Nonlinear Science and Complexity

J.A. Tenreiro Machado • Albert C.J. Luo •
Ramiro S. Barbosa • Manuel F. Silva •
Lino B. Figueiredo
Editors

Nonlinear Science and Complexity

Springer

Editors
J.A. Tenreiro Machado
Dept. Electrical Engineering
Institute of Engineering
of the Polytechnic Institute of Porto
Rua Dr. Antonio Bernardino
de Almeida, 431
4200-072 Porto
Portugal
jtm@isep.ipp.pt

Manuel F. Silva
Dept. Electrical Engineering
Institute of Engineering
of the Polytechnic Institute of Porto
Rua Dr. Antonio Bernardino
de Almeida, 431
4200-072 Porto
Portugal
mss@isep.ipp.pt

Albert C.J. Luo
Department of Mechanical
and Industrial Engineering
Southern Illinois University Edwardsville
62026-1805 Edwardsville, IL
USA
aluo@siue.edu

Ramiro S. Barbosa
Dept. Electrical Engineering
Institute of Engineering
of the Polytechnic Institute of Porto
Rua Dr. Antonio Bernardino
de Almeida, 431
4200-072 Porto
Portugal
rsb@isep.ipp.pt

Lino B. Figueiredo
Dept. Electrical Engineering
Institute of Engineering
of the Polytechnic Institute of Porto
Rua Dr. Antonio Bernardino
de Almeida, 431
4200-072 Porto
Portugal
lbf@isep.ipp.pt

ISBN 978-90-481-9883-2
e-ISBN 978-90-481-9884-9
DOI 10.1007/978-90-481-9884-9
Springer Dordrecht Heidelberg London New York

Library of Congress Control Number: 2010938376

Cover design: eStudio Calamar S.L.

Printed on acid-free paper

Springer is part of Springer Science+Business Media (www.springer.com)

Preface

This book will present recent developments and discoveries in the vital areas of nonlinear science and complexity, to stimulate more research, and to rapidly pass such discoveries to our community. The materials presented in this book include: nonlinear dynamical systems, Lie group analysis and applications, nonlinear fluid mechanics, celestial mechanics, fractional dynamics and applications, mathematic modeling in engineering complexity for a better understanding of physical mechanism and mathematical theory of complex nature and systems.

This book is based on the 2nd Conference on Nonlinear Science and Complexity, NSC'08, that took place at Porto, Portugal, during 28–31 July 2008. This conference succeeds the NSC'06 held at Beijing, China, during 6–12 August 2006. The aim of the conference was to present the fundamental and frontier theories and techniques for modern science and technology, and to stimulate more research interest for exploration of nonlinear science and complexity. The conference focused also on principles, analytical and symbolic approaches, computational techniques in nonlinear physical science and nonlinear mathematics. After peer-reviewed, 105 papers were accepted for presentations from 30 countries. Later 49 papers were selected for publication in the edited book, divided into five groups. The selected manuscripts were further improved and the edited book represents a valuable contribution to the field of nonlinear science and complexity.

- The topic on nonlinear dynamical systems with thirteen papers presents multiple system synchronization, gear transmission systems, cutting dynamics in material process, fuzzy and stochastic dynamical systems, discontinuous systems, scattered in Parts I and VIII.
- The topic on Lie group analysis and application plays an important role in searching closed-form solutions for nonlinear ordinary and partial differential equations. Six papers are selected for publication in Part II of the edited book.
- The topic on nonlinear dynamics of celestial mechanics presented the basic theory and methods in the field. Nine papers are selected for publications in Part III of the book. The interesting results will be useful to scientists in astronomy.
- The topic on mathematical modeling for nonlinear systems in science and engineering is arranged in Parts IV and V with eleven papers. The Bose-Einstein

condensates, boundary layers, incomplete markets and pneumatic systems are presented.
- The topic on fractional dynamics and computational techniques presents the recent development of fractional calculus and numerical computations. In this group, ten papers are selected for publication in Parts VI and VII.

The editors believe that the edited book presents the recent developments and discoveries in nonlinear science and complexity. The materials in this edited book provide important information and tools for students and scientists in the community of nonlinear science and complexity.

J.A. Tenreiro Machado
Albert C.J. Luo
Ramiro S. Barbosa
Manuel F. Silva
Lino B. Figueiredo

Contents

Part VIII Nonlinear Systems

Part I
Nonlinear Dynamics of Continuous and Discontinuous Dynamical Systems

On Synchronization and Its Complexity of Multiple Dynamical Systems

Albert C.J. Luo

Abstract In this paper, the synchronization and its complexity of multiple dynamical systems under certain constraints are presented. The basic concepts of the synchronizations of two or more dynamical systems are introduced. The slave and master systems with the constraints is described through discontinuous dynamical systems, and the sufficient and necessary conditions for the synchronicity of the two systems can be developed from the theory of discontinuous dynamical systems. Finally, the synchronization for multiple slave system with multiple master systems is discussed under any constraints.

Keywords Synchronization · Constraints · Synchronicity · Synchronization complexity

1 Introduction

The investigation on synchronization in dynamical systems should return back to the 17th century. In 1673, Huygens [1] gave the detailed description of the synchronization of two pendulum clocks with weak interaction. In fact, Huygens looked into two modal shapes of vibration. If the coupled pendulums possess small oscillations with the same initial conditions or the initial phase difference is zero, the two pendulums will be synchronized. If the initial phase difference is 180°, the anti-synchronization of two pendulums can be observed. For a general case, the motion of the two pendulums will be combined by the synchronization and anti-synchronization modes of vibration. So far, there are four classes of synchronizations of two or more dynamical systems: (i) identical or complete synchronization, (ii) generalized synchronization, (iii) phase synchronization, (iv) anticipated and lag synchronization and amplitude envelope synchronization. All the synchronizations

---

A.C.J. Luo
Department of Mechanical and Industrial Engineering, Southern Illinois University Edwardsville, Edwardsville, IL 62026-1805, USA

J.A.T. Machado et al. (eds.), *Nonlinear Science and Complexity*,
DOI 10.1007/978-90-481-9884-9_1,

of two or more systems possess at least one constraint for synchronicity, and such synchronizations experience the asymptotic stability characteristics. Once the two or more systems form a synchronization state at one or more specific constraints, such a state should be stable. Such a discussion can be referred to Pikosky et al. [2] and Boccaletti [3]. In this paper, the concepts for synchronization of two or more dynamical systems will be introduced, and the slave and master systems will be described through discontinuous dynamical systems with constraint boundaries. The corresponding synchronization complexity will be discussed.

2 Basic Concepts

Consider two dynamic systems as

$$\dot{\mathbf{x}} = \mathbf{F}(\mathbf{x}, t, \mathbf{p}) \in \Re^n \tag{1}$$

and

$$\dot{\tilde{\mathbf{x}}} = \tilde{\mathbf{F}}(\tilde{\mathbf{x}}, t, \tilde{\mathbf{p}}) \in \Re^{\tilde{n}} \tag{2}$$

where $\mathbf{F} = (F_1, F_2, \ldots, F_n)^T, \mathbf{x} = (x_1, x_2, \ldots, x_n)^T$ and $\mathbf{p} = (p_1, p_2, \ldots, p_k)^T$; $\tilde{\mathbf{F}} = (\tilde{F}_1, \tilde{F}_2, \ldots, \tilde{F}_{\tilde{n}})^T$, $\tilde{\mathbf{x}} = (\tilde{x}_1, \tilde{x}_2, \ldots, \tilde{x}_{\tilde{n}})^T$ and $\tilde{\mathbf{p}} = (\tilde{p}_1, \tilde{p}_2, \ldots, \tilde{p}_{\tilde{k}})^T$. The vector functions $\mathbf{F}$ and $\tilde{\mathbf{F}}$ can be time-dependent or time-independent. Consider a time interval $I_{12} \equiv (t_1, t_2) \subset \Re$ and domains $U_{\mathbf{x}} \subseteq \Re^n$ and $\tilde{U}_{\tilde{\mathbf{x}}} \subseteq \Re^{\tilde{n}}$. For initial conditions $(t_0, \mathbf{x}_0) \in I_{12} \times U_{\mathbf{x}}$ and $(t_0, \tilde{\mathbf{x}}_0) \in I_{12} \times \tilde{U}_{\tilde{\mathbf{x}}}$, the corresponding flows of the two systems are $\mathbf{x}(t) = \mathbf{\Phi}(t, \mathbf{x}_0, t_0, \mathbf{p})$ and $\tilde{\mathbf{x}}(t) = \tilde{\mathbf{\Phi}}(t, \tilde{\mathbf{x}}_0, t_0, \tilde{\mathbf{p}})$ for $(t, \mathbf{x}) \in I_{12} \times U_{\mathbf{x}}$ and $(t, \tilde{\mathbf{x}}) \in I_{12} \times \tilde{U}_{\tilde{\mathbf{x}}}$ with $\mathbf{p} \in U_{\mathbf{p}} \subseteq \Re^k$ and $\tilde{\mathbf{p}} \in U_{\tilde{\mathbf{p}}} \subseteq \Re^{\tilde{k}}$. The semi-group properties of two flows hold (i.e., $\mathbf{\Phi}(t+s, \mathbf{x}_0, t_0, \mathbf{p}) = \mathbf{\Phi}(t, \mathbf{\Phi}(s, \mathbf{x}_0, t_0, \mathbf{p}), s, \mathbf{p})$ and $\mathbf{x}(t_0) = \mathbf{\Phi}(t_0, \mathbf{x}_0, t_0, \mathbf{p})$, $\tilde{\mathbf{\Phi}}(t+s, \tilde{\mathbf{x}}_0, t_0, \tilde{\mathbf{p}}) = \tilde{\mathbf{\Phi}}(t, \tilde{\mathbf{\Phi}}(s, \tilde{\mathbf{x}}_0, t_0, \tilde{\mathbf{p}}), s, \tilde{\mathbf{p}})$ and $\tilde{\mathbf{x}}(t_0) = \tilde{\mathbf{\Phi}}(t_0, \tilde{\mathbf{x}}_0, t_0, \tilde{\mathbf{p}})$).

To investigate the synchronization of the two systems in (1) and (2), the slave and master systems are defined as follows:

Definition 1 A system in (2) is called a *master* system if its flow $\tilde{\mathbf{x}}(t)$ is independent. A system in (1) is called a *slave* system of the master system if its flow $\mathbf{x}(t)$is constrained by the flow $\tilde{\mathbf{x}}(t)$of the master system.

From the foregoing definition, a *slave* system is constrained by a *master* system via specific conditions. Such a phenomenon is called the *synchronization* of the slave and master systems under such a specific conditions. To make this concept clear, a formal definition is given, i.e.,

Definition 2 If a flow $\mathbf{x}(t)$of the slave system in (1) is constrained with a flow $\tilde{\mathbf{x}}(t)$ of a *master* system in (2) through the following function

$$\varphi(\mathbf{x}(t), \tilde{\mathbf{x}}(t), t, \boldsymbol{\lambda}) = 0, \quad \boldsymbol{\lambda} \in \Re^{k_\varphi}, \tag{3}$$

for time $t \in [t_{m_1}, t_{m_2}]$, then the slave system is said to be *synchronized* with the master system in the sense of (3) for time $t \in [t_{m_1}, t_{m_2}]$, denoted by the $(n : \tilde{n})$-dimensional synchronization of the slave and master systems in the sense of (3). If $t_{m_2} \to \infty$, the slave system is said to be *absolutely synchronized* with the master system in the sense of (3) for time $t \in [t_{m_1}, \infty)$. Two special cases are given as follows.

(i) For $n = \tilde{n}$, such a synchronization is called an *equi-dimensional* system synchronization in the sense of (3) for $t \in [t_{m_1}, t_{m_2}]$.
(ii) For $n = \tilde{n}$, such a synchronization is called an *absolute*, equi-dimensional system synchronization in the sense of (3) for $t \in [t_{m_1}, \infty)$.

If $n \neq \tilde{n}$, the $(n : \tilde{n})$-synchronization is called a *non-equi-dimensional* system synchronization. Under a certain rule in (3), it is interesting that a slave system can follow another completely different master system to synchronize. From the proceeding definition, it can be seen that the slave system is synchronized with the master system under a constraint condition. In fact, constraint conditions for such a synchronization phenomenon can be more than one. In other words, the slave system can be synchronized with the master system under multiple constraints. Thus, the definition for the synchronization of a slave system with a master system under multiple constraints is given as follows:

Definition 3 An n-dimensional slave system in (1) is called *to be synchronized* with an $\tilde{n}$-dimensional master system in (2) of the *$(n : \tilde{n}; l)$-type* (or an*$(n : \tilde{n}; l)$-synchronization*) if there are l-linearly independent functions $\varphi^{(j)}(\mathbf{x}(t), \tilde{\mathbf{x}}(t), t, \boldsymbol{\lambda}_j)$ ($j \in \mathbb{L}$ and $\mathbb{L} = \{1, 2, \ldots, l\}$) to make two flows $\tilde{\mathbf{x}}(t)$ and $\mathbf{x}(t)$of the master and slave systems satisfy

$$\varphi^{(j)}(\mathbf{x}(t), \tilde{\mathbf{x}}(t), t, \boldsymbol{\lambda}_j) = 0 \quad \text{for } \boldsymbol{\lambda}_j \in \Re^{n_j} \text{ and } j \in \mathbb{L} \tag{4}$$

for time $t \in [t_{m_1}, t_{m_2}]$. If $t_{m_2} \to \infty$, the synchronization of the slave and master systems is called an *absolute, $(n : \tilde{n}; l)$-synchronization* in the sense of (4) for time $t \in [t_{m_1}, \infty)$. The six special cases are given as follows:

(i) For $l = n$, the synchronization of the slave and master systems is called to *a complete, $(n : \tilde{n}; n)$-synchronization* in the sense of (4) for $t \in [t_{m_1}, t_{m_2}]$.
(ii) For $l = n$ and $t_{m_2} \to \infty$, the synchronization of the slave and master systems is called *an absolute, complete, $(n : \tilde{n}; n)$-synchronization* in the sense of (4) for $t \in [t_{m_1}, \infty)$.
(iii) If $n = \tilde{n} > l$, the synchronization of the slave and master systems is called an equi-dimensional, $(n : n; l)$-synchronization in the sense of (4) for $t \in [t_{m_1}, t_{m_2}]$.
(iv) If $n = \tilde{n} > l$ and $t_{m_1} \to \infty$, the synchronization of the slave and master systems is called an absolute, equi-dimensional, $(n : n; l)$-synchronization in the sense of (4) for $t \in [t_{m_1}, \infty)$.
(v) If $n = \tilde{n} = l$, the synchronization of the slave and master systems is called a complete, equi-dimensional, $(n : n; n)$-synchronization (simply called *a synchronization*) in the sense of (4) for $t \in [t_{m_1}, t_{m_2}]$.

(vi) If $n = \tilde{n} = l$ and $t_{m_1} \to \infty$, the synchronization of the slave and master systems is called an absolute, complete, equi-dimensional, $(n:n;n)$-synchronization (simply called *an absolute synchronization*) in the sense of (4) for $t \in [t_{m_1}, \infty)$.

3 Discontinuous Descriptions

Introduce a new variable vector

$$\mathbf{X} = (\mathbf{x}; \tilde{\mathbf{x}})^T = (x_1, x_2, \ldots, x_n; \tilde{x}_1, \tilde{x}_2, \ldots, \tilde{x}_{\tilde{n}})^T \in \Re^{n+\tilde{n}}. \tag{5}$$

Note that $(\bullet;\bullet) \equiv (\bullet,\bullet)$ is just for a combined vector of two state vectors of the slave and master systems.

Definition 4 A boundary in an $(n+\tilde{n})$-dimensional phase space of the slave and master systems, relative to the constraint conditions in (3), is defined as

$$\begin{aligned}\partial\Omega_{12} &= \bar{\Omega}_1 \cap \bar{\Omega}_2 \\ &= \big\{\mathbf{X}^{(0)} \big| \varphi(\mathbf{X}^{(0)}, t, \boldsymbol{\lambda}) \equiv \varphi(\mathbf{x}^{(0)}(t), \tilde{\mathbf{x}}^{(0)}(t), t, \boldsymbol{\lambda}) = 0, \\ &\qquad \varphi \text{ is } C^r\text{-continuous } (r \geq 1)\big\} \subset \Re^{n+\tilde{n}-1};\end{aligned} \tag{6}$$

and two corresponding domains are defined as

$$\begin{aligned}\Omega_1 &= \big\{\mathbf{X}^{(1)} \big| \varphi(\mathbf{X}^{(1)}, t, \boldsymbol{\lambda}) \equiv \varphi(\mathbf{x}^{(1)}(t), \tilde{\mathbf{x}}^{(1)}(t), t, \boldsymbol{\lambda}) > 0, \\ &\qquad \varphi \text{ is } C^r\text{-continuous } (r \geq 1)\big\} \subset \Re^{n+\tilde{n}}; \\ \Omega_2 &= \big\{\mathbf{X}^{(2)} \big| \varphi(\mathbf{X}^{(2)}, t, \boldsymbol{\lambda}) \equiv \varphi(\mathbf{x}^{(2)}(t), \tilde{\mathbf{x}}^{(2)}(t), t, \boldsymbol{\lambda}) < 0, \\ &\qquad \varphi \text{ is } C^r\text{-continuous } (r \geq 1)\big\} \subset \Re^{n+\tilde{n}}.\end{aligned} \tag{7}$$

On the two domains, a discontinuous dynamical system is defined as

$$\dot{\mathbf{X}}^{(\alpha)} = \mathbb{F}^{(\alpha)}(\mathbf{X}^{(\alpha)}, t, \boldsymbol{\pi}^{(\alpha)}) \quad \text{in } \Omega_\alpha \tag{8}$$

where $\mathbb{F}^{(\alpha)} = (\mathbf{F}^{(\alpha)}; \tilde{\mathbf{F}}^{(\alpha)})^T = (F_1^{(\alpha)}, F_2^{(\alpha)}, \ldots, F_n^{(\alpha)}; \tilde{F}_1^{(\alpha)}, \tilde{F}_2^{(\alpha)}, \ldots, \tilde{F}_{\tilde{n}}^{(\alpha)})^T$ and $\boldsymbol{\pi}^{(\alpha)} = (\mathbf{p}_\alpha, \tilde{\mathbf{p}}_\alpha)^T$. Suppose there is a vector field $\mathbb{F}^{(0)}(\mathbf{X}^{(0)}, t, \boldsymbol{\lambda})$on the boundary to make $\varphi(\mathbf{X}^{(0)}, t, \boldsymbol{\lambda}) = 0$, and the corresponding dynamical system is expressed by

$$\dot{\mathbf{X}}^{(0)} = \mathbb{F}^{(0)}(\mathbf{X}^{(0)}, t, \boldsymbol{\lambda}) \quad \text{on } \partial\Omega_{12}. \tag{9}$$

The domains Ω_α $(\alpha = 1, 2)$ are separated by the boundary $\partial\Omega_{12}$, as shown in Fig. 1. For a point $(\mathbf{x}^{(1)}, \tilde{\mathbf{x}}^{(1)}) \in \Omega_1$ at time t, one obtains $\varphi(\mathbf{x}^{(1)}, \tilde{\mathbf{x}}^{(1)}, t, \boldsymbol{\lambda}) > 0$. However, for a point $(\mathbf{x}^{(2)}, \tilde{\mathbf{x}}^{(2)}) \in \Omega_2$ at time t, one obtains $\varphi(\mathbf{x}^{(2)}, \tilde{\mathbf{x}}^{(2)}, t, \boldsymbol{\lambda}) < 0$. On the boundary $(\mathbf{x}^{(0)}, \tilde{\mathbf{x}}^{(0)}) \in \partial\Omega_{12}$, $\varphi(\mathbf{x}^{(0)}, \tilde{\mathbf{x}}^{(0)}, t, \boldsymbol{\lambda}) = 0$ is required. If the constraint condition is time-independent, the separation boundary determined by the

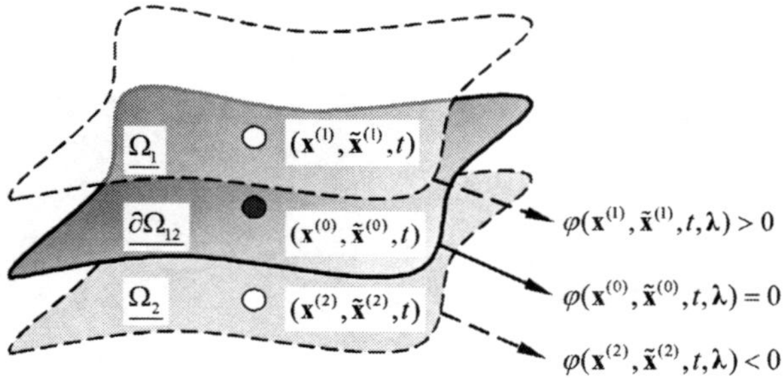

Fig. 1 Constraint boundary and domains in $(n + \tilde{n})$-dimensional state space

constraint condition is invariant. If there are many synchronization conditions, the above definition can be extended as follows.

Definition 5 The jth-boundary in an $(n + \tilde{n})$-dimensional phase space of the slave and master systems, relative to the constraint conditions in (4), is defined as

$$\begin{aligned}\partial\Omega_{(12,j)} &= \bar{\Omega}_{(1,j)} \cap \bar{\Omega}_{(2,j)} \\ &= \big\{\mathbf{X}^{(0,j)} \big| \varphi^{(j)}(\mathbf{X}^{(0,j)}, t, \boldsymbol{\lambda}_j) \equiv \varphi^{(j)}(\mathbf{x}^{(0,j)}(t), \tilde{\mathbf{x}}^{(0,j)}(t), t, \boldsymbol{\lambda}_j) = 0, \\ &\qquad \varphi^{(j)} \text{ is } C^{r_j}\text{-continuous } (r_j \geq 1)\big\} \\ &\subset \Re^{n+\tilde{n}-1};\end{aligned} \tag{10}$$

and the two domains for the jth-boundary are defined as

$$\begin{aligned}\Omega_{(1,j)} &= \big\{\mathbf{X}^{(1,j)} \big| \varphi^{(j)}(\mathbf{X}^{(1,j)}, t, \boldsymbol{\lambda}_j) \equiv \varphi^{(j)}(\mathbf{x}^{(1,j)}(t), \tilde{\mathbf{x}}^{(1,j)}(t), t, \boldsymbol{\lambda}_j) > 0, \\ &\qquad \varphi^{(j)} \text{ is } C^{r_j}\text{-continuous } (r_j \geq 1)\big\} \\ &\subset \Re^{n+\tilde{n}}; \\ \Omega_{(2,j)} &= \big\{\mathbf{X}^{(2,j)} \big| \varphi^{(j)}(\mathbf{X}^{(2,j)}, t, \boldsymbol{\lambda}_j) \equiv \varphi^{(j)}(\mathbf{x}^{(2,j)}(t), \tilde{\mathbf{x}}^{(2,j)}(t), t, \boldsymbol{\lambda}_j) < 0, \\ &\qquad \varphi^{(j)} \text{ is } C^{r_j}\text{-continuous } (r_j \geq 1)\big\} \\ &\subset \Re^{n+\tilde{n}}.\end{aligned} \tag{11}$$

On the two domains, a discontinuous dynamical system is defined as

$$\dot{\mathbf{X}}^{(\alpha_j,j)} = \mathbb{F}^{(\alpha_j,j)}(\mathbf{X}^{(\alpha_j,j)}, t, \boldsymbol{\pi}_j^{(\alpha_j)}) \quad \text{in } \Omega_{(\alpha_j,j)} \tag{12}$$

where $\mathbb{F}^{(\alpha_j,j)} = (\mathbf{F}^{(\alpha_j,j)}; \tilde{\mathbf{F}}^{(\alpha_j,j)})^T = (F_1^{(\alpha_j,j)}, F_2^{(\alpha_j,j)}, \ldots, F_n^{(\alpha_j,j)}; \tilde{F}_1^{(\alpha_j,j)}, \tilde{F}_2^{(\alpha_j,j)}, \ldots, \tilde{F}_{\tilde{n}}^{(\alpha_j,j)})^T$ and $\boldsymbol{\pi}_j^{(\alpha_j)} = (\mathbf{p}_j^{(\alpha_j)}, \tilde{\mathbf{p}}_j^{(\alpha_j)})^T$. Suppose there is a vector field $\mathbb{F}^{(0,j)}(\mathbf{X}^{(0,j)}, t, \boldsymbol{\lambda}_j)$ on the boundary to make $\varphi^{(j)}(\mathbf{X}^{(0,j)}, t, \boldsymbol{\lambda}) = 0$, and the corresponding dynamical system can be expressed by

$$\dot{\mathbf{X}}^{(0,j)} = \mathbb{F}^{(0,j)}(\mathbf{X}^{(0,j)}, t, \boldsymbol{\lambda}_j) \quad \text{on } \partial\Omega_{(12,j)}. \tag{13}$$

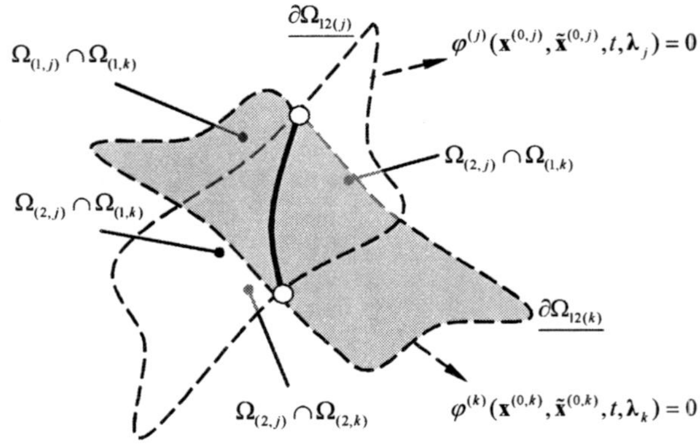

Fig. 2 An intersection of two constraints $\varphi^{(j)} = 0$ and $\varphi^{(k)} = 0$ for $j, k \in \mathbb{L}$ and $j \neq k$

Since l-constraint conditions are linearly independent, any two boundary will be intersected each other. Consider two boundaries $\partial\Omega_{(12,j)}$ and $\partial\Omega_{(12,k)}$, and their intersection is given by

$$\partial\Omega_{(12,jk)} = \partial\Omega_{(12,j)} \cap \partial\Omega_{(12,k)} \subset \Re^{n+\tilde{n}-2} \tag{14}$$

and the domains is separated into four sub-domains

$$\Omega_{(\alpha_j\alpha_k,jk)} = \Omega_{(\alpha_j,j)} \cap \Omega_{(\alpha_k,k)} \subset \Re^{n+\tilde{n}} \quad \text{for } j,k = 1,2,\ldots \text{ and } \alpha_j, \alpha_k = 1,2. \tag{15}$$

Such a partition of the domain in state space is sketched in Fig. 2. The intersection of two constraint conditions in state space is depicted by a dark $(n+\tilde{n}-2)$-manifold. For the l-linearly independent constraints, the state space partition can be determined via such linearly independent constraint conditions. Based on the l-constraint conditions, the corresponding intersection of boundaries is

$$\partial\Omega_{12(\mathbf{J})} = \bigcap_{j=1}^{l} \partial\Omega_{(12,j)} \subset \Re^{n+\tilde{n}-l} \tag{16}$$

which gives an $(n+\tilde{n}-l)$-dimensional manifold. If $n = l$, the intersection manifold is in an $\tilde{n}$-dimensional phase space. The slave system can be completely synchronized with the master system. From the l-constraint conditions, the domain in $(n+\tilde{n})$-dimensional state space is partitioned into the following sub-domains, i.e.,

$$\Omega_{\boldsymbol{\alpha}} = \Omega_{(\alpha_1\alpha_2\cdots\alpha_l)} = \bigcap_{j=1}^{l} \Omega_{(\alpha_j,j)} \subset \Re^{n+\tilde{n}} \quad \text{for } \alpha_j = 1,2 \text{ and } j \in \mathbb{L}. \tag{17}$$

The total domain is a union of all domains

$$\mho = \bigcup_{j=1}^{l} \bigcup_{\alpha_j=1}^{2} \left(\bigcap_{j=1}^{l} \Omega_{(\alpha_j,j)} \right) \subset \Re^{n+\tilde{n}}.$$

From a theory of discontinuous dynamical system in Luo [4, 5], at least the slave system possesses discontinuous vector fields to make the flow stay on the super-surface. The synchronization can keep on the super-surface. The constraint can be used as a super-surface for the synchronization of the slave and master systems. The synchronization, desynchronization and penetration can be treated as sink flow, source flows and passable flows on the boundary, respectively. The corresponding necessary and sufficient conditions can be developed from Luo [4–6].

4 Complexity by System Synchronization

To discuss the synchronization complexity, consider many master systems and many slave systems. A few master and slave systems with constraints can be synchronized.

Definition 6 A $\mathscr{S}$-set of slave systems is defined as

$$\mathscr{S} \equiv \left\{ {}^{(I)}\dot{\mathbf{X}} = {}^{(I)}\mathbf{F}({}^{(I)}\mathbf{X}, t, {}^{(I)}\mathbf{p}) \middle| I = 1, 2, \ldots; {}^{(I)}\mathbf{X} \in \Re^{n_{(I)}}; {}^{(I)}\mathbf{p} \in \Re^{k_{(I)}} \right\} \tag{18}$$

and an $\mathscr{M}$-set of master systems is defined as

$$\mathscr{M} \equiv \left\{ {}^{(\tilde{I})}\dot{\tilde{\mathbf{X}}} = {}^{(\tilde{I})}\mathbf{F}({}^{(\tilde{I})}\tilde{\mathbf{X}}, t, {}^{(\tilde{I})}\tilde{\mathbf{p}}) \middle| \tilde{I} = 1, 2, \ldots; {}^{(\tilde{I})}\tilde{\mathbf{X}} \in \Re^{\tilde{n}_{(\tilde{I})}}; {}^{(\tilde{I})}\tilde{\mathbf{p}} \in \Re^{\tilde{k}_{(\tilde{I})}} \right\}. \tag{19}$$

This definition gives a cluster of slave systems and a cluster of master systems. To investigate the synchronization of the slave and master systems, the slave and master systems can be selected from such $\mathscr{S}$-set of slave systems and $\mathscr{M}$-set of master systems. For any slave system in the $\mathscr{S}$-set of slave systems, it can be synchronized with many master systems in the $\mathscr{M}$-set of master systems with the corresponding constraints. The constraints for such synchronization can be either single or multiple constraints, and the synchronized components for such constraints can be either full or partial components from those slave and master systems. Based on this reason, the subspace set in state space should be defined.

Definition 7 A subspace set of the Ith-slave system is defined as

$$\Omega_{\mathscr{S}} \equiv \left\{ {}^{(I,\mu)}\Omega_{\mathscr{S}} \middle| \mu = 1, 2, \ldots; I = 1, 2, \ldots \right\} \tag{20}$$

where

$$\begin{aligned} {}^{(I,\mu)}\Omega_{\mathscr{S}} \equiv \Big\{ {}^{(I,\mu)}\mathbf{X} \Big| {}^{(I,\mu)}\mathbf{X} = ({}^{(I,\mu)}X_1, {}^{(I,\mu)}X_2, \ldots, {}^{(I,\mu)}X_{\Gamma_\mu})^T; \mu = 1, 2, \ldots; \\ \text{and } \Gamma_\mu < n_{(I)}; {}^{(I,\mu)}X_i \in \{{}^{(I)}X_1, {}^{(I)}X_2, \ldots, {}^{(I)}X_{n_{(I)}}\}; i = 1, 2, \ldots, \Gamma_\mu \Big\} \end{aligned} \tag{21}$$

and a subspace set of the $\tilde{I}$th- master system is defined as

$$\Omega_{\mathscr{M}} \equiv \left\{ {}^{(\tilde{I},\tilde{\mu})}\Omega_{\mathscr{M}} \middle| \tilde{\mu} = 1, 2, \ldots; \tilde{I} = 1, 2, \ldots \right\} \tag{22}$$

where

$$
\begin{aligned}
{}^{(\tilde{I},\tilde{\mu})}\Omega_{\mathscr{M}} \equiv \Big\{ & {}^{(\tilde{I},\tilde{\mu})}\mathbf{X} \Big| {}^{(\tilde{I},\tilde{\mu})}\mathbf{X} = ({}^{(\tilde{I},\tilde{\mu})}X_1, {}^{(\tilde{I},\tilde{\mu})}X_2, \ldots, {}^{(\tilde{I},\tilde{\mu})}X_{\tilde{\Gamma}_{\tilde{\mu}}})^T; \tilde{\mu} = 1, 2, \ldots; \\
& \text{and } \tilde{\Gamma}_{\tilde{\mu}} < \tilde{n}_{(\tilde{I})}; {}^{(\tilde{I},\tilde{\mu})}X_i \in \{{}^{(\tilde{I})}X_1, {}^{(\tilde{I})}X_2, \ldots, {}^{(\tilde{I})}X_{n_{(\tilde{I})}}\}; \\
& i = 1, 2, \ldots, \tilde{\Gamma}_{\tilde{\mu}} \Big\}.
\end{aligned} \tag{23}
$$

From the foregoing definitions of the two subspace sets for slave and master systems, each subspace for the Ith-slave system (or the $\tilde{I}$th-master system) is arbitrarily selected from $n_{(I)}$-components (or $\tilde{n}_{(\tilde{I})}$-components). Based on such phase subspaces for the Ith- slave system and the $\tilde{I}$th-system, the corresponding constraint can be defined for the synchronization of such slave and master systems on the two subspace. Thus the corresponding C-set of the constraints for the slave and master systems is defined as follows.

Definition 8 For two subspaces${}^{(I,\mu)}\Omega_{\mathscr{S}}$ and ${}^{(\tilde{I},\tilde{\mu})}\Omega_{\mathscr{M}}$, a C-set of constraints is defined as

$$
\mathscr{C} \equiv \Big\{ {}^{(I,\tilde{I})}\mathscr{C} \Big| I = 1, 2, \ldots; \tilde{I} = 1, 2, \ldots \Big\} \tag{24}
$$

where

$$
\begin{aligned}
{}^{(I,\tilde{I})}\mathscr{C} \equiv \Big\{ & {}^{(I,\tilde{I})}\varphi_j({}^{(I,\mu)}\mathbf{X}, {}^{(\tilde{I},\tilde{\mu})}\tilde{\mathbf{X}}, t, \boldsymbol{\lambda}_j) = 0 \Big| j = 1, 2, \ldots; \\
& \mu, \tilde{\mu} = 1, 2, \ldots; \boldsymbol{\lambda}_j \in \Re^{n_j} \Big\}.
\end{aligned} \tag{25}
$$

Definition 9 Consider M-slave systems from the $\mathscr{S}$-set of slave systems and $\tilde{M}$-master systems from the $\mathscr{M}$-set of master systems

$$
{}^{(I)}\dot{\mathbf{X}} = {}^{(I)}\mathbf{F}({}^{(I)}\mathbf{X}, t, {}^{(I)}\mathbf{p}) \quad \text{for all } I \in \{1, 2, \ldots, M\}, \tag{26}
$$

$$
{}^{(\tilde{I})}\dot{\tilde{\mathbf{X}}} = {}^{(\tilde{I})}\mathbf{F}({}^{(\tilde{I})}\tilde{\mathbf{X}}, t, {}^{(\tilde{I})}\tilde{\mathbf{p}}) \quad \text{for all } \tilde{I} \in \{1, 2, \ldots, \tilde{M}\}. \tag{27}
$$

There are l-constraints on two subspaces sets $\Omega_{\mathscr{S}}$ and $\Omega_{\mathscr{M}}$,

$$
{}^{(I,\tilde{I})}\varphi_j({}^{(I,\mu)}\mathbf{X}, {}^{(\tilde{I},\tilde{\mu})}\tilde{\mathbf{X}}, t, \boldsymbol{\lambda}_j) = 0 \quad \text{for all } j \in \{1, 2, \ldots, l\} \tag{28}
$$

with $l \le \sum_{I=1}^{M} n_{(I)}$. If all the l-constraints in (28) hold for time $t \in [t_{m_1}, t_{m_2}]$, then the M-slave systems with $\tilde{M}$-master systems are called to be synchronized for time $t \in [t_{m_1}, t_{m_2}]$in the sense of (28).

The foregoing definition gives the synchronization between two clusters of slave and master systems are discussed. For $I = \tilde{I} = 1$, the foregoing definition implies the slave and master systems are one to one. If the two subspace sets of the slave and master systems take all components in state space, and the corresponding constraints in (28) becomes (3) or (4). The synchronicity for such slave and master system was discussed from the theory of discontinuous dynamical systems in Luo [4–6]. To further explain the above definition, one slave system with multiple master systems or one master system with multiple slave systems can be discussed first.

Definition 10 Consider M-slave systems from the $\mathscr{S}$-set of slave systems and a master system from the $\mathscr{M}$-set of master systems

$$^{(I)}\dot{\mathbf{X}} = {}^{(I)}\mathbf{F}({}^{(I)}\mathbf{X}, t, {}^{(I)}\mathbf{p}) \quad \text{for all } I \in \{1, 2, \ldots, M\}, \tag{29}$$

$$^{(1)}\dot{\tilde{\mathbf{X}}} = {}^{(1)}\mathbf{F}({}^{(1)}\tilde{\mathbf{X}}, t, {}^{(1)}\tilde{\mathbf{p}}). \tag{30}$$

There are l-constraints on two subspaces sets $\Omega_{\mathscr{S}}$ and $\Omega_{\mathscr{M}}$,

$$^{(I,1)}\varphi_j({}^{(I,\mu)}\mathbf{X}, {}^{(1,\tilde{\mu})}\tilde{\mathbf{X}}, t, \boldsymbol{\lambda}_j) = 0 \quad \text{for all } j \in \{1, 2, \ldots, l\} \tag{31}$$

with $L \leq \sum_{l=1}^{M} n_{(l)}$. If all the l-constraints in (31) hold for time $t \in [t_{m_1}, t_{m_2}]$, then the M-slave systems with the master system are called to be synchronized for time $t \in [t_{m_1}, t_{m_2}]$ in the sense of (31).

This definition tells that M-slave systems are synchronized with one master system with different constraints. For each $I \in \{1, 2, \ldots, M\}$, the corresponding slave system synchronized with the master system can be discussed. It is of great interest to consider two master systems for M-slave systems under different constraints.

Definition 11 Consider M-slave systems from the $\mathscr{S}$-set of slave systems and two master system from the $\mathscr{M}$-set of master systems

$$^{(I)}\dot{\mathbf{X}} = {}^{(I)}\mathbf{F}({}^{(I)}\mathbf{X}, t, {}^{(I)}\mathbf{p}) \quad \text{for all } I \in \{1, 2, \ldots, M\}, \tag{32}$$

$$^{(1)}\dot{\tilde{\mathbf{X}}} = {}^{(1)}\mathbf{F}({}^{(1)}\tilde{\mathbf{X}}, t, {}^{(1)}\tilde{\mathbf{p}}) \quad \text{and} \quad {}^{(2)}\dot{\tilde{\mathbf{X}}} = {}^{(2)}\mathbf{F}({}^{(2)}\tilde{\mathbf{X}}, t, {}^{(2)}\tilde{\mathbf{p}}). \tag{33}$$

There are l-constraints on two subspaces sets $\Omega_{\mathscr{S}}$ and $\Omega_{\mathscr{M}}$,

$$\left.\begin{aligned} {}^{(I,1)}\varphi_j({}^{(I,\mu)}\mathbf{X}, {}^{(1,\tilde{\mu})}\tilde{\mathbf{X}}, t, \boldsymbol{\lambda}_j) = 0 \\ {}^{(I,2)}\varphi_j({}^{(I,\mu)}\mathbf{X}, {}^{(2,\tilde{\mu})}\tilde{\mathbf{X}}, t, \boldsymbol{\lambda}_j) = 0 \end{aligned}\right\} \quad \text{for all } j \in \{1, 2, \ldots, l\} \tag{34}$$

with $l \leq \sum_{l=1}^{M} n_{(I)}$. If all the l-constraints in (34) hold for time $t \in [t_{m_1}, t_{m_2}]$, then the M-slave systems with the two master systems are called to be synchronized for time $t \in [t_{m_1}, t_{m_2}]$ in the sense of (34).

The foregoing definition gives that the M-slave systems can be synchronized with two master systems under different constraints. If we consider the two master systems to be two parent systems, the slave systems are treated as M-children systems. Further, the synchronicity of the parent and child systems can be called *the similarity of the parent and child systems*. For each child (or slave) system, under certain constraints in (34), the similarity of the two parent systems with the child system can be investigated as the synchronicity of the slave and master systems as discussed. The synchronization of a slave system with multiple master systems under certain constraints can be also discussed, and the corresponding definition is given as follows.

Definition 12 Consider *a* slave system from the $\mathscr{S}$-set of slave systems and $\tilde{M}$-master system from the $\mathscr{M}$-set of master systems

$$^{(1)}\dot{\mathbf{X}} = {}^{(1)}\mathbf{F}({}^{(1)}\mathbf{X}, t, {}^{(1)}\mathbf{p}), \tag{35}$$

$$^{(\tilde{I})}\dot{\tilde{\mathbf{X}}} = {}^{(\tilde{I})}\mathbf{F}({}^{(\tilde{I})}\tilde{\mathbf{X}}, t, {}^{(\tilde{I})}\tilde{\mathbf{p}}) \quad \text{for all } \tilde{I} \in \{1, 2, \ldots, \tilde{M}\}. \tag{36}$$

There are l-constraints on two subspaces sets $\Omega_{\mathscr{S}}$ and $\Omega_{\mathscr{M}}$,

$$^{(1,\tilde{I})}\varphi_j({}^{(1,\mu)}\mathbf{X}, {}^{(\tilde{I},\tilde{\mu})}\tilde{\mathbf{X}}, t, \boldsymbol{\lambda}_j) = 0 \quad \text{for all } j \in \{1, 2, \ldots, l\}. \tag{37}$$

with $l \leq n_{(1)}$. If all the l-constraints in (37) hold for time $t \in [t_{m_1}, t_{m_2}]$, then the slave system with the $\tilde{M}$-master systems are called to be synchronized for time $t \in [t_{m_1}, t_{m_2}]$ in the sense of (37).

The definition gives the slave system controlled by the $\tilde{M}$-master systems under the l-constraints. The synchronization of such systems can also be investigated through the theory of discontinuous dynamical systems in Luo [4–6].

References

1. C. Huygens (Hugenii), *Horologium Oscillatorium* (Apud F. Muguet, Paris, 1673). English Translation, *The Pendulum Clock* (Iowa State University, Ames, 1986)
2. A. Pikovsky, M. Rosenblum, J. Kurths, *Synchronization: A Universal Concept in Nonlinear Science* (Cambridge University Press, Cambridge, 2001)
3. S. Boccaletti, *The Synchronized Dynamics of Complex Systems* (Elsevier, Amsterdam, 2008)
4. A.C.J. Luo, A theory for non-smooth dynamical systems on the connectable domains. Commun. Nonlinear Sci. Numer. Simul. **10**, 1–55 (2005)
5. A.C.J. Luo, *Singularity and Dynamics on Discontinuous Vector Fields* (Elsevier, Amsterdam, 2006)
6. A.C.J. Luo, *Global Transversality, Resonance and Chaotic Dynamics* (Singapore, World Scientific, 2008)

Periodic and Chaotic Motions in a Gear-pair Transmission System with Impacts

Albert C.J. Luo and Dennis O'Connor

Abstract Nonlinear dynamical behaviors of a gear transmission system with impacts are investigated. The transmission system is described through an impact model with possible stick between the two gears. Based on the mapping structures, periodic motions of such a system are predicted analytically. To understand the global dynamical behaviors of the gear transmission system, system parameter maps are developed. Numerical simulations for periodic and chaotic motions are performed from the parameter maps.

Keywords Gear-pair transmission systems · Impact chatter · Stick motion

1 Introduction

Gear transmission systems are extensively used in mechanical engineering and an efficient gear transmission is necessary to save energy in mechanical transmission as discussed in Changenet et al. [1]. From the current principles and theories, impacting chatter is a source to dissipate energy, and the released energy will cause vibration and noise in the system. On the other hand, the reduction of vibration and noise in transmission systems will enhance the corresponding transmission efficiency.

The early investigations of gear transmission systems focused on the mesh geometries, kinematics and strength of teeth as in Buckingham [2, 3]. For low-speed gear systems, the linear model was developed, which gave a reasonable prediction of gear-tooth vibrations. With increasing rotation speed in gear transmission systems, vibrations and noise become serious. Hartog and Mikina [4] used a piecewise linear system without damping to model gear transmission systems, and the symmetric periodic motion in such a system was investigated. Ozguven and Houser [5] gave

A.C.J. Luo · D. O'Connor
Department of Mechanical and Industrial Engineering, Southern Illinois University Edwardsville, Edwardsville, IL 62026, USA

J.A.T. Machado et al. (eds.), *Nonlinear Science and Complexity*,
DOI 10.1007/978-90-481-9884-9_2,

a survey on the mathematical models of gear transmission systems. The piecewise linear model and the impact model were the two main mechanical models to investigate the origin of vibration and noise. In 1984, Pfeiffer [6] presented an impact model of gear transmissions, and the theoretical and experimental investigations on regular and chaotic motions in the gear box were later carried out in Karagiannis and Pfeiffer [7]. One also used a piecewise linear model to investigate the dynamics of gear transmission systems (e.g., Theodossiades and Natsiavas [8]).

To model vibrations in gear transmission systems, Luo and Chen [9] gave an analytical prediction of the simplest, periodic motion through a piecewise linear, impacting system. In addition, the local singularity theory in Luo [10] was used to obtain the corresponding grazing of periodic motions, and chaotic motions were simulated numerically through such a piecewise linear system. the grazing mechanism of the strange fragmentation of such a piecewise linear system was discussed in Luo and Chen [11]. Luo and Chen [12] used the mapping structure technique to analytically predict arbitrary periodic motions of such a piecewise linear system. In this piecewise linear model, it was assumed that impact locations were fixed, and the perfectly plastic impact was considered. Separation of the two gears occurred at the same location as the gear impact. Compared with the existing models, this model can give a better prediction of periodic motions in gear transmission systems, but the related assumptions may not be realistic to practical transmission systems. In this paper, the two gears will be considered to be independent, and impacts between the two gears occur at different locations. This gear transmission system with impact will be modeled by a piecewise linear system with impacts. Luo and O'Connor [13, 14] discussed the mechanism of impacting chatter with stick, and analytical prediction of periodic chatter with/without stick. In this paper, the global nonlinear behaviors of such a gear transmission system will be discussed and parameter maps will be developed. Numerical illustrations will be presented for parameter characteristics of impacting chatter with/without stick.

2 Equations of Motion

To model the gear transmission system, consider a periodically forced oscillator confined between the teeth of a second oscillator, as shown in Fig. 1. Interaction between the two gears causes impacting and sticking together. Since the gears are supported by shafts, each gear m_i $(i = 1, 2)$ is connected to a spring and a damper. The spring stiffness k_i is from the twisting shafts of a gear transmission system, and the damper damping r_i is from lubricating fluids. The free-flying gap between two teeth of the driven gear is d. The external force $B_0 + A_0 \cos \Omega t$ acts on the driving gear m_1 where A_0 and Ω are the amplitude and frequency of the oscillation torque, respectively. B_0 is from the constant torque. The displacements of each mass measured from their equilibriums are expressed by $x^{(1)}$ and $x^{(2)}$. Impacts between two gears are described through the impact law with restitution coefficient e. The equilibrium of the first gear is set at the center of the two teeth of the second gear at

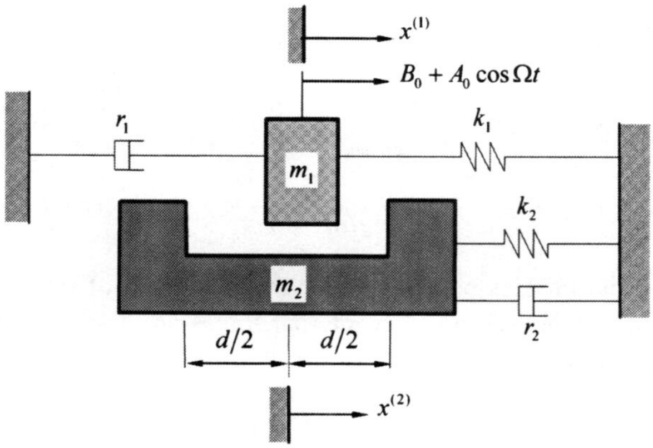

Fig. 1 A mechanical model for a gear transmission

equilibrium. Without any interaction between two gear oscillators, the equations of motion are for $i = 1, 2$

$$\ddot{x}_2^{(i)} + 2\zeta_2^{(i)}\dot{x}_2^{(i)} + (\omega_2^{(i)})^2 x_2^{(i)} = b_2^{(i)} + Q_2^{(i)} \cos \Omega t \tag{1}$$

where

$$\left.\begin{array}{llll} \zeta_2^{(i)} = \dfrac{r_i}{2m_i}, & \omega_2^{(i)} = \sqrt{\dfrac{k_i}{m_i}} & (i = 1, 2); & \\ b_2^{(1)} = \dfrac{B_0}{m_1}, & Q_2^{(1)} = \dfrac{A_0}{m_1}, & b_2^{(2)} = 0, & Q_2^{(2)} = 0 \end{array}\right\} \tag{2}$$

for the mechanical model in Fig. 1. Once $|x_2^{(i)} - x_2^{(\bar{i})}| = \frac{d}{2} (i, \bar{i} \in \{1,2\}$ and $i \neq \bar{i})$, the impact between the two gears occurs. From momentum conservation and the simple impact law, two velocities of the two gears after impacting are given by

$$\dot{x}_2^{(i)+} = I_1^{(i)} \dot{x}_2^{(i)-} + I_2^{(i)} \dot{x}_2^{(\bar{i})-} \tag{3}$$

where the superscripts "−" and "+" represent before and after impact, and the corresponding coefficients are

$$\left.\begin{array}{ll} I_1^{(1)} = \dfrac{m_1 - m_2 e}{m_1 + m_2}, & I_2^{(1)} = \dfrac{(1+e)\, m_2}{m_1 + m_2}, \\ I_1^{(2)} = \dfrac{(1+e)\, m_1}{m_1 + m_2}, & I_1^{(2)} = \dfrac{m_2 - m_1 e}{m_1 + m_2}. \end{array}\right\} \tag{4}$$

Once two gear oscillators stick together, equations of motion are for $i = 1, 2$ and $\alpha = 1, 3$

$$\ddot{x}_\alpha^{(i)} + 2\zeta_\alpha^{(i)}\dot{x}_\alpha^{(i)} + (\omega_\alpha^{(i)})^2 x_\alpha^{(i)} = b_\alpha^{(i)} + Q_\alpha^{(i)} \cos \Omega t \tag{5}$$

where

$$\left.\begin{aligned}&\zeta_\alpha^{(i)}=\frac{r_1+r_2}{2(m_1+m_2)},\qquad \omega_\alpha^{(i)}=\sqrt{\frac{k_1+k_2}{m_1+m_2}},\qquad Q_\alpha^{(i)}=\frac{A_0}{m_1+m_2};\\&b_\alpha^{(1)}=\frac{B_0}{m_1+m_2}\pm\frac{k_2d}{2(m_1+m_2)},\qquad b_\alpha^{(2)}=\frac{B_0}{m_1+m_2}\mp\frac{k_1d}{2(m_1+m_2)}.\end{aligned}\right\}\tag{6}$$

From physics points of view, there is a pair of internal forces during the sticking of two gears, and such internal forces are assumed to be positive in the negative direction, expressed by for $\alpha = 1, 3$

$$\left.\begin{aligned}f_\alpha^{(1)}&=-m_1\ddot{x}_\alpha^{(1)}-r_1\dot{x}_\alpha^{(1)}-k_1x_\alpha^{(1)}+B_0+A_0\cos\Omega t,\\f_\alpha^{(2)}&=-m_2\ddot{x}_\alpha^{(2)}-r_2\dot{x}_\alpha^{(2)}-k_0x_\alpha^{(2)}.\end{aligned}\right\}\tag{7}$$

From the Newton's third law, we have

$$f_\alpha^{(1)}=-f_\alpha^{(2)}.\tag{8}$$

Consider the 2nd gear to be a base reference as in Fig. 1. In region $\alpha = 1$, $f_\alpha^{(1)} > 0$ and $f_\alpha^{(2)} < 0$, but in region $\alpha = 3$, $f_\alpha^{(1)} < 0$ and $f_\alpha^{(2)} > 0$. The stick motion vanishing requires

$$f_\alpha^{(i)}=0\quad \text{for } i=1,2.\tag{9}$$

The stick condition for two gear oscillators is given for $i = 1, 2$ and $\alpha = 1, 3$

$$f_\alpha^{(i)}\operatorname{sgn}(x_\alpha^{(i)}-x_\alpha^{(\bar{i})})>0.\tag{10}$$

Further, the condition for stick vanishing is given by

$$f_\alpha^{(i)}\operatorname{sgn}(x_\alpha^{(i)}-x_\alpha^{(\bar{i})})=0.\tag{11}$$

In region $\alpha = 2$, two gear oscillators do not interfere each other. So $f_2^{(i)} = 0$ holds always.

3 Switching Sets and Mappings

As a result of the two gears impacting, the phase plane for each gear is discontinuous. The phase plane domains and boundaries were mathematically defined in Luo and O'Connor [13]. Based on the connectable domain, the mapping structures were introduced to describe possible motions. For the gear transmission system, equations of motion in the absolute frame are from Luo and O'Connor [14]

$$\dot{\mathbf{x}}_\alpha^{(i)}=\mathbf{F}_\alpha^{(i)}(\mathbf{x}_\alpha^{(i)},t)\tag{12}$$

for $i = 1, 2$ and $\alpha = 1, 2, 3$ with the following vectors

$$\begin{aligned}\mathbf{x}_\alpha^{(i)} &= (x_\alpha^{(i)}, \dot{x}_\alpha^{(i)})^T = (x_\alpha^{(i)}, y_\alpha^{(i)})^T,\\ \mathbf{F}_\alpha^{(i)} &= (\dot{x}_\alpha^{(i)}, F_\alpha^{(i)})^T = (y_\alpha^{(i)}, F_\alpha^{(i)})^T;\end{aligned} \tag{13}$$

where

$$F_\alpha^{(i)} = -2\zeta_\alpha^{(i)}\dot{x}_\alpha^{(i)} - (\omega_\alpha^{(i)})^2 x_\alpha^{(i)} + b_\alpha^{(i)} + Q_\alpha^{(i)} \cos \Omega t, \tag{14}$$

and the superscript "i" represents the ith mass and the subscript "α" represents the α-domain. From discontinuous boundaries in [14], the switching planes based on the two impacting chatter boundaries are defined as

$$\left.\begin{aligned}{}^R\Sigma_{2\infty}^{(i)} &= \left\{(t_k, x_k^{(i)}, \dot{x}_k^{(i)}, \dot{x}_k^{(\bar{i})}) \middle| x_k^{(\bar{i})} = x_k^{(i)} - \frac{d}{2}, \dot{x}_k^{(i)} \neq \dot{x}_k^{(\bar{i})}\right\},\\ {}^L\Sigma_{2\infty}^{(i)} &= \left\{(t_k, x_k^{(i)}, \dot{x}_k^{(i)}, \dot{x}_k^{(\bar{i})}) \middle| x_k^{(\bar{i})} = x_k^{(i)} + \frac{d}{2}, \dot{x}_k^{(i)} \neq \dot{x}_k^{(\bar{i})}\right\}.\end{aligned}\right\} \tag{15}$$

From now on, $x_k^{(i)} \equiv x^{(i)}(t_k)$ and $\dot{x}_k^{(i)} \equiv \dot{x}^{(i)}(t_k)$ on the separation boundary at time t_k. are switching displacement and velocity. The switching phase is defined by $\varphi_k = \text{mod}(\Omega t_k, 2\pi)$. Based on the above definitions of switching planes, four mappings are defined in the absolute frame as

$$\left.\begin{aligned}&P_2 : {}^R\Sigma_{2\infty}^{(i)} \to {}^R\Sigma_{2\infty}^{(i)}, \qquad P_3 : {}^R\Sigma_{2\infty}^{(i)} \to {}^L\Sigma_{2\infty}^{(i)};\\ &P_5 : {}^L\Sigma_{2\infty}^{(i)} \to {}^L\Sigma_{2\infty}^{(i)}, \qquad P_6 : {}^L\Sigma_{2\infty}^{(i)} \to {}^R\Sigma_{2\infty}^{(i)}.\end{aligned}\right\} \tag{16}$$

To investigate stick motions in the gear transmission system, the switching planes for stick are defined as

$$\left.\begin{aligned}\Sigma_{12}^{(i)} &= \left\{(t_k, x_k^{(i)}, \dot{x}_k^{(i)}, \dot{x}_k^{(\bar{i})}) \middle| x_k^{(i)} = {}^R x_{2-}^{(\bar{i})} + \frac{d}{2}, \dot{x}_k^{(i)} = {}^R\dot{x}_{2-}^{(\bar{i})}\right\},\\ \Sigma_{21}^{(i)} &= \left\{(t_k, x_k^{(i)}, \dot{x}_k^{(i)}, \dot{x}_k^{(\bar{i})}) \middle| x_k^{(i)} = {}^R x_{2+}^{(\bar{i})} + \frac{d}{2}, \dot{x}_k^{(i)} = {}^R\dot{x}_{2+}^{(\bar{i})}\right\};\\ \Sigma_{23}^{(i)} &= \left\{(t_k, x_k^{(i)}, \dot{x}_k^{(i)}, \dot{x}_k^{(\bar{i})}) \middle| x_k^{(i)} = {}^L x_{2-}^{(\bar{i})} - \frac{d}{2}, \dot{x}_k^{(i)} = {}^L\dot{x}_{2-}^{(\bar{i})}\right\},\\ \Sigma_{32}^{(i)} &= \left\{(t_k, x_k^{(i)}, \dot{x}_k^{(i)}, \dot{x}_k^{(\bar{i})}) \middle| x_k^{(i)} = {}^L x_{2+}^{(\bar{i})} - \frac{d}{2}, \dot{x}_k^{(i)} = {}^L\dot{x}_{2+}^{(\bar{i})}\right\}.\end{aligned}\right\} \tag{17}$$

The two switching planes can be treated as the same for all mappings. Except for two stick mappings (i.e., P_1 and P_4). the other mappings are the same as in (16). From the stick switching planes, the mappings are defined as

$$\left.\begin{aligned}&P_1 : \Sigma_{21}^{(i)} \to \Sigma_{12}^{(i)}, \qquad P_2 : \Sigma_{12}^{(i)} \to \Sigma_{21}^{(i)}, \qquad P_3 : \Sigma_{12}^{(i)} \to \Sigma_{23}^{(i)};\\ &P_4 : \Sigma_{23}^{(i)} \to \Sigma_{32}^{(i)}, \qquad P_5 : \Sigma_{23}^{(i)} \to \Sigma_{32}^{(i)}, \qquad P_6 : \Sigma_{32}^{(i)} \to \Sigma_{21}^{(i)}.\end{aligned}\right\} \tag{18}$$

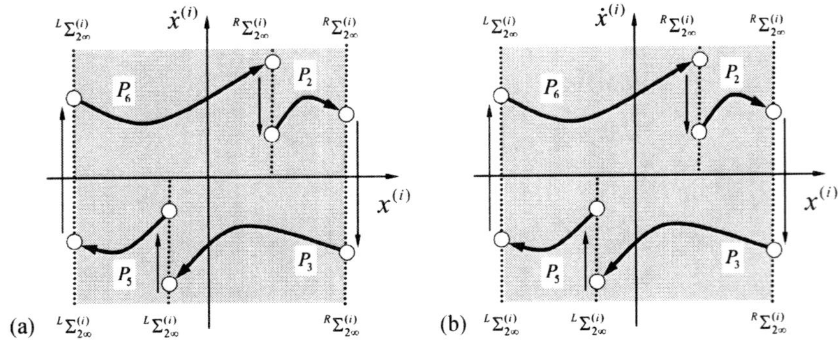

Fig. 2 Basic mappings: (**a**) impacting chatter only and (**b**) with stick switching. The straight line with arrow represents an impact on the boundary

With mixed switching planes, four mappings are defined by

$$\left.\begin{array}{ll} P_2 : \Sigma_{12}^{(i)} \to {}^{R}\Sigma_{2\infty}^{(i)}, & P_2 : {}^{R}\Sigma_{2\infty}^{(i)} \to \Sigma_{21}^{(i)}; \\ P_3 : \Sigma_{12}^{(i)} \to {}^{L}\Sigma_{2\infty}^{(i)}, & P_3 : {}^{R}\Sigma_{2\infty}^{(i)} \to \Sigma_{23}^{(i)}; \end{array}\right\} \tag{19a}$$

$$\left.\begin{array}{ll} P_5 : \Sigma_{23}^{(i)} \to {}^{L}\Sigma_{2\infty}^{(i)}, & P_5 : {}^{L}\Sigma_{2\infty}^{(i)} \to \Sigma_{32}^{(i)}, \\ P_6 : \Sigma_{32}^{(i)} \to {}^{R}\Sigma_{2\infty}^{(i)}, & P_6 : {}^{L}\Sigma_{2\infty}^{(i)} \to \Sigma_{21}^{(i)}. \end{array}\right\} \tag{19b}$$

Among four basic mappings, the two mappings (P_2 and P_5) are local and the other two mappings (P_3 and P_6) are global. The local mapping will map the motion from a switching plane onto itself. However, the global mapping will map the motion from a switching plane to another one. Such mappings are sketched in Fig. 2(a). The corresponding switching planes are labeled. On the impacting chatter boundaries, impacts are expressed by thin straight lines with arrows. The mappings relative to the stick switching planes only are sketched in Fig. 2(b). Only two stick mappings (P_1 and P_2) are new, and the other four mappings are the same as in Fig. 2(a). The mappings based on the sticking and impacting switching planes are presented in Fig. 3(a) and (b).

Set a vector as

$$\mathbf{y}_k \equiv (t_k, x_k^{(i)}, \dot{x}_k^{(i)}, \dot{x}_k^{(\bar{i})})^T. \tag{20}$$

For the impacting maps $P_\sigma\,(\sigma = 1, 2, \ldots, 6)$, $\mathbf{y}_{k+1} = P_\sigma \mathbf{y}_k$ can be expressed by

$$P_\sigma : (t_k, x_k^{(i)}, \dot{x}_k^{(i)}, \dot{x}_k^{(\bar{i})}) \to (t_{k+1}, x_{k+1}^{(i)}, \dot{x}_{k+1}^{(i)}, \dot{x}_{k+1}^{(\bar{i})}). \tag{21}$$

From Appendix in Luo and O'Connor [13, 14], the absolute displacement and velocity for two gear oscillators can be obtained with initial conditions $(t_k, x_k^{(i)}, \dot{x}_k^{(i)})$ and $(t_k, x_k^{(\bar{i})}, \dot{x}_k^{(\bar{i})})$. The final state for time t_{k+1} can be given. The switching planes give $x_\gamma^{(\bar{i})} = x_\gamma^{(i)} \pm \frac{d}{2}$ $(\gamma = k, k+1)$.

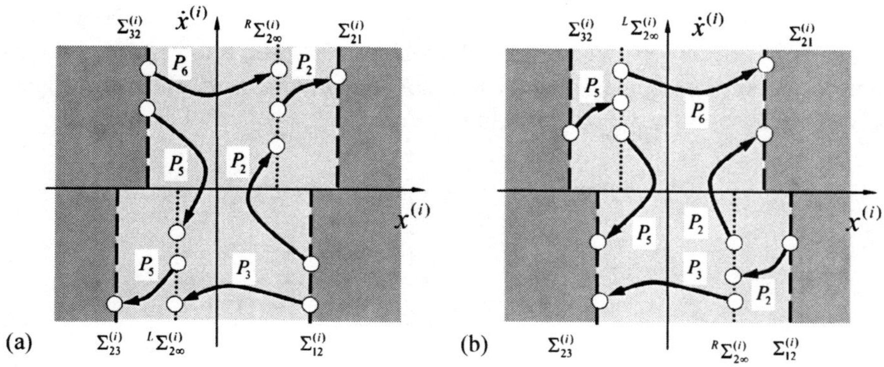

Fig. 3 Mappings between switching planes for stick and impacting

4 Mapping Structures

To describe motions in such a gear transmission system, the notation for mapping actions of basic mappings is introduced as in Luo [10, 15]

$$P_{n_k \cdots n_2 n_1} \equiv P_{n_k} \circ \cdots \circ P_{n_2} \circ P_{n_1} \tag{22}$$

where the mapping $P_{n_j} (n_j \in \{1, 2, \ldots, 6\}, j = 1, 2, \ldots, k)$ is defined in the previous section. Consider a generalized mapping structure as

$$\begin{aligned} &\underbrace{P_{(65^{k_{s4}}4^{k_{s3}}31^{k_{s2}}2^{k_{s1}})\cdots(65^{k_{14}}4^{k_{13}}31^{k_{12}}2^{k_{11}})}}_{s\text{-terms}} \\ &\quad = \underbrace{P_{(65^{k_{s4}}4^{k_{s3}}31^{k_{s2}}2^{k_{s1}})} \circ \cdots \circ P_{(65^{k_{14}}4^{k_{13}}31^{k_{12}}2^{k_{11}})}}_{s\text{-terms}} \end{aligned} \tag{23}$$

where $(k_{\mu\nu} \in \{0, \mathbb{N}\}, \mu = 1, 2, \ldots, s, \nu = 1, 2, 3, 4)$. From the generalized mapping structure, consider a simple mapping structure of periodic motions for impacting chatter. For instance, the mapping structure is

$$P_{65^n 32^m} = P_6 \circ P_{5^n} \circ P_3 \circ P_{2^m} \tag{24}$$

where $m, n \in \{0, \mathbb{N}\}$. Such a mapping structure gives $(m + 1)$-impacts on the right boundary and $(n + 1)$-impacts on the left boundary, which are described by mappings P_2 and P_5, respectively. Through the global mappings P_3 and P_6, the impacting chatters on the two boundaries are connected together. Consider a periodic motion of $P_{65^n 32^m}$ with period $T_1 = k_1 T$ $(k_1 \in \mathbb{N})$. If the mapping structure copies itself, a new mapping structure is:

$$P_{(65^n 32^m)^{2^l}} = P_{(65^n 32^m)^{2^{l-1}}} \circ P_{(65^n 32^m)^{2^{l-1}}}. \tag{25}$$

As $l \to \infty$, a chaotic motion relative to mapping structure $P_{65^n32^m}$ is formed. The prescribed chaos is generated by period-doubling. However, if the grazing bifurcation occurs, such a mapping structure may not be copied by itself. The new mapping structures are combined by the two different mapping structures. For instance,

$$\begin{aligned} P_{65^{n_2}32^{m_2}65^{n_1}32^{m_1}} &= P_{65^{n_2}32^{m_2}} \circ P_{65^{n_1}32^{m_1}}, \\ &\vdots \\ P_{65^{n_l}32^{m_l}\ldots 65^{n_1}32^{m_1}} &= \underbrace{P_{65^{n_l}32^{m_l}} \circ \cdots \circ P_{65^{n_1}32^{m_1}}}_{l\text{-terms}}. \end{aligned} \tag{26}$$

Such a gazing bifurcation will cause the discontinuity of periodic motions, and chaotic motions may exist between periodic motions of $P_{65^{n_l}32^{m_l}\ldots 65^{n_1}32^{m_1}}$ and $P_{65^{n_{l-1}}32^{m_{l-1}}\ldots 65^{n_1}32^{m_1}}$.

For low excitation frequency, the impacting chatter accompanying stick motion exists in the gear transmission system. Consider a simple chatter with stick motion with the following mapping structure

$$P_{645^n312^m} = P_6 \circ P_4 \circ P_{5^n} \circ P_3 \circ P_1 \circ P_{2^m}. \tag{27}$$

From the above mapping structure, m-impacts on the right boundary and n-impacts on the left boundary, which are described by mappings P_2 and P_5, respectively. In addition, both the mth mapping of P_2 and the nth mapping of P_5 map the impacting boundary to the stick boundary, and the stick mappings are P_1 and P_4, respectively. The two global mappings P_3 and P_6 connect the impact and stick boundaries. Similarly, a mapping structure for period-doubling is

$$P_{(645^n312^m)^{2^l}} = P_{(645^n312^m)^{2^{l-1}}} \circ P_{(645^n312^m)^{2^{l-1}}}. \tag{28}$$

Due to grazing bifurcation, the mapping structures are:

$$\begin{aligned} P_{645^{n_2}32^{m_2}65^{n_1}312^{m_1}} &= P_{645^{n_2}312^{m_2}} \circ P_{645^{n_1}312^{m_1}}, \\ &\vdots \\ P_{645^{n_l}312^{m_l}\ldots 645^{n_1}312^{m_1}} &= \underbrace{P_{645^{n_l}312^{m_l}} \circ \cdots \circ P_{645^{n_1}312^{m_1}}}_{l\text{-terms}}. \end{aligned} \tag{29}$$

To help one understand two sorts of mapping structures, the two simple mapping structures are shown in Fig. 4(a) and (b) for the impacting chatter with and without stick of two gear systems. Similarly, the other mapping structures can be discussed through the generalized mapping structure in (25). Periodic and chaotic motions relative to a certain mapping structure can be determined.

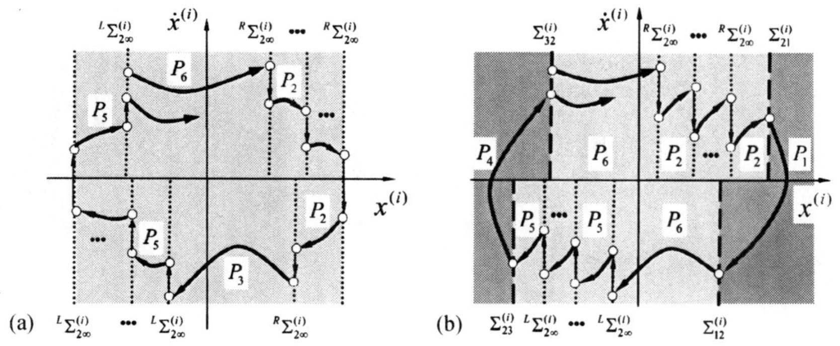

Fig. 4 Mapping structures for (**a**) impacting chatter and (**b**) impacting chatter with stick motion of two gear systems

5 Parameter Maps and Illustrations

The parameter map for excitation frequency Ω versus restitution e are shown in Fig. 5 for parameters ($m_1 = 2, m_2 = 1, r_1 = r_2 = 0.6, k_1 = 30, k_2 = 20, Q_0 = 50.0$ and $d = 1.0$. In Fig. 5(a), the entire range of excitation frequency for two masses experiencing interaction is presented. The zoomed view of the parameter map is given in Fig. 5(b) for $\Omega \in [0, 8]$. The chatter with stick possesses a mapping structure of $P_{645^n312^n}$ for $n = 1, 2, \ldots, 70$. The number of impacting chatters increases with increasing e. The region labeled by "Chatter" represents chatter with stick where the chatter impacts number approaches infinity as $e \to 1$. The region just above the region for the chatters with stick has complex mapping structure. Within the "complex motion" region, chaotic and periodic motions of impacting chatter without stick exist, and the corresponding mapping structures are relative to $P_{65^232^2}$ and $P_{65^332^3}$. In additions, the regions relative to periodic motions of $P_{65^232^2}$ and $P_{65^332^3}$ are labeled. With increasing excitation frequency, symmetric and asymmetric periodic motions with the mapping structure of P_{63} are presented. The larger region is symmetric while the smaller region is asymmetric. For higher excitation frequency, the two

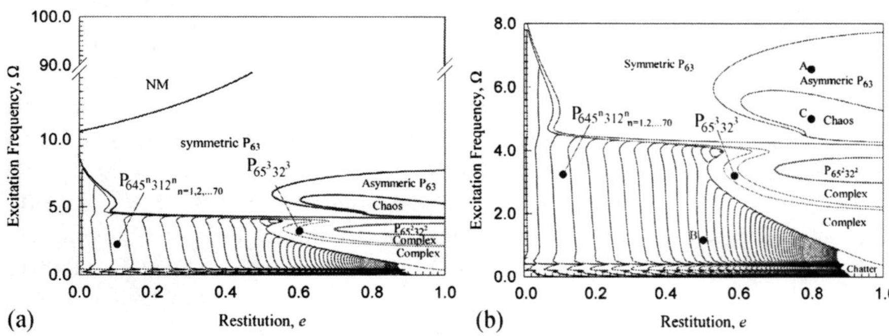

Fig. 5 Parameter map for excitation frequency versus restitution ($m_1 = 2, m_2 = 1, r_1 = r_2 = 0.6, k_1 = 30, k_2 = 20, Q_0 = 50.0$ and $d = 1.0$)

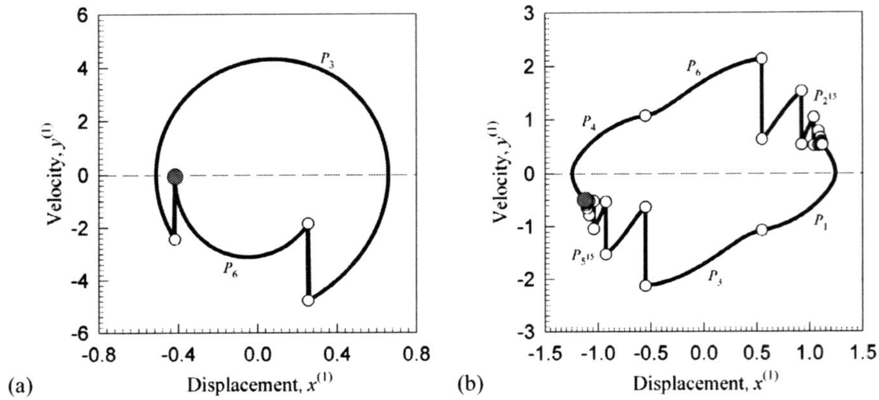

Fig. 6 Phase planes: (**a**) asymmetric impacts $P_{63}(\Omega = 6.6$ and $e = 0.8$; $t_0 \approx 0.8852$, $x_0^{(1)} \approx -0.4179$, $y_0^{(1)} \approx -0.0770$ and $x_0^{(2)} \approx -0.9179$, $y_0^{(2)} \approx -4.0045)$ and (**b**) impact chatter with stick $P_{5^{15}312^{15}64}$ ($\Omega = 1.0$ and $e = 0.5$; $t_0 \approx 2.6000$, $x_0^{(1)} \approx -1.1201$, $y_0^{(1)} \approx -0.5273$ and $x_0^{(2)} \approx -0.6201$, $y_0^{(2)} \approx -0.5271$)

masses will not contact each other, and such a region is labeled by "NM". It means the two gears do not transfer any energy.

To demonstrate motions with specific mapping structures in the parameter map, three sets of excitation frequency and restitution are used, and they are labeled through points A, B and C in Fig. 5(b). At the point "A", $\Omega = 6.6$ and $e = 0.8$ are selected. For this point, the initial conditions are $t_0 \approx 0.8852$, $x_0^{(1)} \approx -0.4179$, $y_0^{(1)} \approx -0.0770$ and $x_0^{(2)} \approx -0.9179$, $y_0^{(2)} \approx -4.0045$. The corresponding phase plane is plotted in Fig. 6(a). The motion starts with just after the driving gear impacts at the right hand side of the driven gear. The next impact takes place at the left hand side of the driven gear and then returns back to the right side again. The asymmetric motion is relative to mapping P_6 and P_3, its twin asymmetric motion will not presented and the detailed discussion can referred to Luo [15]. For parameters (i.e., $\Omega = 1.0$ and $e = 0.5$) labeled "B" in Fig. 5(b), the periodic motion of impacting chatter with stick $P_{5^{15}312^{15}64}$ is plotted in Fig. 6(b) with initial conditions ($t_0 \approx 2.6000$, $x_0^{(1)} \approx -1.1201$, $y_0^{(1)} \approx -0.5273$ and $x_0^{(2)} \approx -0.6201$, $y_0^{(2)} \approx -0.5271$). The driving gear begins at the onset of stick motion relative to P_4 on the left hand side of the driven gear. Crossing the tooth gap from the left to right side of the driven gear is the mapping of P_6. The two gears impact fifteen times (i.e., $P_{2^{15}}$) before a new stick motion is formed on the right side, and the stick motion is described through the mapping of P_1. The second half of the periodic motion can be described in a similar fashion. Finally, the chaotic motion is demonstrated through Poincaré mapping sections at point C (i.e., $\Omega = 5.0$ and $e = 0.8$). The initial conditions are $t_0 \approx 0.0641$, $x_0^{(1)} \approx -1.5161$, $y_0^{(1)} \approx 6.0031$ and $x_0^{(2)} \approx -2.0161$, $y_0^{(2)} \approx 3.5209$. The switching points are plotted in Fig. 7 for ten thousand periods (10^4T) of the excitation forcing. The Poincaré mapping sections of switching points for the 1st and 2nd masses are given in Fig. 7(a) and (b), respectively. The switching points describe the posi-

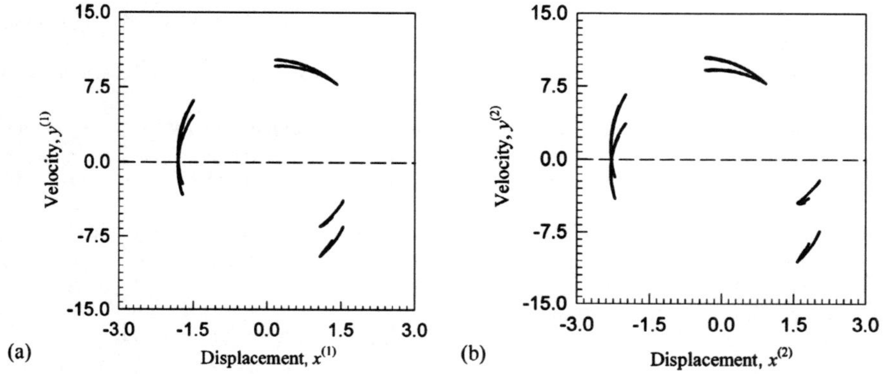

Fig. 7 Poincaré mapping sections for chaos ($\Omega = 5.0$ and $e = 0.8$): (**a**) mass m_1 and (**b**) mass m_2. ($t_0 \approx 0.0641$, $x_0^{(1)} \approx -1.5161$, $y_0^{(1)} \approx 6.0031$ and $x_0^{(2)} \approx -2.0161$, $y_0^{(2)} \approx 3.5209$)

tion and velocity of the driving and driven gears upon impact. The switching points form a strange attractor of chaotic motions for such a gear transmission system. In a similar fashion, the periodic and chaotic motions can be illustrated.

6 Conclusions

Nonlinear dynamical behaviors of a gear transmission system with impacts were investigated through an impact model with possible stick between the two gears. Switching sets and basic mappings were introduced to identify periodic and chaotic motions in such a gear transmission system. To understand the global dynamical behaviors of the gear transmission system, system parameter maps were developed analytically and numerically. Numerical simulations for illustration of periodic and chaotic motions in such a gear transmission system were performed from the parameter maps.

References

1. C. Changenet, X. Oviedo-Marlot, P. Velex, Power loss predictions in geared transmissions using thermal networks-applications to a six-speed manual gearbox. ASME J. Mech. Design **128**(3), 618–625 (2006)
2. E. Buckingham, *Dynamic Loads on Gear Teeth* (American Special Publication, New York, 1931)
3. E. Buckingham, *Analytical Mechanics of Gears* (McGraw-Hill, New York, 1949)
4. J.P.D. Hartog, S.J. Mikina, Forced vibrations with non-linear spring constants. ASME J. Appl. Mech. **58**, 157–164 (1932)
5. H.N. Ozguven, D.R. Houser, Mathematical models used in gear dynamics—a review. J. Sound Vib. **121**(3), 383–411 (1988)
6. F. Pfeiffer, Mechanische systems mit unstetigen ubergangen. Ing. Arch. **54**, 232–240 (1984)

7. K. Karagiannis, F. Pfeiffer, Theoretical and experimental investigations of gear Box. Nonlinear Dyn. **2**, 367–387 (1991)
8. S. Theodossiades, S. Natsiavas, Non-linear dynamics of gear-pair systems with periodic stiffness and backlash. J. Sound Vib. **229**(2), 287–310 (2000)
9. A.C.J. Luo, L.D. Chen, Periodic motion and grazing in a harmonically forced, piecewise linear, oscillator with impacts. Chaos Solitons Fractals **24**, 567–578 (2005)
10. A.C.J. Luo, A theory for non-smooth dynamic systems on the connectable domains. Commun. Nonlinear Sci. Numer. Simul. **10**, 1–55 (2005)
11. A.C.J. Luo, L.D. Chen, The grazing mechanism of the strange attractor fragmentation of a harmonically forced, piecewise, linear oscillator with impacts. IMeChE Part K, J. Multi-Body Dyn. **220**, 35–51 (2006)
12. A.C.J. Luo, L.D. Chen, Arbitrary periodic motions and grazing switching of a forced piecewise-linear, impacting oscillator. ASME J. Vib. Acoust. **129**, 276–285 (2007)
13. A.C.J. Luo, D. O'Connor, Nonlinear dynamics of a gear transmission system, Part I: mechanism of impacting chatter with stick, in *2007 ASME International Design Engineering Conferences and Exposition*, September 4–7, 2007, Las Vegas, Nevada. IDETC2007-34881 (2007)
14. A.C.J. Luo, D. O'Connor, Nonlinear dynamics of a gear transmission system, Part II: periodic impacting chatter and stick, in *2007 ASME International Mechanical Engineering Congress and Exposition*, November 10–16, 2007, Seattle, Washington. IMECE2007-43192 (2007)
15. A.C.J. Luo, *Singularity and Dynamics on Discontinuous Vector Fields* (Elsevier, Amsterdam, 2006)

Analytical Prediction of Interrupted Cutting Periodic Motions in a Machine Tool

Brandon C. Gegg, Steve C.S. Suh, and Albert C.J. Luo

Abstract The methodology for prediction of interrupted cutting periodic motions in a machining system is developed. The interrupted cutting mappings in the vicinity of the system constraints are defined. The criteria for the interrupted cutting periodic motions are developed through the state variables and mapping forms. The periodic interrupted cutting motions in a two-degree-of-freedom model are predicted numerically and analytically via closed form solutions. The chip and tool-piece seizure in the machine-tool system is also discussed. The bifurcations are caused by interactions of continuous dynamical systems in the neighborhood of the boundary.

Keywords Contact mechanism · Cutting · Cutting dynamics · Discontinuity · Friction · Interrupted cutting · Machine-tool · Machine tool vibration · Manufacturing · Tool-piece · Workpiece

1 Introduction

Researchers have continually worked to improve the performance of machining systems. Understanding the underlying dynamics of machining systems is necessary for machining limits to be expanded. A basic representation of the machine-tool system can be described by three situations: (i) the tool not contacting with work-piece, (ii) the tool contacting the work-piece *without* cutting, and (iii) the tool contacting the work-piece *with* cutting. A lot of research has been conducted for cutting only.

B.C. Gegg (✉) · S.C.S. Suh
Department of Mechanical Engineering, Texas A&M University, College Station, TX 77843-3123, USA
e-mail: brandon-gegg@tamu.edu

A.C.J. Luo
Department of Mechanical Engineering, Southern Illinois University at Edwardsville, Edwardsville, IL 62026-1805, USA

J.A.T. Machado et al. (eds.), *Nonlinear Science and Complexity*,
DOI 10.1007/978-90-481-9884-9_3,

Merchant [1] in 1945, as among the earliest researchers, developed the shear angle solution from the principle of minimum work. In 2006, the intermittent loss of cutting was presented by Chandiramani and Pothala [2]. In 2008, Gegg et al. [3] presented the loss of contact between the work-piece and tool-piece. A few researches have been completed in this area. The multiple-discontinuities (e.g., the cutting and thrust forces, elastic deformation and stagnation effects) are a natural occurrence in machining systems. For example, the friction forces are typically dependent on the relative velocity between the chip and tool rake face. Some of the earliest studies of discontinuous systems were found in 1930s. For instance, Hartog [4] investigated the forced vibration with Coulomb and viscous damping in theory and experiment. In 1994, Wiercigroch [5] studied the stick-slip phenomena for such a special case. In 2005, Luo [6] developed a general theory for the local singularity of non-smooth dynamical systems on connectable domains (also see, Luo [7]). In 2006, Luo and Gegg [8] applied such a general theory of discontinuous systems to a forced, dry-friction oscillator. The stick and non-stick motions and grazing phenomenon with respect to a friction (velocity) boundary were presented through the vector fields of the oscillator. Machine-tool systems contain multiple discontinuities, which can be analyzed through such a methodology. The discontinuities for the most basic machine-tool systems are considered as: (i) displacement boundaries (i.e., loss of contact with the work-piece); force boundaries (onset of cutting), and (ii) velocity boundaries (i.e., chip/tool rake and work-piece/tool flank stick-slip). The existence of multiple boundaries is dependent on the complexity of the model.

2 Mechanical Model

Consider a machine-tool model given by an oscillator with two-degrees of freedom for regular and interrupted cutting, as shown in Fig. 1. The corresponding parameters are labeled. The (x, y)-coordinate system measures the deflection of the tool from the equilibrium point (X_{eq}, Y_{eq}) in the $(\mathbf{e}_x, \mathbf{e}_y)$ global coordinate system, as shown in Fig. 1. The equation of motion for such a machine-tool model is from

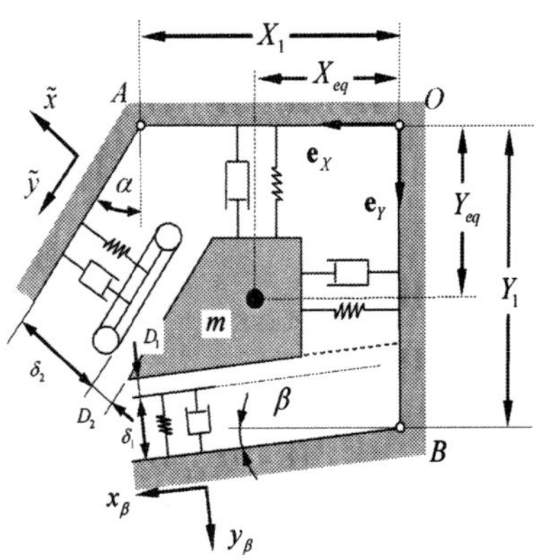

Fig. 1 Geometric characteristics of cutting processing for the machine tool

Gegg [3].

$$\begin{bmatrix} 1 & 0 \\ 0 & 1 \end{bmatrix} \begin{Bmatrix} \ddot{x} \\ \ddot{y} \end{Bmatrix} + \begin{bmatrix} B_{11}^{(i)} & B_{12}^{(i)} \\ B_{21}^{(i)} & B_{22}^{(i)} \end{bmatrix} \begin{Bmatrix} \dot{x} \\ \dot{y} \end{Bmatrix} + \begin{bmatrix} K_{11}^{(i)} & K_{12}^{(i)} \\ K_{21}^{(i)} & K_{22}^{(i)} \end{bmatrix} \begin{Bmatrix} x \\ y \end{Bmatrix}$$

$$= \begin{Bmatrix} A_x^{(i)} \\ A_y^{(i)} \end{Bmatrix} \cos(t) + \begin{Bmatrix} C_x^{(i)} \\ C_y^{(i)} \end{Bmatrix} \quad \text{for } i = 1, 2, 3, 4; \tag{1}$$

where all the coefficients are listed in Appendix. For a special case of interrupted cutting (i.e., chip and tool-piece seizure),

$$\left.\begin{aligned} &\tilde{y} = \tilde{y}_0 - V(t - t_0), \qquad \dot{\tilde{y}} = V, \qquad \ddot{\tilde{y}} = 0, \\ &\ddot{\tilde{x}} + 2d\dot{\tilde{x}} + \omega^2 \tilde{x} = A_0 \cos(t) + B_0 t + C_0; \end{aligned}\right\} \tag{2}$$

where $V = \bar{V}/\Omega$. $\bar{V}$ is the chip velocity in the $\tilde{y}$-coordinate system and Ω is the excitation frequency of the periodical force acting on the tool-piece by contact with the work-piece.

$$\left.\begin{aligned} &\tilde{y} = (x - x_0)\sin\alpha + (y - y_0)\cos\alpha, \\ &\tilde{w} = \dot{x}\sin(\alpha) + \dot{y}\cos(\alpha). \end{aligned}\right\} \tag{3}$$

A more detailed explanation of this model was given in Gegg et al. [3]. The machine-tool motion described by (1)–(2) can be modeled by the discontinuous system theory as in Luo [6, 7]. The state and vector fields are defined as,

$$\mathbf{D}_1 \triangleq (D_1, \dot{D}_1)^T \equiv (u, v)^T \quad \text{and} \quad \mathbf{F}_1 \triangleq (v, F_{D_1}^{(i)}(\mathbf{D}_1, t))^T, \tag{4}$$

$$\mathbf{D}_2 \triangleq (D_2, \dot{D}_2)^T \equiv (r, s)^T \quad \text{and} \quad \mathbf{F}_2 \triangleq (s, F_{D_2}^{(i)}(\mathbf{D}_2, t))^T, \tag{5}$$

$$\mathbf{D}_3 = \tilde{\mathbf{y}} = (\tilde{y}, \tilde{w})^T \quad \text{and} \quad \mathbf{F}_{\tilde{y}}^{(\kappa)}(\tilde{\mathbf{x}}, \tilde{\mathbf{y}}, t) \triangleq (\tilde{w}, F_{\tilde{y}}^{(\kappa)}(\tilde{\mathbf{x}}, \tilde{\mathbf{y}}, t))^T$$
$$(\kappa \in \{0, 3, 4\}), \tag{6}$$

$$\mathbf{D}_4 = (p, q)^T \quad \text{and} \quad \mathbf{F}_{D_4}^{(\kappa)}(\tilde{\mathbf{x}}, \tilde{\mathbf{y}}, t) \triangleq (q, F_{D_4}^{(\kappa)}(\tilde{\mathbf{x}}, \tilde{\mathbf{y}}, t))^T \quad (\kappa \in \{2, 3\}), \tag{7}$$

where $\mathbf{x} = (x, \dot{x})^T$, $\mathbf{y} = (y, \dot{y})^T$, $\tilde{\mathbf{x}} = (\tilde{x}, \dot{\tilde{x}})^T$ and

$$\left.\begin{aligned} &u = (X_{eq} + x)\sin\beta + (Y_1 - Y_{eq} - y)\cos\beta - \delta_1, \\ &v = \Omega(\dot{x}\sin\beta - \dot{y}\cos\beta); \end{aligned}\right\} \tag{8}$$

$$\left.\begin{aligned} &r = (Y_{eq} + y)\sin\alpha + (X_1 - X_{eq} - x)\cos\alpha - \delta_2, \\ &s = \Omega(\dot{y}\sin\alpha - \dot{x}\cos\alpha); \end{aligned}\right\} \tag{9}$$

$$p = L_c - (\tilde{y}_0 - \tilde{y}) \quad \text{and} \quad q = \dot{\tilde{y}} - V. \tag{10}$$

The corresponding forces are given by

$$F_{D_1}^{(i)}(\mathbf{D}_1,t) = \ddot{D}_1 = \Omega^2(\ddot{x}^{(i)}\sin\beta + \ddot{y}^{(i)}\cos\beta)$$
$$= \Omega^2[F_x^{(i)}(\mathbf{x},\mathbf{y},t)\sin\beta + F_y^{(i)}(\mathbf{x},\mathbf{y},t)\cos\beta], \tag{11}$$
$$F_{D_2}^{(i)}(\mathbf{D}_2,t) = \ddot{D}_2 = \Omega^2(\ddot{x}^{(i)}\sin\alpha + \ddot{y}^{(i)}\cos\alpha)$$
$$= \Omega^2[F_x^{(i)}(\mathbf{x},\mathbf{y},t)\sin\alpha + F_y^{(i)}(\mathbf{x},\mathbf{y},t)\cos\alpha], \tag{12}$$
$$F_{\tilde{y}}^{(i)}(\tilde{\mathbf{x}},\tilde{\mathbf{y}},t) = F_{D_3}^{(i)}(\tilde{\mathbf{x}},\tilde{\mathbf{y}},t) = F_{D_4}^{(i)}(\tilde{\mathbf{x}},\tilde{\mathbf{y}},t)$$
$$= \ddot{\tilde{y}}^{(i)}(t) = \ddot{x}^{(i)}\sin\alpha + \ddot{y}^{(i)}\cos\alpha$$
$$= F_x^{(i)}(\mathbf{x},\mathbf{y},t)\sin\alpha + F_y^{(i)}(\mathbf{x},\mathbf{y},t)\cos\alpha. \tag{13}$$

Since cutting processing, there are many dynamical states. From such cutting dynamical states, the machine-tool dynamical systems can be described by a discontinuous dynamical system with many boundaries. Discontinuous systems theory requires that the appropriate phase planes be partitioned to identify the discontinuities for the machine-tool system, as sketched in Fig. 2. The domains are defined as

$$\Omega_1 = \{(x,y,\dot{x},\dot{y}) | u(x,y,\dot{x},\dot{y}) \in (0,\infty)\}, \tag{14}$$
$$\Omega_2 = \{(x,y,\dot{x},\dot{y}) | u(x,y,\dot{x},\dot{y}) \in (-\infty,0), r(x,y,\dot{x},\dot{y}) \in (0,\infty),$$
$$p(x,y,\dot{x},\dot{y}) \in (-\infty,0), \tilde{w}(x,y,\dot{x},\dot{y}) \in (-\infty,V)\}, \tag{15}$$
$$\Omega_3 = \{(x,y,\dot{x},\dot{y}) | u(x,y,\dot{x},\dot{y}) \in (-\infty,0), r(x,y,\dot{x},\dot{y}) \in (-\infty,0),$$
$$p(x,y,\dot{x},\dot{y}) \in (0,L_0), \tilde{w}(x,y,\dot{x},\dot{y}) \in (-\infty,V)\}, \tag{16}$$
$$\Omega_4 = \{(x,y,\dot{x},\dot{y}) | u(x,y,\dot{x},\dot{y}) \in (-\infty,0), r(x,y,\dot{x},\dot{y}) \in (-\infty,0),$$
$$\tilde{w}(x,y,\dot{x},\dot{y}) \in (V,\infty)\} \tag{17}$$

and

$$\partial\Omega_{12} = \{(x,y,\dot{x},\dot{y}) | \varphi_{12} = \varphi_{21} = u(x,y,\dot{x},\dot{y}) = 0\}; \tag{18}$$
$$\partial\Omega_{24} = \{(x,y,\dot{x},\dot{y}) | \varphi_{24}^{(1)} = \varphi_{42}^{(1)} = r(x,y,\dot{x},\dot{y}) = 0 \text{ if } \tilde{w} > V,$$
$$\varphi_{24}^{(2)} = \tilde{w}(x,y,\dot{x},\dot{y}) - V = 0 \text{ if } r < 0\}; \tag{19}$$
$$\partial\Omega_{32} = \{(x,y,\dot{x},\dot{y}) | \varphi_{32}^{(1)} = \varphi_{23}^{(1)} = r(x,y,\dot{x},\dot{y}) = 0 \text{ if } \tilde{w} < V,$$
$$\varphi_{32}^{(2)} = p(x,y,\dot{x},\dot{y}) = 0 \text{ if } r < 0\}; \tag{20}$$
$$\partial\Omega_{34} = \{(x,y,\dot{x},\dot{y}) | \varphi_{34} = \varphi_{43} = \tilde{w}(x,y,\dot{x},\dot{y}) - V = 0 \text{ if } r < 0\}. \tag{21}$$

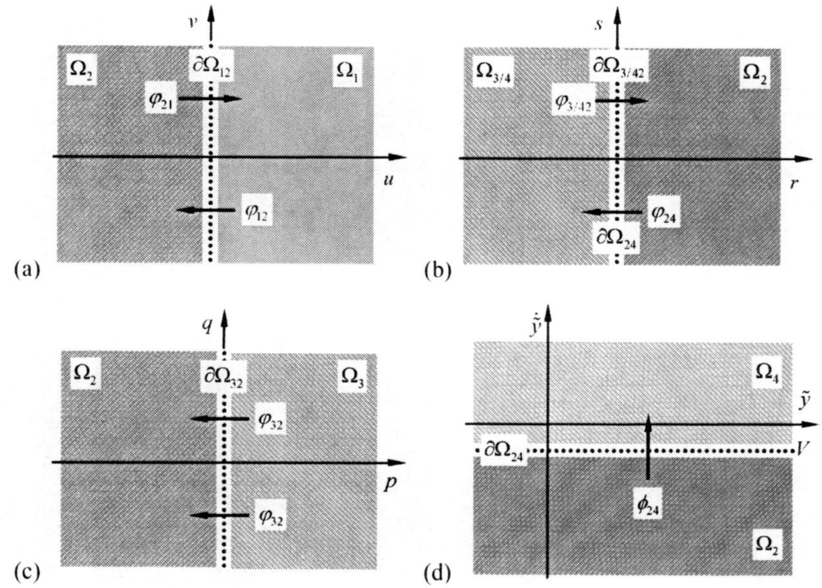

Fig. 2 Partitions in phase space for the displacement and velocity discontinuities of this machine–tool system: **(a)** (u, v), **(b)** (r, s), **(c)** (p, q) and **(d)** $(\tilde{y}, \dot{\tilde{y}})$

3 Motion Switchability Conditions

From Luo [6, 7], the passable motion is guaranteed for $t_n \subset (t_s, t_e)$ by

$$\left[\mathbf{n}^T_{\partial\Omega_{ij}} \cdot \mathbf{F}^{(i)}_{D_i}(\mathbf{D}_i, t_{n-})\right] \times \left[\mathbf{n}^T_{\partial\Omega_{ij}} \cdot \mathbf{F}^{(j)}_{D_i}(\mathbf{D}_i, t_{n+})\right] > 0, \tag{22}$$

(for $i \neq j$ and $i, j = 1, 2, 3, 4$); where the normal vector for the boundaries are

$$\mathbf{n}_{\partial\Omega_{12}} = \nabla\varphi_{12} = \left(\frac{\partial\varphi_{12}}{\partial u}, \frac{\partial\varphi_{12}}{\partial v}\right)^T_{(u_m, v_m)} = (1, 0)^T, \tag{23}$$

$$\mathbf{n}_{\partial\Omega_{24}} = \begin{cases} \nabla\varphi^{(1)}_{24} = (\frac{\partial\varphi^{(1)}_{24}}{\partial r}, \frac{\partial\varphi^{(1)}_{24}}{\partial s})^T_{(r_m, s_m)} = (1, 0)^T & \text{if } \tilde{w} > V, \\ \nabla\varphi^{(2)}_{24} = (\frac{\partial\varphi^{(2)}_{24}}{\partial \tilde{y}}, \frac{\partial\varphi^{(2)}_{24}}{\partial \tilde{w}})^T_{(\tilde{y}_m, \tilde{w}_m)} = (0, 1)^T & \text{if } r < 0; \end{cases} \tag{24}$$

$$\mathbf{n}_{\partial\Omega_{32}} = \begin{cases} \nabla\varphi^{(1)}_{32} = (\frac{\partial\varphi^{(1)}_{32}}{\partial p}, \frac{\partial\varphi^{(1)}_{32}}{\partial q})^T_{(p_m, q_m)} = (1, 0)^T & \text{if } r < 0, \\ \nabla\varphi^{(2)}_{32} = (\frac{\partial\varphi^{(2)}_{32}}{\partial r}, \frac{\partial\varphi^{(2)}_{32}}{\partial s})^T_{(r_m, s_m)} = (1, 0)^T & \text{if } \tilde{w} < V; \end{cases} \tag{25}$$

respectively. In addition,

$$\mathbf{n}_{\partial\Omega_{34}} = \nabla\varphi_{34} = \left(\frac{\partial\varphi_{34}}{\partial \tilde{y}}, \frac{\partial\varphi_{34}}{\partial \tilde{w}}\right)^T_{(\tilde{y}_m, \tilde{w}_m)} = (0, 1)^T. \tag{26}$$

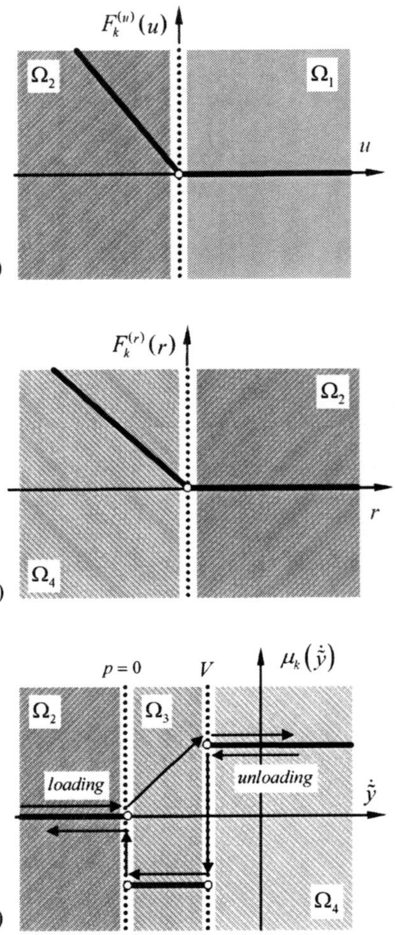

Fig. 3 Force definitions for this machine-tool system: (**a**) domain Ω_1 to domain Ω_2 force condition, (**b**) domain Ω_2 to domain Ω_4 force condition and (**c**) loading and unloading paths

The corresponding normal comments of vector fields are

$$\mathbf{n}_{\partial\Omega_{12}}^T \bullet \mathbf{F}_{D_1}^{(j)}(\mathbf{D}_1, t_{n-}) = v, \quad \text{for } j = 1, 2; \tag{27}$$

$$\left.\begin{array}{ll} \mathbf{n}_{\partial\Omega_{24}}^T \bullet \mathbf{F}_{D_2}^{(j)}(\mathbf{D}_2, t_{n-}) = s & \text{if } \tilde{w} > V, \\ \mathbf{n}_{\partial\Omega_{24}}^T \bullet \mathbf{F}_{D_3}^{(j)}(\mathbf{D}_3, t_{n-}) = F_{\tilde{y}}^{(j)}(\tilde{\mathbf{x}}, \tilde{\mathbf{y}}, t) & \text{if } r < 0, \end{array}\right\} \quad \text{for } j = 2, 4; \tag{28}$$

$$\left.\begin{array}{ll} \mathbf{n}_{\partial\Omega_{32}}^T \bullet \mathbf{F}_{D_4}^{(j)}(\mathbf{D}_4, t_{n-}) = q & \text{if } \tilde{w} < V, \\ \mathbf{n}_{\partial\Omega_{32}}^T \bullet \mathbf{F}_{D_2}^{(j)}(\mathbf{D}_2, t_{n-}) = s & \text{if } r < 0, \end{array}\right\} \quad \text{for } j = 2, 3; \tag{29}$$

$$\mathbf{n}_{\partial\Omega_{34}}^T \bullet \mathbf{F}_{D_3}^{(j)}(\mathbf{D}_3, t_{n-}) = F_{\tilde{y}}^{(j)}(\tilde{\mathbf{x}}, \tilde{\mathbf{y}}, t) \quad \text{for } j = 0, 3, 4. \tag{30}$$

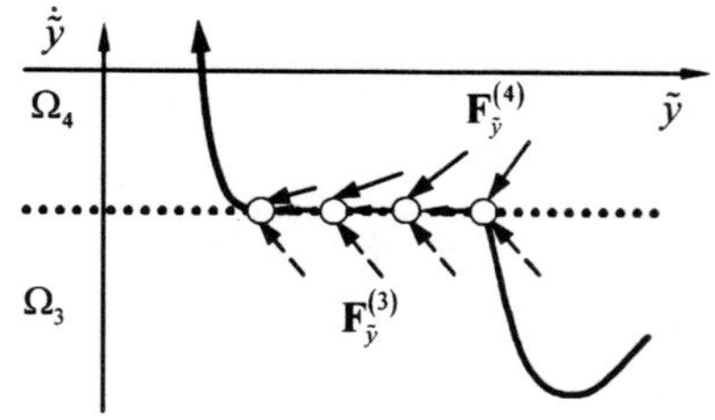

Fig. 4 Vector fields for chip seizure (stick motion)

The non-passable motion and vanishing of the non-passable motion is guaranteed for $t_m = [t_i, t_{i+1}]$ by

$$\left[\mathbf{n}_{\partial\Omega_{ij}}^T \bullet \mathbf{F}_{D_i}^{(i)}(\mathbf{D}_i, t_{m-})\right] \times \left[\mathbf{n}_{\partial\Omega_{ij}}^T \bullet \mathbf{F}_{D_i}^{(j)}(\mathbf{D}_i, t_{m+})\right] \leq 0. \tag{31}$$

The motion switches from domain Ω_1 to domain Ω_2 and (for $i \neq j$ and $i, j = 1, 2, 3, 4$), respectively. The force conditions at the point of switching are shown in Fig. 3. The stiffness force ($F_k^{(\gamma)}$ for $\gamma = u, r$) as domain Ω_1 to domain Ω_2 and domain Ω_2 to domain Ω_4 are shown in Fig. 3(a) and (b); respectively. The kinematic friction coefficient distribution switching from domain Ω_2 to domain Ω_4 jumps past domain Ω_3 on a loading path, see Fig. 3(c). The unloading path begins in domain Ω_4 moves through domain Ω_3 and ends at Ω_2. The vector fields at the domain switching points, with respect to the boundary, define whether the motion will be passable or non-passable, which is sketched in Fig. 4.

4 Mappings Structure

The motion of the machine tool is tracked through the phase plane by the mappings $P_{\alpha_{jk}}$ (for $j = 1, 2, 3, 4$; $k = 1, 2, 3, 4$; $\alpha = 0, 1, 2, 3, 4$), where j is the initial boundary, k is the final boundary and α is the domain, as shown in Fig. 5. The mappings describe the following cases: (i) vibration of the tool with no contact of the work-piece ($\alpha = 1$); (ii) the tool in contact with the work-piece but no cutting ($\alpha = 2$); (iii) the tool in contact with the work-piece with cutting where $\dot{z} < 0$ ($\alpha = 3$); (iv) the tool in contact with the work-piece with cutting where $\dot{z} > 0$ ($\alpha = 4$); and (v) the tool in contact with the tool in the special case where the chip tool rake face seizure occurs, $\dot{z} \equiv 0$ ($\alpha = 0$). The governing equation for each mapping can be represented by

$$f_m^{(\alpha)}(\tilde{\mathbf{x}}_0, \tilde{\mathbf{y}}_0, t_0, t) = 0 \quad (m = 1, 2, 3, 4). \tag{32}$$

The action of one mapping given a set of initial conditions $(\tilde{\mathbf{x}}_i, \tilde{\mathbf{y}}_i, t_i)$ yields a set of final conditions $(\tilde{\mathbf{x}}_{i+1}, \tilde{\mathbf{y}}_{i+1}, t)$ for this machine-tool system. The mappings can be combined to describe the trajectory of periodic orbit in the phase plane. For example, consider the mappings $P_{3_{33}}$ and $P_{4_{33}}$ in series; which can be simplified

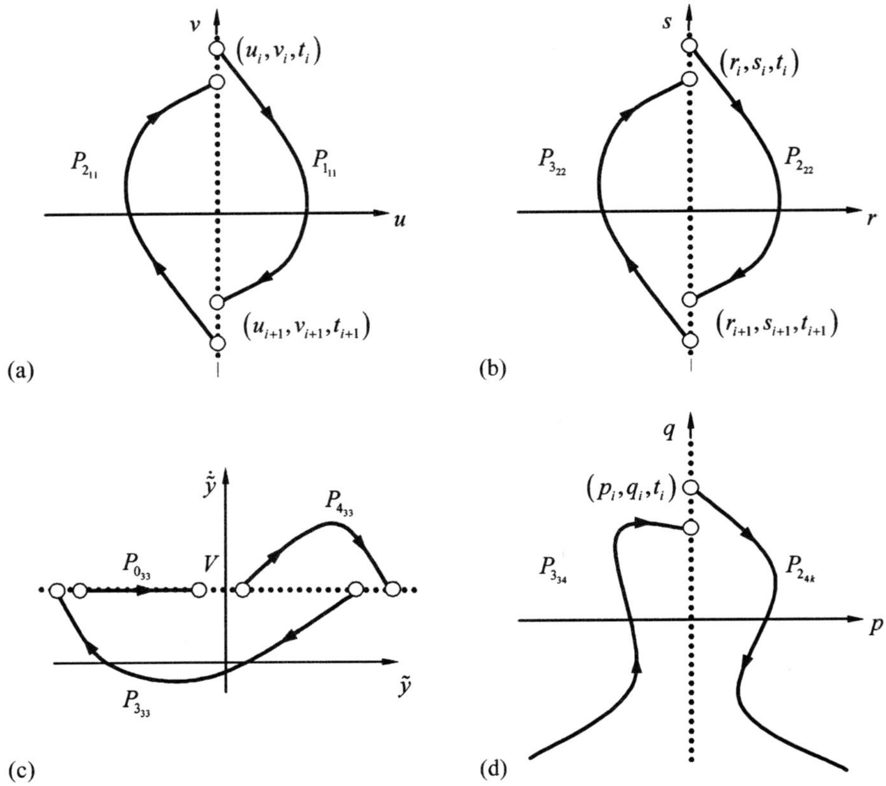

Fig. 5 Mappings: **(a)** $\mathbf{D}_1$, **(b)** $\mathbf{D}_2$, **(c)** $\mathbf{D}_3$ and **(d)** $\mathbf{D}_4$, phase planes

using the notation $P_{43} = P_{4_{33}} \circ P_{3_{33}}$. The governing equations for this motion is

$$\left.\begin{aligned} f_k^{(3)}(\tilde{\mathbf{x}}_i, \tilde{\mathbf{y}}_i, t_i; \tilde{\mathbf{x}}_{i+1}, \tilde{\mathbf{y}}_{i+1}, t_{i+1}) = 0 \\ f_k^{(4)}(\tilde{\mathbf{x}}_{i+1}, \tilde{\mathbf{y}}_{i+1}, t_{i+1}; \tilde{\mathbf{x}}_{i+2}, \tilde{\mathbf{y}}_{i+2}, t_{i+2}) = 0 \end{aligned}\right\} \quad (k = 1, 2, 3, 4). \tag{33}$$

The periodicity conditions are

$$(\tilde{\mathbf{x}}_i, \tilde{\mathbf{y}}_i) = (\tilde{\mathbf{x}}_{i+2}, \tilde{\mathbf{y}}_{i+2}) \quad \text{and} \quad t_2 = t_1 + 2N\pi/\Omega. \tag{34}$$

Equations (33) and (34) can be solved through traditional numerical techniques. Varying a system parameter gives a description of how the orbits change.

5 Numerical and Analytical Predictions

The numerical and analytical predictions of the interrupted periodic cutting motions for this two degree of freedom oscillator with dry friction, subject to

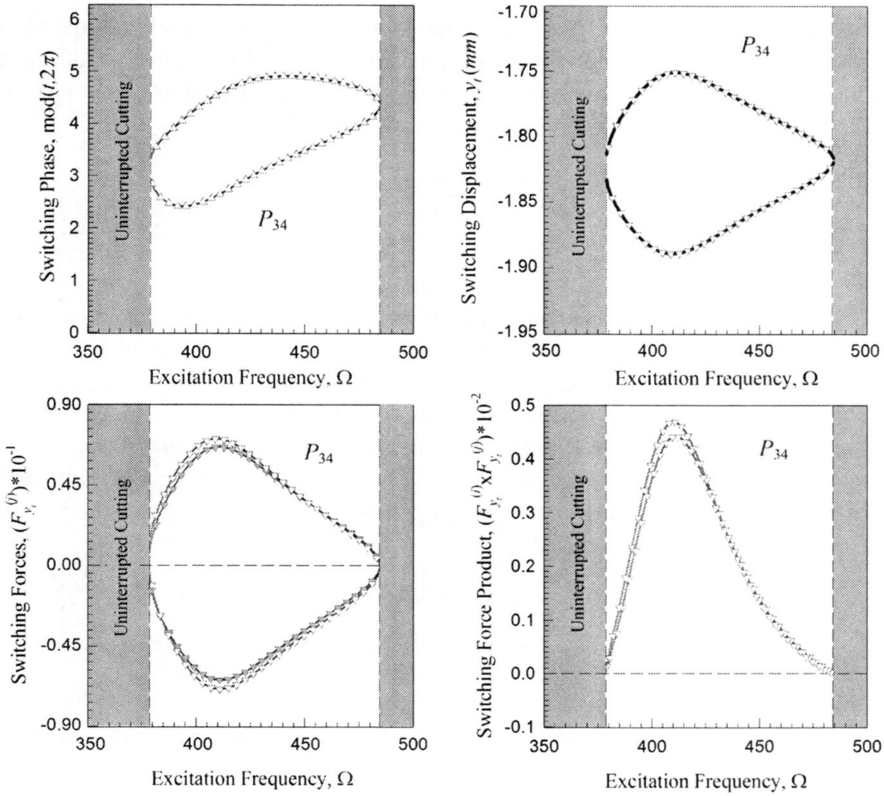

Fig. 6 Numerical and analytical predictions of (**a**) switching phase (mod($\Omega t_i, 2\pi$)), (**b**) switching displacement ($\tilde{y}$), (**c**) switching forces ($F_{\tilde{y}}^{(3)}$ and $F_{\tilde{y}}^{(4)}$) and (**d**) switching force product $F_{\tilde{y}}^{(3)} \times F_{\tilde{y}}^{(4)}$ for interrupted periodic motions

a periodic force, is presented over the range of excitation frequency of $\Omega \in [379.0, 484.4]$ rad/s. The dynamical system parameters are $m = 10$ kg, $d_x = 740$ Ns/mm, $d_y = 630$ Ns/mm, $k_x = k_y = 560$ kN/mm, $d_1 = d_2 = 0$ Ns/mm, and the external force and geometry parameters are $\delta_1 = \delta_2 = 10^{-3}$ m, $L_c = 0.5 \times 10^{-3}$ m, $\bar{V} = -20$ mm/s, $\mu = 0.7$, $\alpha = \frac{\pi}{4}$ rad, $\beta = 0.1$ rad, $\eta = \frac{\pi}{4}$ rad, $A = 500$ N, $X_1 = Y_1 = 10^{-3}$ m, $X_{eq} = Y_{eq} = 5 \times 10^{-3}$ m. The switching phase mod $(\Omega t_i, 2\pi)$ and switching displacement y_t versus excitation frequency (Ω) are illustrated in Fig. 6(a) and (b), respectively. The numerical and analytical predictions are illustrated by the solid curves and triangular symbol in Fig. 6, respectively. The most useful information is found in Fig. 6(c) and (d), where the switching forces ($F_{\tilde{y}}^{(3)}$ and $F_{\tilde{y}}^{(4)}$) and switching force products ($F_{\tilde{y}}^{(3)} \times F_{\tilde{y}}^{(4)}$) versus excitation frequency (Ω) are shown.

The periodic motion observed through a range of excitation frequency $\Omega \in [379.0, 484.4]$ rad/s is the mapping structure $P_{43} = P_4 \circ P_3$. Outside the neighborhood of the interval $\Omega \in [379.0, 484.4]$ rad/s the periodic motions do not intersect the discontinuity (or pure cutting occurs, no interruptions). The predictions of Fig. 6 are completed via the closed form solution to (1) and (2). The contact conditions are $(x_1^*, y_1^*) = (0.3941, -4.4638)$ mm, and $(x_2^*, y_2^*) = (0.2720, -2.9126)$ mm.

6 Summary

The phase planes are introduced to analyze the vector fields at the boundaries of the machine-tool system as a discontinuous dynamical system. The basic mappings are introduced for developing a mapping structure of periodic motions in the machine-tool system. Through such mapping structures periodic motions of the machine-tool system can be predicted analytically. For illustration of this methodology, one of periodic motions varying with excitation frequency is predicted analytically and numerically. For this system, there are many interesting phenomena, and the further investigation will be conducted.

Appendix

The dynamical system parameters for the machine-tool system, in the case the tool does not contact the work-piece, domain Ω_1 are

$$B_{11}^{(1)} = \frac{1}{m\Omega}d_x, \qquad B_{12}^{(1)} = B_{21}^{(1)} = 0, \qquad B_{22}^{(1)} = \frac{1}{m\Omega}d_y; \tag{A.1}$$

$$K_{11}^{(1)} = \frac{1}{m\Omega^2}k_x, \qquad K_{12}^{(1)} = K_{21}^{(1)} = 0, \qquad K_{22}^{(1)} = \frac{1}{m\Omega^2}k_y; \tag{A.2}$$

and

$$C_x^{(1)} = C_y^{(1)} = A_x^{(1)} = A_y^{(1)} = 0. \tag{A.3}$$

The dynamical system parameters for this machine-tool system, in the case the tool contacts the work-piece where no cutting occurs, domain Ω_2 are

$$\left.\begin{aligned} B_{11}^{(2)} &= \frac{1}{m\Omega}[d_x + d_1 \sin^2\beta], & B_{12}^{(2)} &= -\frac{1}{m\Omega}d_1 \cos\beta \sin\beta, \\ B_{21}^{(2)} &= -\frac{1}{m\Omega}d_1 \cos\beta \sin\beta, & B_{22}^{(2)} &= \frac{1}{m\Omega}[d_y + d_1 \cos^2\beta]; \end{aligned}\right\} \tag{A.4}$$

$$\left.\begin{aligned} K_{11}^{(2)} &= \frac{1}{m\Omega^2}[k_x + k_1 \sin^2\beta], & K_{12}^{(2)} &= -\frac{1}{m\Omega^2}k_1 \cos\beta \sin\beta, \\ K_{21}^{(2)} &= -\frac{1}{m\Omega^2}k_1 \cos\beta \sin\beta, & K_{22}^{(2)} &= \frac{1}{m\Omega^2}[k_y + k_1 \cos^2\beta]; \end{aligned}\right\} \tag{A.5}$$

and

$$\left.\begin{aligned} C_x^{(2)} &= \frac{1}{m\Omega^2}\{k_1[x_1^* \sin\beta - y_1^* \cos\beta]\sin\beta\},\\ C_y^{(2)} &= \frac{1}{m\Omega^2}\{-k_1[x_1^* \sin\beta - y_1^* \cos\beta]\cos\beta\},\\ A_x^{(2)} &= \frac{A}{m\Omega^2}\sin\eta, \qquad A_y^{(2)} = \frac{\mathrm{A}}{\mathrm{m}\Omega^2}\cos\eta. \end{aligned}\right\} \tag{A.6}$$

The dynamical system parameters for this machine-tool system, in the case the tool contacts the work-piece where cutting occurs where $\dot{z} < 0$ and $D_4 > 0$, domain Ω_3 and $\dot{z} > 0$, domain Ω_4;

$$\left.\begin{aligned} B_{11}^{(j)} &= \frac{1}{m\Omega}[d_x + d_1 \sin^2\beta + d_2\cos\alpha(\cos\alpha + (-1)^{j-1}\mu\sin\alpha)],\\ B_{12}^{(j)} &= \frac{1}{m\Omega}[-d_1\cos\beta\sin\beta - d_2\sin\alpha(\cos\alpha + (-1)^{j-1}\mu\sin\alpha)],\\ B_{21}^{(j)} &= \frac{1}{m\Omega}[-d_1\sin\beta\cos\beta - d_2\cos\alpha(\sin\alpha + (-1)^{j}\mu\cos\alpha)],\\ B_{22}^{(j)} &= \frac{1}{m\Omega}[d_y + d_1\cos^2\beta + d_2\sin\alpha(\sin\alpha + (-1)^{j}\mu\cos\alpha)]; \end{aligned}\right\} \tag{A.7}$$

$$\left.\begin{aligned} K_{11}^{(j)} &= \frac{1}{m\Omega^2}[k_x + k_1 \sin^2\beta + k_2\cos\alpha(\cos\alpha + (-1)^{j-1}\mu\sin\alpha)],\\ K_{12}^{(j)} &= \frac{1}{m\Omega^2}[-k_1\cos\beta\sin\beta + k_2\sin\alpha(\cos\alpha + (-1)^{j-1}\mu\sin\alpha)],\\ K_{21}^{(j)} &= \frac{1}{m\Omega^2}[-k_1\cos\beta\sin\beta - k_2\cos\alpha(\sin\alpha + (-1)^{j}\mu\cos\alpha)],\\ K_{22}^{(j)} &= \frac{1}{m\Omega^2}[k_y + k_1\cos^2\beta + k_2\sin\alpha(\sin\alpha + (-1)^{j}\mu\cos\alpha)]; \end{aligned}\right\} \tag{A.8}$$

and

$$\begin{aligned} C_x^{(j)} &= \frac{1}{m\Omega^2}\{k_1[x_1^*\sin\beta - y_1^*\cos\beta]\sin\beta\\ &\quad + k_2[x_2^*\cos\alpha - y_2^*\sin\alpha][\cos\alpha + (-1)^{j-1}\mu\sin\alpha]\}, \end{aligned} \tag{A.9}$$

$$\begin{aligned} C_y^{(j)} &= \frac{1}{m\Omega^2}\{-k_1[x_1^*\sin\beta - y_1^*\cos\beta]\cos\beta\\ &\quad + k_2[-x_2^*\cos\alpha + y_2^*\sin\alpha][\sin\alpha + (-1)^{j}\mu\cos\alpha]\}, \end{aligned} \tag{A.10}$$

$$A_x^{(j)} = \frac{A}{m\Omega^2}\sin\eta, \qquad A_y^{(j)} = \frac{A}{m\Omega^2}\cos\eta \tag{A.11}$$

for $j = 3, 4$; respectively. The parameters for the machine-tool where the chip adheres to the tool-piece rake face ($\dot{z} \equiv 0$) are

$$d = \frac{1}{2m\Omega}[d_2 + d_1 \sin^2(\alpha + \beta) + d_x \cos^2 \alpha + d_y \sin^2 \alpha], \tag{A.12}$$

$$\omega^2 = \frac{1}{m\Omega^2}[k_1 \sin^2(\alpha + \beta) + k_2 + k_x \cos^2 \alpha + k_y \sin^2 \alpha], \tag{A.13}$$

$$A_0 = \frac{A}{m\Omega^2} \sin(\eta - \alpha), \tag{A.14}$$

$$B_0 = \frac{V}{m\Omega^2}[k_1 \cos(\alpha + \beta) \sin(\alpha + \beta) + (k_x - k_y) \cos \alpha \sin \alpha], \tag{A.15}$$

$$\begin{aligned} C_0 = \frac{1}{m\Omega^2}(&\{[d_1 V - k_1(Vt_0 + \tilde{y}_0)] \cos(\alpha + \beta) + k_1[x_1^* \sin \beta - y_1^* \cos \beta]\} \\ &\times \sin(\alpha + \beta) + [V(d_x - d_y) + (Vt_0 + \tilde{y}_0)(k_y - k_x)] \\ &\times \cos \alpha \sin \alpha + k_2 \tilde{x}_2^*). \end{aligned} \tag{A.16}$$

References

1. M.E. Merchant, Mechanics of the metal cutting process. I. Orthogonal cutting and a type 2 chip. J. Appl. Phys. **16**(5), 267–275 (1945)
2. N.K. Chandiramani, T. Pothala, Dynamics of 2-dof regenerative chatter during turning. J. Sound Vib. **290**, 488–464 (2006)
3. B.C. Gegg, C.S. Suh, A.C.J. Luo, Stick and non-stick periodic motions of a machine tool in the cutting process, in *MSEC ICMP2008/DYN-72052*, Evanston, Illinois (2008)
4. J.P. Den Hartog, Forced vibrations with Coulomb and viscous damping. Trans. Am. Soc. Mech. Eng. **53**, 107–115 (1931)
5. M. Wiercigroch, A note on the switch function for the stick-slip phenomenon. J. Sound Vib. **175**(5), 700–704 (1994)
6. A.C.J. Luo, A theory for non-smooth dynamical systems on connectable domains. Commun. Nonlinear Sci. Numer. Simul. **10**, 1–55 (2005)
7. A.C.J. Luo, *Singularity and Dynamics of Discontinuous Vector Fields* (Elsevier, Amsterdam, 2006)
8. A.C.J. Luo, B.C. Gegg, On the mechanism of stick and non-stick, periodic motions in a forced linear oscillator including dry friction. ASME J. Vib. Acoust. **128**, 97–105 (2006)

Control of Hopf Bifurcation and Chaos as Applied to Multimachine System

Majdi M. Alomari and Benedykt S. Rodanski

Abstract Based on bifurcation theory and center manifold theory, both linear and nonlinear controllers are used to control a Hopf bifurcation and chaos. The second system of the IEEE second benchmark model of Subsynchronous Resonance (SSR) is considered. The system can be mathematically modeled as a set of first order nonlinear ordinary differential equations with the compensation factor ($\mu = X_c/X_L$) as a control parameter. So, bifurcation theory can be applied to nonlinear dynamical systems, which can be written as $dx/dt = F(x; \mu)$. The dynamics of the damper winding, automatic voltage regulator (AVR), and power system stabilizer (PSS) on SSR in power system are included. Both linear and nonlinear controllers are used to control the Hopf bifurcation and chaos. The results show that linear controller can only delay the inception of a bifurcation to some desired value of the bifurcation parameter. On the other hand, when the control objective is set to stabilize the periodic solution, nonlinear controller must be used.

Keywords Subsynchronous resonance · Hopf bifurcation · Control of chaos · Nonlinear controller

1 Introduction

The phenomenon of subsynchronous resonance occurs mainly in series capacitor-compensated transmission systems. In power systems series compensation is considered as a powerful technique based on economic and technical considerations for increasing effectively the power transfer capability as well as improving the stability of these systems. However, this introduces problems as well as with the benefits, namely the electromechanical interaction between electrical resonant circuits of the transmission system and the torsional natural frequencies of the turbine-generator

M.M. Alomari (✉) · B.S. Rodanski
University of Technology, Sydney (UTS), PO Box 123, Broadway, NSW 2007, Australia
e-mail: majdi.m.alomari@eng.uts.edu.au

J.A.T. Machado et al. (eds.), *Nonlinear Science and Complexity*,
DOI 10.1007/978-90-481-9884-9_4,

rotor. This phenomenon is called subsynchronous resonance (SSR), and it can cause shaft fatigue and possible damage or failure. SSR has been studied extensively since 1970, when a major transmission network in southern California experienced shaft failure to its turbine-generator unit with series compensation. Actually, power systems have rich bifurcation phenomena. Recently, power system dynamics has been studied using the nonlinear dynamics point of view, which utilizes the bifurcation theory. Bifurcation is used to indicate a qualitative change in the features of a system, such as the number and types of solution upon a small variation in the parameters of a system.

Harb et al. [1] applied a bifurcation analysis together with the method of multiple scales and Floquet theory to the CHOLLA # 4 turbine-generator system. Tomim et al. [2] proposed an index that identifies Hopf bifurcation points in power systems susceptible to subsynchronous resonance. Abed and Fu [3, 4] illustrated how the static feedback controller (u) can be chosen to suppress discontinuous bifurcations of fixed points such as subcritical Hopf bifurcations. They showed that subcritical Hopf bifurcation is converted to supercritical Hopf bifurcation by using a nonlinear static feedback. Nayfeh et al. [5] used a nonlinear state feedback controller in the form of $u = Kx^3$ to change the subcritical to a supercritical Hopf bifurcation. Also, they used this controller to reduce the amplitude of the limit cycle born near the bifurcation value as the controller gain value increases.

We focus on the torsional interaction effect, which results from the interaction of the electrical subsynchronous mode with the torsional mode. We use bifurcation theory and chaos to investigate the complex dynamics of the considered system. The type of the Hopf bifurcation is determined by numerical integration of the system, with specific amount of initial disturbances, slightly before and after the bifurcation value. On further increase of the compensation factor, the system experiences chaos via torus attractor. Chaos is a bounded steady-state behavior that is not an equilibrium solution or a periodic solution or a quasiperiodic solution [6].

2 System Description

The system considered is the two different machine infinite bus system, shown in Fig. 1. The two machines have a common torsional mode connected to a single series compensated transmission line. The model and the parameters are provided in the second system of the IEEE second benchmark model. The electro-mechanical systems for the first and second units are shown in Fig. 2. The first unit consists of exciter (EX.), generator (Gen.1), low-pressure (LP1) and high-pressure (HP1) turbine sections. And the second unit consists of generator (Gen.2), low-pressure (LP2) and high-pressure (HP2) turbine sections. Every section has its own angular momentum constant M and damping coefficient D, and every pair of successive masses have their own shaft stiffness constant K, as shown in Fig. 2. The data for electrical and mechanical system are provided in [7]. Replacement of these generators with a single equivalent generator will change the resonance characteristics and therefore is not justified. Consequently, each generator is represented in its own rotor frame of reference and suitable transformation is made.

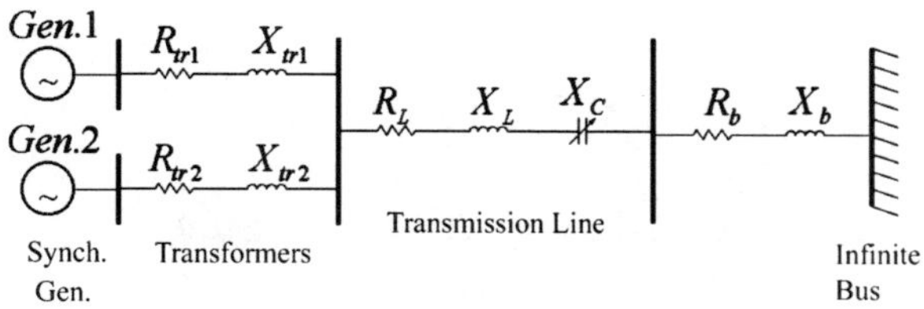

Fig. 1 Electrical system (two different machine infinite bus system)

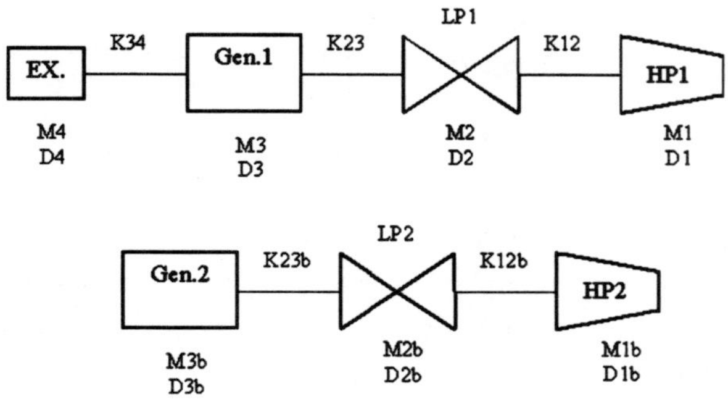

Fig. 2 Electro-mechanical systems for the first and second units

3 Mathematical Model

The mathematical model of the electrical and mechanical system will be presented in this section. Actually, the electrical system includes the dynamic nonlinear mathematical model of a synchronous generator and that of the transmission line. The generator model considered in this study includes five equations, d-axis stator winding, q-axis stator winding, d-axis rotor field winding, q-axis rotor damper winding and d-axis rotor damper winding equations. Each mass of the mechanical system can be modeled by a second order ordinary differential equation (swing equation), which is presented in state space model as two first order ordinary differential equations. Using the direct and quadrature d–q axes and Park's transformation, we can write the complete mathematical model that describes the dynamics of the system. The mathematical model of the electrical and mechanical system is provided in [8].

4 System Response without Controller

In this section we investigate the case of adding damper windings, automatic voltage regulator (AVR) and power system stabilizer (PSS) to the first generator. Figure 3 shows the block diagram of the use of AVR together with the PSS [9].

The operating point stability regions in the δ_{r1} plane together with two Hopf bifurcation points are depicted in Fig. 4. We observe that the power system has a stable

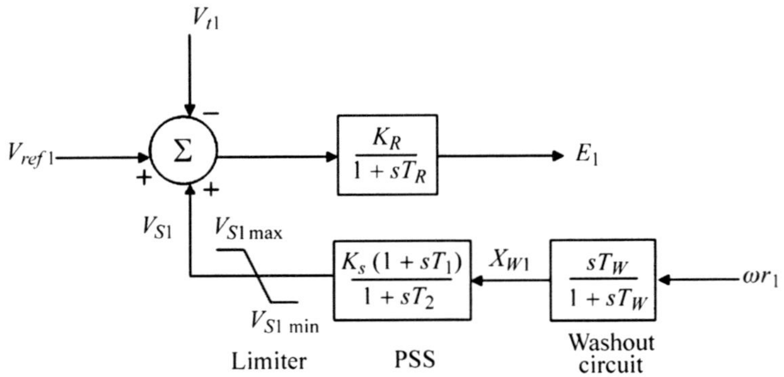

Fig. 3 Block diagram of the use of AVR and PSS to the first generator

Fig. 4 Bifurcation diagram showing variation of the first generator rotor angle δ_{r1} with the compensation factor μ (for the case of no controller)

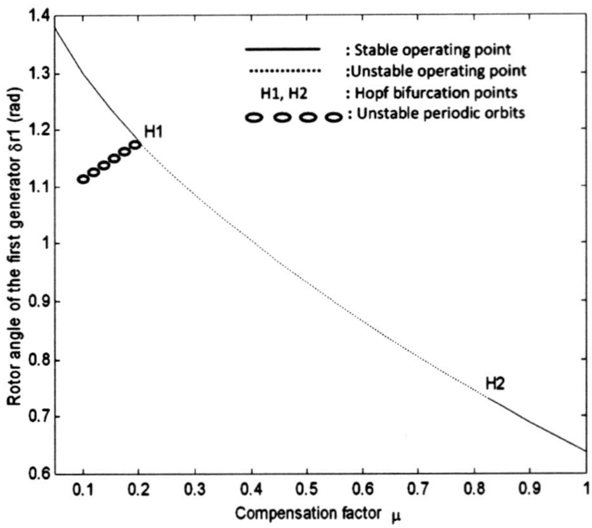

operating point to the left of $H_1 \approx 0.198377$ and to the right of $H_2 \approx 0.824135$, and has an unstable operating point between H_1 and H_2. The operating point loses stability at a Hopf bifurcation point, namely $\mu = H_1$. It regains stability at a reverse Hopf bifurcation, namely $\mu = H_2$. In this case a pair of complex conjugate eigenvalues will transversally cross from left half to right half of the complex plane, and then back to the left half. To determine whether the limit cycles created due to the Hopf bifurcation are stable or unstable, we obtain the time response of the system by numerical integration with small disturbance slightly before H_1. Figure 5 shows the response of the system with 7% initial disturbance on the speed of the generator at $\mu = 0.182265$, which is less than H_1. It can be observed that the system is unstable. Therefore, the type of this Hopf bifurcation is subcritical. So, the periodic solution emanating at the bifurcation point is unstable.

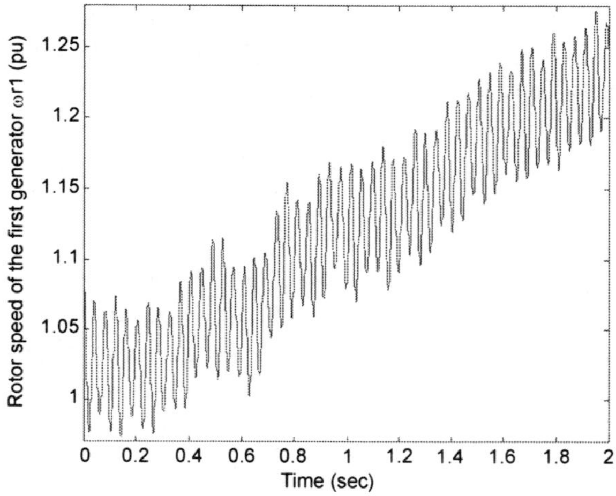

Fig. 5 Rotor speed of the generator at $\mu = 0.182265$ with 7% initial disturbance in rotor speed of generator (for the case of no controller)

5 Control of Hopf Bifurcation and Chaos

Consider the nonlinear dynamical system presented in the form:

$$\begin{aligned} \dot{x} &= f(x, u; \mu), \\ y &= g(x). \end{aligned} \tag{1}$$

Where x is a state variables vector, f is the field vector, μ is the control parameter of the system, y is the system outputs and u is the system state feedback control inputs. At any value of compensation factor μ, the operating points (equilibrium solutions) are obtained by setting the derivatives of the state variables in the system equal to zero.

$$F(x_e, u_e; \mu) = 0. \tag{2}$$

Where u_e represents the control input value when the system is at the equilibrium. The stability of the equilibrium solution is studied by examination of the eigenvalues of the Jacobean matrix evaluated at the operating point. Consider the system undergoing a Hopf bifurcation at the considered equilibrium point. That is, the critical eigenvalues of A cross imaginary axis at $\pm j\beta$, while all other eigenvalues have strictly negative real part.

5.1 Control of Critical Modes

Consider the linearization of the system as follows:

$$\begin{aligned} \delta\dot{x} &= D_x f(x_e, u_e; \mu)\delta x + D_u f(x_e, u_e; \mu)\delta u = A\delta x + B\delta u, \\ \delta y &= D_x g(x_e)\delta x = C^T \delta x. \end{aligned} \tag{3}$$

Let λ_j and λ_k denote the critical eigenvalues with corresponding normalized (left) eigenvectors P_j and P_k. Utilizing model control approaches [10], it may be possible to change these eigenvalues without affecting the others. This is achieved by considering the feedback law [11]:

$$\delta u = -\sum_{l=j,k} (k_l (p_l)^T)\delta x. \tag{4}$$

Where the model feedback gains k_j and k_k are determined such that the moved eigenvalues are at $\hat{\lambda}_j$ and, $\hat{\lambda}_k$. So, if it is desired to move eigenvalues $\lambda_{j,k} = \pm j\beta$ to $\hat{\lambda}_{j,k} = -\varepsilon \pm j\beta$, the modal feedback gains can be easily calculated as:

$$\begin{aligned} k_j &= \frac{(\lambda_j - \hat{\lambda}_j)(\lambda_j - \hat{\lambda}_k)}{(p_j)^T B(\lambda_j - \lambda_k)} = \frac{\varepsilon}{(p_j)^T B}\left(1 - \frac{j\varepsilon}{2\beta}\right), \\ k_k &= \frac{(\lambda_k - \hat{\lambda}_k)(\lambda_k - \hat{\lambda}_j)}{(p_k)^T B(\lambda_k - \lambda_j)} = \frac{\varepsilon}{(p_k)^T B}\left(1 + \frac{j\varepsilon}{2\beta}\right). \end{aligned} \tag{5}$$

We emphasize that modal feedback gains calculation to alter the critical modes is only possible if they are controllable. Besides, $(I_j)^T B$ and its complex conjugate $(I_k)^T B$ must be nonzero.

5.2 Control of Hopf Bifurcation

The bifurcation point is transferred to the origin via simple change of coordinates, with $m = \mu - \mu_o$ and $w = u - u_e$. By utilizing appropriate similarity linear transformations, the Jacobian is transformed to diagonal form, with a 2×2 block for the critical complex eigenvalues. The system can be expressed in the form:

$$\begin{aligned} \dot{x_c} &= J_c X_c + f_c(x_c, x_s, w; m), \\ \dot{x_s} &= J_s X_s + f_s(x_c, x_s, w; m). \end{aligned} \tag{6}$$

Where J_s is a matrix whose eigenvalues all have negative real parts (i.e., J_s is a stable matrix) and the matrix J_c has the form:

$$J_c = \begin{bmatrix} 0 & -\beta \\ \beta & 0 \end{bmatrix} \tag{7}$$

The functions f_c, f_s and their derivatives vanish at the origin. By the center manifold theorem [12] it can be ascertain that in the vicinity of origin (i.e., x_s, m, w: small) a smooth invariant manifold $x_s = h\ (x_c, m, w)$ for (6) exists. This center manifold is tangent to the eigenspace of the linearized system J_c, with $h(0, m, w) = h'(0, m, w) = 0$. Substituting the manifold constraint into the first part of (6), the bifurcation equations can be obtained as:

$$\dot{x}_c = J_c x_c + f_c(x_c, h(x_c, m, w), m, w). \tag{8}$$

For the purpose of notational simplicity, let $x_c = [x\ z]^T$, the bifurcation equations can now be considered as:

$$\dot{x} = f(x, z, m; w), \qquad \dot{z} = g(x, z, m; w). \tag{9}$$

Where $f(0, 0, m, w) = g(0, 0, w, m) = 0$, and the Jacobian evaluated near the origin with $m = 0$, is J_c given by (7). The control w is of feedback type with $w(0, 0) = 0$. Applying the Taylor expansion of (9) results in:

$$\begin{aligned} \dot{x} &= (f_{mx}m + f_w w_x)x - (\beta - f_w w_z - f_{mz} - f_{mz}m)z, \\ \dot{z} &= (\beta + g_w w_x + g_{mx}m)x + (g_{mz}m + g_w w_z)z. \end{aligned} \tag{10}$$

The characteristic polynomial of the Jacobian matrix can be obtained as:

$$\begin{aligned} &\lambda^2 - \lambda[f_w w_x + g_w w_z + m(f_{mx} + g_{mz})] \\ &\quad + \beta^2 - \beta[f_w w_z - g_w w_x + m(g_{mx} - f_{mz})] \\ &\quad - m[f_w(w_z g_{mx} - w_x g_{mz}) + g_w(w_x f_{mz} - w_z f_{mx})] = 0. \end{aligned} \tag{11}$$

The roots of the characteristic polynomial are $\lambda_{1,2}(m, w) = \alpha(m, w) \pm j\omega(m, w)$, where

$$\begin{aligned} \alpha(m, w) &= \frac{1}{2}[f_w w_x + g_w w_z + m(f_{mx} + g_{mz})], \\ \omega(m, w) &= \beta - \frac{f_w w_z - g_w w_x - m(g_{mx} - f_{mz})}{2} \\ &\quad + m\left[\frac{(f_{mx} - g_{mz})(f_w w_x - g_w w_z) + 2(f_w w_z g_{mx} + g_w w_x f_{mz})}{\beta}\right]. \end{aligned} \tag{12}$$

In the case of no control effort (i.e. $w = 0$), the bifurcation parameter dependent eigenvalues can be evaluated as:

$$\lambda_{1,2}(m) = \frac{m}{2}(f_{mx} + g_{mz}) \pm j\left(\beta + \frac{m}{2}(g_{mx} - f_{mz})\right). \tag{13}$$

Which has complex poles at $\pm j\beta$ for $m = 0$. The transversality condition requires that:

$$\frac{d}{dm}[\mathrm{Re}(\lambda(m))]_{m=0} \neq 0 \rightarrow \alpha_1 = (f_{mx} + g_{mz}) \neq 0. \tag{14}$$

Assume that the stability coefficient S be defined as:

$$\begin{aligned} S &= \frac{1}{16}(f_{xxx} + g_{xxz} + f_{xzz} + g_{zzz}) \\ &\quad + \frac{1}{16}[f_{xz}(f_{xx} + f_{zz}) - g_{xz}(g_{xx} + g_{zz}) - f_{xx}g_{xx} + f_{zz}g_{zz}]. \end{aligned} \tag{15}$$

The Hopf bifurcation theorem establishes that in these circumstances, if the genericity condition $S|_{x=z=m=0} \neq 0$, is also satisfied a curve of periodic solutions bifurcates from the origin into $m < 0$ provided $S\alpha_1$ is positive or into $m > 0$ if $S\alpha_1$ is negative. If α_1 is negative, the origin is stable for $m > 0$ and unstable for $m < 0$; conversely for α_1 positive, the origin is stable for $m < 0$ and unstable for $m > 0$. The periodic solutions on the side of $m = 0$ for which they exist, are stable if the origin is unstable and vice versa. On other words, for $\alpha_1 > 0$, a supercritical Hopf bifurcation occurs if $S < 0$; the origin is stable for $m < 0$ and unstable for $m > 0$. As m passes through zero, the stable periodic solutions bifurcate into $m < 0$. On the other hand, with α_1 positive, if $S > 0$, the origin is stable and a subcritical Hopf bifurcation is displayed, with unstable periodic orbits bifurcating into $m < 0$. In case of $\alpha_1 < 0$, situation is similar with the sign of m changed.

To study the possibility of rendering a subcritical Hopf bifurcation supercritical, the effects of control on S in (15) must also be investigated. For nonzero control effort, the new stability coefficient S_w to be evaluated at origin will be:

$$S_w = S + \frac{1}{16}[(w_{xxx} + w_{xzz})f_w + (w_{zzz} + w_{xxz})g_w] + \frac{1}{16\beta}[w_{xz}(f_w^2 - g_w^2)(w_{xx} + w_{zz}) + f_w g_w (w_{zz}^2 - w_{xx}^2)]. \tag{16}$$

From S_w and $\lambda_{1,2}$ (m, w) it is clear that only the feedback of critical variables up to cubic terms, may have any effect on the existence of a Hopf bifurcation or changing its stability attitude. The elimination of subcritical bifurcation requires that α (m, w) in (12) be always negative. For a system with controllable modes $((I)^T b \neq 0)$, this can be achieved by modal control approaches proposed by (4) and (5), linear control law can only delay the bifurcation point to the quested value of the bifurcation parameter (that is affect on the location of eigenvalues). On the other hand, nonlinear control (quadratic and/or cubic) can change the subcritical Hopf bifurcation to supercritical.

The quadratic feedback control law:

$$w = k_1 x^2 + k_2 z^2 \tag{17}$$

will change the subcritical Hopf bifurcation to supercritical, provided the critical modes are controllable and the feedback gains k_1 and k_2 are chosen such that

$$\frac{f_w g_w}{4\beta}(k_2^2 - k_1^2) < -S. \tag{18}$$

Where S is defined by (15). The quadratic feedback will make the stability coefficient negative and hence changing the bifurcation to supercritical.

On the other hand, the cubic feedback control law:

$$w = k_1 x^3 + k_2 z^3 \tag{19}$$

will change the subcritical Hopf bifurcation to supercritical, where the feedback gains k_1 and k_2 are chosen such that

$$\frac{6}{16}(k_1 f_w + k_2 g_w) < -S. \tag{20}$$

The cubic feedback will change the sign of the stability coefficient of (16) resulting in a supercritical Hopf bifurcation.

5.3 Numerical Simulation Results

For instance, consider the nonlinear dynamical system presented in the form:

$$\dot{x} = F(x; \mu) + u. \tag{21}$$

5.3.1 Linear Controller

The case of including the dynamics of the two axes damper windings, AVR and PSS is considered. So, in this case we have 27 ordinary nonlinear differential equations. It is easy to show that the critical modes of the linearized system around the operating point are controllable. Therefore, linear controller can be used to delay the occurrence of bifurcation to some desired value. Equations (4) and (5) can be used, with $\varepsilon = 0.45$, to identify the linear control:

$$\begin{aligned} u = [&4.25, -7.01, 7.40, -4.05, 8.22, 3.34, -5.65, 6.26, -2.80, 5.33, 2.95, 2.74,\\ &-20.74, -21.25, -36.38, 19.85, -18.24, -20.37, -28.25, 19.84, 34.93,\\ &17.25, 32.67, 12.34, 15.64, 41.45, 22.34]\delta x \end{aligned} \tag{22}$$

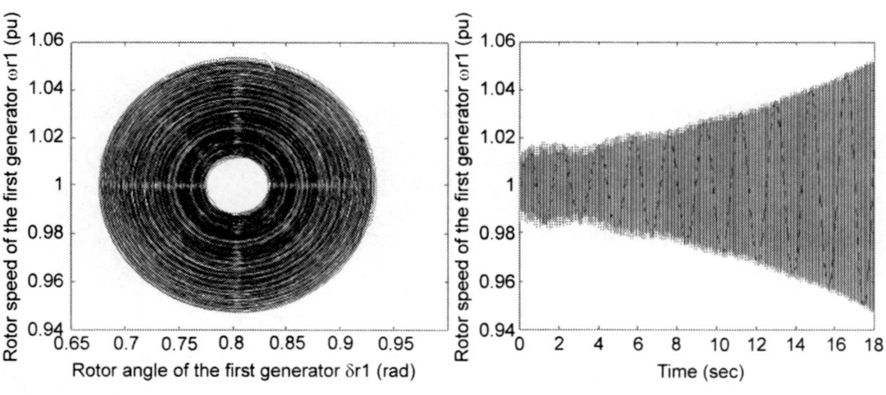

Fig. 6 Two-dimensional projection of the phase portrait onto ω_{rn}–δ_{rn} plane (*left*) and the time histories of the corresponding rotor speed of generator (*right*) at $\mu = 0.35$ (for the case of adding damper windings, AVR and PSS without controller)

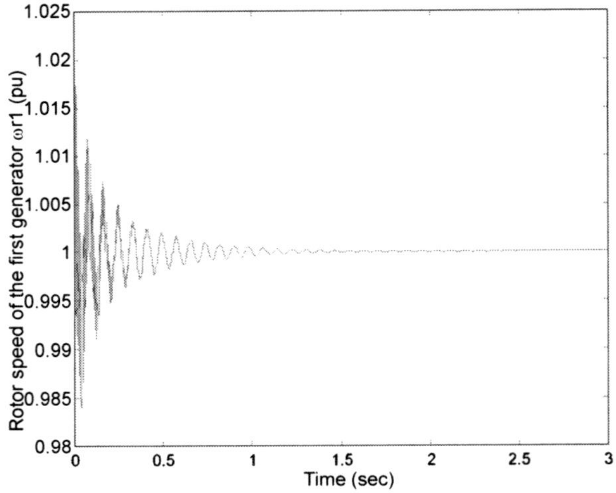

Fig. 7 Rotor speed of the generators at $\mu = 0.35$ with 2% initial disturbance in rotor speed of generators

that will move the critical eigenvalues to $-\varepsilon \pm j\beta$ without affecting the other eigenvalues which are in the left-half plane. This controller will delay the bifurcation point (H_1) to a new value of $\mu = H_{1C} = 0.401520$. As mentioned before, linear control law can only delay the bifurcation point to some desired value of the bifurcation parameter. Figure 6 shows the response of the system with 1% initial disturbance on the speed of the generator at $\mu = 0.35$ without controller. It can be observed that the system is unstable. Figure 7 shows the system response after a 2% initial disturbance in generator rotor speed at $\mu = 0.35$ when the linear feedback controller is applied. It can be observed that, the system has been stabilized.

5.3.2 Nonlinear Controller

The case of including the dynamics of the two axes damper windings, AVR and PSS with linear controller is considered in this section. To facilitate the use of previous results, the center manifold needs to be approximated. The new variables

$$y = V^{-1} \cdot \delta x \tag{23}$$

are now introduced, where V is the transformation identified such that the Jacobian of

$$\dot{y} = V^{-1} \cdot \dot{x} \tag{24}$$

evaluated near the bifurcation point is:

$$J = \begin{bmatrix} J_c & 0 \\ 0 & J_s \end{bmatrix} = \begin{bmatrix} 0 & -\beta & 0 & \cdots & \cdots & 0 \\ \beta & 0 & 0 & \cdots & \cdots & 0 \\ 0 & 0 & -|\lambda_1| & 0 & \cdots & 0 \\ \vdots & \vdots & 0 & \ddots & & \vdots \\ \vdots & \vdots & \vdots & & \ddots & 0 \\ 0 & 0 & 0 & \cdots & 0 & -|\lambda_{25}| \end{bmatrix}. \tag{25}$$

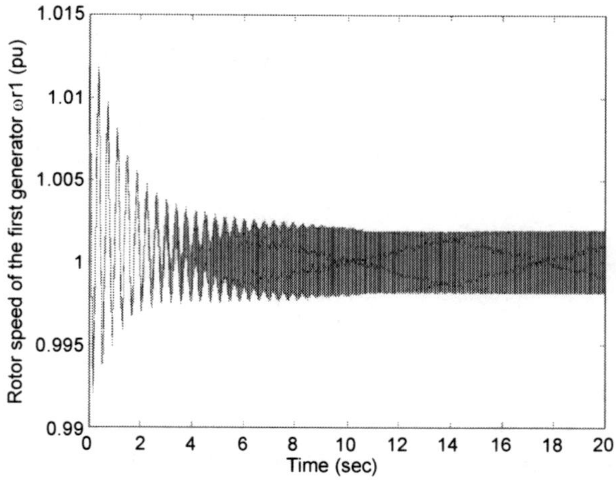

Fig. 8 Rotor speed of the first generator at $\mu = H \approx 0.401525$ with 2% initial disturbance in rotor speed of generator (for the case of adding nonlinear (quadratic) controller)

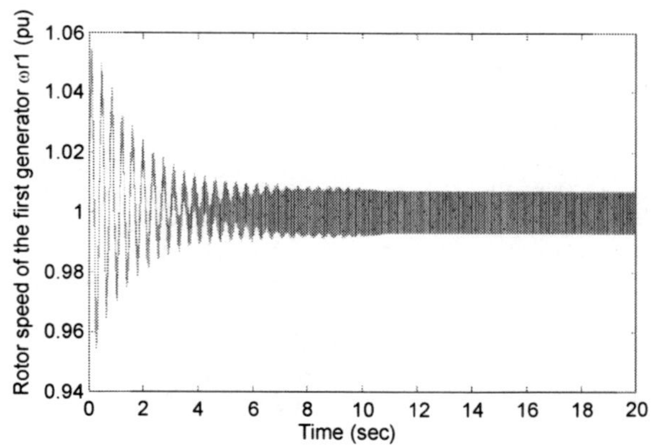

Fig. 9 Rotor speed of the first generator at $\mu = H \approx 0.401525$ with 2% initial disturbance in rotor speed of generator (for the case of adding nonlinear (cubic) controller)

Substituting the approximate center manifold constraint h in (6), it can be seen that it must satisfy the differential equation

$$h^{'}(x_c) \cdot \dot{x_c} - \dot{x_s} = 0. \tag{26}$$

By utilizing the center manifold theorem and using (25), the bifurcation equations in the form of (8) can be identified. On the other hand, by using (14) and (15) α_1 is calculated to be 0.7426 and the stability coefficient S is 0.1235. So, for the control established by quadratic feedback of critical variables $w = k_1 x^2 + k_2 z^2$, the condition which is going to be achieved must be met will be

$$1.245(k_2^2 - k_1^2) < -0.1235. \tag{27}$$

With $k_1 = 1$ and $k_2 = 0.25$, this will correspond to a negative value; resulting in stabilized oscillatory responses. Hence, the Hopf bifurcation will be changed from subcritical to supercritical. Figure 8 shows the system response after a 2% initial disturbance in generator rotor speed at $\mu = 0.401525$, which is greater than H_{1C}.

It can be observed that the system routes to a periodic solution giving rise to oscillations. Hence, the type of this Hopf bifurcation is supercritical. So, the periodic solution emanating at the bifurcation point is stable.

The same type of results in the quadratic feedback controller can be obtained by cubic feedback of critical variables $w = k_1 x^3 + k_2 z^3$. In this case, the condition that must be met will be:

$$2.254k_1 + 3.165k_2 < -0.1235. \tag{28}$$

By choosing the feedback gains as $k_1 = 1$ and $k_2 = -0.8$, this coefficient will be negative to render the subcritical Hopf bifurcation supercritical. Figure 9 shows the system response after a 2% initial disturbance in generator rotor speed at $\mu = H \approx 0.401525$, which is greater than H_{1C}. It can be observed that the system routes to a periodic solution giving rise to oscillations. Hence, the type of this Hopf bifurcation is supercritical. So, the periodic solution emanating at the bifurcation point is stable.

References

1. A.M. Harb, A.H. Nayfeh, L. Mili, Bifurcation control for mitigating subsynchronous oscillations in power systems, in *14th PSCC*, Seville, 24–28 June 2002
2. M.A. Tomim, A.C. Zambroni de Souza, P.P. Carvalho Mendes, G. Lambert-Torres, Identification of Hopf bifurcation in power systems susceptible to subsynchronous resonance, in *IEEE Bologna Power Tech Conference*, Bologna, Italy, June 23–26, 2003
3. E.H. Abed, J.H. Fu, Local feedback stabilization and bifurcation control, place. I. Hopf bifurcation. Syst. Control Lett. **7**, 11–17 (1986)
4. E.H. Abed, J.H. Fu, Local feedback stabilization and bifurcation control. II. Stationary bifurcation. Syst. Control Lett. **8**, 467–473 (1987)
5. A.H. Nayfeh, A.M. Harb, C.-M. Chin, Bifurcations in power system model. Int. J. Bifurc. Chaos **6**(3), 497–512 (1996)
6. A.H. Nayfeh, B. Balachandran, *Applied Nonlinear Dynamics* (Wiley, New York, 1995)
7. IEEE SSR Working Group, Second benchmark model for computer simulation of subsynchronous resonance. IEEE Trans. Power Appar. Syst. **PAS-104**(5), 1057–1064 (1985)
8. M.M. Alomari, B.S. Rodanski, The effects of machine components on bifurcation and chaos as applied to multimachine system, in *CHAOS2008, Chaotic Modeling and Simulation International Conference*, Chania, Crete, Greece, 3–6 June 2008
9. K.R. Padiyar, M.K. Geetha, K.U. Rao, A novel power flow controller for controlled series compensation, in *IEE, AC and DC Power Transmission*, 29 April–3 May. Conference Publication, No. 423 (1996), pp. 329–334
10. J. Van de Vegte, *Feedback Control Systems*, 3rd edn. (Prentice Hall, Englewood Cliffs, 1995)
11. S.A. Shahrestani, D.J. Hill, Control of nonlinear bifurcating systems, The University of Sydney, NSW, 2006
12. J. Guckenheimer, P. Holmes, *Nonlinear Oscillations, Dynamical Systems, and Bifurcations of Vector Fields*, 3rd edn. (Springer, New York, 1990)

Part II
Lie Group Analysis and Applications in Nonlinear Sciences

Group-Invariant Solutions of Fractional Differential Equations

**R.K. Gazizov, A.A. Kasatkin,
and S.Y. Lukashchuk**

Abstract In this paper, the results of symmetry analysis for some nonlinear fractional differential equations are presented. Applications of the symmetries for constructing exact solutions are illustrated.

Keywords Exact solutions · Lie transformations groups · Nonlinear fractional differential equations

1 Introduction

In recent years, the fractional calculus is increasingly used as effective tool to describe physical, chemical, and biological processes in complex irregular and heterogeneous systems (see for example [6]). Nevertheless, both analytical and numerical methods of solving equations with fractional order derivatives are not developed well enough. Most of existing methods allows one to find solutions only for limited classes of linear equations and for isolated examples of nonlinear equations (see Refs. [4, 5, 7–10] and references therein).

It is well known that modern group analysis can be effectively used to find exact solutions of ordinary and partial differential equation (see for example [3]).

Nevertheless, this approach is not widely applied yet to symmetry properties investigation of fractional differential equations (FDE). For example in [1] an admitted group of dilations is found for linear wave-diffusion equation of fractional order

R.K. Gazizov (✉) · A.A. Kasatkin · S.Y. Lukashchuk
Ufa State Aviation Technical University, Karl Marx str. 12, Ufa, Russia
e-mail: gazizov@mail.rb.ru

A.A. Kasatkin
e-mail: alexei_kasatkin@mail.ru

S.Y. Lukashchuk
e-mail: lsu@mail.rb.ru

J.A.T. Machado et al. (eds.), *Nonlinear Science and Complexity*,
DOI 10.1007/978-90-481-9884-9_5,

and these transformations are used to construct the automodel solution, but not all admitted groups of this equation are found there.

Recently, in [2], we have adapted methods of Lie continuous group for symmetry analysis of FDE and proposed prolongation formulas for fractional derivatives. Some examples of constructing symmetries of FDE and using these symmetries for constructing exact solutions of equations under consideration are also presented in [2]. In this paper we solve the problem of group classification for a wide class of FDEs with one independent variable (Sect. 3) and construct solutions of some equations using admitted groups (Sect. 4). Required formulas from fractional calculus and from [2] are presented in Sect. 2.

2 Transformation Groups and Symmetries of FDE

We consider fractional differential equation

$$D_x^\alpha y = f(x, y), \quad 0 < \alpha < 1, \tag{1}$$

where y is a function of independent variable x, D_x^α is the Riemann-Liouville fractional derivative operator:

$$D_x^\alpha y(x) = \frac{1}{\Gamma(1-\alpha)} \frac{d}{dx} \int_0^x \frac{y(t)}{(x-t)^\alpha} dt. \tag{2}$$

One-parameter group of transformations

$$\begin{aligned} \bar{x} &= \varphi(x, y, a), & \bar{y} &= \psi(x, y, a); \\ \varphi|_{a=0} &= x, & \psi|_{a=0} &= y, \end{aligned} \tag{3}$$

has the infinitesimal operator

$$X = \xi(x, y)\frac{\partial}{\partial x} + \eta(x, y)\frac{\partial}{\partial y}, \tag{4}$$

where

$$\xi(x, y) = \left.\frac{\partial \varphi}{\partial a}\right|_{a=0}, \qquad \eta(x, y) = \left.\frac{\partial \psi}{\partial a}\right|_{a=0}.$$

It means that the transformations (3) may be rewritten as infinitesimal transformations

$$\bar{x} = x + a\xi(x, y) + o(a), \qquad \bar{y} = y + a\eta(x, y) + o(a). \tag{5}$$

We consider transformations which conserve the structure of fractional derivative operator (2). In (2) the lower limit of the integral is fixed and, therefore, the equation $x = 0$ should be invariant with respect to such transformations. This invariance condition arrives to

$$\xi(x, y(x))|_{x=0} = 0. \tag{6}$$

When transformations (5) with (6) are applied to operator of fractional differentiation $D_x^\alpha y$, infinitesimal transformation of fractional derivative is obtained:

$$D_{\bar{x}}^\alpha \bar{y} = D_x^\alpha y + a\zeta_\alpha + o(a),$$

where ζ_α is given by the following expression (see [2])

$$\zeta_\alpha = D_x^\alpha \eta + D_x^\alpha (D_x(\xi)y) + \xi D_x^{\alpha+1} y - D_x^{\alpha+1}(\xi y). \tag{7}$$

Using Leibnitz's rule

$$D_x^\alpha (u(x)v(x)) = \sum_{n=0}^{\infty} \binom{\alpha}{n} D_x^{\alpha-n} u(x)\, D_x^n v(x), \quad \alpha > 0, \tag{8}$$

one can see that

$$\zeta_\alpha = D_x^\alpha(\eta) - \alpha D_x(\xi) D_x^\alpha(y) - \sum_{n=1}^{\infty} \binom{\alpha}{n+1} I_x^{n-\alpha}(y) D_x^{n+1}(\xi).$$

By chain rule, one can obtain

$$\zeta_\alpha = \partial_x^\alpha(\eta) + [\eta_y - \alpha D_x(\xi)] D_x^\alpha(y) - y\partial_x^\alpha(\eta_y) + \mu$$
$$+ \sum_{n=1}^{\infty} \left[\binom{\alpha}{n} \partial_x^n(\eta_y) - \binom{\alpha}{n+1} D_x^{n+1}(\xi) \right] I_x^{n-\alpha}(y), \tag{9}$$

$$\mu = \sum_{n=2}^{\infty} \sum_{m=2}^{n} \sum_{k=2}^{m} \sum_{r=0}^{k-1} \binom{\alpha}{n} \binom{n}{m} \binom{k}{r} \frac{1}{k!} \frac{x^{n-\alpha}}{\Gamma(n+1-\alpha)} [-y]^r$$
$$\times D_x^m [y^{k-r}] \frac{\partial^{n-m+k} \eta(x,y)}{\partial x^{n-m} \partial y^k}.$$

Here μ contains only nonlinear combinations of $y', y'', \ldots$.

Example 1 Symmetries of equation $D_x^\alpha y = 0$. In this case determining equation [3]

$$\zeta_\alpha |_{D_x^\alpha y = 0} = 0$$

by virtue of (9) can be rewritten in the form

$$\partial_x^\alpha(\eta) - y\partial_x^\alpha(\eta_y) + \mu + \sum_{n=1}^{\infty} \left[\binom{\alpha}{n} \partial_x^n(\eta_y) - \binom{\alpha}{n+1} D_x^{n+1}(\xi) \right] I_x^{n-\alpha}(y) = 0.$$

Variables $x, y, y', y'', \ldots$ (contained in $\mu, D_x^{n+1}\xi$) and $I_x^{n-\alpha}(y)$ are considered independent here. Splitting with respect to $I_x^{n-\alpha}(y)$ leads to infinite overdetermined system of linear fractional differential equations

$$\partial_x^\alpha(\eta) - y\partial_x^\alpha(\eta_y) + \mu = 0,$$

$$\binom{\alpha}{n}\partial_x^n(\eta_y) - \binom{\alpha}{n+1}D_x^{n+1}\xi = 0, \quad n \in \mathbb{N}.$$

Further splitting allows to solve this system and to find admitted operators

$$X_1 = x\frac{\partial}{\partial x}, \qquad X_2 = y\frac{\partial}{\partial y},$$

$$X_3 = x^2\frac{\partial}{\partial x} + (\alpha - 1)xy\frac{\partial}{\partial y}, \qquad X_4 = x^{\alpha-1}\frac{\partial}{\partial y}.$$

Remark 1 Prolongation formula (7) is valid for arbitrary order α, including negative (in this case it gives prolongation on fractional integrals). It also can be generalized for two independent variables (see [2]).

Remark 2 In literature some alternative definitions of fractional derivative are considered, which are useful for applications. One of them is *Caputo fractional derivative* defined by

$$^C D_x^\alpha y(x) = \frac{1}{\Gamma(m-\alpha)}\int_0^x \frac{y^{(m)}(t)}{(x-t)^{\alpha+1-m}}dt. \tag{10}$$

For Caputo fractional derivatives prolongation formula can be written in the form (for $0 < \alpha < 1$)

$$^C\zeta_\alpha = {}^C D_x^\alpha(\eta) - \alpha D_x(\xi)^C D_x^\alpha(y) - \sum_{n=1}^{\infty}\binom{\alpha}{n+1} I_x^n({}^C D_x^\alpha(y))D_x^{n+1}(\xi).$$

3 Group Classification of Equations $D_x^\alpha y = f(x, y)$

Consider the equation

$$D_x^\alpha(y) = f(x, y), \quad 0 < \alpha < 1. \tag{11}$$

with an independent variable x, dependent variable y, and arbitrary function $f(x, y)$.

Problem of *group classification* is to determine all 'non-equivalent' equations with non-trivial symmetry group. Here 'non-equivalent' means that the equations are not connected by *equivalence transformations*.

Transformation of the variables x, y

$$\bar{x} = \Phi(x, y, a), \qquad \bar{y} = \Psi(x, y, a) \tag{12}$$

is called equivalence transformation for (11) if the equation has the same form (maybe with changed function f) in new variables $\bar{x}, \bar{y}$:

$$D_{\bar{x}}^{\alpha}\bar{y} = \bar{f}(\bar{x}, \bar{y}). \tag{13}$$

By analogue with constructing symmetries, infinitesimal approach may be used for constructing equivalence transformations. For (11), equivalence transformations have the form

$$\begin{cases} \bar{x} = \dfrac{a_1 x}{1 - a_3 x}, \\ \bar{y} = a_2(1 - a_3 x)^{1-\alpha} y + \nu\left(\dfrac{a_1 x}{1 - a_3 x}\right), \\ \bar{f} = \dfrac{a_2}{a_1^{\alpha}}(1 - a_3 x)^{1+\alpha} f + D_{\omega}^{\alpha}\nu(\omega)|_{\omega = a_1 x(1 - a_3 x)^{-1}}, \end{cases} \tag{14}$$

where $a_1 > 0, a_2 \neq 0, a_3$ and $\nu(x)$ are arbitrary.

Admitted operators of (11) are sought in the form (4), where coefficients ξ and η are to be found from determining equation

$$\zeta_{\alpha}|_{D^{\alpha}y=f} - \xi f_x - \eta f_y = 0.$$

Splitting this equation lets us find that

$$\xi(x) = C_1 x + C_2 x^2, \qquad \eta(x, y) = (\alpha - 1)C_2 xy + C_3 y + q(x),$$

and the determining equation reduces to the form

$$\begin{aligned} &D_x^{\alpha} q(x) - q(x) f_y + [C_3 - \alpha C_1 - (\alpha + 1)C_2 x] f(x, y) \\ &\quad - (C_1 x + C_2 x^2) f_x(x, y) - [C_3 + (\alpha - 1)C_2 x] y f_y(x, y) = 0. \end{aligned}$$

From this equation the following classification result can be obtained.

Equation

$$D^{\alpha} y = f(x, y), \quad 0 < \alpha < 1,$$

has no symmetries for arbitrary function $f(x, y)$. Symmetries exists in the following cases (up to equivalence transformations (14)):

(1) $f(x, y) = y\Psi(x)$.
Admitted operators:

$$Z_1 = y\frac{\partial}{\partial y}, \qquad Z_2 = q_0(x)\frac{\partial}{\partial y},$$

where q_0 satisfies $D_x^{\alpha} q_0 = q_0 \Psi(x)$.
Additional extensions:

$$\text{(a) } \Psi(x) = \frac{k}{x^{\alpha}}: \quad Z_3 = x\frac{\partial}{\partial x}.$$

Here Z_2 has the form $Z_2 = x^\gamma \frac{\partial}{\partial y}$, where $k = \frac{\Gamma(\gamma+1)}{\Gamma(\gamma+1-\alpha)}$.

$$\text{(b) } \Psi(x) = \pm \frac{1}{x^{2\alpha}}: \quad Z_3 = x^2 \frac{\partial}{\partial x} + (\alpha - 1)xy \frac{\partial}{\partial y}.$$

If $\Psi(x) = x^{-2\alpha}$, then $Z_2 = x^{\alpha-1} e^{-1/x} \frac{\partial}{\partial y}$.
If $\Psi(x) = -x^{-2\alpha}$, the solution of $D_x^\alpha q_0 = q_0 \Psi(x)$ can not be constructed.

$$\text{(c) } \Psi(x) = 0: \quad Z_3 = x^2 \frac{\partial}{\partial x} + (\alpha - 1)xy \frac{\partial}{\partial y},$$

$$Z_4 = x \frac{\partial}{\partial x}.$$

Here $Z_2 = x^{\alpha-1} \frac{\partial}{\partial y}$.

(2) $f(x, y) = x^{-1-\alpha} \Psi(yx^{1-\alpha})$.
Admitted operator is

$$Z_1 = x^2 \frac{\partial}{\partial x} + (\alpha - 1)xy \frac{\partial}{\partial y}.$$

Additional extensions (except above-mentioned):

$$\text{(a) } \Psi(z) = e^{\pm z}: \quad Z_2 = x \frac{\partial}{\partial x} + (\alpha - 1)y \frac{\partial}{\partial y} \pm \alpha x^{\alpha-1} \frac{\partial}{\partial y}.$$

$$\text{(b) } \Psi(z) = z^\lambda: \quad Z_2 = x \frac{\partial}{\partial x} - \frac{1 - \lambda(1-\alpha)}{1-\lambda} y \frac{\partial}{\partial y}, \quad \lambda \neq 0, 1.$$

(3) $f(x, y) = x^{\beta-\alpha} \Psi(y/x^\beta)$.
Admitted operator is

$$Z_1 = x \frac{\partial}{\partial x} + \beta y \frac{\partial}{\partial y}.$$

(4) $f(x, y) = x^{-1-\alpha} e^{\mp 1/x} \Psi(yx^{1-\alpha} e^{\pm 1/x})$:
Admitted operator is

$$Z_1 = x^2 \frac{\partial}{\partial x} + (\alpha - 1)xy \frac{\partial}{\partial y} \pm y \frac{\partial}{\partial y}.$$

4 Exact Solutions of Equation $D^\alpha y = f(x, y)$

In this section we use obtained symmetries of equations considered in Sect. 3 to construct their solutions.

(1) Equation

$$D^\alpha y = x^{-1-\alpha}\Psi(yx^{1-\alpha})$$

is invariant under the group of projective transformations generated by

$$X_3 = x^2\frac{\partial}{\partial x} + (\alpha - 1)xy\frac{\partial}{\partial y}.$$

Invariant solution has the form $y = Cx^{\alpha-1}$, where $\Psi(C) = 0$. Therefore, if $C \neq 0$ is the root of $\Psi(C) = 0$, then $y = Cx^{\alpha-1}$ is nontrivial invariant solution.

(2) Equation

$$D^\alpha y = x^{-1-\alpha}e^{-\frac{1}{x}}\Psi(yx^{1-\alpha}e^{\frac{1}{x}})$$

admits operator

$$x^2\frac{\partial}{\partial x} + (\alpha - 1)xy\frac{\partial}{\partial y} + y\frac{\partial}{\partial y}$$

and invariant solution is

$$y = Cx^{\alpha-1}e^{-1/x},$$

where $C \neq 0$ is the root of $C = \Psi(C)$.

(3) Solution of the equation

$$D^\alpha y = x^{\beta-\alpha}\Psi(y/x^\beta), \tag{15}$$

invariant under dilations group generated by $x\frac{\partial}{\partial x} + \beta y\frac{\partial}{\partial y}$, has the form

$$y(x) = cx^\beta, \quad c = \text{const},$$

where c is determined by equation

$$c\frac{\Gamma(\beta+1)}{\Gamma(\beta+1-\alpha)} = \Psi(c).$$

In particular, for $\beta = -\alpha$, $\Psi(z) = b - cz^2$, (15) takes the form

$$D^\alpha y + ay^2 = \frac{b}{x^{2\alpha}}$$

(fractional Riccati equation, [2]) and admitted operator is

$$X = x\frac{\partial}{\partial x} - \alpha y\frac{\partial}{\partial y}.$$

Hence, invariant solution has the form

$$y(x) = \frac{c}{x^\alpha},$$

where c satisfies the square equation

$$ac^2 + \frac{\Gamma(1-\alpha)}{\Gamma(1-2\alpha)}c - b = 0.$$

(4) Equation

$$D^\alpha y = Cx^{\beta-\alpha}\left(\frac{y}{x^\beta}\right)^{\frac{\beta+1}{\beta+1-\alpha}}. \tag{16}$$

admits two-dimensional symmetry algebra (see case (2) (b) of Sect. 3)

$$Z_1 = x^2\frac{\partial}{\partial x} + (\alpha-1)xy\frac{\partial}{\partial y}, \qquad Z_2 = x\frac{\partial}{\partial x} + \beta y\frac{\partial}{\partial y}.$$

Invariant solution under Z_2-generated group is sought as $y = \gamma x^\beta$.
If $\beta \le -1$, no fractional derivative of order $\alpha \in (0,1)$ exists for function γx^β and there are no invariant solutions. If $\beta > -1$, solution exists and it is given by formula

$$y = \gamma x^\beta, \quad \text{where } \gamma = \left(C\,\frac{\Gamma(\beta+1-\alpha)}{\Gamma(\beta+1)}\right)^{\frac{\alpha-\beta-1}{\alpha}} = \text{const}. \tag{17}$$

Operator Z_1 generates the group of projective transformations

$$\bar{x} = \frac{x}{1-ax}, \qquad \bar{y} = y(1-ax)^{1-\alpha}.$$

Applied to (17), it gives the one-parameter family of solutions of (16):

$$y = \gamma\frac{x^\beta}{(1-ax)^{\beta+1-\alpha}}, \quad \beta > -1,\ \beta \ne \alpha - 1.$$

References

1. E. Buckwar, Y. Luchko, Invariance of a partial differential equation of fractional order under the lie group of scaling transformations. J. Math. Anal. Appl. **227**, 81–97 (1998)
2. R.K. Gazizov, A.A. Kasatkin, S.Y. Lukashchuk, Continuous transformation groups of fractional differential equations. Vestn. USATU 9 **3**(21), 125–135 (2007) (in Russian)
3. N.H. Ibragimov (ed.), *CRC Handbook of Lie Group Analysis of Differential Equations, vols. 1,2,3* (CRC Press, Boca Raton, 1996)
4. A.A. Kilbas, H.M. Srivastava, J.J. Trujillo, *Theory and Applications of Fractional Differential Equations*. North-Holland Mathematics Studies, vol. 204 (Elsevier, Amsterdam, 2006)
5. F. Mainardy, P. Paradisi, R. Gorenflo, Probability distributions generated by fractional diffusion equations, in *Econophysics: an Emerging Science*, ed. by J. Kertesz, I. Kondor (Kluwer Academic, Dordrecht, 1999)
6. R. Metzler, J. Klafter, The random walks guide to anomalous diffusion: a fractional dynamics approach. Phys. Rep. **339**, 1–77 (2000)

7. K.S. Miller, B. Ross, *An Introduction to the Fractional Calculus and Fractional Differential Equations* (Wiley, New York, 1993)
8. K.B. Oldham, J. Spanier, *The Fractional Calculus* (Academic Press, San Diego, 1974)
9. I. Podlubny, *Fractional Differential Equations. An Introduction to Fractional Derivatives, Fractional Differential Equations, Some Methods of Their Solution and Some of Their Applications* (Academic Press, San Diego, 1999)
10. S. Samko, A. Kilbas, O. Marichev, *Fractional Integrals and Derivatives: Theory and Applications* (Gordon and Breach Science, Yverdon, 1993)

Type-II Hidden Symmetries for Some Nonlinear Partial Differential Equations

Maria Luz Gandarias

Abstract The Type-II hidden symmetries are extra symmetries in addition to the inherited symmetries of the differential equations when the number of independent and dependent variables is reduced by a Lie point symmetry. In this paper we analyze the connection between one of the methods analyzed in Abraham-Shrauner and Govinder (J. Nonlinear Math. Phys. 13:612, 2006) and the weak symmetries of the partial differential equation in order to determine the source of these hidden symmetries. We have considered some of the models presented in Abraham-Shrauner and Govinder (J. Nonlinear Math. Phys. 13:612, 2006), as well as a linear three-dimensional wave equation considered in Abraham-Shrauner et al. (J. Phys. A, Math. Theor., 39:5739, 2006).

Keywords Hidden symmetries · Weak symmetries · Partial differential equations

1 Introduction

There is no existing general theory for solving nonlinear partial differential equations (PDE's) and it happens that many PDE's of physical importance are nonlinear. Lie classical symmetries admitted by nonlinear PDE's are useful for finding invariant solutions.

If a PDE is invariant under a Lie group, the number of independent variables can be reduced by one. The reduced equation loses the symmetry used to reduce the number of variables and may lose other Lie symmetries depending on the structure of the associated Lie algebra. If a PDE loses (gains) a symmetry in addition to the one used to reduce the number of independent variables of the PDE, the PDE possesses a Type I (Type II) hidden symmetry [1]. It has been noted [2] that these Type II hidden symmetries do not arise from contact symmetries or nonlocal symmetries

M.L. Gandarias (✉)
Department of Mathematics, University of Cádiz, Cádiz, Spain
e-mail: marialuz.gandarias@uca.es

J.A.T. Machado et al. (eds.), *Nonlinear Science and Complexity*,

DOI 10.1007/978-90-481-9884-9_6,

since the transformations to reduce the number of variables involve only variables. Thus the origin of these hidden symmetries must be in point symmetries [1]. In [1] B. Abraham-Shrauner and K.S. Govinder have identified a common provenance for the Type II hidden symmetries of differential equations reduced from PDE's that covers the PDE's studied. In [1] it was pointed out that the crucial point is that the differential equation that is reduced from a PDE and possesses a Type II hidden symmetry is also a reduced differential equation from one or more other PDE's. The inherited symmetries from these other PDE's are a larger class of Lie point symmetries that includes the Type II hidden symmetries. The Type II hidden symmetries are actually inherited symmetries from one or more of the other PDE's. The crucial question [1] is whether we can identify the PDE's from which the Type II hidden symmetries are inherited. In [1] two methods were proposed: some PDE's may be constructed by calculating the invariants by reverse transformations and some PDE's may be identified by inspection.

The *weak symmetries* were introduced in Olver and Rosenau [4]. Their approach consists in calculating the symmetries of the basic equation supplemented by certain differential constraints, chosen in order to weaken the invariance criterion of the basic system and to provide us with the larger Lie-point symmetry groups for the augmented system. In this way one obtains an overdetermined nonlinear system of equations and the solution set is, in this case, quite larger than the corresponding to classical symmetries.

In this paper, we consider weak symmetries of PDE's with special differential constraint in order to determine the source of these Type II hidden symmetries. The main result is that we can identify the PDE from which the Type II hidden symmetries are inherited by using as differential constraint the *side condition* from which the reduction has been derived.

In [1] the investigation was confined to hidden symmetries of PDE's for which the number of independent variables is reduced by Lie symmetries.

2 Weak Symmetries for the Model Equation

We begin by considering the model equation introduced in [1]

$$u_{xxx} + u(u_t + cu_x) = 0 \tag{1}$$

where c is a constant and the subscripts denote differentiation with respect to the variable indicated. Applying the Lie classical method to (1) leads to a four-parameter Lie group. Associated with this Lie group we have a Lie algebra which can be represented by the following generators [1]:

$$\begin{aligned} &\mathbf{v}_1 = \partial_x, \qquad \mathbf{v}_2 = \partial_t, \\ &\mathbf{v}_3 = (x + 2ct)\partial_x + 3t\partial_t, \qquad \mathbf{v}_4 = ct\partial_x + t\partial_t + u\partial_u. \end{aligned} \tag{2}$$

If we reduce (1) by using the generator $c\mathbf{v}_1+\mathbf{v}_2$ we get $u=w(z)$, $z=x-ct$ and the reduced ODE is

$$w_{zzz}=0 \tag{3}$$

which admits a seven-parameter Lie group. The associated Lie algebra can be represented by the following generators

$$\begin{aligned} &\mathbf{w}_1=\partial_z, \qquad \mathbf{w}_2=\partial_w, \qquad \mathbf{w}_3=z^2\partial_w, \qquad \mathbf{w}_4=z\partial_z, \\ &\mathbf{w}_5=z\partial_w, \qquad \mathbf{w}_6=w\partial_w, \qquad \mathbf{w}_7=\frac{1}{2}z^2\partial_z+zw\partial_w. \end{aligned} \tag{4}$$

The inherited symmetries are $\mathbf{v}_2\to\mathbf{w}_1$, $\mathbf{v}_3\to\mathbf{w}_4$, $\mathbf{v}_4\to\mathbf{w}_6$, all of which can be inferred by looking at the Lie algebra of (1). The other symmetries are Type II symmetries [1]. Two possible methods have been identified in [1] for finding possible PDE's the symmetries of which are inherited in the transformations $w=u$, $z=x-ct$ in (1). The first method proposed is to *guess* possible PDE's, evaluate their Lie point symmetries and then check if the group generators reduce to (4). Some PDE's that reduce to (3) by using the variables z and w and were proposed, by guessing, in [1] are

$$u_{xxx}=0, \qquad u_{ttt}=0, \qquad u_{xxt}=0, \qquad u_{xtt}=0. \tag{5}$$

We propose to have as differential constraint the side condition from which the reduction has been derived and to derive weak symmetries, that is, Lie classical symmetries of the original equation and the side condition. The PDE from which the hidden symmetries are inherited is the original PDE in which we substitute the side condition from which the reduction has been derived.

We are going to derive some weak symmetries of the model equation (1), choosing as side condition the differential constraint

$$cu_x+u_t=0, \tag{6}$$

which is associated to the generator $\mathbf{v}_1+c\mathbf{v}_2$ that has been used to derive the reduction

$$u=w(z), \qquad z=x-ct.$$

Applying Lie classical method to (1) with the side condition (6) we get:

$$\xi=\xi(t,x), \qquad \tau=\tau(t), \qquad \phi=\alpha(x,t)u+\beta(x,t),$$

where $\alpha(x,t)=\xi_x(x,t)u+g_1(t)$ and $\xi(x,t)$ and $\beta(x,t)$ must satisfy $\xi_{xxx}=\beta_{xxx}=0$. To apply the method in practice we use the MACSYMA package [3].

This yields the following generators

$$\begin{aligned}
&\mathbf{u}_1 = f_1(t)\partial_x, \qquad \mathbf{u}_2 = f_2(t)\partial_u, \qquad \mathbf{u}_3 = f_3(t)\partial_t, \qquad \mathbf{u}_4 = f_4(t)x^2\partial_u,\\
&\mathbf{u}_5 = f_5(t)x\partial_x, \qquad \mathbf{u}_6 = f_6(t)x\partial_u, \qquad \mathbf{u}_7 = f_7(t)u\partial_u,\\
&\mathbf{u}_8 = f_8(t)\left(\frac{1}{2}x^2\partial_x + xu\partial_u\right),
\end{aligned} \tag{7}$$

with $f_i(t), i = 1,\dots,8$, arbitrary functions. However, by appropriate choice of polynomials in t for $f_i(t)$ (and also taking combinations) the group generators reduce to the seven generators (4).These generators (7) have been derived in [1] by considering the classical Lie symmetries of the first equation of (5), namely $u_{xxx} = 0$. By interchanging x and t the symmetries of $u_{ttt} = 0$ can also be given by (7).

It was pointed out in [1] that symmetry $\mathbf{w}_7$ is not inherited by the other two equations derived by guessing, namely $u_{xxt} = 0$, $u_{xtt} = 0$. Nevertheless we prove that $\mathbf{w}_7$ is inherited as a weak symmetry of any of (5) with the side condition

$$u_t + cu_x = 0. \tag{8}$$

The crucial point is that $\mathbf{u}_7$ is a Lie symmetry of any of (5) in which we have substituted the side condition (8), and this equation is precisely $u_{xxx} = 0$ or $u_{ttt} = 0$.

Until now we have assumed that the PDE are all reduced by using the same variables as the original PDE (1). This does not have to be the case, we now consider the following equation introduced in [1]

$$u_{xxx} + u_{xx}\left(u_x + \frac{t}{x}u_t\right) = 0. \tag{9}$$

The generators of the classical symmetries are

$$\begin{aligned}
&\mathbf{v}_1 = x\partial_x, \qquad \mathbf{v}_2 = t\partial_t,\\
&\mathbf{v}_3 = \frac{x}{t}\partial_u, \qquad \mathbf{v}_4 = x\log(t)\partial_x + t\log(t)\partial_t + u\partial_u.
\end{aligned} \tag{10}$$

If we reduce (9) by using $\mathbf{v}_1 + \mathbf{v}_2$ then the new independent variable is $z = \frac{x}{t}$ with the dependent variable unchanged. We consider the side condition corresponding to this reduction which is

$$xu_x + tu_t = 0. \tag{11}$$

Then one requires that the group transformation leaves invariant the set of solutions of (9) and of the side condition (11) we obtain the Lie generators (7). The same happens if we consider the following example introduced in [1]

$$u_{xxx} + u_{xx}(xu_{xx} + tu_{tx}) = 0, \tag{12}$$

the generators of the classical symmetries are

$$\begin{aligned} &\mathbf{v}_1 = x\partial_x, \qquad \mathbf{v}_2 = t\partial_t, \qquad \mathbf{v}_3 = t\partial_x, \\ &\mathbf{v}_4 = tx\partial_x + t^2\partial_t + tu\partial_u, \qquad \mathbf{v}_5 = x\partial_u, \qquad \mathbf{v}_f = f(t)\partial_u. \end{aligned} \tag{13}$$

If we reduce (12) by using $\mathbf{v}_4$ then the new independent variable is $z = \frac{x}{t}$ with the new dependent variable w and $u = tw(z)$. We consider the side condition corresponding to this reduction which is

$$xu_x + tu_t = u. \tag{14}$$

Then one requires that the group transformation leaves invariant the set of solutions of (12) and of the side condition (14) we obtain the Lie generators (7).

3 Linear Three-Dimensional Wave Equation

In [2] the existence of an extra symmetry besides the inherited symmetries of the linear three-dimensional wave equation has been analyzed. The linear three-dimensional wave equation is

$$u_{xx} + u_{yy} + u_{zz} - u_{tt} = 0, \tag{15}$$

where u is the wave function, x, y and z are the spatial coordinates and t is the time normalized by the wave speed. The Lie group generators of (15) which appeared in [2] are 16 group generators and an infinite-dimensional generator corresponding to the linear equations. After reducing (15) by the scaling generator and by the rotation generator the resultant PDE is

$$4v(1-v)w_{vv} - 4vsw_{vs} + (1-s^2)w_{ss} + (4-6v)w_s - 2sw_s = 0, \tag{16}$$

with $v = \frac{x^2}{t^2} + \frac{y^2}{t^2}$, $s = \frac{z}{t}$. The three Lie point symmetries of (16) are

$$\mathbf{w}_1 = 2sv\partial_v + (s^2-1)\partial_s, \qquad \mathbf{w}_2 = w\partial_w, \qquad \mathbf{w}_\infty = F_w(v,s)\partial_w, \tag{17}$$

The reduction by $\mathbf{w}_1$ is the ODE

$$\sigma w_{\sigma\sigma} + w_\sigma = 0 \tag{18}$$

with $\sigma = \frac{v}{1-s^2}$. The associated Lie algebra can be represented by the following generators

$$\begin{aligned} &\mathbf{u}_1 = k_1\sigma w\partial_\sigma, && \mathbf{u}_2 = k_2(\sigma\log(\sigma)\partial_\sigma + w^2\partial_w), \\ &\mathbf{u}_3 = k_3(\sigma\log^2(\sigma)\partial_\sigma + w\log(w)), && \mathbf{u}_4 = k_4 w\partial_w, \\ &\mathbf{u}_5 = k_5\log(\sigma)\partial_w, && \mathbf{u}_6 = k_6\partial_w, \\ &\mathbf{u}_7 = k_7\sigma\partial_\sigma, && \mathbf{u}_8 = k_8 v\log(\sigma)\partial_\sigma, \end{aligned} \tag{19}$$

with $k_i, i = 1, \ldots, 8$, arbitrary constants, which already appeared in [2]. The ODE has two inherited symmetries and six Type-II hidden symmetries. In order to determine the other possible PDEs the inherited symmetries of which include all the symmetries in (18) we consider the PDE equation obtained considering the PDE (16) in $(2+1)$ dimensions and the side condition from which the reduction was derived. This side condition associated to generator $\mathbf{w_1}$ is

$$2vsw_v + (s^2 - 1)w_s = 0. \tag{20}$$

Applying the classical method to equation (16) with the side condition (20) we get the following generators:

$$\begin{aligned}
&\mathbf{v}_1 = F_1(s)vw\partial_v, && \mathbf{v}_2 = F_2(s)(vs\log(v)\partial_v + w^2\partial_w),\\
&\mathbf{v}_3 = F_3(s)(v\log^2(v)\partial_v + w\log(w)), && \mathbf{v}_4 = F_4(s)w\partial_w,\\
&\mathbf{v}_5 = F_5(s)\log(v)\partial_w, && \mathbf{v}_6 = F_6(s)\partial_w,\\
&\mathbf{v}_7 = F_7(s)v\partial_v, && \mathbf{v}_8 = F_8(s)v\log(v)\partial_v,
\end{aligned} \tag{21}$$

with $F_i(s), i = 1, \ldots, 8$, arbitrary functions, which already appeared in [1].

The PDE the inherited symmetries of which include all the symmetries in (19) is

$$v^2 w_{vv} + vw_v = 0. \tag{22}$$

It was pointed out in [2] that the determination of (22) is subtle and can be guessed from (18) however the PDE (22) from which the hidden symmetries are inherited is the original PDE in which we substitute the side condition (20) from which the reduction has been derived.

Acknowledgements The support of DGICYT project MTM2006-05031, Junta de Andalucía group FQM-201 and project P06-FQM-01448 are gratefully acknowledged.

References

1. B. Abraham-Shrauner, K.S. Govinder, J. Nonlinear Math. Phys. **13**, 612 (2006)
2. B. Abraham-Shrauner, K.S. Govinder, J.A. Arrigo, J. Phys. A, Math. Theor. **39**, 5739 (2006)
3. B. Champagne, W. Hereman, P. Winternitz, Comput. Phys. Commun. **66**, 319 (1991)
4. P.J. Olver, P. Rosenau, Phys. Lett. A **144**, 107 (1986)

Nonclassical and Potential Symmetries for a Boussinesq Equation with Nonlinear Dispersion

M.S. Bruzón and M.L. Gandarias

Abstract In this paper we consider a generalized Boussinesq equation which includes nonlinear dispersion. For this equation nonclassical and potential symmetries are derived. We prove that the nonclassical method applied to this equation leads to new symmetries, which can not be obtained by Lie classical method. We also have written the equation in a conserved form and a new class of nonlocal symmetries have been obtained.

Keywords Partial differential equation · Nonclassical symmetries · Potential symmetries

1 Introduction

The Boussinesq equation, which belongs to the KdV family of equations and describes motions of long waves in shallow water under gravity propagating in both directions, is given by

$$u_{tt} = u_{xx} + cu_{xxxx} + (u^2)_{xx} = 0. \tag{1}$$

Here $u = u(x,t)$ is a sufficiently often differentiable function, which for $c = -1$ gives the good Boussinesq or well-posed equation, while for $c = 1$ the bad or ill-posed classical equation [3, 4].

In [7] Clarkson obtained some nonclassical symmetry reductions and exact solutions for a Boussinesq equation. Gandarias and Bruzón [9] studied the classical and the nonclassical method for another Boussinesq equation.

M.S. Bruzón (✉) · M.L. Gandarias
Department of Mathematics, University Cádiz, Puerto Real, Cádiz 11510, Spain
e-mail: matematicas.casem@uca.es

M.L. Gandarias
e-mail: marialuz.gandarias@uca.es

J.A.T. Machado et al. (eds.), *Nonlinear Science and Complexity*,
DOI 10.1007/978-90-481-9884-9_7,

In [10] Rosenau extended the Boussinesq equations to include nonlinear dispersion to the effect that the new equations support compact and semi-compact solitary structures in higher dimensions,

$$u_{tt} = au_{xx} + (u^{m+1})_{xx} + b[u(u^m)_{xx}]_{xx}, \tag{2}$$

where a and b are arbitrary constants. Equation (2) describes for $a = 0$ the vibrations of a purely an harmonic lattice and support travelling structures with a compact support [10, 11].

One of the most useful point transformations are those which form a continuous group. Lie classical symmetries admitted by nonlinear partial differential equations (PDE's) are useful for finding invariant solutions. In [5] we studied similarity reductions of the generalized Boussinesq equation (2), with a, b, m arbitrary constants and $m \neq 0$. Motivated by the fact that symmetry reductions for many PDE's are known that are not obtained by using the classical Lie method there have been several generalizations of the classical Lie group method for symmetry reductions. Bluman and Cole [1], in their study of symmetry reductions of the linear heat equation, proposed the so-called nonclassical method of group-invariant solutions. In [2] Bluman introduced a method to find a new class of symmetries for a PDE when it can be written in a conserved form. These symmetries are nonlocal symmetries which are called *potential* symmetries. In [8] Gandarias introduced a new class of symmetries for a PDE, which can be written in the form of conservation laws. These symmetries, called *nonclassical potential* symmetries are realized as nonclassical symmetries of an associated system.

2 Nonclassical Symmetries

The basic idea of the method is that the PDE (2) is augmented with the invariance surface condition

$$\Phi \equiv \xi u_x + \tau u_t - \eta = 0, \tag{3}$$

which is associated to the vector field

$$V = \xi(x,t,u)\partial_x + \tau(x,t,u)\partial_t + \eta(x,t,u)\partial_u. \tag{4}$$

By requiring that both, (2) and (3), are invariant under the transformation with infinitesimal generator (4), an overdetermined nonlinear system of equations for the infinitesimals $\xi(x,t,u)$, $\tau(x,t,u)$ and $\eta(x,t,u)$ is obtained. The number of determining equations arising in the nonclassical method is smaller than for the classical method, consequently the set of solutions is, in general, larger than for the classical method. However, the associated vector fields do not form a vector space.

To obtain nonclassical symmetries of (2) we apply the algorithm described in [7] for calculating the determining equations and we use the MACSYMA program symmgrp.max [6]. We can distinguish two different cases:

In the case $\tau \neq 0$, without loss of generality, we may set $\tau(x,t,u) = 1$, and we obtain a set of sixteen determining equations for the infinitesimals $\xi(x,t,u)$ and $\eta(x,t,u)$. Solving this system we obtain

1. If a, b and m are arbitrary constants

$$\xi = k_1, \qquad \eta = 0,$$

   where k_1 and k_2 are constants.
2. If m is arbitrary and $a = 0$, $\xi = \frac{k_1}{t+k_2}$, $\eta = -\frac{2u}{m(t+k_2)}$, where k_1 and k_2 are constants.
3. If a is arbitrary and $m = -1$, $\xi = \frac{x+k_1}{t+k_2}$, $\eta = -\frac{2u}{t+k_2}$, where k_1 and k_2 are constants.
4. If $a = 0$ and $m = -1$, $\xi = \frac{-x+k_1k_3}{k_1(t+k_2)}$, $\eta = \frac{2(k_1+2)u}{k_1(t+k_2)}$, where k_1, k_2 and k_3 are constants.

By comparing these symmetries with the symmetries obtained by the classical method given in [5] we can observe that *the nonclassical method applied to* (2) *gives only rise to the classical symmetries.*

In the case $\tau = 0$, without loss of generality, we may set $\xi = 1$ and we obtain one overdetermined system for the infinitesimal η.

The complexity of this system is the reason why we cannot solve it in general. Thus we proceed, by making ansatz on the form of $\eta(x,t,u)$, to solve the system.

For $b = -\frac{1}{m^2}$, with $m = 1, 2$, choosing $\eta = \eta(x,u)$, we find that the infinitesimal generators take the form:

$$\xi = 1, \qquad \tau = 0, \qquad \eta = \frac{u \cosh x}{\sinh x}. \tag{5}$$

It is easy to check that these generators do not satisfy the Lie classical determining equations. Therefore we obtain the nonclassical symmetry reduction

$$z = t, \qquad u = h(t) \sinh x,$$

where $h(t)$ satisfies the following linear second-order ODE's:

- For $m = 1$

$$h'' - a\,h = 0. \tag{6}$$

- For $m = 2$

$$h'' + 2h^3 - a\,h = 0. \tag{7}$$

The solutions of (6) yield the following exact solutions of (2),

If $a > 0$,

$$u = (k_1 \exp(\sqrt{a}t + k_2 \exp(-\sqrt{a}t)) \sinh x. \tag{8}$$

If $a < 0$,

$$u = (k_1 \cos(\sqrt{-a}\,t) - k_2 \sin(\sqrt{-a}\,t)) \sinh x. \tag{9}$$

If $a=0$,

$$u=(k_1 t+k_2)\sinh x. \tag{10}$$

After multiplying (7) by $2y'$ and integrating once with respect to z we get

$$(h')^2=-h^4+ah^2. \tag{11}$$

This equation is solvable in terms of the Jacobian elliptic functions.

We remark that, when $b=-1$ and $m=1$ (2) does not admit any classical symmetry but translations. Consequently (8), (9), (10), which *are not* travelling waves reductions, can not be obtained by Lie classical symmetries.

For $m=1$, choosing $\eta=\eta(x,t)$, we find the following infinitesimal generators,

$$\xi=1, \qquad \tau=0, \qquad \eta=x\psi_2(t)+\psi_1(t), \tag{12}$$

where $\psi_1(t)$ and $\psi_2(t)$ satisfy

$$\frac{d^2\psi_2}{dt^2}-6\psi_2^2=0, \tag{13}$$

$$\frac{d^2\psi_1}{dt^2}-6\psi_1\psi_2=0, \tag{14}$$

respectively. In this case we obtain the nonclassical symmetry reduction

$$z=t, \qquad u=x^2\psi_2(t)+x\psi_1(t)+\psi_0(t),$$

where $\psi_2(t)$ satisfies the Weierstrass elliptic function equation (13) and $\psi_1(t)$ satisfies the Lamé equation (14) [7].

3 Classical Potential Symmetries

In order to find potential symmetries of (2) we write the equation in a conserved form and the associated auxiliary system is given by

$$\begin{cases} v_x=-u_t, \\ v_t=au_x+(u^{m+1})_x+b[u(u^m)_{xx}]_x. \end{cases} \tag{15}$$

If $(u(x),v(x))$ satisfies (15), then $u(x)$ solves the generalized Boussinesq equation. The basic idea for obtaining *classical potential* symmetries is to require that the infinitesimal generator

$$X=\xi(x,t,u,v)\partial_x+\tau(x,t,u,v)\partial_t+\phi_1(x,t,u,v)\partial_u+\phi_2(x,t,u,v)\partial_v \tag{16}$$

leaves invariant the set of solutions of (15). This yields to an overdetermined, non linear system of equations for the infinitesimals $\xi(x,t,u,v)$, $\tau(x,t,u,v)$, $\phi_1(x,t,u,v)$ and $\phi_2(x,t,u,v)$. We obtain classical potential symmetries if

$$(\xi_v)^2+(\tau_v)^2+(\phi_{1,v})^2\neq 0. \tag{17}$$

The classical method applied to (15) leads to the classical symmetries.

4 Nonclassical Potential Symmetries

The basic idea for obtaining *nonclassical potential* symmetries is that the potential system (15) is augmented with the invariance surface conditions

$$\xi u_x + \tau u_t - \phi_1 = 0, \qquad \xi v_x + \tau v_t - \phi_2 = 0, \tag{18}$$

which is associated with the vector field

$$X_1 = \xi(x,t,u,v)\partial_x + \tau(x,t,u,v)\partial_t + \phi_1(x,t,u,v)\partial_u + \phi_2(x,t,u,v)\partial_v. \tag{19}$$

By requiring that both (15) and (18) are invariant under the transformations with infinitesimal generator (19) one obtains an overdetermined, nonlinear system of equations for the infinitesimals $\xi(x,t,u,v)$, $\tau(x,t,u,v)$, $\phi_1(x,t,u,v)$ and $\phi_2(x,t,u,v)$.

In the case $\tau \neq 0$, without loss of generality, we may set $\tau(x,t,v) = 1$. The nonclassical method applied to (15) yields to the classical symmetries.

In the case $\tau = 0$, without loss of generality, we may set $\xi = 1$ and we obtain overdetermined non linear system of equations for the infinitesimals ϕ_1 and ϕ_2 which is solve by making ansatz on the form of $\phi_1(x,t,u,v)$ and $\phi_2(x,t,u,v)$. In this way we have found one solution.

For $a = 0$ and $m = -1$ we obtain the infinitesimal generators

$$\xi = 1, \qquad \tau = 0, \qquad \phi_1 = ku\psi(v), \qquad \phi_2 = \omega(x,v),$$

where k is constant and ω and ψ satisfies $-k\,\psi\,\omega + \frac{\partial\omega}{\partial x} + \omega\frac{\partial\omega}{\partial v} = 0$.

In the case that $\omega = \omega(v)$ the infinitesimal generators are:

$$\xi = 1, \qquad \tau = 0, \qquad \phi_1 = u\frac{d\omega}{d\,v}, \qquad \phi_2 = \omega(v).$$

We obtain the nonclassical potential symmetry reduction

$$z = t, \qquad u = \exp\left(kx\frac{d\omega}{d\,v}\right)h_1(t)$$

and v is given by $\int \frac{dv}{\omega(v)} = kx + h_2(t)$.

5 Concluding Remarks

We have considered the generalized Boussinesq equation (2). We have proved that the nonclassical method with $\tau \neq 0$ applied to (2) gives only rise to classical symmetries; for $\tau = 0$ leads to new symmetries, which can not be obtained by Lie classical method.

We also have written the equation in a conserved form and a new class of nonlocal symmetries have been obtained.

Acknowledgements The support of DGICYT project MTM2006-05031, Junta de Andalucía group FQM-201 and project P06-FQM-01448 are gratefully acknowledged.

References

1. G.W. Bluman, J. Cole, Phys. J. Math. Mech. **18**, 1025–1042 (1969)
2. G.W. Bluman, S. Kumei, J. Math. Phys. **5**, 1019–1023 (1980)
3. M.J. Boussinesq, C. R. Acad. Sci. Paris **72**, 755–759 (1871)
4. M.J. Boussinesq, J. Math. Pure Appl. Ser. **7**, 55–108 (1872)
5. M.S. Bruzón, M.L. Gandarias, J. Ramírez, *Proceedings of the International Conference SPT* (World Scientific, Cala Gonone, 2001)
6. B. Champagne, W. Hereman, P. Winternitz, Comput. Phys. Commun. **66**, 319–340 (1991)
7. P.A. Clarkson, Chaos Solitons Fractals **5**, 2261–2301 (1995)
8. M.L. Gandarias, in *Proceedings and Lecture Notes*. CRM, vol. 25 (2000), pp. 161–165
9. M.L. Gandarias, M.S. Bruzón, J. Nonlinear Math. Phys. **5**, 8–12 (1998)
10. P. Rosenau, J. Phys. Lett. A **275**, 193–203 (2000)
11. P. Rosenau, Phys. Rev. Lett. **73**, 1737–1741 (1994)

Application of the Composite Variational Principle to Shallow Water Equations

Emrullah Yasar and Teoman Ozer

Abstract In this study, we derive new local conservation laws of the one-layer shallow water equations in the plane flow which are belong to the family of evolution type equations by the composite variational principle of view in the development of the study (N.H. Ibragimov, A new conservation theorem. J. Math. Anal. Appl. 333:311–328, 2007).

Keywords Adjoint equation · Conservation laws · Shallow water equations · Symmetries

1 Introduction

Conservation laws has been used in many research areas such as physical interpretation of basic properties of the given system e.g. energy, momentum, center of mass, spin [2, 3] etc., investigation of stability theory [4], integrability of differential equations [5]. This means that, finding the conservation laws of the given system has a key role.

E. Noether [6] established first systematically relationship between conservation laws and continuous symmetry of the equation. However this relationship is satisfied by only the Euler-Lagrange type equations. For instance Noether's approach can not applied to the evolution-type equations. In the literature, in order to overcome of this problem several methods was suggested by some authors. The direct

E. Yasar (✉)
Department of Mathematics, Faculty of Arts and Sciences, Uludag University, 16059 Bursa, Turkey
e-mail: eyasar@uludag.edu.tr

T. Ozer
Faculty of Civil Engineering, Division of Mechanics, Istanbul Technical University, 34469 Istanbul, Turkey
e-mail: tozer@itu.edu.tr

J.A.T. Machado et al. (eds.), *Nonlinear Science and Complexity*,
DOI 10.1007/978-90-481-9884-9_8,

method (in fact, this method has been known since more than 200 years, see [7]), the characteristic method [8], the variational approach (multiplier approach) for arbitrary functions as well as on the solution space [2], symmetry conditions on the conserved quantities [9], the direct construction formula approach [10], the partial Noether approach [11].

Very recently, Ibragimov proposed a general approach on finding conservation laws of single or systems of partial differential equations (PDEs). The short brief of the method [1] with step by step is as follows (Because of our purposes we give the all instructions for two dependent and independent variables as well as for first order PDEs. See, also [12–14]):

(1) Let $x^n = (x^1, x^2)$ be two independent variables with $x^1 = t$, $x^2 = x$ and $u^\alpha = (u^1, u^2)$ be two dependent variables with $u^1 = u$, $u^2 = v$. Since
$$F \equiv u_t^\alpha = \Delta(x^\alpha, u^\alpha, u_x^\alpha, \ldots), \quad \alpha = 1, 2 \tag{1}$$
evolution equations have not the usual Lagrangian, formal Lagrangian is formed. Formal Lagrangian is multiplication of a new adjoint variable, $v^\beta = (v^1, v^2)$, with a given equation, where $v^1 = v$, $v^2 = w$. Namely,
$$L = vF. \tag{2}$$

(2) With this formal Lagrangian,
$$F^* = \frac{\delta L}{\delta u} \tag{3}$$
adjoint equation is created. Here, $\frac{\delta}{\delta u}$ is variational derivative with $\frac{\delta}{\delta u} = \frac{\partial}{\partial u} - D_i(\frac{\partial}{\partial u_i}) + D_i D_j(\frac{\partial}{\partial u_{ij}}) + \cdots$.

(3) Original equation F and adjoint equation F^* are being together the Euler-Lagrange type equations. Indeed,
$$\frac{\delta L}{\delta u} = F^*, \qquad \frac{\delta L}{\delta v} = F. \tag{4}$$

(4) Adjoint equation F^* inherits symmetries of the original equation. If,
$$X = \xi^i \frac{\partial}{\partial x^i} + \eta \frac{\partial}{\partial u} \tag{5}$$
is admitted by the original equation then adjoint equation has the following symmetry:
$$Y = X + \eta^* \frac{\partial}{\partial v}.$$

(5) Symmetries and formal Lagrangian of the given equation always satisfies
$$X(L) + L D_i(\xi^i) = 0 \tag{6}$$
invariance condition.

(6) If formal Lagrangian and symmetries are substituted in the
$$C^i = v^\beta \left[\xi^i F_\beta + (\eta^\alpha - \xi^j u_j^\alpha) \frac{\partial F_\beta}{\partial u_i^\alpha} \right], \quad i = 1, 2. \tag{7}$$

(7) If (3), i.e., adjoint equation is self adjoint then local conservation laws are obtained . Otherwise nonlocal conservation laws are derived.

2 Symmetry Group Analysis of the Shallow Water Equations in the Plane Flow

For the plane geometry, if viscous effects are unimportant, the motion of current in plane flow is

$$\begin{aligned} h_t + uh_x + hu_x &= 0, \\ u_t + uu_x + h_x &= 0, \end{aligned} \tag{8}$$

where $h(x,t)$ is the thickness of the current, $u(x,t)$ is the depth-averaged horizontal fluid speed in the current. We consider a one-parameter Lie group of infinitesimal transformations in (x,t,u,h) given

$$\begin{aligned} x^* &= x^*(x,t,u,h;\varepsilon), \qquad & t^* &= t^*(x,t,u,h;\varepsilon), \\ u^* &= u^*(x,t,u,h;\varepsilon), \qquad & h^* &= h^*(x,t,u,h;\varepsilon), \end{aligned} \tag{9}$$

where ε is the group parameter. We then require that this transformation leave the set of solutions of (8) invariant. This leads to an over determined linear system of equations for infinitesimals $\xi^x(x,t,u,h)$, $\xi^t(x,t,u,h)$, $\eta^u(x,t,u,h)$ and $\eta^h(x,t,u,h)$. The associated Lie algebra of infinitesimals:

$$X = \xi^x \frac{\partial}{\partial x} + \xi^t \frac{\partial}{\partial t} + \eta^u \frac{\partial}{\partial u} + \eta^h \frac{\partial}{\partial h}.$$

Applying the classical method to (8) yields a system of equations that leads to a five parameter Lie group. Associated with this Lie group, we have a Lie algebra that can be represented by the generators

$$\begin{aligned} & X_1 = \frac{\partial}{\partial x}, \qquad X_2 = \frac{\partial}{\partial t}, \qquad X_3 = x\frac{\partial}{\partial x} + t\frac{\partial}{\partial t}, \\ & X_4 = t\frac{\partial}{\partial x} + \frac{\partial}{\partial u}, \qquad X_5 = x\frac{\partial}{\partial x} + u\frac{\partial}{\partial u} + 2h\frac{\partial}{\partial h}. \end{aligned} \tag{10}$$

3 Derivation of Conservation Laws

We write the Lagrangian (2) for (8) in the following form:

$$L = v(h_t + uh_x + hu_x) + w(u_t + uu_x + h_x), \tag{11}$$

where v and w is the adjoint variables. With this Lagrangian we have

$$\frac{\delta L}{\delta u} = -(w_t + uw_x + hv_x), \qquad \frac{\delta L}{\delta v} = h_t + uh_x + hu_x \tag{12}$$

and

$$\frac{\delta L}{\delta h} = -(w_x + v_t + uv_x), \qquad \frac{\delta L}{\delta w} = u_t + uu_x + h_x. \tag{13}$$

It follows from (12)–(13) that the Euler-Lagrange equations for the Lagrangian (11) provide the plane flow equations (8) and the following adjoint equations for the new dependent variables v, w

$$\begin{aligned} w_t + u w_x + h v_x &= 0, \\ w_x + v_t + u v_x &= 0. \end{aligned} \tag{14}$$

Let $w = h,\ v = u$ in (14) then we yield plane flow of one-layer shallow water equations

$$\begin{aligned} h_t + h u_x + h_x u &= 0, \\ u_t + h_x + u u_x &= 0. \end{aligned} \tag{15}$$

Therefore, plane flow of one-layer shallow-water equations is self-adjoint.

Let us find the conservation laws furnished for instance by the symmetry X_4 in (10). Applying the formula (7) to the Galilean symmetry $X_4 = t\frac{\partial}{\partial x} + \frac{\partial}{\partial u}$ where $\xi^1 = t$, $\xi^2 = 0$, $\eta^1 = 1$, $\eta^2 = 0$, $v = u$, $w = h$ and to the Lagrangian (11), we obtain the following conserved vectors:

$$\begin{aligned} C_4^1 &= v\left(tF_1 + (1 - tu_x)\frac{\partial F_1}{\partial u_x} - th_x\frac{\partial F_1}{\partial h_x}\right) + w\left(tF_2 - th_x\frac{\partial F_2}{\partial h_x} + (1 - tu_x)\frac{\partial F_2}{\partial u_x}\right) \\ &= 2uh + t(uh_t + hu_t), \\ C_4^2 &= v\left((1 - tu_x)\frac{\partial F_1}{\partial u_t} - th_x\frac{\partial F_1}{\partial h_t}\right) + w\left(-th_x\frac{\partial F_2}{\partial h_t} + (1 - tu_x)\frac{\partial F_2}{\partial u_t}\right) \\ &= -tuh_x - thu_x + h. \end{aligned}$$

Thus,

$$\begin{aligned} C_4^1 &= 2uh + t(uh_t + hu_t), \\ C_4^2 &= -tuh_x - thu_x + h. \end{aligned}$$

If one substitute these quantities in the $D_i(C^i) = 0$ and after some simplifications we yield

$$C_4^1 = uh, \qquad C_4^2 = h. \tag{16}$$

In a similar way, applying the formula (7) to the dilation symmetry $X_3 = x\frac{\partial}{\partial x} + t\frac{\partial}{\partial x}$ in (10) where $\xi^1 = x$, $\xi^2 = t$, $\eta^1 = 0$, $\eta^2 = 0$ and to the Lagrangian (11), we obtain the following flux of the conservation law

$$\begin{aligned} C_3^1 &= v\left(xF_1 - (tu_t + xu_x)\frac{\partial F_1}{\partial u_x} - (th_t + xh_x)\frac{\partial F_1}{\partial h_x}\right) \\ &\quad + w\left(xF_2 - (tu_t + xu_x)\frac{\partial F_2}{\partial u_x} - (th_t + xh_x)\frac{\partial F_2}{\partial h_x}\right) \\ &= xuh_t - 2thuu_t - tu^2h_t + xhu_t - thh_t. \end{aligned}$$

The operator X_3 provides the following density

$$C_3^2 = v\left(tF_1 - (tu_t + xu_x)\frac{\partial F_1}{\partial u_t} - (th_t + xh_x)\frac{\partial F_1}{\partial h_t}\right)$$
$$+ w\left(tF_2 - (tu_t + xu_x)\frac{\partial F_2}{\partial u_t} - (th_t + xh_x)\frac{\partial F_2}{\partial h_t}\right)$$
$$= tu^2h_x + 2thuu_x - xuh_x - xhu_x + thh_x.$$

If we substitute these quantities in the $D_i(C^i) = 0$ and after some simplifications we have

$$\begin{aligned} C_3^1 &= hu^2 - thh_t, \\ C_3^2 &= thh_x + uh. \end{aligned} \tag{17}$$

Summing up, following the above procedures we can easily obtain conserved vectors for each symmetry. For instance we get the following conserved vectors corresponding the operator X_5

$$\begin{aligned} C_5^1 &= u^2h + \frac{1}{2}h^2, \\ C_5^2 &= uh. \end{aligned} \tag{18}$$

4 Conclusion

In conclusion, we have applied composite variational approach to the shallow water equations in the plane flow. First, we obtained Lie-point symmetries. Then we constructed adjoint equations by applying formal Lagrangian to variational derivative. It is seen that, adjoint (14) are self adjoint and they admit symmetries of (8). Equations (8) and (14) together is a member of family of the Euler-Lagrange type equations. Then, the corresponding local conservation laws are given for the first time by using the composite variational method, which is a powerful tool for PDEs.

Acknowledgements This work is part of the PhD thesis of the first author Emrullah Yasar.

References

1. N.H. Ibragimov, J. Math. Anal. Appl. **333**, 311–328 (2007)
2. P.J. Olver, *Application of Lie Groups to Differential Equations* (Springer, New York, 1993)
3. A. Zee, *Quantum Field Theory* (Princeton University Press, Princeton, 2003)
4. A.M. Lyapunov, *Stability of Motion* (Academic Press, New York, 1966)
5. R.J. LeVeque, *Numerical Methods for Conservation Laws*. Lect. in Math. (Birkhäuser, Basel, 1992)
6. E. Noether, Nacr. Konig. Gesell. Wissen, Gottingen, Math.-Phys. Kl. **2**, 235–257 (1918)
7. P.S. Laplace, *Celestial Mechanics* (New York, 1966), p. 1798 (English translation)
8. H. Steudel, Z. Naturforsch. A **17**, 129–132 (1962)
9. A.H. Kara, F.M. Mahomed, Int. J. Theor. Phys **39**, 23–40 (2000)
10. S.C. Anco, G.W. Bluman, Eur. J. Appl. Math. **13**, 545–566 (2002)

11. A.H. Kara, F.M. Mahomed, Nonlinear Dyn. **45**, 367–383 (2006)
12. R. Khamitova, Commun. Nonlinear Sci. Numer. Simul. (2008). doi:10.1016/j.cnsns.2008.08.009
13. E. Yasar, Nonlinear Dyn. **54**(4), 307–312 (2008)
14. E. Yasar, T. Ozer, Conservation laws for one-layer shallow water wave systems. Nonlinear Anal., Real World Appl. **11**, 838–848 (2010)

Conserved Forms of Second Order-Ordinary Differential Equations

C. Muriel and J.L. Romero

Abstract In this paper we prove that λ-symmetries of any second-order ordinary differential equation can be used to construct an integrating factor of the equation, and that the associated conserved form can be derived from the algorithm of reduction associated to the λ-symmetry.

Keywords Ordinary differential equations · λ-symmetries · Integrating factors · First integrals

1 Introduction

In the literature it is usual to consider integrating factors only for first-order ordinary differential equations. In 1874 Sophus Lie proved that a first-order ordinary differential equation can be solved by quadrature if the Lie point symmetries of the equation are known. It is well-known that the first-order ordinary differential equation

$$M(x,u) + N(x,u)u_x = 0 \tag{1}$$

possesses a one-parameter group with infinitesimal generator $v = \xi(x,u)\partial_x + \eta(x,u)\partial_u$ if and only if the function

$$\mu = \frac{1}{\xi M + \eta N} \tag{2}$$

is an integrating factor with $\xi M + \eta N \neq 0$.

For equations of order $n > 1$, it may appear integrating factors that do not proceed from Lie point symmetries. For example, the exact second-order equation

$$u_{xx} = D_x((x + x^2)e^u) \tag{3}$$

C. Muriel (✉) · J.L. Romero
Department of Mathematics, University of Cádiz, 11510 Puerto Real, Cádiz, Spain
e-mail: concepcion.muriel@uca.es

J.A.T. Machado et al. (eds.), *Nonlinear Science and Complexity*,
DOI 10.1007/978-90-481-9884-9_9, © Springer Science+Business Media B.V. 2011

admits no Lie point symmetry [12, p. 182] but admits the trivial integrating factor $\mu(x,u,u_x)=1$.

Several methods have been developed to find integrating factors for equations of arbitrary order n [1, 4, 6].

An interesting approach to this problem is based on variational derivatives [1, 4]:

$$\frac{\delta}{\delta u}=\frac{\partial}{\partial u}-D_x\frac{\partial}{\partial u_x}+D_x^2\frac{\partial}{\partial u_{xx}}+\cdots, \tag{4}$$

where D_x denotes the total derivative operator with respect to x. Let us denote by $u^{(n)}=(u,u_1,\ldots,u_n)$, where u_i is the i-th order derivative of u with respect to the independent variable x. The integrating factors $\mu(x,u^{n-1)})$ of an n-th order equation

$$M(x,u^{(n-1)})+N(x,u^{(n-1)})u_n=0 \tag{5}$$

are determined by the following equation:

$$\frac{\delta}{\delta u}(\mu(M+Nu_n))=0. \tag{6}$$

For first-order equations, the corresponding equation (6) becomes the single linear partial differential equation $(\mu M)_u-(\mu N)_x=0$, which always has an infinite number of solutions. For $n=2$, the determining equation (6) gives an over-determined system of two second-order linear partial differential equations [4]. These systems become more complicated for higher orders.

Other methods to find determining equations for integrating factors use specific *ansatzes* for the integrating factor depending on the form of the equation [1, 3]. Different systems of determining equations for integrating factors can also be derived dealing with the adjoint-symmetries of the equation [1, 2].

In this paper we apply the theory of λ-symmetries [7] to the problem of finding integrating factors of second-order equations, including equations without Lie point symmetries. We first associate a λ-symmetry to a known integrating factor and then we consider the converse problem: given a λ-symmetry we show how to construct an integrating factor.

The technique we present here is specially useful for equations that do not have Lie point symmetries. For any second-order equation it is possible to calculate a λ-symmetry by solving a quasi-linear first-order partial differential equation. Once this λ-symmetry has been calculated, an integrating factor and the associated conserved form can be derived dealing with first-order ordinary differential equations.

2 λ-Symmetries Associated to Integrating Factors

λ-symmetries for ordinary differential equations were first introduced in [7], motivated by the existence of equations without Lie point symmetries that can be integrated or reduced in order. In fact, most of the known methods of order reduction can be derived by the algorithm of reduction associated to λ-symmetries [9].

There exist integrating factors of equations of order $n > 1$ that do not possess Lie point symmetries. Our next result proves that there always exists at least one λ-symmetry associated to a given integrating factor.

Theorem 1 *If a second-order ordinary differential equation*

$$u_{xx} = M(x, u, u_x) \tag{7}$$

admits an integrating factor $\mu = \mu(x, u, u_x)$:

$$\mu \cdot (u_{xx} - M(x, u, u_x)) = D_x(G(x, u, u_x)) \tag{8}$$

then the vector field $v = \partial_u$ *is a* λ*-symmetry of* (7) *for function* $\lambda = -\frac{G_u}{G_{u_x}}$.

Proof The first λ-prolongation of $v = \partial_u$ for $\lambda = -\frac{G_u}{G_{u_x}}$ is given by

$$v^{[\lambda,(1)]} = \partial_u - \frac{G_u(x, u, u_x)}{G_{u_x}(x, u, u_x)}\partial_{u_x}. \tag{9}$$

It is clear that x and $G(x, u, u_x)$ are two independent invariants of $v^{[\lambda,(1)]}$. By Theorem 2 in [8], $D_x(G(x, u, u_x))$ is also an invariant of $v^{[\lambda,(2)]}$. By applying $v^{[\lambda,(2)]}$ to both members in (8) we get

$$\mu \cdot v^{[\lambda,(2)]}(u_{xx} - M(x, u, u_x)) = -v^{[\lambda,(2)]}(\mu) \cdot (u_{xx} - M(x, u, u_x)). \tag{10}$$

When u_{xx} is replaced by $M(x, u, u_x)$ in (10) we obtain

$$v^{[\lambda,(2)]}(u_{xx} - M(x, u, u_x)) = 0 \quad \text{when } u_{xx} = M(x, u, u_x). \tag{11}$$

This proves the result. □

Example 1 Let us consider again the exact equation (3). Since $\mu(x, u, u_x) = 1$ is an integrating factor of (3), the vector field $v = \partial_u$ is a λ-symmetry of (3) for function $\lambda(x, u, u_x) = -(x + x^2)e^u$. We want to remark that this equation does not admit Lie point symmetries.

3 Integrating Factors Associated to λ-Symmetries

In Theorem 1 it has been proved that any integrating factor has associated a λ-symmetry. The question that naturally arises is how to recover an integrating factor from a previously known λ-symmetry.

Let us assume that a given second-order ordinary differential equation

$$u_{xx} = M(x, u, u_x) \tag{12}$$

admits some λ-symmetry v. By introducing canonical coordinates, if it is necessary, it can be assumed that $v = \partial_u$ for some function $\lambda = \lambda(x, u, u_x)$. Let $w(x, u, u_x)$ be a first-order invariant of $v^{[\lambda,(1)]}$. The set $\{x, w, D_x(w)\}$ is a complete system of invariants of $v^{[\lambda,(2)]}$ [8]. Since v is a λ-symmetry of (12), the subvariety

$$\mathcal{L} = \{(x, u, u_x) : u_{xx} = M(x, u, u_x)\} \tag{13}$$

is invariant for $v^{[\lambda,(2)]}$. Then $\mathcal{L}$ is, locally, the solution set of an invariant function $\Delta_R(x, w, D_x w)$ (Proposition 2.18 in [11]). By considering w as a new variable depending on x, $\Delta_R(x, w, w_x) = 0$ is the reduced equation associated to the λ-symmetry v. Let $D_x(G(x, w)) = 0$ be a conserved form of this reduced first-order equation. Locally, the subvariety $\mathcal{L}$ is the solution set of function $D_x(G(x, w(x, u, u_x)))$. By Proposition 2.10 in [11] there exists some function μ such that

$$\mu \cdot (u_{xx} - M(x, u, u_x)) = D_x(G(x, w(x, u, u_x))). \tag{14}$$

By comparing the coefficients of u_{xx} we conclude that

$$\mu(x, u, u_x) = G_w(x, w(x, u, u_x)) \cdot w_{u_x}(x, u, u_x) \tag{15}$$

is an integrating factor of (12) and $D_x(G(x, w(x, u, u_x))) = 0$ is the associated conserved form.

We emphasize this result in the next theorem:

Theorem 2 *If a second-order ordinary differential equation* (12) *admits the vector field* $v = \partial_u$ *as* λ*-symmetry, then an integrating factor is given by*

$$\mu(x, u, u_x) = G_w(x, w(x, u, u_x)) \cdot w_{u_x}(x, u, u_x) \tag{16}$$

where

- $w(x, u, u_x)$ *is a first-order invariant of* $v^{[\lambda,(1)]}$,
- $G(x, w)$ *is a conserved form of the first-order reduced equation associated to the* λ*-symmetry* v.

4 Algorithm to Calculate Integrating Factors Based on λ-Symmetries

As a consequence of Theorem 2, the next algorithm can be followed to find integrating factors of a given a second-order equation

$$u_{xx} = M(x, u, u_x). \tag{17}$$

1. Find one particular solution $\lambda = \lambda(x, u, u_x)$ of the quasi-linear first-order partial differential determining equation:
$$\lambda_x + u_x \lambda_u + M \lambda_{u_x} = M_u + M_{u_x} \lambda - \lambda^2. \tag{18}$$
For this function λ, the vector field $v = \partial_u$ becomes a λ-symmetry of (17).
2. Calculate any particular solution $w = w(x, u, u_x)$ of the equation
$$w_u + w_{u_x} \lambda(x, u, u_x) = 0. \tag{19}$$
This function w is a first-order invariant of $v^{[\lambda,(1)]}$.
3. Write (17) in terms of $\{x, w, D_x w\}$ and find some conserved form $D_x(G(x, w)) = 0$ of the reduced equation (first-order equation).

4. An integrating factor of (17) is given by (16). The associated conserved form of (17) is given, without additional computations, by $D_x(G(x, w(x, u, u_x))) = 0$.

Example 2 The second-order equation

$$u_{xx} = \frac{u_x^2}{u} + \left(u + \frac{x}{u}\right)u_x - 1 \tag{20}$$

is an instance of a Painlevé-type equation [5] which has no Lie point symmetries.

1. It can be checked that function $\lambda = u + \frac{x}{u} + \frac{u_x}{u}$ is a particular solution of the corresponding (18). In consequence, $v = \partial_u$ is a λ-symmetry of (20) for $\lambda = u + \frac{x}{u} + \frac{u_x}{u}$.
2. A first-order invariant of $v^{[\lambda,(1)]}$ is given by

$$w = \frac{u_x}{u} - u + \frac{x}{u}. \tag{21}$$

3. In terms of $\{x, w, w_x\}$, (20) becomes $w_x = 0$. A conserved form $D_x(G(x, w)) = 0$ of this reduced equation is defined by $G(x, w) = w$.
4. Since $w_{u_x} = \frac{1}{u}$ and $G_w(x, w) = 1$, we obtain the integrating factor defined by (16):

$$\mu(x, u, u_x) = \frac{1}{u} \tag{22}$$

and the associated conserved form (14):

$$D_x\left(\frac{u_x}{u} - u + \frac{x}{u}\right) = 0. \tag{23}$$

Remark 1 The integrating factor (22) can also be derived by the method based on variational derivatives. It can be checked that (22) is a particular solution of the corresponding second-order system of partial differential equations (6.6.17–18) in [4]:

$$\begin{aligned}
&(u_x u^2 - u + u_x^2 + x u_x)\mu_{u_x u_x} + u u_x \mu_{u u_x} + u \mu_{x u_x} \\
&\quad + 2(u^2 + 2u_x + x)\mu_{u_x} + 2u\mu_u + 2\mu = 0, \\
&(u_x^2 u^3 - u_x u^2 + u_x^3 u + u_x^2 x u)\mu_{u u_x} + (u_x u^3 - u^2 + u_x^2 u + u_x x u)\mu_{x u_x} \\
&\quad + 2u^2 u_x \mu_{xu} + (u^2 u_x^2)\mu_{uu} + u^2 \mu_{xx} + (-u_x^3 + u^2 u_x^2 - x u_x^2 + u u_x)\mu_{u_x} \\
&\quad + (u^2 + u_x^2 u)\mu_u + (u^3 + 2u_x u + xu)\mu_x + (u - u_x^2)\mu = 0.
\end{aligned} \tag{24}$$

In this case the form of solution (22) is simple and it could have been found as solution of the system (24). However, for another equations that admit more complicated integrating factors, it is not easy to find solutions for this type of systems (Example 2 in [10]).

5 Conclusions

We have presented an alternative approach to the problem of determining integrating factors for second-order ordinary differential equations based on λ-symmetries.

The method may simplify the computations derived by other methods. A λ-symmetry of a second-order equation can be calculated by means of a particular solution of a quasi-linear first-order partial differential equation. By using the algorithm of reduction associated to the λ-symmetry, the integrating factor and the associated first integral can be determined dealing with first-order ordinary differential equations.

The method is specially interesting when it is applied to equations without Lie point symmetries, as it is shown with a particular example of a Painlevé-type equation.

The analogous problem for ordinary differential equations of arbitrary has been considered in a separate paper [10].

References

1. W. Bluman, C. Anco, *Symmetry and Integration Methods for Differential Equations* (Springer, New York, 2002)
2. W. Bluman, C. Anco, Euro Int. App. Math. **9**, 245–259 (1998)
3. E.S. Cheb-Terrab, A.D. Roche, J. Symb. Comput. **27**, 501–519 (1999)
4. N.H. Ibragimov, *A Practical Course in Differential Equations and Mathematical Modelling* (ALGA, Karlskrona, 2006)
5. E. Ince, *Ordinary Differential Equations* (Longmans, Green & Co, London, 1927)
6. P.G.L. Leach, S.E. Bouquet, J. Nonlinear Math. Phys. **9**(2), 73–91 (2002)
7. C. Muriel, J.L. Romero, IMA J. Appl. Math. **66**(2), 111–125 (2001)
8. C. Muriel, J.L. Romero, Theor. Math. Phys. **133**(2), 1565–1575 (2002)
9. C. Muriel, J.L. Romero, J. Lie Theory **13**, 167–188 (2003)
10. C. Muriel, J.L. Romero, J. Nonlinear Math. Phys. **15**(3), 290–299 (2008)
11. P.J. Olver, *Applications of Lie Groups to Differential Equations* (Springer, New York, 1993)
12. P.J. Olver, *Equivalence, Invariants and Symmetry* (Cambridge University Press, Cambridge, 1995)

Analytical Investigation of a Two-Phase Model Describing a Three-Way-Catalytic Converter

J. Volkmann and N. Migranov

Abstract The two-phase-model describes the thermal and chemical behaviour of a Three-Way-Catalytic converter (TWC). According to this model energy, mass balance equations for gas and solid phase are taking into account. These equations were investigated by group analysis and separation of variables method in order to construct solutions.

Keywords Group analysis · Separation of variables · Partial differential equations · Catalytic converter · Two-phase-model

1 Introduction

The article focuses on Lie symmetry analysis of equations describing a two-phase model for a three-way-catalytic converter. The origins of this model approach are found in investigations of Liu/Amundson [1] and Eigenberger [2, 3]. Based on the work written by Nieken [4] balance equations for energy and mass describing the temperature and the concentration of several compounds are considered. The momentum balance is negated. Mass flux density in [kg/s] and pressure are significant parameters which are determinable experimentally. A more detailed description of this model can be found in the literature (e.g. [1–3, 5, 6]). These model equations can be regarded as

J. Volkmann (✉)
Research Center for Mathematics and Physics, Bashkir State Pedagogical University, 3A, October Revolution Street, Ufa 45 0000, Russia
e-mail: Jo.Volkmann@fh-wolfenbuettell.de

N. Migranov
Department of Theoretical Physics, Bashkir State Pedagogical University, 3A, October Revolution Street, Ufa 45 0000, Russia
e-mail: ufangm@yahoo.co.uk

J.A.T. Machado et al. (eds.), *Nonlinear Science and Complexity*,
DOI 10.1007/978-90-481-9884-9_10,

- Balance of energy in the gas phase:

$$\epsilon \varrho_g c_{p,g} \frac{\partial T_g}{\partial t} = -\epsilon \varrho_g c_{p,g} v_g \frac{\partial T_g}{\partial z} + \lambda \epsilon \frac{\partial^2 T_g}{\partial z^2} + \alpha a_v (T_s - T_g), \tag{1}$$

- Balance of energy in the solid phase:

$$(1-\epsilon)\varrho_s c_{p,s} \frac{\partial T_s}{\partial t} + (1-\epsilon)\lambda_s \frac{\partial^2 T_s}{\partial z^2} - \alpha a_v (T_s - T_g) + a_x \sum_{j=1}^{J} (-\Delta H_j) R^j = 0, \tag{2}$$

- Balance of mass in the gas phase:

$$\epsilon \varrho_g \frac{\partial w_{i,g}}{\partial t} = -\epsilon \varrho_g v_g \frac{\partial w_{i,g}}{\partial z} - \epsilon D_{eff} \frac{\partial^2 w_{i,g}}{\partial z^2} - \varrho_g \beta_j a_x (w_{i,g} - w_{i,s}),$$

- Balance of mass in the solid phase:

$$0 = \varrho_g \beta_j a_x (w_{i,g} - w_{i,s}) + a_x \sum_{i=1}^{J} M_i R^i, \tag{3}$$

with initial- and boundary conditions (see [6]).

Here the following parameters are used: ϵ is the porosity of a porous medium describing the fraction of void space in the material where the void may contain, for example, air or water. It is defined by the ratio $\frac{V_v}{V_T}$ where V_v is the volume of void-space and V_T is the total or bulk volume of material, including the solid and void components. v_g the velocity of the gas, ϱ the density of the gas (index g) and solid (index s). Furthermore c_p is the specific heat capacity of the gas (index g) and solid (index s). λ represents the heat conductivity, α the heat exchanging coefficient with the dimension [W/(m^2 K)]. $a_V = \frac{O_{mon}}{Adz}$ is a geometry factor including the surface O_m of the catalytic converter. $a_x = \frac{O_{mon,act}}{Adz}$ is another geometry factor taking into account the active layer inside. ΔH_j represents the enthalpy of the reaction and R^j describes the reaction velocity. Furthermore D_{eff} is an effective diffusion coefficient, M_i the mol mass of the substance i with the dimension [kg/kmol]. In the factor $G_z = \frac{\dot{m}}{A} = \frac{\epsilon \varrho_g A v_g}{A} = \epsilon \varrho_g v_g$ the mass flow is taken into account. L represents the length of the component.

The variables $T_g, T_s, w_{i,g}$ describe the temperature of the gas, of the solid and the mass fraction which is the ratio of the mass of substance i and the whole mass respectively.

2 The Group Theoretical Approach

In this chapter the system of equations

$$\frac{\partial T_g}{\partial t} = A_1 \frac{\partial T_g}{\partial z} + A_2 \frac{\partial^2 T_g}{\partial z^2} + A_3 (T_s - T_g), \tag{4}$$

$$\frac{\partial T_S}{\partial t} = B_1 \frac{\partial^2 T_s}{\partial z^2} + B_2 - B_3(T_s - T_g), \tag{5}$$

$$\frac{\partial w_{i,g}}{\partial t} = -C_1 \frac{\partial w_{i,g}}{\partial z} + C_2 \frac{\partial^2 w_{i,g}}{\partial z^2} - F. \tag{6}$$

is investigated by symmetry analysis written in dimensionless variables. A closure look shows that (4) and (5) are coupled and (6) is decoupled. So the investigation starts with (4) and (5). Using the *Mathematica* package *MathLie* [21] one can find the following infinitesimals $\phi^1 = -\frac{B_2 k_3 - B_3 k_3 T_g + B_3 \mathcal{F}(t,x) + \frac{\partial \mathcal{F}}{\partial t} - B_1 \frac{\partial^2 \mathcal{F}}{\partial z^2}}{B_3}$, $\phi^2 = k_3 T_s + \mathcal{F}(t,x)$, $\xi^1 = k_1$, $\xi^2 = k_2$, where $\mathcal{F}(t,x)$ satisfies

$$\begin{aligned} &-A_3 B_2 k_3 - (-A_3 + B_3\epsilon)\frac{\partial \mathcal{F}}{\partial t} + \epsilon \frac{\partial^2 \mathcal{F}}{\partial t} + A_1 B_3 \frac{\partial \mathcal{F}}{\partial x} + A_2 \frac{\partial^2 \mathcal{F}}{\partial x \partial t} + A_3 B_1 \frac{\partial^2 \mathcal{F}}{\partial x^2} \\ &- A_2 B_3 \epsilon \frac{\partial^3 \mathcal{F}}{\partial x^2 \partial t} - B_1 \epsilon \frac{\partial^3 \mathcal{F}}{\partial x^2 \partial t} - A_1 B_1 \frac{\partial^3 \mathcal{F}}{\partial x^3} + A_2 B_1 \epsilon \frac{\partial^4 \mathcal{F}}{\partial x^4} = 0. \end{aligned} \tag{7}$$

For the further calculations an expression for $\mathcal{F}$ is chosen in such way that $-A_3 B_2 k_3 + (-A_3 + B_3\epsilon)\frac{\partial \mathcal{F}}{\partial t} + \epsilon \frac{\partial^2 \mathcal{F}}{\partial t^2} = 0$ has to be solved. The solution of this equation is

$$\mathcal{F} = -\frac{A_3 B_2 k_3 t}{A_3 - B_3\epsilon} + \frac{\exp(-t\frac{-A_3 + B_3\epsilon}{\epsilon})\epsilon k_4}{A_3 - B_3\epsilon} + k_5. \tag{8}$$

This expression has also the property to annul (7). By substituting (8) into the infinitesimals it is

$$\begin{aligned} \phi^1 &= \frac{A_3(B_2 B_3 k_3 t + B_3 k_3 T_g - e^{-B_3 t + \frac{A_3}{\epsilon} t} k_5 - B_3 k_4)}{B_3(A_3 - B_3\epsilon)} \\ &\quad + \frac{B_3\epsilon(B_2 k_3 + B_3(-k_3 T_g + k_4))}{B_3(A_3 - B_3\epsilon)}, \\ \phi^2 &= -\frac{A_3 + B_2 k_3 t}{A_3 - B_3\epsilon} + k_3 T_s + \frac{\exp(-t\frac{-A_3 + B_3\epsilon}{\epsilon})\epsilon k_5}{A_3 - B_3\epsilon} + k_4, \\ \xi^1 &= k_1, \qquad \xi^2 = k_2. \end{aligned}$$

In this case a five parametric group is found. The generators of the related algebra read

$$\begin{aligned} V_1 &= \partial_x, \qquad V_2 = \partial_t, \\ V_3 &= \frac{A_3(B_2 t + T_g) + (B_2 - B_3 T_g)\epsilon}{A_3 - B_3\epsilon}\partial_{T_g} - \frac{A_3 B_2 t - A_3 T_s + B_2 \epsilon T_s}{A_3 - B_3\epsilon}\partial_{T_s}, \\ V_4 &= -\partial_{T_g} + \partial_{T_s}, \qquad V_5 = \frac{e^{-B_3 t + \frac{A_3 t}{\epsilon}} A_3}{B_3(-A_3 + B_3\epsilon)}\partial_{T_g} - \frac{e^{-B_3 t + \frac{A_3 t}{\epsilon}}\epsilon}{-A_3 + B_3\epsilon}\partial_{T_s}. \end{aligned}$$

Commutator table and the properties of this algebra can be found in [6] .

3 Reductions and Solutions

The next part of the investigation of system (4) and (5) is devoted to calculate reductions of several special cases of generators and to solve them.

1. Using the generator $V_1 = \partial_x$ the similarity variables are $t = \zeta$, $T_g = F_1$, $T_s = F_2$. The related ordinary differential equations read

$$\epsilon F_1'(\zeta) = A_3(F_1(\zeta) + F_2(\zeta)), \qquad B_3(F_1(\zeta) + F_2(\zeta)) + F_2'(\zeta) = B_2,$$

with the solution

$$F_1 = \frac{A_3 B_2(-1+\exp(\frac{\zeta(A_3-B_3\epsilon)}{\epsilon}))(-A_3\zeta - \frac{B_3\exp(\frac{\zeta(-A_3+B_3\epsilon)}{\epsilon})\epsilon^2}{A_3-B_3\epsilon})}{(A_3-B_3\epsilon)(-A_3+B_3\epsilon)}$$
$$+\frac{A_3 B_2(A_3\exp(\frac{\zeta(A_3-B_3\epsilon)}{\epsilon}) - B_3\epsilon)(-\zeta + \frac{\exp(\frac{\zeta(-A_3+B_3\epsilon)}{\epsilon})\epsilon}{-A_3+B_3\epsilon})}{(A_3-B_3\epsilon)^2}$$
$$+\frac{(A_3\exp(\frac{\zeta(A_3-B_3\epsilon)}{\epsilon}) - B_3\epsilon)C_1}{A_3-B_3\epsilon} + \frac{A_3(-1+\exp(\frac{\zeta(A_3-B_3\epsilon)}{\epsilon})C_2)}{A_3-B_3\epsilon},$$

$$F_2 = \frac{B_2(-A_3+B_3\exp(\frac{\zeta(A_3-B_3\epsilon)}{\epsilon})\epsilon)(-A_3\zeta - \frac{B_3\epsilon^2\exp(\frac{\zeta(-A_3+B_3\epsilon)}{\epsilon})}{A_3-B_3\epsilon})}{(-A_3+B_3\epsilon)^2}$$
$$+\frac{\epsilon A_3 B_2 B_3(-1+\exp(\frac{\zeta(A_3-B_3\epsilon)}{\epsilon}))(-\zeta + \frac{\epsilon\exp(\frac{\zeta(-A_3+B_3\epsilon)}{\epsilon})}{-A_3+B_3\epsilon})}{(A_3-B_3\epsilon)(-A_3+B_3\epsilon)}$$
$$+\frac{B_3(-1+\exp(\frac{\zeta(A_3-B_3\epsilon)}{\epsilon})\epsilon C_1)}{-A_3+B_3\epsilon} + \frac{(-A_3+B_3\exp(\frac{\zeta(A_3-B_3\epsilon)}{\epsilon}))C_2}{-A_3+B_3\epsilon}.$$

For this case the following initial conditions are valid: $T_g(t=0) = T_g^0$, $T_s(t=0) = T_s^0$. This leads to the following expressions for the constants:

$$C_1 = T_g^0 - \frac{A_3 B_2 \epsilon}{(A_3-B_3\epsilon)(-A_3+B_3\epsilon)},$$
$$C_2 = T_s^0 + \frac{B_2 B_3 \epsilon^2}{(A_3-B_3\epsilon)(-A_3+B_3\epsilon)}.$$

2. The next generator which is considered is $V_2 = \partial_t$. The similarity variables in this case are $x = \zeta$, $T_g = F_1$, $T_s = F_2$. The related ordinary differential equations are

$$A_1\frac{dF_1}{d\zeta} = A_3(F_1(\zeta) + F_2(\zeta)) + A_2\epsilon\frac{d^2F_1}{d\zeta^2}, \tag{9}$$

$$B_3(F_1(\zeta) + F_2(\zeta)) = B_2 + B_1\frac{d^2F_2}{d\zeta}. \tag{10}$$

Equation (10) can be solved for F_1 and the result can be differentiated twice and can be substituted into (9):

$$A_2\epsilon\left(\frac{B_1}{B_3}\frac{d^4F_2}{d\zeta^4}-\frac{d^2F_2}{d\zeta^2}\right)-A_1\left(\frac{B_1}{B_3}\frac{d^3F_2}{d\zeta^3}-\frac{dF_2}{d\zeta}\right)$$
$$+A_3\left(\frac{B_1}{B_3}\frac{d^2F_2}{d\zeta^2}+\frac{B_2}{B_3}-F_2\right)+A_3F_2=0. \qquad (11)$$

The substitution $\frac{dF_2}{d\zeta}=G$ allows to reduce of the order of (11). After dividing this so obtained equation by $A_2\epsilon\frac{B_1}{B_3}$ and introducing the parameter $\kappa=\frac{A_1}{A_2\epsilon}$ it is

$$\frac{d^3G}{d\zeta^3}-\kappa\frac{d^2G}{d\zeta^2}-\frac{A_3\kappa}{A_1}\frac{dG}{d\zeta}-\frac{B_3}{B_1}\frac{dG}{d\zeta}+\frac{B_3\kappa}{B_1}G+\frac{A_3B_2\kappa}{A_1B_1}=0. \qquad (12)$$

With the assumption $\kappa\ll 1$ the solution can be expressed in a convergent series concerning the parameter κ:

$$G(\zeta)=u_1(\zeta)+\kappa u_2(\zeta). \qquad (13)$$

Equation (13) can be inserted into (12) and the coefficients for each power of κ can be written out to the following system of equations:

$$\kappa^0:\quad -\frac{B_3}{B_1}\frac{du_1}{d\zeta}+\frac{d^3u_1}{d\zeta^2}=0, \qquad (14)$$

$$\kappa:\quad \frac{A_3B_2}{A_1B_1}+\frac{B_3}{B_1}u_1+\frac{A_3}{A_1}\frac{du_1}{d\zeta}-\frac{B_3}{B_1}\frac{du_2}{d\zeta}-\frac{d^2u_1}{d\zeta^2}+\frac{d^3u_2}{d\zeta^3}=0. \qquad (15)$$

The solution of (14) is

$$u_1(\zeta)=\sqrt{\frac{B_1}{B_3}}\exp\left(\sqrt{\frac{B_3}{B_1}}\zeta\right)K_1-\sqrt{\frac{B_1}{B_3}}\exp\left(-\sqrt{\frac{B_3}{B_1}}\zeta\right)K_2+K_3,$$

where K_1, K_2, K_3 are constants of integration. After inserting this into (15) an expression for u_2 is found:

$$u_2=\frac{(A_3B_2+A_1B_3K_3)\zeta}{A_1B_3}+\exp\left(\sqrt{\frac{B_3}{B_1}}\zeta\right)$$
$$\times\left(-\frac{A_3B_1K_1\zeta}{2A_1B_3}+\frac{\sqrt{B1}(3A_3B_1K_1+4A_1B_3C_1)}{4A_1B_3^{\frac{3}{2}}}\right)$$
$$+\exp\left(\sqrt{\frac{B_3}{B_1}}\zeta\right)\left(-\frac{A_3B_1K_2\zeta}{2A_1B_3}+\frac{\sqrt{B1}(3A_3B_1K_2+4A_1B_3C_2)}{4A_1B_3^{\frac{3}{2}}}\right)$$
$$+C_3.$$

Here C_1, C_2, C_3 are also constants of integration. After the substitution of these solutions $u_1(\zeta)$ and $u_2(\zeta)$ into (13) and one quadrature it can be found

$$F_2(x)=\frac{1}{4A_1B_3^2}\Big[2B_3x(A_3B_2x\kappa+A_1B_3(K3x\kappa+2(1+\kappa)C_3))$$
$$+B_1\big(A_3(5B_1(K_1+K_2)\kappa+2\sqrt{B_1}\sqrt{B_3}(-K_1+K_2)x\kappa)$$

$$+4A_1B_3(1+\kappa)(C_1+C_2)\big)\cosh\left(\frac{\sqrt{B_3}x}{\sqrt{B_1}}\right)$$
$$+B_1\big(A_3(5B_1(K_1-K_2)\kappa-2\sqrt{B_1}\sqrt{B_3}(K_1+K_2)x\kappa)$$
$$+4A_1B_3(1+\kappa)(C_1-C_2)\big)\sinh\left(\frac{\sqrt{B_3}x}{\sqrt{B_1}}\right)\Bigg].$$

Now a third term is taken into account using the ansatz $G=u_1(\zeta)+\kappa u_2(\zeta)+\frac{1}{2}\kappa^2u_3(\zeta)$. Inserting this into (12) and writing out the coefficients one finds:

$$\kappa^0:\quad -\frac{B_3}{B_1}\frac{du_1}{d\zeta}+\frac{d^3u_1}{d\zeta^2}=0.$$
$$\kappa^1:\quad \frac{A_3B_2}{A_1B_1}+\frac{B_3}{B_1}u_1+\frac{A_3}{A_1}\frac{du_1}{d\zeta}-\frac{B_3}{B_1}\frac{du_2}{d\zeta}-\frac{d^2u_1}{\zeta^2}+\frac{d^3u_2}{d\zeta^3}=0.$$
$$\kappa^2:\quad \frac{B_3u_2(\zeta)}{B_1}+\frac{A_3}{A_1}\frac{du_2}{d\zeta}-\frac{B_3}{2B_1}\frac{du_3}{d\zeta}-\frac{d^2u_2}{d\zeta^2}+\frac{1}{2}\frac{d^3u_3}{d\zeta^3}=0.$$

The solutions for the functions u_1 and u_2 are the same as in that previous case. For u_3 it is

$$u_3=\frac{(A_3B_2+A_1B_3K3)\zeta^2}{A_1B_3}+\frac{2\zeta(A_3^2B_1B_2+A_1A_3B_1B_3K3+A_1^2B_3^2C_3)}{A_1^2B_3^2}$$
$$+\exp\left(\frac{\sqrt{B_3}\zeta}{\sqrt{B_1}}\right)\left(\frac{A_3^2B_1^{\frac{3}{2}}K_1\zeta^2}{4A_1^2B_3^{\frac{3}{2}}}+\zeta\Omega_1+\Psi_{1,4}\right)$$
$$+\exp\left(-\frac{\sqrt{B_3}\zeta}{\sqrt{B_1}}\right)\left(-\frac{A_3^2B_1^{\frac{3}{2}}K_2\zeta^2}{4A_1^2B_3^{\frac{3}{2}}}-\zeta\Omega_2-\Psi_{2,5}\right)+C_6$$

with

$$\Omega_i=\frac{(-A_3^2B_1^2K_i-A_1A_3B_1^{\frac{3}{2}}\sqrt{B_3}K_i-A_1A_3B_1B_3C_i)}{A_1^2B_3^2},\qquad i=1,2$$
$$\Psi_{i,j}=\frac{\sqrt{B_1}}{4A_1^2B_3^{\frac{5}{2}}}(5A_3^2B_1^2K_i+6A_1A_3B_1^{\frac{3}{2}}\sqrt{B_3}K_i+6A_1A_3B_1B_3C_i+4A_1^2B_3^2C_j)$$

with $i=1,2$ and $j=4,5$ in the expression $\Psi_{i,j}$.

4 Conclusion

In this paper a two phase model describing a catalytic convert was investigated using balance equations for the gas and solid phase for energy and mass.

The equations in dimensionless form were investigated by the group theoretical approach. Due to the fact that this system was decoupled the equations for the temperature can be investigated first by group theoretical approach. A five-dimensional algebra was found.

For special generators the reduction was done in order to transform the original problem to ordinary differential equations. In one case a closed representation for the solution was found describing the behaviour of the temperature. In a second case a small parameter was introduced and methods of perturbation theory were applied in order to construct solutions.

In further investigations the chemistry has to be taken into account as well as the coupling of all equations.

References

1. S.L. Liu, N.R. Amundson, Stability of adiabatic packed bed reactors. JEC Found. **1**, 200–208 (1962)
2. G. Eigenberger, On the dynamic behaviour of the catalytic fixed-bed reactor in the region of multiple steady states I: The invluence of heat conduction in two phase models. Chem. Eng. Sci. **27**, 1909–1915 (1972)
3. G. Eigenberger, On the dynamic behaviour of the catalytic fixed-bed reactor in the region of multiple steady states II: The invluence of the boundary conditions in the catalyst phase. Chem. Eng. Sci. **27**, 1917–1924 (1972)
4. U. Nieken, *Abluftreinigung in katalytischen Festbettreaktoren bei periodischer Strömungsumkehr, Fortschritt–Bericht VDI, Reihe 3, Verfahrenstechnik* (Düsseldorf, 1993)
5. Th. Kirchner, *Experimentelle Untersuchungen und dynamische Simulation der Autoabgaskatalysi zur Verbesserung des Kaltstartverhaltens, Fortschritt-Bericht VDI, Reihe 12, Verfahrenstechnik/Fahrzeugtechnik* (Stuttgart, 1997)
6. J. Volkmann, N. Migranov, *Analytical investigation of a two-phase model describing a three-way-catalytic converter*, in preparation
7. G.J. Taylor, The dispersion of matter in turbulent flow trough a pipe. Proc. R. Soc. Lond. A **223**, 446 (1954)
8. A.D. Polyanin, A.J. Zhurov, A.V. Vyaz'min, Exact solutions of nonlinear heat- and mass-transfer equations. Theor. Found. Chem. Eng. **34**(5), 451–464 (2000)
9. A.D. Polyanin, A.J. Zhurov, A.V. Vyaz'min, *Exact solutions of heat and mass transfer equations*. www.mat.unb.br/~matcont/19_6.ps
10. N. Ibragimov, *Transformation Groups Applied to Mathematical Physics* (Dordrecht, Reidel, 1985)
11. N. Ibragimov, *CRC Handbook of Lie Group Analysis of Differential Equations*, vol. 1 (CRC Press, Boca Raton, 1994)
12. N. Ibragimov, *CRC Handbook of Lie Group Analysis of Differential Equations*, vol. 2 (CRC Press, Boca Raton, 1995)
13. N. Ibragimov, *CRC Handbook of Lie Group Analysis of Differential Equations*, vol. 3 (CRC Press, Boca Raton, 1996)
14. L.V. Ovsiannikov, *Group Analysis of Differential Equations* (Academic Press, New York, 1982)
15. G. Bluman, St.C. Anco, *Symmetry and Integration Methods for Differential Equations* (Springer, New York, 2002)
16. P.J. Olver, *Application of Lie Groups to Differential Equations* (Springer, New York, 1986)
17. A.N. Tikhonov, A.A. Samarskii, *Equations of Mathematical Physics* (Dover, New York, 1990)

18. J. Volkmann, *Mathematical Models and their Investigation of Catalytic Converters and Particulate Traps*, Preprint (2007)
19. H.S. Carslaw, J.C. Jaeger, *Conduction of Heat in Solids* (Clarendon Press, Oxford, 1984)
20. N. Ibragimov, *A Practical Course in Differential Equations and Mathematical Modelling: Classical and New Methods, Nonlinear Mathematical Models, Symmetry and Invariance Principles* (ALGA, Karlskrona, 2006)
21. G. Baumann, *Symmetry Analysis of Differential Equations with Mathematica* (Telos/Springer, New York, 2000)
22. R. Schmid, *Investigation of Lie Algebras*, Private communication, Ulm, 2002

Part III
Celestial Mechanics and Dynamical Astronomy: Methods and Applications

The Role of Invariant Manifolds in the Formation of Spiral Arms and Rings in Barred Galaxies

M. Romero-Gómez, E. Athanassoula,
J.J. Masdemont, and C. García-Gómez

Abstract We propose a new theory to explain the formation of spiral arms and of all types of outer rings in barred galaxies. We have extended and applied a technique used in celestial mechanics to compute transfer orbits. Thus, our theory is based on the chaotic orbital motion driven by the invariant manifolds associated to the periodic orbits around the hyperbolic equilibrium points. In particular, spiral arms and outer rings are related to the presence of heteroclinic or homoclinic orbits. Thus, R_1 rings are associated to the presence of heteroclinic orbits, while R_1R_2 rings are associated to the presence of homoclinic orbits. Spiral arms and R_2 rings, however, appear when there exist neither heteroclinic nor homoclinic orbits. We examine the parameter space of three realistic, yet simple, barred galaxy models and discuss the formation of the different morphologies according to the properties of the galaxy model. The different morphologies arise from differences in the dynamical parameters of the galaxy.

Keywords Galactic dynamics · Invariant manifolds · Spiral structure · Ring structure

M. Romero-Gómez (✉) · E. Athanassoula
Laboratoire d'Astrophysique de Marseille, Observatoire Astronomique de Marseille Provence, 38, rue Frederic Joliot-Curie, 13388 Marseille, France
e-mail: merce.romerogomez@oamp.fr

J.J. Masdemont
I.E.E.C. & Dep. Mat. Aplicada I, Universitat Politècnica de Catalunya, Av. Diagonal 647, 08028 Barcelona, Spain

C. García-Gómez
D.E.I.M., Universitat Rovira i Virgili, Av. Països Catalans 26, 43007 Tarragona, Spain

J.A.T. Machado et al. (eds.), *Nonlinear Science and Complexity*,
DOI 10.1007/978-90-481-9884-9_11,

1 Introduction

Bars are very common features in disk galaxies. According to Eskridge et al. [1] in the near infrared 56% of the galaxies are strongly barred and 6% are weakly barred. A large fraction of barred galaxies show either spiral arms emanating from the ends of the bar or spirals that end up forming outer rings (Elmegreen & Elmegreen [2]; Sandage & Bedke [3]).

Spiral arms are believed to be density waves (Lindblad [4]), that propagate outwards towards the principal Lindblad resonances, where they damp down (Toomre [5]). Some mechanism for replenishment is, therefore, needed (see for example Lindblad [6]; Toomre [5, 9]; Toomre & Toomre [7]; Sanders & Huntley [8]; Athanassoula [10] for more details). Schwarz [11–13] relates, however, the rings with the position of the principal resonances. There are different types of outer rings and they can be classified according to the relative orientation of the principal axes of the inner and outer rings (Buta [14]). If the two axes are perpendicular, the outer ring has an eight-shape and it is called R_1 ring. If they are parallel, it is called R_2 ring. There are galaxies where both types of rings are present, in which case the outer ring is simply called R_1R_2 ring.

Our approach is from the dynamical systems point of view. We first note that both spiral arms and (inner and outer) rings emanate from, or are linked to, the ends of the bar, where the unstable equilibrium points of a rotating system are located. We also note that no common theory for the formation of both features has been presented, so far. We therefore study in detail the neighbourhood of the unstable points and we find that spiral arms and rings are flux tubes driven by the invariant manifolds associated to the planar periodic orbits around the unstable equilibrium points.

2 Results

Our results show that the dynamics around the hyperbolic equilibrium points are essentially dominated by the planar and vertical families of periodic orbits and invariant tori. We also computed the invariant manifolds associated to both invariant objects (Romero-Gómez et al. [15]) and proved that for a bar potential, the motion around the unstable equilibrium points is essentially driven by the invariant manifolds associated to the planar periodic orbits.

One of our goals is to check separately the influence of each of the main free parameters in the bar model on the shape of the invariant manifolds of the unstable planar periodic orbits. In order to do so, we make families of models in which only one of the free parameters is varied, while the others are kept fixed. For each model, we compute the family of the planar periodic orbits around the hyperbolic equilibrium point, and for each periodic orbit, we compute the stable and unstable invariant manifolds. Our results show that only the bar pattern speed and the bar strength have an influence on the shape of the invariant manifolds, and thus, on the morphology of the galaxy (Romero-Gómez et al. [16]).

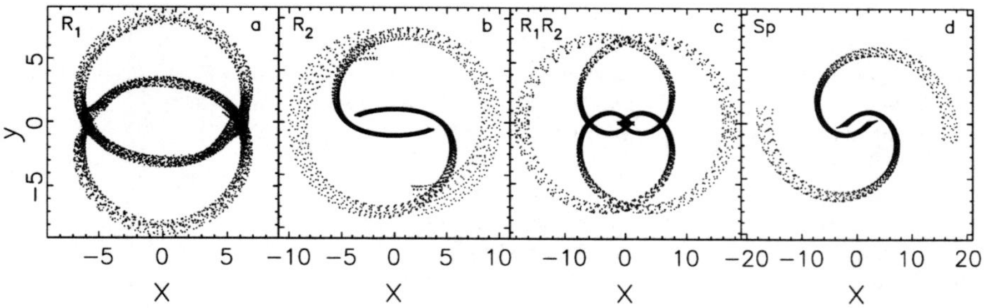

Fig. 1 Rings and spiral arms morphologies in barred galaxies. (**a**) rR_1 ring. (**b**) R_2 ring. (**c**) R_1R_2 ring. (**d**) Spiral arms

Our results also show that the morphologies obtained do not depend on the type of bar potential we use, but on the presence of homoclinic or heteroclinic orbits. If heteroclinic orbits exist, then the ring of the galaxy is classified as rR_1 (Romero-Gómez et al. [17], see Fig. 1a). The inner branches of the invariant manifolds associated to the periodic orbits around the unstable Lagrangian points outline an inner ring that encircles the bar and is elongated along it. The outer branches of the same invariant manifolds form an outer ring whose principal axis is perpendicular to the bar major axis. If the model does not have either heteroclinic or homoclinic orbits and only transit orbits are present, the barred galaxy will present two spiral arms emanating from the ends of the bar. The outer branches of the unstable invariant manifolds will spiral out from the ends of the bar and they will not return to its vicinity (Fig. 1d). If the outer branches of the unstable invariant manifolds intersect in configuration space with each other, then they form the characteristic shape of R_2 rings (Fig. 1b). That is, the trajectories outline an outer ring whose principal axis is parallel to the bar major axis. If only homoclinic orbits exist, the inner branches of the invariant manifolds form an inner ring, while the outer branches outline both types of outer rings, thus the barred galaxy presents an R_1R_2 ring morphology (Fig. 1c).

Acknowledgements This work partially supported by the Spanish MCyT-FEDER Grant MTM2006-00478 and AYA2007-60366, the French grant ANR-06-BLAN-0172, and the Marie Curie Research Training Network, Astronet. MRG acknowledges her "Becario MAE-AECI".

References

1. P.B. Eskridge, J.A. Frogel, R.W. Podge, A.C. Quillen, R.L. Davies, D.L. DePoy, M.L. Houdashelt, L.E. Kuchinski, S.V. Ramírez, K.V. Sellgren, D.M. Terndrup, G.P. Tiede, Astron. J. **119**, 536–544 (2000)
2. D.M. Elmegreen, B.G. Elmegreen, Mon. Not. R. Astron. Soc. **201**, 1021–1034 (1982)
3. A. Sandage, J. Bedke, *The Carnegie Atlas of Galaxies* (Carnegie Inst., Washington, 1994)
4. B. Lindblad, Stockh. Obs. Ann. **22**(5) (1963)
5. A. Toomre, Astrophys. J. **158**, 899–914 (1969)
6. P.O. Lindblad, Stockh. Obs. Ann. **21**(4) (1960)

7. A. Toomre, J. Toomre, Astrophys. J. **178**, 623–666 (1972)
8. R.H. Sanders, J.M. Huntley, Astrophys. J. **209**, 53–65 (1976)
9. A. Toomre, The structure and evolution of normal galaxies, in *Proc. of the Advanced Study Institute*, ed. by S.M. Fall, D. Lynden-Ball (Cambridge, 1981)
10. E. Athanassoula, Phys. Rep. **114**, 319–403 (1984)
11. M.P. Schwarz, Astrophys. J. **247**, 77–88 (1981)
12. M.P. Schwarz, Mon. Not. R. Astron. Soc. **209**, 93–109 (1984)
13. M.P. Schwarz, Mon. Not. R. Astron. Soc. **212**, 677–686 (1985)
14. R. Buta, Astrophys. J. Suppl. **96**, 39–116 (1995)
15. M. Romero-Gómez, J.J. Masdemont, C. García-Gómez, E. Athanassoula, Commun. Nonlinear Sci. Numer. Simul. **14**, 4123–4138 (2009)
16. M. Romero-Gómez, E. Athanassoula, J.J. Masdemont, C. García-Gómez, Astron. Astrophys. **472**, 63–75 (2007)
17. M. Romero-Gómez, J.J. Masdemont, E. Athanassoula, C. García-Gómez, Astron. Astrophys. **453**, 39–45 (2006)

Continuous and Discrete Concepts for Detecting Transport Barriers in the Planar Circular Restricted Three Body Problem

Michael Dellnitz, Kathrin Padberg, Robert Preis, and Bianca Thiere

Abstract In the last two decades, the mathematical analysis of material transport has received considerable interest in many scientific fields, in particular in astrodynamics. In this contribution we will focus on the numerical detection and approximation of transport barriers in the solar system. For this we consider and combine several techniques for the mathematical treatment of transport processes—using both continuous concepts from dynamical systems theory and discrete ideas from graph theory. For the demonstration of our results we consider the planar circular restricted three body problem with Sun and Jupiter as primaries, a simple model for describing the motion of asteroids in the solar system.

Keywords Transport barriers · Dynamical systems · Almost invariant sets · Invariant manifolds · Expansion

1 Introduction

The transport of material constitutes an important aspect of many natural systems. During the last two decades different mathematical concepts have been developed to get a better understanding of the mechanisms of particle transport and to estimate

M. Dellnitz (✉) · R. Preis · B. Thiere
Department of Mathematics, University of Paderborn, 33095 Paderborn, Germany
e-mail: dellnitz@math.upb.de

R. Preis
e-mail: robsy@math.upb.de

B. Thiere
e-mail: thiere@math.upb.de

K. Padberg
Institute for Transport and Economics, and Center for Information Services and High Performance Computing, Technische Universität Dresden, 01062 Dresden, Germany
e-mail: padberg@vwi.tu-dresden.de

J.A.T. Machado et al. (eds.), *Nonlinear Science and Complexity*,
DOI 10.1007/978-90-481-9884-9_12, © Springer Science+Business Media B.V. 2011

transport rates and probabilities [1–3]. Areas of application cover many scientific fields, such as fluid dynamics, ocean dynamics, molecular dynamics, physical chemistry, and astrodynamics [1, 4–6]. In this context, the analysis of transport phenomena in the solar system has received considerable scientific interest [7], in particular since recent sightings of asteroids. The numerical analysis of material transport in the solar system will be the core of this contribution.

Whereas in Dellnitz et al. [8, 9] the computation of transport rates and probabilities played a major role, here we focus on the analysis of the macroscopic structure of the underlying dynamical system, and, in particular, on the detection and approximation of barriers to particle transport.

For this we consider and combine several techniques for the mathematical treatment of transport processes—using both continuous concepts from dynamical systems theory (e.g. invariant manifolds, finite-time Lyapunov exponents) and discrete ideas from graph theory (e.g. graph partitioning, graph based expansion). Our methods are based on the set-oriented approach [10, 11] for the analysis and approximation of complicated dynamical behavior [12–14] and extend the ideas described in Dellnitz et al. [8, 9] and Padberg [15]. For the demonstration of our results we consider an appropriate Poincaré map in the planar circular restricted three body problem (PCRTBP) with Sun and Jupiter as primaries.

In the following, we give a brief overview of our methods and results, a more detailed treatment can be found in Padberg et al. [16].

2 Methods and Results

We consider the motion of a particle (asteroid) in the field of the Sun and Jupiter as described by the PCRTBP. We fix a certain energy level and consider a Poincaré section M, reducing the system to a two-dimensional time-discrete map $f : M \to M$ on a subset M of $\mathbb{R}^2$. f is area and orientation preserving.

Within the set-oriented framework, we first approximate the recurrent set within M by covering it with a finite collection of boxes $\mathcal{B} = \{B_1, \ldots, B_n\}$. This provides us with a convenient discretization of the region of interest and will be the basis for the following computations.

For details on the model under consideration and the computational framework we refer to Szebehely [17] and Dellnitz et al. [8]. The set-oriented algorithms are implemented in the software package GAIO [11].

2.1 *Finite-Time Lyapunov Exponents*

An established geometrical approach for the analysis of transport phenomena relies on the approximation of stable and unstable manifolds of hyperbolic period points of the map f. Their transversal intersection gives rise to complicated dynamical behavior and explains transport in terms of lobe dynamics [1].

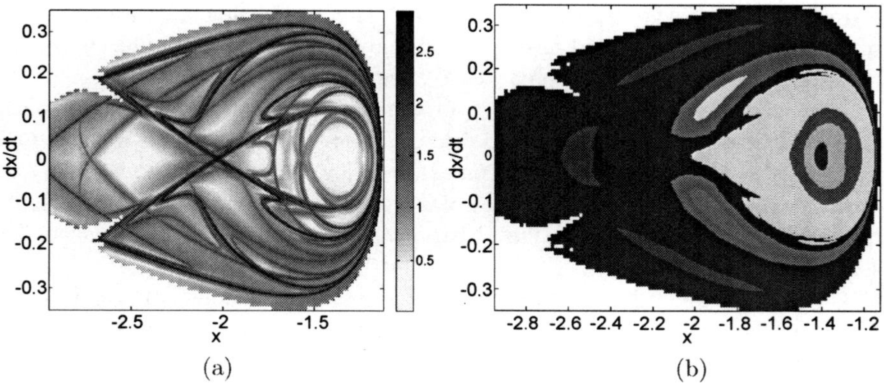

Fig. 1 (**a**) Approximation of transport barriers (here, the stable and unstable manifolds of a hyperbolic fixed point and of further hyperbolic periodic points of low period) using a set-oriented FTLE approach. *Dark areas* correspond to high values of the FTLE, computed using three iterations of f in forward and backward time. The relevant region in phase space is covered by a collection of small boxes. (**b**) Nine almost invariant sets obtained via a graph partitioning approach. Parts of the transport barriers obtained via the FTLE approach bound very well the almost invariant sets (e.g. *the large light region*). In addition, graph partitioning picks up regular (and truly invariant) regions

Finite-time Lyapunov exponents (FTLE) [18–22] are increasingly used (especially in nonautonomous systems) for the approximation of transport barriers and invariant manifolds. This quantity measures how much a small initial perturbation evolves under the (linearized) dynamics and it is expected to be large in the vicinity of invariant manifolds of hyperbolic objects. So local maxima or ridges in the scalar FTLE field typically define boundaries between regions that are characterized by a minimal exchange of particles [22]. For obvious reasons such regions are known as almost invariant sets [12] and play a crucial role in the macroscopic analysis of dynamical systems, in particular with respect to transport.

Here we compute a set-oriented approximation of the scalar FTLE field [15] by measuring the maximum divergence of small perturbations in the initial conditions under repeated application of f.

In our example system the approach detects the major transport barriers corresponding to stable and unstable manifolds of a hyperbolic fixed point and of further low-period hyperbolic periodic points of f. Moreover, the FTLE approach highlights the complicated homoclinic and heteroclinic tangles that provide the basis for the transport mechanism (see Fig. 1(a)). Light colors in the FTLE field can be related to regular regions such as invariant tori.

2.2 Graph Partitioning Techniques

So far, we have discussed how to compute major transport barriers of the underlying system. In this paragraph we will point out how to find regions of interest—almost invariant sets—for the computation of transport rates and probabilities, using a graph based instead of a geometric concept. Interestingly, almost invariant sets are often bounded by stable and unstable manifolds of the hyperbolic objects [8, 26] such that we expect the two approaches to compare well.

Almost invariant sets are usually approximated using a probabilistic approach [12], where—based on a set-oriented discretization—the dynamics is reduced to a finite-state Markov process. In this context, we view the boxes as the vertices and transitions between them as the edges of a weighted, directed graph. The weights of the edges are determined by the transitions between boxes, as, for instance, described in Dellnitz et al. [8].

Graph based algorithms can then be used to find a reasonable partition of phase space into almost invariant regions [13], i.e. sets of vertices that are highly coupled within themselves and only loosely coupled with other parts, and to compute transport rates between sets of interest [8].

As the graph partitioning problem for most commonly used cost functions is NP-complete, efficient heuristic methods have been developed for a number of different applications, see e.g. Preis [23]. For our computations we use the tool GADS [24], which connects the graph partitioning software library PARTY [23] and GAIO [11] to efficiently interlock graph based techniques with set-oriented methods [25].

Dominant almost invariant regions in the Sun-Jupiter problem approximated via graph partitioning techniques are shown in Fig. 1(b). The boundaries of some of the detected regions appear to be decomposed of branches of invariant manifolds of hyperbolic periodic points of the Poincaré map, visible as ridges in the FTLE field in Fig. 1(a) (e.g. the large light region in Fig. 1(b)). Moreover, graph partitioning also picks up regular regions such as invariant tori, which are also partly bounded by transport barriers obtained from the FTLE approach (e.g. the family of sets located in the large light region in Fig. 1(b)). So the geometric information related to invariant manifolds and high values of FTLE appears to be well coded in the graph. This is in good agreement with the studies in Dellnitz et al. [8].

2.3 Graph Based Expansion

Motivated by the similarity of the results obtained by the two different—continuous and discrete—methods explained above we now focus on another approach which, again, is only based on the analysis of the graph, i.e. without any geometric information.

The graph partitions give us the positions of some boundaries between the parts as a by-product. However, we would like to approximate the boundaries directly, in order to get more detailed information about the transport mechanism at work. Thus, we explore the use of graph expansion values, which is a well known notion in graph theory. It, roughly speaking, measures how much any set of vertices expands in a local neighborhood of the graph, and hence, has conceptual similarities to the FTLE approach. In our studies we therefore use this notion in order to detect invariant manifolds in the underlying dynamical system, i.e. regions in the vicinity of such transport barriers are expected to have a large graph expansion.

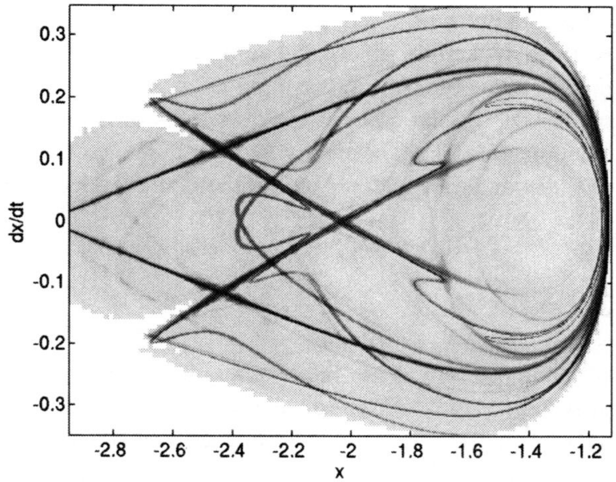

Fig. 2 Graph based expansion approach. Here vertices that induce particularly expansive subgraphs are *highlighted*. The results compare very well to the FTLE approach

We use several variations of the neighborhood structure (neighborhood subgraph with a small radius) in order to compute an expansion value for each vertex. A detailed definition and the development of heuristics to extract the transport barriers directly from the graph can be found in Padberg et al. [16]. A test computation using this approach for the same previous system can be seen in Fig. 2. Here we colored each box (i.e. each vertex v) according to the expansion value of the respective neighborhood subgraph of radius 4 induced by v. This corresponds to measuring the expansion of discrete trajectories (with respect to the initial vertex v) of length 4 on the graph. The approximated structures match very well the transport barriers obtained via the FTLE approach described above.

A natural extension of the graph expansion method is to use the multi-level structure of the set-oriented approach. For instance, the box covering can be iteratively

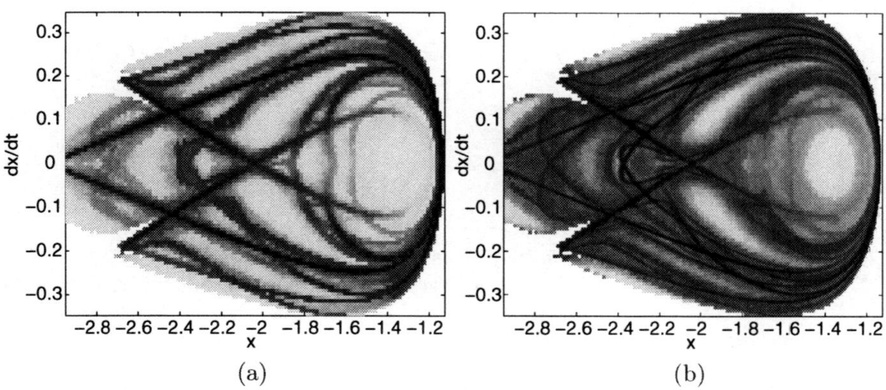

Fig. 3 Adaptive graph expansion. Here boxes inducing high values of graph expansion have been successively refined, resulting in a detailed approximation of the transport barriers. Figure (**a**) shows the initial coarse box covering whereas in (**b**) boxes with high values of graph expansion are refined

refined in areas of high graph expansion, see Fig. 3. Again the major transport barriers are picked up nicely and compare well to the other results. The adaptive approach allows for an increasingly detailed approximation of the transport barriers while keeping the computational costs at an acceptable level. For instance, in the example considered here only 34 % of the boxes are needed for the same resolution of the boundaries as compared to the non-adaptive approach.

3 Discussion

As demonstrated above the combination of geometrical and graph based methods provides a powerful tool for the qualitative and quantitative analysis of transport in dynamical systems. Both the FTLE approach and the graph partitioning ansatz define consistent almost invariant regions. This is in good agreement with related work on almost invariant sets and invariant manifolds [26]. The application of graph based expansion compares well to the FTLE approach and confirms that the reduction of the dynamical system f to a discrete graph with a finite-state Markov process retains all relevant information from the dynamics. Moreover, the graph based expansion ansatz is computationally inexpensive compared to partitioning methods and it can probably be used to obtain an initial guess for the solution of graph partitioning problems. Furthermore, the multi-level structure of the set-oriented approach allows for the development of adaptive methods. For a more detailed discussion of local expansion concepts for the approximation of transport barriers in dynamical systems we refer to Padberg et al. [16].

In the astrodynamical application considered here, the results allow us to draw conclusions about transport of particles between the Jupiter region and a neighborhood of the Sun. Based on the approximation of the relevant sets we can now compute transition probabilities and estimate for instance the risk of an asteroid impact, as discussed in Dellnitz et al. [8, 9].

Acknowledgements This research was partly supported by the EU funded Marie Curie Research Training Network *AstroNet*.

References

1. S. Wiggins, *Chaotic Transport in Dynamical Systems* (Springer, New York, 1992)
2. R.S. MacKay, J.D. Meiss, I.C. Percival, Transport in Hamiltonian systems. Physica D **13**, 55–81 (1984)
3. V. Rom-Kedar, S. Wiggins, Transport in two-dimensional maps. Arch. Ration. Mech. Anal. **109**, 239–298 (1990)
4. H. Aref, The development of chaotic advection. Phys. Fluids **14**, 1315–1325 (2002)
5. J.D. Meiss, Symplectic maps, variational principles, and transport. Rev. Mod. Phys. **64**, 795–848 (1992)
6. S. Wiggins, The dynamical systems approach to Lagrangian transport in oceanic flows. Annu. Rev. Fluid Mech. **37**, 295–328 (2005)

7. B.J. Gladman, J.A. Burns, M. Duncan, P. Lee, H.F. Levison, The exchange of impact ejecta between terrestrial planets. Sciences **271**, 1387–1392 (1996)
8. M. Dellnitz, O. Junge, W.S. Koon, F. Lekien, M.W. Lo, J.E. Marsden, K. Padberg, R. Preis, S.D. Ross, B. Thiere, Transport in dynamical astronomy and multibody problems. Int. J. Bifurc. Chaos **15**, 699–727 (2005)
9. M. Dellnitz, O. Junge, M.W. Lo, J.E. Marsden, K. Padberg, R. Preis, S. Ross, B. Thiere, Transport of mars-crossers from the quasi-Hilda region. Phys. Rev. Lett. **94**, 231102 (2005)
10. M. Dellnitz, A. Hohmann, A subdivision algorithm for the computation of unstable manifolds and global attractors. Numer. Math. **75**, 293–317 (1997)
11. M. Dellnitz, G. Froyland, O. Junge, The algorithms behind GAIO—Set oriented numerical methods for dynamical systems, in *Ergodic Theory, Analysis, and Efficient Simulation of Dynamical Systems*, ed. by B. Fiedler (Springer, Berlin, 2001), pp. 145–174
12. M. Dellnitz, O. Junge, On the approximation of complicated dynamical behavior. SIAM J. Numer. Anal. **36**, 491–515 (1999)
13. M. Dellnitz, R. Preis, Congestion and almost invariant sets in dynamical systems, in *Proceedings of SNSC'01*, ed. by F. Winkler (Springer, Berlin, 2003), pp. 183–209
14. G. Froyland, M. Dellnitz, Detecting and locating near-optimal almost-invariant sets and cycles. SIAM J. Sci. Comput. **24**, 1839–1863 (2003)
15. K. Padberg, Numerical analysis of transport in dynamical systems, PhD thesis, Universität Paderborn, Germany (2005)
16. K. Padberg, B. Thiere, R. Preis, M. Dellnitz, Local expansion concepts for detecting transport barriers in dynamical systems. Commun. Nonlinear Sci. Numer. Simul. **14**(12), 4176–4190 (2009)
17. V. Szebehely, *Theory of Orbits* (Academic Press, New York, 1967)
18. G. Haller, Finding finite-time invariant manifolds in two-dimensional velocity fields. Chaos **10**, 99–108 (2000)
19. G. Haller, Distinguished material surfaces and coherent structures in three-dimensional fluid flows. Physica D **149**, 248–277 (2001)
20. G. Haller, A. Poje, Finite-time transport in aperiodic flows. Physica D **119**, 352–380 (1998)
21. G. Haller, G. Yuan, Lagrangian coherent structures and mixing in two-dimensional turbulence. Physica D **147**, 352–370 (2000)
22. S.C. Shadden, F. Lekien, J.E. Marsden, Definition and properties of Lagrangian coherent structures from finite-time Lyapunov exponents in two-dimensional aperiodic flows. Physica D **212**, 271–304 (2005)
23. R. Preis, Analyses and design of efficient graph partitioning methods, PhD thesis, Universität Paderborn, Germany (2000)
24. R. Preis, GADS—Graph algorithms for dynamical systems, Technical Report (2004)
25. M. Dellnitz, K. Padberg, R. Preis, Integrating multilevel graph partitioning with hierarchical set oriented methods for the analysis of dynamical systems, Technical report, Preprint 152, DFG Priority Program: Analysis, Modeling and Simulation of Multiscale Problems (2004)
26. G. Froyland, K. Padberg, Almost-invariant sets and invariant manifolds—connecting probabilistic and geometric descriptions of coherent structures in flows. Physica D **238**, 1507–1523 (2009)

Low-Energy Transfers in the Earth–Moon System

Elisa Maria Alessi, Gerard Gómez, and Josep J. Masdemont

Abstract The aim of this work is to compute low-energy trajectories in the Earth–Moon system within the framework of the Circular Restricted Three-Body Problem. It is known that this model admits five equilibrium points, in a proper reference system. We look for connection paths between the neighbourhood of a given collinear libration point and one of the primaries. We focus on the point L_1 and on the point L_2, whose linear behaviour is of type center $\times$ center $\times$ saddle. We consider Lindstedt–Poincaré series expansion as main procedure to compute invariant stable manifolds associated with periodic and quasi-periodic orbits around L_1 and L_2. It turns out that direct lunar transfers are allowed only from certain regions on the Moon's surface and that the most advantageous connections between the Earth and a libration point orbit take place at the local maxima of the function distance between the Earth and a given stable manifold.

Keywords Earth–Moon system · Libration points · Low-energy transfers

1 Introduction

In this work, we apply the tools of the Dynamical Systems Theory and the model of the Circular Restricted Three-Body Problem (CR3BP) to the Earth–Moon system in

E.M. Alessi (✉) · G. Gómez
IEEC & Departament de Matemàtica Aplicada i Anàlisi, Universitat de Barcelona, Gran Via 585, Barcelona 08007, Spain
e-mail: elim@maia.ub.es

G. Gómez
e-mail: gerard@maia.ub.es

J.J. Masdemont
IEEC & Departament de Matemàtica Aplicada I, ETSEIB, Universitat Politècnica de Catalunya, Diagonal 647, Barcelona 08028, Spain
e-mail: josep@barquins.upc.edu

J.A.T. Machado et al. (eds.), *Nonlinear Science and Complexity*,
DOI 10.1007/978-90-481-9884-9_13, © Springer Science+Business Media B.V. 2011

order to construct low-energy transfers from the Moon's surface to a libration point orbit and from a Low Earth Orbit (LEO) to a libration point orbit.

Recently, worldwide space agencies are designing missions aiming at the development of the appropriate technology needed for the future lunar exploration. We mention SELENE (JAXA), Chang'e 1 (China), Chandrayaan-1 (India) and Lunar Reconnaissance Orbiter (NASA) as examples of such an effort. This renewed enthusiasm is motivated by the study of the lunar surface and environment, of the lunar origin and evolution and of the Earth environment. On top of that, the Moon might represent an ideal departure point for interplanetary missions.

In this framework, we believe that the collinear equilibrium points L_1 and L_2 may stand for optimal rendezvous locations. We can imagine to position a station in orbit around one of these points and to move there either from the Earth or the Moon. We recall that the idea of a lunar L_1 space hub has already been considered, in particular as a construction and repair facility [1].

The key point of our research is to take advantage of the central and hyperbolic invariant manifolds associated with each collinear libration point. With respect to the lunar rescue orbits, we analyze from which regions of the Moon's surface we can reach a nominal libration point orbit on a trajectory belonging to the associated stable manifold, the angle and the velocity of departure and the transfer time. Regarding the departure from a given LEO, it turns out that the stable manifold passes quite far from the Earth and thus the transfer is established on two trajectories, one linking the LEO with the manifold and the other lying on the manifold.

2 The Model

The Circular Restricted Three-Body Problem [2] studies the behaviour of a particle P with infinitesimal mass m_3 which moves under the gravitational attraction of two primaries, P_1 and P_2, of masses m_1 and m_2, which follow a circular orbit around their common centre of mass.

To describe the motion of the particle, we introduce a synodical reference system $\{O, x, y, z\}$ which rotates around the z-axis with a constant angular velocity ω equal to the mean motion n of the primaries. The origin of the reference frame is set at the barycenter of the system, and the x-axis on the line which joins the primaries, oriented in the direction of the largest primary. In this way we work with m_1 and m_2 fixed on the x-axis. See Fig. 1.

The set of units is chosen in such a way that the unit of length is defined as the distance between the primaries and the unit of time is defined by imposing the mean motion n of the relative orbit of the primaries to be unitary. Moreover, we define the mass ratio μ as $\mu = \frac{m_2}{m_1+m_2}$.

With these assumptions, the CR3BP equations of motion can be written as

$$\ddot{x} - 2\dot{y} = \frac{\partial \Omega}{\partial x} = x - \frac{(1-\mu)}{r_1^3}(x-\mu) - \frac{\mu}{r_2^3}(x+1-\mu),$$

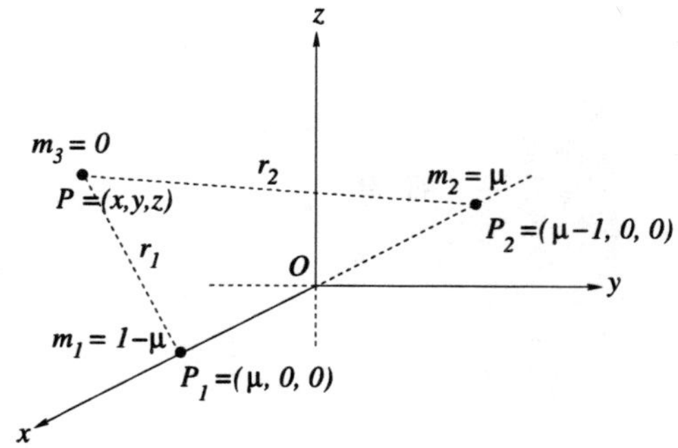

Fig. 1 The circular restricted three-body problem in the synodical reference system with adimensional units

$$
\begin{aligned}
\ddot{y} + 2\dot{x} &= \frac{\partial \Omega}{\partial y} = y - \frac{(1-\mu)}{r_1^3} y - \frac{\mu}{r_2^3} y, \\
\ddot{z} &= \frac{\partial \Omega}{\partial z} = -\frac{(1-\mu)}{r_1^3} z - \frac{\mu}{r_2^3} z,
\end{aligned}
\tag{1}
$$

where

$$
\Omega(x, y, z) = \frac{1}{2}(x^2 + y^2) + \frac{1-\mu}{r_1} + \frac{\mu}{r_2} + \frac{1}{2}(1-\mu)\mu, \tag{2}
$$

and r_1 and r_2 are the distances from P to P_1 and P_2, respectively.

System (1) has a first integral, called Jacobi integral, which is given by

$$
C = 2\Omega - (\dot{x}^2 + \dot{y}^2 + \dot{z}^2), \tag{3}
$$

where C is the so-called Jacobi constant.

3 Collinear Points Dynamics

In the synodical reference system, there exist five equilibrium points, called libration points. Three of them, the collinear ones, are in the line joining the primaries and they are usually denoted as L_1, L_2 and L_3.

The linear behaviour associated with the dynamics in a neighbourhood of the collinear points is of type centre × centre × saddle. The centre × centre part determines the neutral behaviour of the flow of the CR3BP around them, in particular it defines their central manifold. On the other hand, the saddle component of the linear approximation determines the instability of the equilibrium points.

The central manifold is filled with periodic and quasi-periodic solutions. We consider halo type orbits, that is, three-dimensional periodic orbits symmetric with respect to the $\{y = 0\}$ plane and quasi-periodic Lissajous orbits lying on invariant tori.

Due to the hyperbolic character, each type of periodic and quasi-periodic orbits has a stable and an unstable invariant manifold. They look like tubes (see Fig. 2) of asymptotic trajectories tending to, or departing from, the corresponding orbit. When

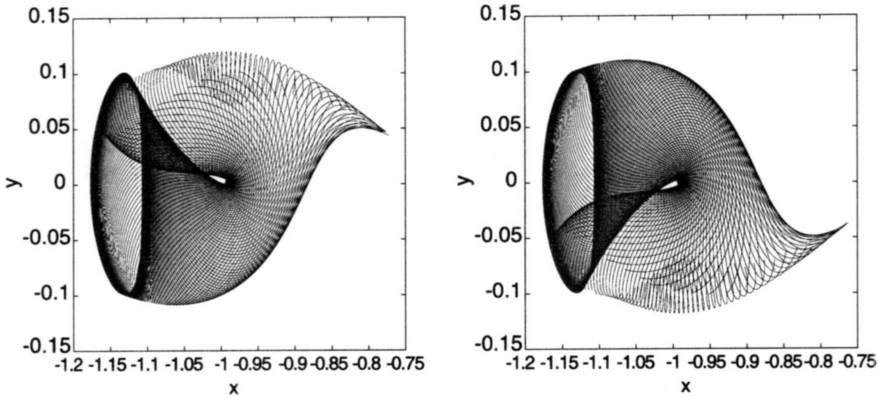

Fig. 2 Negative branch of the unstable invariant manifold (*left*) and of the stable invariant manifold (*right*) associated with a given halo orbit around the point L_2 in the Earth–Moon system. *The dot* appearing in the center of *both plots* represents the Moon

going forwards in time, the trajectories on the stable manifold approach exponentially the periodic/quasi-periodic orbit, while those on the unstable manifold leave it exponentially. Each manifold has two branches, a positive and a negative one.

4 Methodology

The computation of all different kinds of orbits, as well as their invariant stable and unstable manifolds, can be done in different ways. In this work, we determine halo and Lissajous orbits together with the corresponding stable manifolds by an order 25 Lindstedt–Poincaré procedure, which takes in consideration high order terms in the equations of motion and produces initial conditions for the invariant objects with high degree of accuracy [3]. In particular, it provides semi-analytical expressions for the invariant objects in terms of suitable amplitudes, α_1, α_2, α_3 and α_4, and phases, ϕ_1 and ϕ_2, by series expansions.

The formal series solution is of the type

$$x(t) = \sum e^{[(i-j)\theta_3]}\big[x_{ijkm}^{pq}\cos(p\theta_1 + q\theta_2) + \bar{x}_{ijkm}^{pq}\sin(p\theta_1 + q\theta_2)\big]\alpha_1^i\alpha_2^j\alpha_3^k\alpha_4^m,$$

$$y(t) = \sum e^{[(i-j)\theta_3]}\big[y_{ijkm}^{pq}\cos(p\theta_1 + q\theta_2) + \bar{y}_{ijkm}^{pq}\sin(p\theta_1 + q\theta_2)\big]\alpha_1^i\alpha_2^j\alpha_3^k\alpha_4^m,$$

$$z(t) = \sum e^{[(i-j)\theta_3]}\big[z_{ijkm}^{pq}\cos(p\theta_1 + q\theta_2) + \bar{z}_{ijkm}^{pq}\sin(p\theta_1 + q\theta_2)\big]\alpha_1^i\alpha_2^j\alpha_3^k\alpha_4^m,$$

where $\theta_1 = \omega t + \phi_1$, $\theta_2 = \nu t + \phi_2$, $\theta_3 = \lambda t$, $\omega = \sum \omega_{ijkm}\alpha_1^i\alpha_2^j\alpha_3^k\alpha_4^m$, $\nu = \sum \nu_{ijkm}\alpha_1^i\alpha_2^j\alpha_3^k\alpha_4^m$, $\lambda = \sum \lambda_{ijkm}\alpha_1^i\alpha_2^j\alpha_3^k\alpha_4^m$ and summations are extended over all $i, j, k, m \in \mathbb{N}$ and $p, q \in \mathbb{Z}$.

The two frequencies ω and ν are, respectively, the in-plane and the out-of-plane frequencies, while λ determines the hyperbolic motion. The values of α_3 and α_4

characterize the size of the orbit and are, respectively, the in-plane and the out-of-plane amplitudes. The solution restricted to the central manifold (bounded orbits) is obtained setting $\alpha_1 = \alpha_2 = 0$. Setting $\alpha_1 = 0$ and $\alpha_2 \neq 0$ ($\alpha_2 = 0$, $\alpha_1 \neq 0$) we get their stable (unstable) manifold.

Lissajous orbits are characterized by two frequencies, which tend to the frequencies related to both centers, when the amplitudes tend to zero. Halo orbits appear when the two frequencies are equal, that is, when the in-plane motion and the out-of-plane one have the same period. As a consequence, the in-plane and out-of-plane amplitudes α_3 and α_4 are no longer independent.

5 Rescue Orbits

As first application, we perform a numerical globalisation of the stable invariant manifolds associated with orbits of the halo and the Lissajous families around the collinear points L_1 and L_2, until they reach the Moon's surface (considered as a sphere). The initial conditions considered are associated with equally spaced values of the phase ϕ_1 (and ϕ_2 in the Lissajous case) in the range $[0, 2\pi]$.

If an orbit reaches the Moon's surface backwards in time, we compute the latitude φ and the longitude λ corresponding to the arrival point, the physical velocity of arrival, the physical transfer time and the arrival angle ϑ, defined as the angle between the velocity vector and the Moon's surface normal vector.

We keep track of the number of loops that the transfer orbit performs around the Moon before reaching it. We are interested in almost direct transfers and we do not see operational advantages in trajectories winding around the Moon indefinitely. For this purpose we compute the number of minima of the r_2 function along the orbits. If we get more than 5 minima, then we discard such trajectory and we proceed to explore the next point of the manifold. We remark that we reject the minima associated with the loops exhibited by the trajectories before leaving the neighbourhood of the periodic/quasi-periodic orbit.

5.1 Results

As general result, we have found that not all the orbits of a stable manifold can get to the Moon. It depends on the size of the arrival orbit considered, on the branch of the manifold, and on the phase/phases associated with the trajectory of the invariant manifold.

The main difference between the two explorations, the halo and the Lissajous one, derives from the fact that an hyperbolic invariant manifold associated with a halo orbit is a two-dimensional object, while the one associated with a Lissajous orbit is three-dimensional. As a consequence, the intersection of the stable manifold of a quasi-periodic orbit with the surface of the Moon covers a two-dimensional region instead of being a curve.

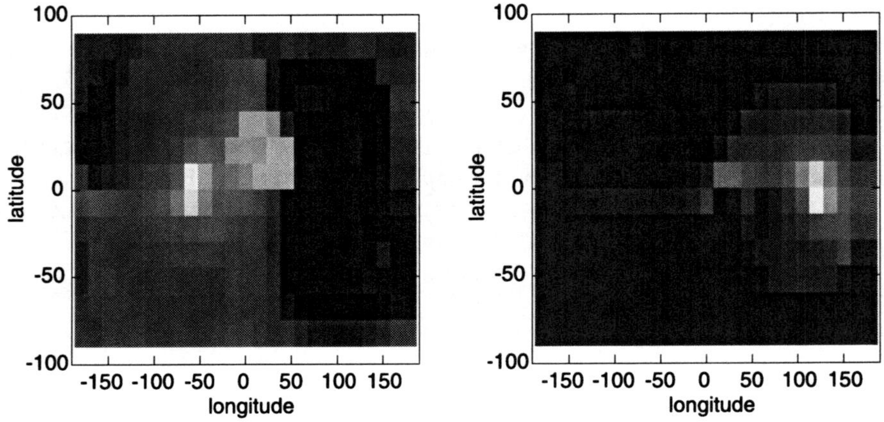

Fig. 3 Number of opportunities of departure from the Moon's surface per unit of length of the arrival halo orbit and per unit of area element. *A lighter shade of gray* corresponds to a greater chance. *On the left*, the L_1 case; *on the right*, the L_2 one

If we consider a nominal halo orbit either around L_1 or around L_2, we cannot complete a rescue transfer starting at an arbitrary point of the surface of the Moon if it does not perform at least one loop around the Moon. As we increase the number of loops, the area of the rescue region on the surface of the Moon increases. In fact, if we allow of at least 3 minima, one can reach the halo families departing from any point of the surface of the Moon.

On the other hand, the points of allowed departure are not uniformly distributed on the Moon's surface, that is, there exist regions where we have more chances to take off joining the stable invariant manifold associated with a given halo/Lissajous orbit. This is illustrated in Fig. 3 for the halo case, where a lighter shade of gray corresponds to a greater probability of departure.

In both cases, the modulus of the velocity at the departure from the surface of the Moon is almost equal to the escape velocity of the Moon (about 2.4 km/s), as expected from the conservation of the Jacobi constant. Concerning the transfer time, the results for L_1 and L_2 are very similar. For direct trajectories, the transfer time is approximately of 10 days in both cases.

6 LEO Transfers

As mentioned before, the hyperbolic manifolds in the CR3BP do not provide direct transfers to the greatest primary. Because of this, we construct transfers between a nominal Lissajous orbit around the point L_1 and a Keplerian orbit around the Earth by means of the associated stable invariant manifold and of an additional arc trajectory. In this way, we deal with two manoeuvres, one to insert into the hyperbolic manifold from the arc segment, say Δv_1, and one to inject into the arc segment from the LEO orbit, say Δv_2.

For each square Lissajous orbit considered, we take a well-defined number of initial conditions on the negative branch of the corresponding stable invariant manifold. They are associated with equally spaced values of the phases ϕ_1 and ϕ_2 in the range $[0, 2\pi]$ and given by an order 25 Lindstedt–Poincaré series expansion. As departure location, we fix a sphere of radius R from the Earth's surface (LEO sphere).

Each initial condition on the manifold is propagated backwards in time until the trajectory gets to a certain distance with respect to the Earth. At this point we apply Δv_1. This first correction to the velocity vector is performed starting from a Two-Body Problem (2BP) approximation refined imposing two constraints, that is, to reach the given LEO sphere tangentially.

6.1 Results

We have considered two cases: in the first one, the insertion manoeuvre on the manifold takes place at the local minima (perigees) of the function distance between the Earth and the point chosen on the given stable manifold; in the second case, at the local maxima (apogees).

We have chosen a LEO sphere of radius 360 km from the Earth's surface. In this configuration, the procedure adopted has always converged quite fast. It would quite likely do for different values of R.

As general result, it turned out that greater the initial distance from the Earth cheaper the total cost, though longer the journey. The cost of the two manoeuvres in the perigee case is of about 4.3 km/s, of about 3.6 km/s in the apogee situation. The former transfer spends approximately 1 day on the arc segment, the latter about 3.5 days. The most expensive manoeuvre takes place at the LEO sphere, being Δv_1 at least 1.5 km/s smaller than Δv_2.

7 Conclusions

We have seen that transfers between the Moon and a libration point orbit either around the collinear libration point L_1 or around the collinear libration point L_2 are feasible from any point on the Moon's surface if we allow of not direct paths.

To go from a nominal LEO to a given square Lissajous orbit around the point L_1 by means of the associated stable invariant manifold is possible only by considering an additional trajectory, that is, the connection needs two manoeuvres. It would interesting to see what happens when considering all the points on a given stable invariant manifold, also on the positive branch.

Acknowledgements This work has been supported by the Spanish grants MTM2006-05849 (E.M.A., G.G.) and MTM2006-00478 (J.J.M.) and by the Astronet Marie Curie fellowship (E.M.A.).

References

1. M.W. Lo, S.D. Ross, The lunar L_1 gateway: portal to the stars and beyond. In: *AIAA Space Conference*, Albuquerque, New Mexico, USA, 28–30 August 2001 (2001)
2. V. Szebehely, *Theory of Orbits* (Academic Press, New York, 1967)
3. J.J. Masdemont, Dyn. Syst. **20**, 59–113 (2005)

Gravitational Potential of a Massive Disk. Dynamics Around an Annular Disk

E. Tresaco, A. Elipe, and A. Riaguas

Abstract This article studies the main features of the dynamics around an annular disk. The first part addresses the difficulties finding a usable expression of the gravitational potential of a massive disk that will be used later on to define the differential equations of motion of our dynamical system.

The second part of the article describes the dynamics of a particle orbiting a massive annular disk by means of a description of the main families of periodic orbits, their bifurcations and linear stability.

Keywords Potential theory · Elliptic functions · Periodic orbits

1 The Massive Disk and Its Potential Function

Our goal are the dynamics of an infinitesimal particle moving under the gravitational field of a massive bidimensional annular disk but first we need to deal with finding a proper expression for the potential of a massive disk.

We will consider now a circular plate of radius a in the plane Oxy of a Cartesian coordinate system and with its center at the origin of coordinates, and assuming a total mass M and surface density σ, that is, $M = \pi\sigma a^2$.

E. Tresaco (✉)
Dpto. Matemática Aplicada, Universidad de Zaragoza, Zaragoza, Spain
e-mail: etresaco@unizar.es

A. Elipe
IUMA/Dpto. Matemática Aplicada, Universidad de Zaragoza, Zaragoza, Spain
e-mail: elipe@unizar.es

A. Riaguas
Dpto. Matemática Aplicada, Universidad de Valladolid, Soria, Spain
e-mail: andresrg@mac.uva.es

J.A.T. Machado et al. (eds.), *Nonlinear Science and Complexity*,
DOI 10.1007/978-90-481-9884-9_14,

For any given point in space P of coordinates (x, y, z) where we want to compute the potential, we will make use of the following quantities

$$r^2 = x^2 + y^2, \qquad p^2 = (a+r)^2 + z^2, \qquad k^2 = \frac{4ar}{p^2},$$

$$R^2 = x^2 + y^2 + z^2, \qquad q^2 = (a-r)^2 + z^2, \qquad n^2 = \frac{4ar}{(a+r)^2}.$$

The potential due to the gravitational attraction of the disk is

$$U(P) = -G \int_D \frac{dm}{R},$$

where D denotes the disk, and R the distance from a differential mass element, dm, to the point P. As we consider this body as bidimensional, if ds denotes the differential element of surface then $\sigma ds = dm$, therefore

$$U(P) = -G \int_D \frac{\sigma}{R} ds.$$

This potential is called a *single layer potential* with essential discontinuities at the boundary of the circular plate, but otherwise a continuous function.

Its gradient is a continuous function everywhere except at points in the circular plate. It is not defined for points in the boundary of the circular plate and has a step discontinuity at points in the plate but outside of its boundary.

This potential has been already derived and published, see Ref. [1, 2]. Nevertheless, its closed expression involves elliptic integrals and the expressions given for the potential can not be evaluated in significant areas of space where the potential is otherwise a well defined function or in such a way that produce wrong evaluations when numerically computed. We summarize here the approach derived for two of these papers that have been very useful to develop our formulation of the problem.

In a paper by Krogh, Ng and Snyder [1] closed form expressions are given for the gravitational potential

$$U_{KNS}(r, z) = 2G\sigma\left(z\left(\frac{\pi}{2} + \frac{\pi}{2}\,\mathrm{sign}(a-r)\right) - pE(k) - \frac{a^2 - r^2}{p}K(k) - \frac{(a-r)z^2}{(a+r)p}\Pi(n^2, k)\right). \tag{1}$$

The authors also point to several formulas found in the Byrd and Friedman book [3] and the computational approach in Ref. [4] required to overcome difficulties in computing the function force, specially its z component. This formula (1) matches the expressions (2) using a single formula.

Following Lass and Blitzer [2] the expression of the potential if $r < a$ is:

$$U_{LB}(r, z) = 2G\sigma\left(\pi|z| - pE(k) - \frac{a^2 - r^2}{p}K(k) - \frac{(a-r)z^2}{(a+r)p}\Pi(n^2, k)\right), \tag{2}$$

with $E(k)$, $K(k)$ and $\Pi(n^2, k)$ the complete elliptic integrals of first, second and third kind respectively. For values $r > a$ the expression is valid by removing the term $2G\sigma\pi|z|$.

Both formulas (1), (2) do not represent the potential function for all the points in space for which the potential function has a real finite value. For the cases $r < a$ and $r > a$ are valid but fail to be of use for $r = a$ and $z \neq 0$ because then $n = 1$, and the third elliptic integral is not bounded for these values rendering those formulas useless for the analysis of dynamics.

These formulas (1), (2) fail due to the evaluation of the term containing the elliptic integral of third kind. In order to save this we will make use of the following transformations

$$1 - n^2 = 1 - \frac{4ar}{(a+r)^2} = \frac{(a-r)^2}{(a+r)^2}, \qquad \frac{k}{n} = \frac{a+r}{p}$$

and the relation [413.01] found in the Byrd and Friedman book [3]

$$\Pi(n^2, k^2) = \frac{n\pi\,\Lambda_0(\xi, n)}{2\sqrt{(n^2-k^2)(1-n^2)}}, \qquad \xi = \arcsin\sqrt{\frac{n^2-k^2}{n^2(1-k^2)}} = \arcsin\frac{|z|}{q}$$

to rewrite the potential as

$$\begin{aligned} U(x,y,z) = 2G\sigma\Big(&-pE(k) - \frac{a^2-r^2}{p}K(k) + |z|\Big(\frac{\pi}{2} + \frac{\pi}{2}\operatorname{sign}(a-r)\Big) \\ &- z\frac{\pi}{2}\operatorname{sign}(a-r)\Lambda_0(\phi, k)\Big). \end{aligned} \tag{3}$$

Finally, it is possible to replace the Heuman's lambda function by a combination of elliptic integrals (see [150.03] in Ref. [3]) and reformulate again U after some simplifications as

$$\begin{aligned} U = 2G\sigma\Big(&-pE(k) - \frac{a^2-r^2}{p}K(k) + |z|\Big(\frac{\pi}{2} + \frac{\pi}{2}\operatorname{sign}(a-r)\Big) \\ &- z\operatorname{sign}(a-r)(E(k)F(\phi,k') + K(k)E(\phi,k') - K(k)F(\phi,k'))\Big) \end{aligned} \tag{4}$$

where $k' = \sqrt{1-k^2}$, $\phi = \arcsin\frac{z}{q}$, and $F(\phi, k')$ and $E(\phi, k')$ are the incomplete elliptic integrals of first and second kind respectively. For the computation of the elliptic integrals we have used the algorithms provided by Carlson [5].

Under this form, the potential function and the force function derived from it can be properly evaluated at any point in space where they are defined.

2 Dynamics Around a Circular Annulus

We will study the dynamics of an infinitesimal particle moving under the gravitational field of a massive bidimensional circular annulus. We will consider an annular ring of radius a and b in the plane Oxy and centered at the origin of coordinates. Therefore, the potential created by a continuous massive annulus is

$$U(x,y,z;a,b) = U(x,y,z;a) - U(x,y,z;b) \tag{5}$$

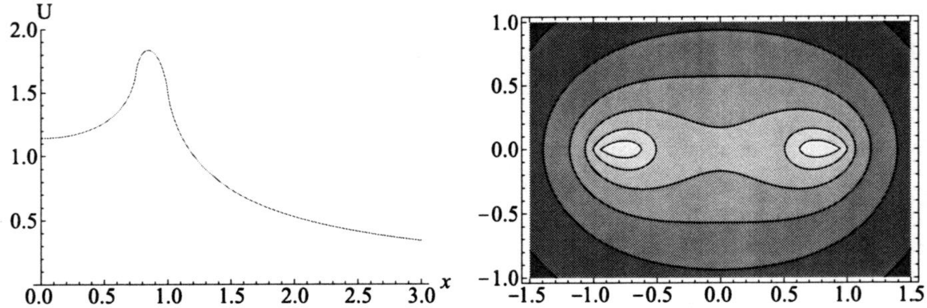

Fig. 1 *Left*: Potential function $U = U(x)$. *Right*: Equipotential curves

where $U(x, y, z; a)$ and $U(x, y, z; b)$ are the potential created two circular concentric plates of radius a and b (see (4)). Note that now $G\sigma = GM/(\pi(a^2 - b^2)) = \mu/(\pi(a^2 - b^2))$.

The potential is symmetric with respect to all axis Ox, Oy and Oz due to the cylindric symmetry of the problem. Figure 1(left) shows the potential function U along the Ox-axis (for $y = z = 0$). We see a local minimum at the central equilibrium point $x = 0$, and that the potential tends to 0, like $1/r$ at the large distances. Figure 1 (right) depicts equipotential curves in the vertical plane for x. It is clear that the origin is an unstable point.

In this preliminary study we mainly focus on planar orbits in the fundamental planes. Polar orbits in the xz-plane, and Equatorial orbits in the xy-plane.

For axis symmetric systems it is natural to employ cylindrical coordinates (r, λ, z). It follows that the angular momentum Λ is conserved, so that term of the kinetic energy can be added to U to form an effective potential $W(r, z)$.

First, we analyze the equilibrium points in the planar case $z = 0$, where the effective potential is now a radial function, therefore we are in presence of an integrable problem thanks to the energy and angular momentum integrals.

By Newton's law, the movement in the equatorial plane is governed by

$$\ddot{r} = \frac{d}{dr}\left(-U(r) - \frac{\Lambda^2}{2r^2}\right) \equiv -W'(r)$$

and the partial expression with respect to r is derived as

$$\frac{\partial U}{\partial r} = -\frac{4G\sigma}{r}\left(\left(\sqrt{R^2 + a^2 + 2ar}\left(1 - \frac{1}{2}{k_a}^2\right)K(k_a) - E(k_a)\right)\right.$$
$$\left. - \left(\sqrt{R^2 + b^2 + 2br}\left(1 - \frac{1}{2}{k_b}^2\right)K(k_b) - E(k_b)\right)\right). \qquad (6)$$

Relative equilibria are given by the critical points of the effective potential,

$$U'(r) = \frac{\Lambda^2}{r^3}.$$

The origin is the unique equilibrium solution in case of angular momentum equal to zero. Therefore, the dynamics is reduced to the linear movement along a diameter

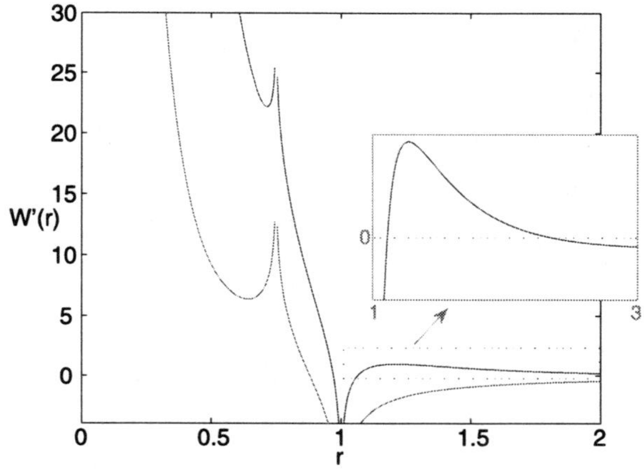

Fig. 2 Critical points of the effective potential for different values of Λ

of the annulus. The critical point inside the annular disk is not consider because it has no physical meaning.

For $\Lambda \neq 0$ it has been observed that there are no critical points inside, and a particle placed there will tend to collapse with the annulus. On the other hand, the dynamics outside depends on the value of the angular momentum, as it can be seen in Fig. 2.

We have found a critical value $\Lambda^*(a, b) > 1$ such that, for smaller values there is no critical points meanwhile for greater values we find two critical points corresponding to a maximum and minimum of the effective potential. And so leading to one stable and one unstable circular orbits. As we increase the angular momentum value, one of the critical points tends to the annulus and the other goes to infinite.

Referring to the dynamics when movement is reduced to z-axis, plotting the phase portrait we will see that the origin is again the only equilibrium point. Energy value at this point is $E_0 = -2\mu/(a + b)$, and only for energies satisfying $E_0 < E < 0$ we will have periodic orbits. These orbits will be vertical oscillations along the z-axis.

Finally we carry out the numerical computation of planar families of periodic orbits. For this purpose we start plotting some Poincaré sections in order to get a preliminary information of the dynamics of the system, nevertheless, these representations although a very useful tool to get approximated values of periodic orbits, are not a systematic way of searching periodic orbits.

Other procedures well suited for this aim are, computation of periodic orbits through the Poincaré map (see Ref. [6]) and the algorithm derived by Henrard and Deprit [7]. Both methods provide numeric continuation of one-parameter family of periodic orbits for a conservative dynamical systems.

We will now show some of the most relevant families that we have found. To compute them we have chosen as continuation parameter the x coordinate of the orbit. Note that the continuation procedures imply the calculus of the variational equations, thus providing information on the linear stability trough the stability index k without much additional effort.

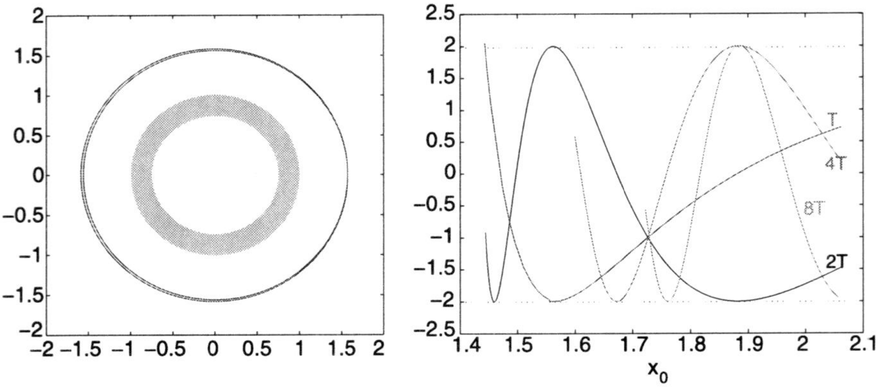

Fig. 3 *Left*: Orbit resulting form the period-doubling bifurcation $2T$. *Right*: Stability index evolution

When restricted to the xy-plane containing the annulus, it was immediately discovered a family of trivial circular periodic orbits outside the ring. The evolution of the stability index for this family shows that when the radius of the orbits goes to infinity it tends asymptotically to 2, while the family ends in a collision with the annulus when decreasing its orbital energy.

It also shows a critical value ($k = -2$), this point indicates a bifurcation with a family of doubling period. This new double-period family has been continued leading again to another doubling bifurcation. Repeating the same procedure we find successive doubling period families whose trace pattern can be seen in Fig. 3. The values of the stability of the orbits of multiple period are given by the formula

$$k_m = 2\cos\left(m \arccos\left(\frac{k}{2}\right)\right), \qquad |k| \leq 2.$$

Lets focus now on orbits perpendicular to the plane of the annulus. We have studied a wide range of polar families although we summarize here only the continuation of two of these families.

First, the family of 8-shape orbits, it consists of 2-arc symmetric periodic orbits centered in the origin of the annulus. Its stability evolution (see Fig. 4) shows that when the orbits radius increases the family members are unstable, while when the orbit radius decreases the family stays in a stable region until it crosses the boundary value, leading to bifurcations with new families of periodic orbits.

The closest critical point ($k = 2$), presents a Pitchfork bifurcation and two new families of stable orbits appear. These new families becomes more asymmetric as energy decreases, until they end with a collision with the annulus.

On the other hand, the other bifurcation point ($k = 2$), describes a monodromy matrix of type 4 (see Ref. [8]) leading to a bifurcation with a non-symmetric family of the same period. This new family consists of inclined-arc stable orbits with center at the origin.

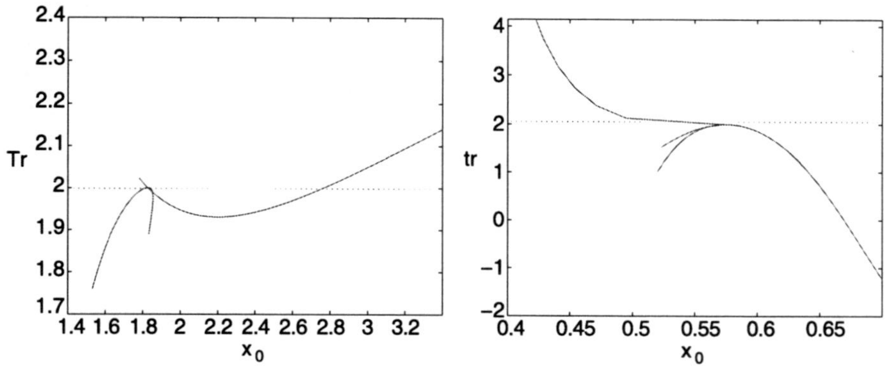

Fig. 4 *Left*: 8-shaped orbit trace evolution. *Right*: Single-line orbits trace evolution

Finally, we describe the behavior of the family of single-line open orbits. It is a stable family that originates out a bifurcation with vertical oscillation, and ends with a collision orbit with the annulus.

Its trace evolution shows a bifurcation with a doubling-period family. Note that the doubling period family, at the bifurcation point with the singular family, have all unit eigenvalues (and thus $k = 2$, giving birth to new families.

Plotting a Poincaré section at this critical value the bifurcation can be easily identified. It is observed a central point that corresponds to the single-arc orbit, surrounding by four isles belonging to a new bifurcated stable family, while the other four hyperbolic points are related to a new unstable family. The evolution of the stability index of these bifurcated branches is depicted in Fig. 4.

Acknowledgements This paper has been supported by the Spanish Ministry of Education and Science, Project AYAP2008-05572.

References

1. F.T. Krough, E.W. Ng, W.V. Snyder, The gravitational field of a disk. Celest. Mech. **26**, 395–405 (1982)
2. H. Lass, L. Blitzer, The gravitational potential due to uniform disks and rings. Celest. Mech. **30**, 225–228 (1983)
3. P.F. Byrd, M.D. Friedman, *Handbook of Elliptic Integrals for Engineers and Scientists* (Springer, New York, 1971), xvi+358
4. R. Bulirsch, Numerical calculation of elliptic integrals and elliptic functions. Numer. Math. **7**, 78–90 (1965)
5. B.C. Carlson, Computing elliptic integrals by duplication. Numer. Math. **33**, 1–16 (1979)
6. D.J. Scheeres, Satellite dynamics about asteroids: computing Poincaré maps for the general case, in *Hamiltonian Systems with Three or More Degrees of Freedom (S'Agaró, 1995)*. NATO Adv. Sci. Inst. Ser. C, Math. Phys. Sci., vol. 533 (Kluwer Academic, Dordrecht, 1999), pp. 554–557
7. A. Deprit, J. Henrard, Natural families of periodic orbits. Astron. J. **72**, 158–172 (1967)
8. M. Hénon, Exploration numérique du problème restreint. Ii Masses égales, stabilité des orbites périodiques, Ann. Astrophys. **28**, 992 (1965)

An Accounting Device for Biasymptotic Solutions: The Scattering Map in the Restricted Three Body Problem

Amadeu Delshams, Josep J. Masdemont, and Pablo Roldán

Abstract We compute the scattering map (see explanation below) in the Spatial Restricted Three Body Problem using a combination of analytical and numerical techniques.

Keywords Restricted three-body problem · Homoclinic and heteroclinic trajectories · Scattering maps · Quasi-periodic motions · Invariant tori · Normally hyperbolic invariant manifolds · Normal forms · Arnold diffusion

The modern geometric theory of dynamical systems originated in the work of Poincaré between 1880 and 1910 on ordinary differential equations and celestial mechanics, particularly in his essay "Sur le problème des trois corps et les équations de la dynamique" [1] in which he discovered homoclinic and heteroclinic orbits, leading to the observation of deterministic chaos.

Many of the tools developed by Poincaré, including the return map, linearization, normal forms, and invariant manifolds, have been successfully used in a variety of celestial mechanics' problems. In effect, he laid out a geometric programme to study the three body problem that is still today being pursued. This fact, combined with a longstanding tradition of mathematical works going back to Laplace, Lagrange, and Poisson, has rendered celestial mechanics a testbed for modern dynamical systems research such as KAM theory and Arnold diffusion.

It is in this spirit that we apply a modern dynamical systems tool known as the *scattering map* to the classical problem of three bodies. We hope that our work on the restricted three body problem can be translated to similar problems, e.g. in chemistry.

Let Λ_1 and Λ_2 be two normally hyperbolic invariant manifolds for a flow, such that the stable manifold of Λ_1 intersects the unstable manifold of Λ_2 transversally

A. Delshams · J.J. Masdemont · P. Roldán (✉)
Departament de Matemàtica Aplicada 1, Universitat Politècnica de Catalunya, Av. Diagonal 647, 08028 Barcelona, Spain
e-mail: pablo.roldan@upc.edu

J.A.T. Machado et al. (eds.), *Nonlinear Science and Complexity*,
DOI 10.1007/978-90-481-9884-9_15,

along a manifold Γ. The scattering map from Λ_2 to Λ_1 is the map that, given an asymptotic orbit in the past, associates the corresponding asymptotic orbit in the future through a heteroclinic orbit.

The scattering map was first introduced by A. Delshams, R. de la Llave and T.M.-Seara as a tool to prove the existence of unbounded energy orbits in geodesic flows perburbed by a periodic potential [2] or a quasi-periodic potential [3]. Later, the scattering map was used to prove the existence of Arnold diffusion in a priori unstable systems [4]. These models are all close to integrable, so they are studied using purely analytical, perturbative methods.

We recently computed the scattering map in the planar restricted three body problem using non-perturbative techniques [5], and we showed that it is a (nontrivial) integrable twist map.

In this paper [5] we compute the scattering map in a problem with three degrees of freedom using also non-perturbative techniques. Specifically, we compute the scattering map between the normally hyperbolic invariant manifolds Λ_1 and Λ_2 associated to the equilibrium points L_1 and L_2 in the spatial Hill's problem.

In the planar problem, for each energy level (in a certain range) there is a unique Lyapunov periodic orbit around $L_{1,2}$. In the spatial problem, this periodic orbit is replaced by a three-dimensional invariant manifold practically full of invariant 2D tori. There are heteroclinic orbits between Λ_1 and Λ_2 connecting these invariant tori in rich combinations. Hence the scattering map in the spatial problem is more complicated, and it allows nontrivial transition chains.

Scattering maps have application to e.g. mission design in Astrodynamics, and to the construction of diffusion orbits in the spatial Hill's problem.

References

1. H.J. Poincaré, Sur le problème des troi corps et les équations de la dynamique. Acta Math. **13**, 1–270 (1890)
2. A. Delshams, R. de la Llave, T.M. Seara, A geometric approach to the existence of orbits with unbounded energy in generic periodic perturbations by a potential of generic geodesic flows of $\mathbf{T}^2$. Commun. Math. Phys. **209**, 353–392 (2000)
3. A. Delshams, R. de la Llave, T.M. Seara, Orbits of unbounded energy in quasi-periodic perturbations of geodesic flows. Adv. Math. **202**, 64–188 (2006)
4. A. Delshams, R. de la Llave, T.M. Seara, A geometric mechanism for diffusion in Hamiltonian systems overcoming the large gap problem: heuristics and rigorous verification on a model. Mem. Am. Math. Soc. **179** (2006), viii+141
5. A. Delshams, J. Masdemont, P. Roldán, Computing the scattering map in the spatial Hill's problem. Discrete Contin. Dyn. Syst. Ser. B **10**, 455–483 (2008)

Optimal Capture Trajectories Using Multiple Gravity Assists

Stefan Jerg, Oliver Junge, and Shane D. Ross

Abstract Graph theoretic methods of optimal control in the presence of uncertainty are applied to a celestial mechanics problem. We find a fuel-efficient spacecraft trajectory which starts at infinity and is captured by the smaller member of a binary system, e.g., a moon of Jupiter, using multiple gravity assists.

Keywords Optimal control · Three-body problem · Celestial mechanics · Gravity assist · Shortest path · Periapsis map · Symplectic map · Graph partitioning

1 Introduction

For low energy spacecraft trajectories such as multi-moon orbiters for the Jupiter system, multiple gravity assists by moons could be used in conjunction with ballistic capture to drastically decrease fuel usage. In this paper, we consider a spacecraft initially in a large orbit around Jupiter. Our goal is to use small impulsive controls to direct the spacecraft into a capture orbit about Callisto, the outermost icy moon of Jupiter. We consider the role of uncertainty, which is critical for space trajectories which are designed using chaotic dynamics. Our model is a family of symplectic

S. Jerg (✉) · O. Junge
Technische Universität München, Zentrum Mathematik, Boltzmannstr. 3, 85747 Garching, Germany
e-mail: jerg@ma.tum.de
url: http://www-m3.ma.tum.de

O. Junge
e-mail: junge@ma.tum.de

S.D. Ross
Engineering Science and Mechanics, Virginia Polytechnic Institute and State University, Blacksburg, VA 24061, USA
e-mail: sdross@vt.edu
url: http://www.esm.vt.edu

J.A.T. Machado et al. (eds.), *Nonlinear Science and Complexity*,
DOI 10.1007/978-90-481-9884-9_16,

twist maps which approximate the spacecraft's motion in the planar circular restricted three-body problem [1]. The maps capture well the dynamics of the full equations of motion; the phase space contains a connected chaotic zone where intersections between unstable resonant orbit manifolds provide the template for lanes of fast migration between orbits of different semimajor axes.

2 The Keplerian Map

The example system we consider is the *Keplerian map* [1],

$$\begin{pmatrix} \omega_{n+1} \\ K_{n+1} \end{pmatrix} = \begin{pmatrix} \omega_n - 2\pi(-2K_{n+1})^{-3/2} \pmod{2\pi} \\ K_n + \mu f(\omega_n; C_J, \bar{K}) \end{pmatrix} \tag{1}$$

of the cylinder $\mathcal{A} = S^1 \times \mathbb{R}$ onto itself. This two-dimensional symplectic twist map is an approximation of a Poincaré map of the planar restricted three-body problem, where the surface of section is at periapsis in the space of orbital elements. The map models a spacecraft on a near-Keplerian orbit about a central body of unit mass, where the spacecraft is perturbed by a smaller body of mass μ. The interaction of the spacecraft with the perturber is modeled as an impulsive kick at periapsis passage, encapsulated in the kick function f.

This map can be used for preliminary design of low energy trajectories which involve multiple gravity assists. A trajectory sent from Earth to the Jovian system,

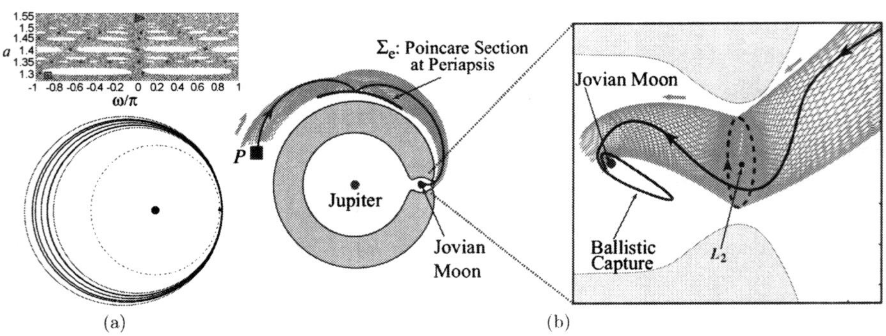

Fig. 1 (**a**) *Upper panel*: a phase space trajectory where the initial point is marked with *a triangle* and the final point with *a square*. *Lower panel*: the configuration space projections in an inertial frame for this trajectory. Jupiter and Callisto are shown at their initial positions, and Callisto's orbit is *dashed*. The uncontrolled spacecraft migration is from larger to smaller semimajor axes, keeping the periapsis direction roughly constant in inertial space. Both the spacecraft and Callisto orbit Jupiter in a counter-clockwise sense. The parameters used are $\mu = 5.667 \times 10^{-5}$, $C_J = 2.995$, $\bar{a} = -1/(2\bar{K}) = 1.35$, appropriate for a spacecraft in the Jupiter-Callisto system. (**b**) A spacecraft P inside a tube of gravitational capture orbits will find itself going from an orbit about Jupiter to an orbit about a Moon. The spacecraft is initially inside a tube whose boundary is the stable invariant manifold of a periodic orbit about L_2. The three-dimensional tube, made up of individual trajectories, is shown as projected onto configuration space. The final intersection of the tube with Σ_e, a Poincaré map at periapsis in the exterior realm

just grazing the orbit of the outermost icy moon Callisto, can migrate using little or no fuel from orbits with large apoapses to smaller ones. This is shown in Fig. 1(a) in both the phase space and the inertial configuration space. From orbits slightly larger than Callisto's, the spacecraft can be captured into an orbit around the moon. The set of all capture orbits is a solid cylindrical tube in the phase space [2, 3], as shown in Fig. 1(b). Followed backward in time this solid tube intersects transversally our Keplerian map, interpreted as a Poincaré surface-of-section. The resulting region is an *exit* from jovicentric orbits exterior to Callisto.

We can consider the location of an exit in the (ω, K)-plane as a target region for computing optimal capture trajectories. The details of the capture orbit around the moon are not considered here, but can be handled by other means at a finer scale [4].

3 Control Problem Formulation

We are interested in studying the dynamics of the Keplerian map (1) subjected to control. We define a family of controlled Keplerian maps $F : \mathcal{A} \times U \to \mathcal{A}$

$$F\left(\begin{pmatrix}\omega_n \\ K_n\end{pmatrix}, u_n\right) = \begin{pmatrix}\omega_{n+1} \\ K_{n+1}\end{pmatrix} = \begin{pmatrix}\omega_n - 2\pi(-2K_{n+1})^{-3/2} \pmod{2\pi} \\ K_n + \mu f(\omega_n) + \alpha u_n\end{pmatrix}, \tag{2}$$

where $u_n \in U = [-u_{\max}, u_{\max}]$, $u_{\max} \ll 1$, and the parametric dependence of f is understood. The term $\alpha = \alpha(C_J, \bar{K})$ is approximated as constant, where

$$\alpha = \sqrt{\frac{1}{\bar{a}}\left(\frac{1+\bar{e}}{1-\bar{e}}\right)}, \quad \text{with } \bar{e} = \sqrt{1 - \left(\frac{C_J - \bar{a}}{2\bar{a}^{3/2}}\right)^2} \text{ and } \bar{a} = -\frac{1}{2\bar{K}}. \tag{3}$$

Note that $F(\cdot, u_n)$ is area-preserving for any u_n. Physically, our control is modeled as a small impulsive thrust maneuver performed at periapsis n changing the speed by u_n. This increases K_n by an energy αu_n in addition to the natural dynamics term $\mu f(\omega_n)$.

Our goal is to control trajectories from a subset $S \subset \mathcal{A}$ to a target region $O \subset \mathcal{A}$. Additionally, we would like to minimize the total ΔV, while maintaining a reasonable transfer time. We model these requirements by considering the cost function $g : \mathcal{A} \times U \to [0, \infty)$,

$$g(a_n, u_n) = \frac{1}{2}|u_n|/u_{\max} + \frac{1}{2}\left(-\frac{1}{2K_n}\right)^{\frac{3}{2}},$$

where $a_n = (\omega_n, K_n)$ and our goal is to minimize the cost given by g that we accumulate along a controlled trajectory.

3.1 Optimal Feedback

Standard methods for solving this (time discrete) optimal control problem include algorithms like *value* or *policy iteration* [5] which compute (approximations to) the *optimal value* function of the problem and a corresponding (approximate) optimal stabilizing *feedback* $u : \mathcal{A} \to U$. For a general *shortest path problem* on a continuous state space, as in our case, a more efficient technique has been proposed [6–8]:

For given $a \in \mathcal{A}$ and $\mathbf{u} \in U^{\mathbb{N}}$ there is a unique associated trajectory $(a_n(a, \mathbf{u}))_{\mathbf{n} \in \mathbb{N}}$ of (2). Let $\mathcal{U}(a) = \{\mathbf{u} \in U^{\mathbb{N}} : a_n(a, \mathbf{u}) \to O \text{ as } n \to \infty\}$ and $S = \{a \in \mathcal{A} : \mathcal{U}(a) \neq \emptyset\}$ the *stabilizable subset* $S \subset \mathcal{A}$. The total cost along a controlled trajectory is given by $J(a, \mathbf{u}) = \sum_{n=0}^{\infty} g(a_n(a, \mathbf{u}), u_n) \in [0, \infty]$.

The construction of the feedback is based on (an approximation to) the *optimal value function* $V : S \to [0, \infty]$, $V(x) = \inf_{\mathbf{u} \in \mathcal{U}(\mathbf{a})} J(a, \mathbf{u})$, which satisfies the *optimality principle*

$$V(a) = \inf_{u \in U} \{g(a, u) + V(F(a, u))\}. \tag{4}$$

The right hand side of this equation can be interpreted as an operator, acting on the function V, the *dynamic programming operator* L. If $\tilde{V}$ is an approximation to V, then one defines the feedback by

$$u(a) = \operatorname*{argmin}_{u \in U} \{g(a, u) + \tilde{V}(F(a, u))\}, \tag{5}$$

whenever this minimum exists.

3.2 Discretization

We are going to approximate V by functions which are piecewise constant. Let $\mathcal{P}$ be a partition of $\mathcal{A}$, i.e. a collection of pairwise disjoint subsets which covers the state space $\mathcal{A}$. For a state $a \in \mathcal{A}$ we let $\rho(a)$ denote the element in the partition which contains a. Let $\mathbb{R}^{\mathcal{P}}$ be the subspace of the space $\mathbb{R}^{\mathcal{A}}$ of all real valued functions on $\mathcal{A}$ which are piecewise constant on the elements of the partition $\mathcal{P}$. The map $\varphi : \mathbb{R}^{\mathcal{A}} \to \mathbb{R}^{\mathcal{P}}$, $\varphi[v](a) = \inf_{a' \in \rho(a)} v(a')$, is a projection onto $\mathbb{R}^{\mathcal{P}}$. We define the *discretized dynamic programming operator* $L_{\mathcal{P}} : \mathbb{R}^{\mathcal{P}} \to \mathbb{R}^{\mathcal{P}}$ by $L_{\mathcal{P}} = \varphi \circ L$. This operator has a unique fixed point $V_{\mathcal{P}}$ which satisfies $V_{\mathcal{P}}(O) = 0$—the approximate (optimal) value function. One can show [8] that the fixed point equation $V_{\mathcal{P}} = L_{\mathcal{P}}[V_{\mathcal{P}}]$ is equivalent to the *discrete optimality principle*

$$V_{\mathcal{P}}(P) = \min_{P' \in \mathcal{F}(P)} \{\mathcal{G}(P, P') + V_{\mathcal{P}}(P')\},$$

where $V_{\mathcal{P}}(P) = V_{\mathcal{P}}(a)$ for any $a \in P \in \mathcal{P}$, the map $\mathcal{F}$ is given by

$$\mathcal{F}(P) = \{P' \in \mathcal{P} : P' \cap f(P, U) \neq \emptyset\} \tag{6}$$

and the cost function $\mathcal{G}$ by

$$\mathcal{G}(P, P') = \inf\{g(a, u) \mid a \in P, F(a, u) \in P', u \in U\}. \tag{7}$$

Note that the approximate value function $V_{\mathcal{P}}(P)$ is the length of the shortest path from P to $\rho(O)$ in the weighted directed graph $(\mathcal{P}, E)$, where the set of edges is defined by $E = \{(P, P') : P' \in \mathcal{F}(P)\}$ and the edge (P, P') is weighted by $\mathcal{G}(P, P')$. As such, it can be computed by, e.g., Dijkstra's algorithm.

In general, parameter uncertainties, modelling errors and small disturbances of the current state a_n may lead to a perturbed state $\tilde{a}_{n+1}$. Grüne and Junge [8] propose a generalization of the graph construction outlined above in order to cope with general disturbances. The following example computation is based on this general approach.

4 Low Energy Multiple Gravity Assists

We consider Jupiter-Callisto system with state space $\mathcal{A} = [-\pi, \pi] \times [-0.4630, -0.03]$ which includes a start region corresponding to spacecraft initially in a large orbit around Jupiter. The target region O is the exit region leading to capture orbits around the moon. We use $u_{\max} = 5$ m/s (in normalized units). The computation of the value function is based on a partition of $\mathcal{A}$ into 2^{20} boxes of equal size (2^{10} boxes in each direction). We use 25 test points on an equidistant grid in each box in state space as well as 65 equally spaced points in the control range $[-u_{\max}, u_{\max}]$ in order to compute the weighted graph. We consider the discretization as uncertainty [8] which corresponds to an additive perturbation of ≈ 1000 km in each time step. Figure 2 shows the resulting approximate value function $\tilde{V}$ and a feedback trajectory starting from the initial point $a_0 = [\omega, K] = [0.036, -0.048]$ in the start region. The corresponding orbit in configuration space is also shown in Fig. 3.

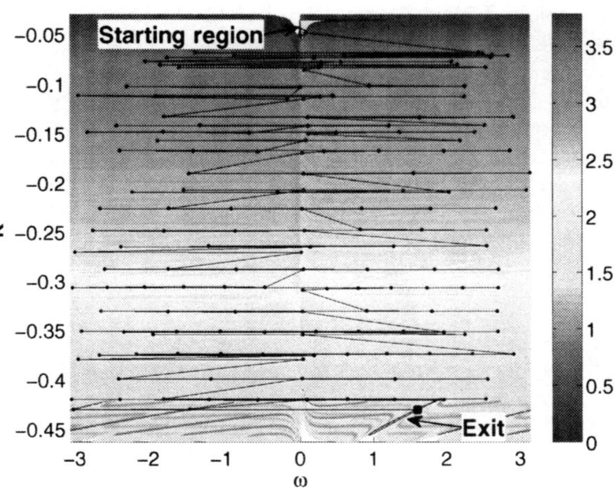

Fig. 2 The optimal value function and a feedback trajectory for the Keplerian map with $(\mu, C_J, \bar{a}) = (5.667 \times 10^{-5}, 2.995, 1.35)$. The initial point contained in the start region (*gray*) is marked by *a triangle* and the final point, which is contained in the exit region (*magenta*), by a *square*

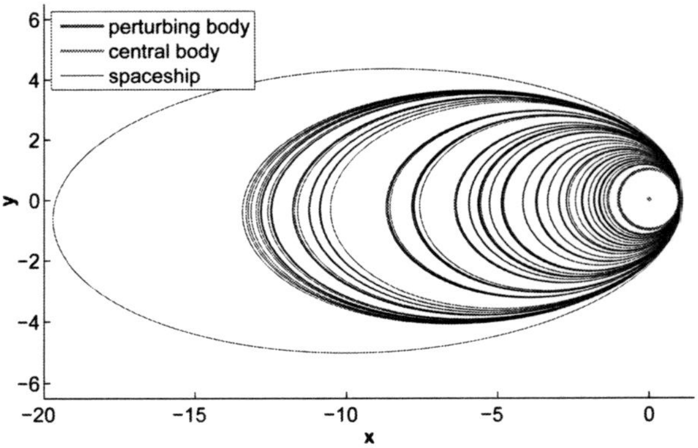

Fig. 3 Projection onto configuration space of the controlled trajectory of Fig. 2 in an inertial frame (normalized units). The spacecraft migration is from larger to smaller semimajor axes, keeping the periapsis direction roughly constant in inertial space

5 Conclusion

We applied a new feedback construction for discrete time optimal control problems with continuous state space which is based on graph theoretic methods to a celestial mechanics problem. We found a fuel-efficient spacecraft trajectory which starts in a large orbit around Jupiter and is captured by the smaller member of a binary system, e.g., a moon of Jupiter, using multiple gravity assists.

References

1. S.D. Ross, D.J. Scheeres, Multiple gravity assists, capture, and escape in the restricted three-body problem. SIAM J. Appl. Dyn. Syst. **6**, 576–596 (2007)
2. W.S. Koon, M.W. Lo, J.E. Marsden, S.D. Ross, Heteroclinic connections between periodic orbits and resonance transitions in celestial mechanics. Chaos **10**, 427–469 (2000)
3. W.S. Koon, M.W. Lo, J.E. Marsden, S.D. Ross, *Dynamical systems, the three-body problem and space mission design* (2008). http://www.shaneross.com/books
4. M.E. Paskowitz, D.J. Scheeres, Robust capture and transfer trajectories for planetary satellite orbiters. J. Guid. Control Dyn. **29**, 342–353 (2006)
5. D.P. Bertsekas, *Dynamic Programming and Optimal Control*, vol. 2 (Athena Scientific, Belmont, 1995)
6. O. Junge, H.M. Osinga, A set oriented approach to global optimal control. ESAIM Control Optim. Calc. Var. **10**, 259–270 (2004)
7. L. Grüne, O. Junge, A set oriented approach to optimal feedback stabilization. Syst. Control Lett. **54**, 169–180 (2005)
8. L. Grüne, O. Junge, Global optimal control of perturbed systems. J. Optim. Theory Appl. **136**, 411–429 (2008)

New Periodic Orbits in the Solar Sail Three-Body Problem

J.D. Biggs, T. Waters, and C. McInnes

Abstract We identify displaced periodic orbits in the circular restricted three-body problem, where the third (small) body is a solar sail. In particular, we consider solar sail orbits in the Earth-Sun system which are high above the ecliptic plane. It is shown that periodic orbits about surfaces of artificial equilibria are naturally present at linear order. Using the method of Lindstedt-Poincaré, we construct nth order approximations to periodic solutions of the nonlinear equations of motion. In the second part of the paper we generalize to the solar sail elliptical restricted three-body problem. A numerical continuation, with the eccentricity, e, as the varying parameter, is used to find periodic orbits above the ecliptic, starting from a known orbit at $e = 0$ and continuing to the required eccentricity of $e = 0.0167$. The stability of these periodic orbits is investigated.

Keywords Displaced periodic orbits · Solar sail · Restricted three body problem

1 Introduction

While the concept of the solar sail has been with us for some time, it is only with recent advances in materials and structures that their use is being seriously considered. A solar sail consists essentially of a large mirror, which uses the momentum change due to photons reflecting off the sail for its propulsion. A natural setting to consider the orbital dynamics of a solar sail is the restricted three-body problem, with

J.D. Biggs (✉) · C. McInnes
Department of Mechanical Engineering, University of Strathclyde, Glasgow, UK
e-mail: james.biggs@strath.ac.uk

C. McInnes
e-mail: colin.mcinnes@strath.ac.uk

T. Waters
Department of Mathematics, University of Portsmouth, Portsmouth, UK
e-mail: thomas.waters@port.ac.uk

J.A.T. Machado et al. (eds.), *Nonlinear Science and Complexity*,
DOI 10.1007/978-90-481-9884-9_17,

the Earth and Sun as the two primaries and the third (small) body a solar sail. We begin here with an analysis of the solar sail circular restricted three-body problem (CRTBP).

There has been some work already carried out regarding the solar sail CRTBP; in McInnes [1] surfaces of equilibrium points are described and in Baoyin and McInnes [2], the authors describe periodic orbits about equilibrium points on the axis joining the primary masses in the solar sail CRTBP.

We investigate periodic orbits high above the ecliptic plane in the solar sail CRTBP. These orbits could potentially be utilised practically for the continued observation, and communication, with the poles. It is shown that periodic orbits exist at linear order, and that these linear solutions can be utilised to find higher order approximations to periodic solutions of the non-linear system using the method of Lindstedt-Poincaré [3]. These approximate orbits are then fine-tuned using a differential corrector to find initial conditions that yield periodic solutions to the full non-linear model [3].

Following this we generalize the problem to the solar sail ERTBP [4] in the Earth-Sun system. A numerical continuation method is used, with the eccentricity e as the varying parameter, to find a periodic orbit above the ecliptic, starting from a known orbit in the solar sail CRTBP. The stability of these periodic orbits are investigated and it is shown that they are unstable and that a bifurcation occurs at $e = 0$.

2 Equations of Motion in the Rotating Frame

We consider a rotating coordinate system in which the primary masses are fixed on the x-axis with the origin at the centre of mass, the z-axis is the axis of rotation and the y-axis completes the triad. We choose our units to set the gravitational constant, the sum of the primary masses, the distance between the primaries, and the magnitude of the angular velocity of the rotating frame to be unity. We shall denote by $\mu = 3 \times 10^{-6}$ the dimensionless mass of the smaller body m_2, the Earth, and therefore the mass of the larger body m_1, the Sun, is given by $1 - \mu$. Denoting by $\boldsymbol{r}, \boldsymbol{r}_1$ and $\boldsymbol{r}_2$ the position of the sail w.r.t. the origin, m_1 and m_2 respectively, the solar sail's equations of motion in the rotating frame are

$$\frac{d^2\boldsymbol{r}}{dt^2} + 2\boldsymbol{\omega} \times \frac{d\boldsymbol{r}}{dt} = \boldsymbol{a} - \boldsymbol{\omega} \times (\boldsymbol{\omega} \times \boldsymbol{r}) - \nabla V \equiv \boldsymbol{F}, \tag{1}$$

with $\boldsymbol{\omega} = \widehat{\boldsymbol{z}}$ and $V = -[(1-\mu)/r_1 + \mu/r_2]$ where $r_i = |\boldsymbol{r}_i|$. These differ from the classical equations of motion in the CRTBP by the radiation pressure acceleration term

$$\boldsymbol{a} = \beta \frac{(1-\mu)}{r_1^2} (\widehat{\boldsymbol{r}}_1 . \boldsymbol{n})^2 \boldsymbol{n}, \tag{2}$$

where β is the sail lightness number, and is the ratio of the solar radiation pressure acceleration to the solar gravitational acceleration. Here $\boldsymbol{n}$ is the unit normal of the sail and describes the sail's orientation. We define $\boldsymbol{n}$ in terms of two angles γ and ϕ w.r.t. the rotating coordinate frame,

$$\boldsymbol{n} = (\cos(\gamma)\cos(\phi), \cos(\gamma)\sin(\phi), \sin(\gamma)), \tag{3}$$

where γ, ϕ are the angles the normal makes with the x–y and x–z plane respectively.

Equilibria are given by specifying the lightness number β and the sail angles ϕ and γ, and solving $\boldsymbol{F} = 0$ in (1). To simplify the search for orbits we assume $\phi = 0$ so the equilibrium (and sail normal) is in the x–z plane. In Fig. 1(i) we show some

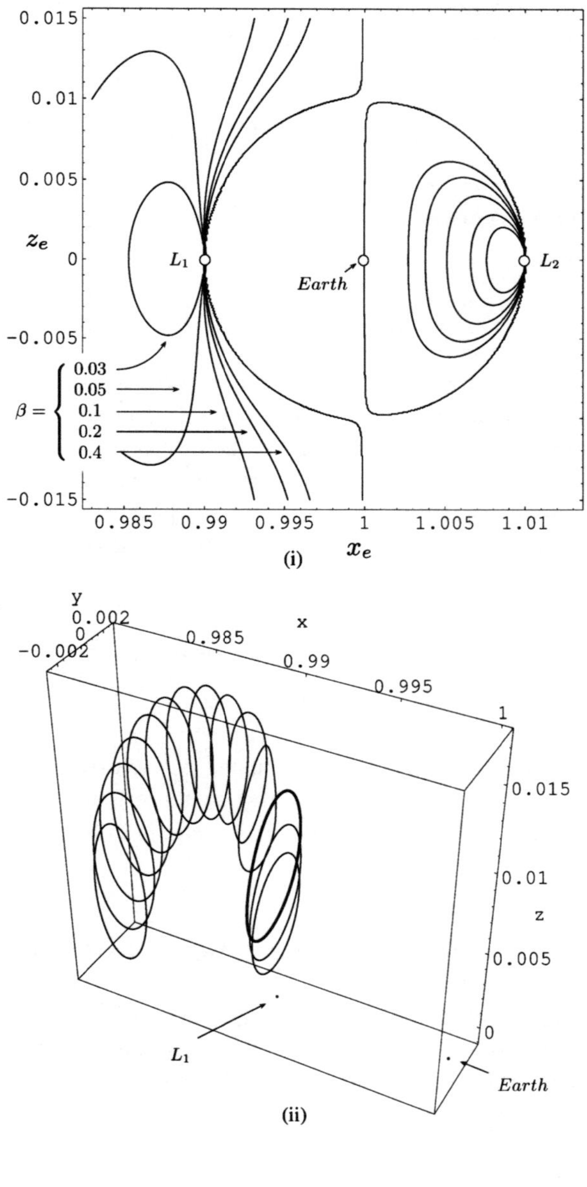

Fig. 1 **(i)** Surfaces of equilibrium points in the x_e–z_e parameter space. *Each curve* is specified by a constant value of β, and the position of the equilibrium point along *the curve* is given by γ. **(ii)** A family of orbits with $\beta = 0.05$. Each orbit has the same amplitude and is about a different equilibrium point along the β level curve shown in Fig. 1(**i**), each equilibrium point being defined by a different γ value. For reference the Earth (to scale) and L_1 are shown

of the equilibria near the Earth for low β values. To put the analysis in this paper well within the near-term we will consider very modest β values of about 0.05.

3 Linearised System

We linearise about the equilibrium points (in the x–z plane) by making the transformation $\mathbf{r} \to \mathbf{r}_e + \delta\mathbf{r}$, Taylor expanding $\mathbf{F}$ about $\mathbf{r}_e$, and neglecting the terms quadratic in $\delta\mathbf{r}$. We assume the orientation of the sail will remain fixed under perturbation of the sail position, in which case γ, ϕ and β are constants. Letting $\delta\mathbf{r} = (\delta x, \delta y, \delta z)^T$ and $\mathbf{X}(t) = (\delta\mathbf{r}, \delta\dot{\mathbf{r}})^T$, our linear system is $\dot{\mathbf{X}}(t) = A\mathbf{X}(t)$ with

$$A = \left(\begin{array}{c|c} 0 & I \\ \hline M & \Omega \end{array}\right), \qquad M = \begin{pmatrix} a & 0 & b \\ 0 & c & 0 \\ d & 0 & e \end{pmatrix}, \qquad \Omega = \begin{pmatrix} 0 & 2 & 0 \\ -2 & 0 & 0 \\ 0 & 0 & 0 \end{pmatrix}, \tag{4}$$

where a dot denotes differentiation w.r.t. t,

$$a = (\partial_x F^x)|_e, \qquad b = (\partial_z F^x)|_e, \qquad c = (\partial_y F^y)|_e,$$
$$d = (\partial_x F^z)|_e, \qquad e = (\partial_z F^z)|_e,$$

and $b \neq d$. The eigenvalues of A are either in pairs of pure imaginary conjugates or real and of opposite sign. Thus equilibria in the x–z plane will have the dynamical structure of centres and saddles, akin to the classical collinear Lagrange points.

If we label the eigenvectors associated with complex eigenvalues $\lambda_a i$ as $\mathbf{u}_a + \mathbf{w}_a i$ with $a = 1, 2$, and the eigenvectors associated with the real eigenvalues $\lambda_r, -\lambda_r$ as $\mathbf{v}_1, \mathbf{v}_2$, then the general solution of the linear system (4) is

$$\begin{aligned} \mathbf{X}(t) = {} & \cos(\lambda_1 t)[A\mathbf{u}_1 + B\mathbf{w}_1] + \sin(\lambda_1 t)[B\mathbf{u}_1 - A\mathbf{w}_1] \\ & + \cos(\lambda_2 t)[C\mathbf{u}_2 + D\mathbf{w}_2] + \sin(\lambda_2 t)[D\mathbf{u}_2 - C\mathbf{w}_2] \\ & + E e^{\lambda_r t}\mathbf{v}_1 + F e^{-\lambda_r t}\mathbf{v}_2. \end{aligned} \tag{5}$$

The linear order solution contains periodic solutions in *both* linear frequencies. By setting $E = F = 0$ we may switch off the real modes, and by setting either $A = B = 0$ or $C = D = 0$ we have periodic solutions in the frequency of our choice.

4 High-Order Approximations to Periodic Orbits

The linear solutions given in the previous section will only closely approximate the motion of the sail given in (1) for small amplitudes. For larger amplitude periodic orbits, we compute high-order approximations using the method of Linstedt-Poincaré [3].

We let ε be a perturbation parameter and expand each coordinate as $x = x_e + \varepsilon x_1 + \varepsilon^2 x_2 + \cdots$ etc. We rescale the time coordinate $\tau = \omega t$ with $\omega = 1 + \varepsilon\omega_1 + \cdots$, and group together the powers of ε in the high-order Taylor expansion of $\mathbf{F}$. We choose our linear solution to be

$$x_1 = kA_y \cos(\lambda\tau + \xi), \qquad y_1 = A_y \sin(\lambda\tau + \xi), \qquad z_1 = mA_y \cos(\lambda\tau + \xi), \quad (6)$$

where λ can be λ_1 or λ_2, k, m are given in terms of components of the eigenvectors and A_y, ξ are free parameters. We use these linear solutions to build up non-linear approximations to periodic orbits one order at a time in the following way:

At each order of ε, the system to be solved will be

$$\begin{aligned} x_n'' - 2y_n' - ax_n - bz_n &= g_1(x_{n-1}, y_{n-1}, z_{n-1}, x_{n-2}, \ldots), \\ y_n'' + 2x_n' - cy_n &= g_2(x_{n-1}, y_{n-1}, z_{n-1}, x_{n-2}, \ldots), \\ z_n'' - dx_n - ez_n &= g_3(x_{n-1}, y_{n-1}, z_{n-1}, x_{n-2}, \ldots), \end{aligned} \quad (7)$$

where prime denotes differentiation w.r.t. τ. The left hand side is the same form as the linear system (4), and on the right hand side the previous orders' solutions act as forcing terms. We use the freedom in ω_n to switch off the resonant or secular terms in the inhomogeneous part, that is those components on the right hand side of the form (6), and what remains is a series of trigonometric subharmonics up to order n.

In calculating the solution at nth order, we find two sets of solutions depending on whether n is even or odd. When n is even, the nth order solutions have the form (letting $T = \lambda\tau + \xi$)

$$\begin{aligned} x_n &= p_{n0} + p_{n2}\cos(2T) + \cdots + p_{nn}\cos(nT), \\ y_n &= q_{n2}\sin(2T) + \cdots + q_{nn}\sin(nT), \\ z_n &= s_{n0} + s_{n2}\cos(2T) + \cdots + s_{nn}\cos(nT), \end{aligned} \quad (8)$$

with $\omega_{n-1} = 0$. When n is odd, the solutions at nth order have the form

$$\begin{aligned} x_n &= p_{n3}\cos(3T) + \cdots + p_{nn}\cos(nT), \\ y_n &= q_{n1}\sin(T) + q_{n3}\sin(3T) + \cdots + q_{nn}\sin(nT), \\ z_n &= s_{n1}\cos(T) + s_{n3}\cos(3T) + \cdots + s_{nn}\cos(nT), \end{aligned} \quad (9)$$

and ω_{n-1} solves

$$\frac{2\lambda\beta_{n1}}{(c+\lambda^2)} + \frac{b\gamma_{n1}}{(e+\lambda^2)} - \alpha_{n1} = 0. \quad (10)$$

Here α_{nj}, β_{nj} and γ_{nj} are the coefficients of the cos, sin and cos terms in the functions g_1, g_2 and g_3 respectively at order n given in (7), and the coefficients p_{nj}, q_{nj}

and s_{nj} are given by

$$
\begin{aligned}
&-(a+j^2\lambda^2)p_{nj} - 2j\lambda q_{nj} - b s_{nj} - \alpha_{nj} = 0,\\
&q_{nj} = \frac{-2j\lambda p_{nj} - \beta_{nj}}{(c+j^2\lambda^2)}, \qquad s_{nj} = \frac{-d p_{nj} - \gamma_{nj}}{(e+j^2\lambda^2)},
\end{aligned} \tag{11}
$$

with the exception of $q_{n0} = 0$ and $p_{n1} = 0$. With these high order approximations, we may find approximate initial data from which to integrate the system of (1). However these will not evolve to exactly periodic trajectories, as they are only approximations to periodic solutions. Thus we use a differential corrector to adjust the initial data so as to close the orbit. An example of a family of periodic orbits found using this method is shown in Fig. 1(ii).

5 The Solar Sail ERTBP

In this section the generalisation to the solar sail ERTBP [4] is considered where the equations of motion are expressed in the rotating-pulsating frame [5]:

$$
\begin{aligned}
x'' - 2y' &= \frac{1}{1+e\cos f}\left(\frac{\partial\Omega}{\partial x} + a_x\right),\\
y'' + 2x' &= \frac{1}{1+e\cos f}\left(\frac{\partial\Omega}{\partial y} + a_y\right),\\
z'' + z &= \frac{1}{1+e\cos f}\left(\frac{\partial\Omega}{\partial z} + a_z\right),
\end{aligned} \tag{12}
$$

where

$$
\Omega = \frac{1}{2}(x^2+y^2+z^2) + \frac{(1-\mu)}{|\boldsymbol{r}_1|} + \frac{\mu}{|\boldsymbol{r}_2|}
$$

and where a_x, a_y, a_z are the components of the solar sail acceleration $\boldsymbol{a} = (a_x, a_y, a_z)^T$ and where $(\cdot)'$ denotes differentiation with respect to the true anomaly f. The pulsating-rotating frame is convenient as the true anomaly appears in the equations of motion as the independent variable and therefore we do not need to solve Kepler's equations. We note that when $e = 0$, (12) are equal to (1). The solar sail ERTBP require a separate analysis as the time appears explicitly in the equations of motion through the true anomaly f and are therefore non-autonomous. Therefore, as the function $\cos f$ in (12) is periodic, any periodic solution will have to be an integer multiple of this period. As such we search for a 1 year periodic orbit ($f = 2\pi$) above the ecliptic plane.

The continuation algorithm used to find periodic orbits above the ecliptic in the solar sail ERTBP is based on a monodromy variant of Newton's method [6]. The initial orbit which will serve as a starter in the numerical continuation is given using the methods in Sect. 4. If e is incremented by a suitably small value, the trajectory remains close enough for Newton's method to converge to a periodic orbit. This process is repeated until a closed orbit is found with the required $e = 0.0167$.

The Newton method starts with an orbit $\mathbf{X}(t)$ initialized at $t = 0$ on a surface of section. In our case we require that the orbit be exactly 1 year so the return map in the rotating-pulsating frame is defined by a T-map of period $f = 2\pi$. The Newton method provides an iterative improvement to the choice of initial conditions for a periodic orbit [6]:

$$\mathbf{X}^*(0) = \mathbf{X}(0) + (I - M)^{-1}[\mathbf{X}(T) - \mathbf{X}(0)] \tag{13}$$

where $\mathbf{X}^*(0)$ is the improved initial condition, I is the identity matrix and M is the monodromy matrix. One of the problems encountered with this Newton method is that the determinant of $(I - M)$ maybe zero and therefore the inverse is not well defined. This problem is resolved by using the Moore-Penrose pseudo inverse. To compute the monodromy matrix M, let $\boldsymbol{\Gamma}(t)$ denote a periodic orbit with period $T = 2\pi$ which satisfies the condition $\boldsymbol{\Gamma}(T) = \boldsymbol{\Gamma}(0)$, by letting $\mathrm{x} = \mathbf{X}(t) - \boldsymbol{\Gamma}(t)$, we may linearize the nonlinear system about this periodic orbit, resulting in the variational equations

$$\dot{\mathrm{x}} = A(t)\mathrm{x}$$

where

$$A(t) = A(t+T) = \left.\frac{\partial f}{\partial \mathbf{X}}\right|_{\mathbf{X}(t)=\boldsymbol{\Gamma}(t)}.$$

Recasting the variational equations in terms of the state transition matrix (or principle fundamental matrix) $\Phi = \partial \mathrm{X}(t)/\partial \mathrm{X}(0)$, we have

$$\dot{\Phi}(t) = A(t)\Phi(t), \qquad \Phi(0) = I$$

where Φ is a 6×6 matrix, then $M = \Phi(t)$. Using Newton's method we obtain initial conditions that yield a 1 year periodic orbit above the ecliptic in the solar sail ERTBP:

$$\begin{aligned}
&x(0) = 0.99026089328, \qquad y(0) = 0.00000002532,\\
&z(0) = 0.01497820749,\\
&x'(0) = 0.00000000062, \qquad y'(0) = 0.00306117311,\\
&z'(0) = -0.00000003900, \qquad f(0) = 0,\\
&\gamma = 0.809196, \qquad \delta = 0.
\end{aligned} \tag{14}$$

6 Stability of Periodic Orbits in the Solar Sail RTBP

The stability of the periodic orbits in Sects. 4 and 5 are determined using Floquet theory [7]: Let the eigenvalues of the monodromy matrix M be denoted by λ_i and the corresponding characteristic (Floquet) exponents α_i defined as $\lambda_i = e^{\alpha_i T}$. Floquet theory then states that the orbit $\mathbf{\Gamma}(t)$ is stable at linear order if and only if the real parts of all the characteristic exponents are less than or equal to zero. In the circular case the characteristic exponents are of the form $\{0, 0, \alpha_i, \bar{\alpha}_i, \pm\alpha_r\}$. However, the characteristic exponents computed in the elliptic case $0 < e \leq 0.0167$ are of the form $\{\alpha_j, \bar{\alpha}_j, \alpha_i, \bar{\alpha}_i, \pm\alpha_r\}$ which is consistent with periodic orbits in the classical ERTBP [8]. This implies that in the solar sail ERTBP there is a bifurcation at $e = 0$, in the sense that the characteristic exponents have changed form. However, in each case the periodic orbit is unstable and will require active control to maintain. This active control could be implemented through small variations in the sail's orientation.

7 Conclusion

In this paper we initially consider periodic orbits above the ecliptic in the solar sail circular restricted three-body problem, where periodic orbits about equilibria are present at linear order. Using the method of Lindstedt-Poincaré, we construct nth order approximations to periodic solutions of the nonlinear equations of motion and use these to compute high amplitude orbits above the ecliptic plane. Following this we generalise to the elliptical restricted three-body problem. A numerical continuation, with the eccentricity, e, as the varying parameter, is used to find a periodic orbit above the ecliptic, starting from a known orbit at $e = 0$ and continuing to the required eccentricity of $e = 0.0167$. The stability of these periodic orbits is investigated and they are shown to be unstable.

References

1. C.R. McInnes, *Solar Sailing: Technology, Dynamics and Mission Applications* (Springer Praxis, London, 1999)
2. H. Baoyin, C. McInnes, Celest. Mech. Dyn. Astron. **94**, 155–171 (2006)
3. T. Waters, C. McInnes, J. Guid. Control Dyn. **30**(3), 687–693 (2007)
4. H. Baoyin, C. McInnes, J. Guid. Control Dyn. **29**(3), 538–543 (2006)
5. V. Szebehely, *Theory of Orbits: The Restricted Problem of Three Bodies* (Academic Press, New York, 1967)
6. R. Marcinek, E. Pollak, J. Chem. Phys. **100**(8), 5894–5904 (1994)
7. R. Grimshaw, *Nonlinear Ordinary Differential Equations* (Blackwell Scientific, Oxford, 1990)
8. R. Broucke, AIAA J. **7**(6), 1003–1009 (1969)

A Review of Invariant Manifold Dynamics of the CRTBP and Some Applications

Josep J. Masdemont

Abstract In this short review we show how the invariant manifolds of quasiperiodic orbits about libration point regimes play a crucial role to study the dynamics in some astrodynamical and astronomical problems, and how they can be used for practical purposes. Some discussion about their computation is also given.

Keywords Normally hyperbolic invariant manifolds · Lindstedt Poincaré · Normal forms · Quasiperiodic orbits · Libration point orbits

1 The Models

The classical Circular Restricted Three Body Problem (CRTBP) considers the motion of an infinitesimal particle (spacecraft) under the gravitational attraction of two point like masses called primaries (Sun and Earth) [24]. In the (non-inertial) synodic coordinate system the equation of motion are,

$$\ddot{X} - 2\dot{Y} = \frac{\partial \Omega_R}{\partial X}, \qquad \ddot{Y} + 2\dot{X} = \frac{\partial \Omega_R}{\partial Y}, \qquad \ddot{Z} = \frac{\partial \Omega_R}{\partial Z}, \tag{1}$$

where, $\Omega_R(X, Y, Z) = (X^2 + Y^2)/2 + (1 - \mu)/r_1 + \mu/r_2 + \mu(1 - \mu)/2$, and r_1, r_2 denote the distances from the spacecraft to the primaries. $r_1^2 = (X - \mu)^2 + Y^2 + Z^2$ (dist. to Earth), and $r_2^2 = (X + 1 - \mu)^2 + Y^2 + Z^2$ (dist. to Sun).

When the mass parameter is small another model, such us the Hill model, can be considered. It s obtained from the CRTBP by a translation of the origin of coordinates to the small primary followed by a rescaling of the coordinates by a factor $\mu^{1/3}$. The choice of the scale factor guarantees that, in the new coordinates, the

J.J. Masdemont (✉)
Departament de Matemàtica Aplicada I, Universitat Politècnica de Catalunya, Diagonal 647 (ETSEIB), 08028 Barcelona, Spain
e-mail: josep@barquins.upc.edu

J.A.T. Machado et al. (eds.), *Nonlinear Science and Complexity*,
DOI 10.1007/978-90-481-9884-9_18,

gravitational force of the secondary is of the same order as the Coriolis force as well as of the centrifugal force. Expanding the result in powers of $\mu^{1/3}$ one obtains the equations of the model [2, 24].

$$\ddot{X} - 2\dot{Y} = \frac{\partial \Omega_H}{\partial X}, \qquad \ddot{Y} + 2\dot{X} = \frac{\partial \Omega_H}{\partial Y}, \qquad \ddot{Z} = \frac{\partial \Omega_H}{\partial Z}, \tag{2}$$

where, $\Omega_H(X, Y, Z) = 3X^2/2 - Z^2/2 + 1/\sqrt{X^2 + Y^2 + Z^2}$.

We note that (1) and (2) have the same form except for the definition of Ω_*. From now on whenever we obtain the same expression for both models just changing the subscript of Ω, or of any other magnitude, the subscript corresponding to the model will be removed. For instance, both models have a Jacobi integral given by,

$$C_B = 2\,\Omega(X, Y, Z) - (\dot{X}^2 + \dot{Y}^2 + \dot{Z}^2). \tag{3}$$

Let us also say that introducing momenta $P_X = \dot{X} - Y$, $P_Y = \dot{Y} + X$ and $P_z = \dot{Z}$, both the CRTBP and Hill models admit a Hamiltonian representation with Hamiltonian,

$$H_B = \frac{1}{2}(P_X^2 + P_Y^2 + P_Z^2) + Y P_X - X P_Y + \frac{1}{2}(X^2 + Y^2) - \Omega(X, Y, Z).$$

The CRTBP has three collinear equilibrium points on the X axis separated by the two primaries. Hill's problem has two equilibrium points on the X axis symmetric with respect to the primary at a distance $1/\sqrt[3]{3}$.

In this review we focus on libration orbits about the equilibrium points L_1 and L_2, the nearest to the smallest primary, which are of major interest for space mission design. Let us write the equations of motion (1) centered in one of these points and at the same time scaling the longitude in such a way that the new distance from the origin (equilibrium point) to the closest primary be equal to one.

Considering (x, y, z) the new coordinates about L_1 or L_2 the CRTBP and Hill equations can be expanded in the form,

$$\left\{ \begin{aligned} \ddot{x} - 2\dot{y} - (1 + 2c_2)x &= \frac{\partial}{\partial x} \sum_{n \geq 3} c_n \rho^n P_n\left(\frac{x}{\rho}\right), \\ \ddot{y} + 2\dot{x} + (c_2 - 1)y &= \frac{\partial}{\partial y} \sum_{n \geq 3} c_n \rho^n P_n\left(\frac{x}{\rho}\right), \\ \ddot{z} + c_2 z &= \frac{\partial}{\partial z} \sum_{n \geq 3} c_n \rho^n P_n\left(\frac{x}{\rho}\right), \end{aligned} \right. \tag{4}$$

where $\rho^2 = x^2 + y^2 + z^2$, P_n is the Legendre polynomial of degree n, and c_n are constants which depend only on the model and the selected equilibrium point [20]. Again introducing momenta $p_x = \dot{x} - y$, $p_y = \dot{y} + x$, $p_z = \dot{z}$, this system is Hamiltonian with Hamiltonian,

$$H = \frac{1}{2}(p_x^2 + p_y^2 + p_z^2) + y p_x - x p_y - \sum_{n \geq 2} c_n \rho^n P_n\left(\frac{x}{\rho}\right). \tag{5}$$

2 Lindstedt Poincaré and Normal Forms Techniques

A convenient procedure to look for periodic and quasiperiodic motion, as well as invariant manifolds, about a libration point is to use a Lindstedt Poincaré technique. To compute invariant tori and their associated hyperbolic invariant manifolds we look for a formal expansions of the solutions of (4) in the form,

$$\begin{cases} x(t) = \sum e^{(i-j)\theta_3}[x_{ijkm}^{pq} \cos \Phi_{pq} + \bar{x}_{ijkm}^{pq} \sin \Phi_{pq}]\alpha_1^i \alpha_2^j \alpha_3^k \alpha_4^m, \\ y(t) = \sum e^{(i-j)\theta_3}[y_{ijkm}^{pq} \cos \Phi_{pq} + \bar{y}_{ijkm}^{pq} \sin \Phi_{pq}]\alpha_1^i \alpha_2^j \alpha_3^k \alpha_4^m, \\ z(t) = \sum e^{(i-j)\theta_3}[z_{ijkm}^{pq} \cos \Phi_{pq} + \bar{z}_{ijkm}^{pq} \sin \Phi_{pq}]\alpha_1^i \alpha_2^j \alpha_3^k \alpha_4^m, \end{cases} \tag{6}$$

where $\Phi_{pq} = \Phi_{pq}(\theta_1, \theta_2) = p\,\theta_1 + q\,\theta_2$ with $\theta_1 = \omega t + \phi_1$, $\theta_2 = \nu t + \phi_2$, $\theta_3 = \lambda t$ and,

$$\omega = \sum \omega_{ijkm}\, \alpha_1^i \alpha_2^j \alpha_3^k \alpha_4^m, \qquad \nu = \sum \nu_{ijkm}\, \alpha_1^i \alpha_2^j \alpha_3^k \alpha_4^m,$$
$$\lambda = \sum \lambda_{ijkm}\, \alpha_1^i \alpha_2^j \alpha_3^k \alpha_4^m.$$

Here α_1 and α_2 represent hyperbolic (unstable and stable amplitudes), α_3 and α_4 represent central (in-plane and out-of-plane amplitudes) and θ_1 and θ_2 phases which characterize the dynamics in a neighborhood of the point. Summation in principle is extended over all $i, j, k, m \in \mathbb{N}$ and $p, q \in \mathbb{Z}$. Nevertheless many coefficients are known to be zero. See [15, 20] for how the coefficients of the expansion are determined as well as how tests of practical convergence can be performed.

2.1 Normal Forms

Another way to semi-analytically obtain the orbits about a libration point is using normal form techniques. The process of computing the type of normal forms that we are going to use is described in detail in [12, 15] and references therein. Here we only summarize the main details of the methodology.

The procedure starts considering the Hamiltonian (5) and performs a (complex) symplectic change of variables such that the second order part is written in the form,

$$H_2(q_1, q_2, q_3, p_1, p_2, p_3) = \sqrt{-1}\nu_0\, q_1 p_1 + \sqrt{-1}\omega_0\, q_2 p_2 + \lambda_0\, q_3 p_3. \tag{7}$$

The type of normal form we are interested to complete up to a high order is known as a reduction to the center manifold [12]. It is accomplished applying the Lie series method [7] and using suitable transformations given by the time one flow of a given Hamiltonian which is know as a generating function (see [10] and references therein).

Let us assume that we have a generating function $G(q, p)$. Applying the cited transformation to the Hamiltonian H one obtains a new Hamiltonian $\hat{H}$ given by,

$$\hat{H} \equiv H + \{H, G\} + \frac{1}{2!}\{\{H, G\}, G\} + \frac{1}{3!}\{\{\{H, G\}, G\}, G\} + \cdots,$$

where $\{f, g\}$ denotes the Poisson bracket of the f, and g functions.

The convenience of the method is clear when H and G are expanded in homogeneous polynomials since, when P and Q are two homogeneous polynomials of degree r and s respectively, then $\{P, Q\}$ is a homogeneous polynomial of degree $r + s - 2$. Thus, if we assume that G is an homogeneous polynomial of degree 3, G_3, then the homogeneous terms of $\hat{H}$ are,

$$\hat{H}_2 = H_2, \qquad \hat{H}_3 = H_3 + \{H_2, G_3\},$$
$$\hat{H}_4 = H_4 + \{H_3, G_3\} + \frac{1}{2!}\{\{H_2, G_3\}, G_3\}, \ldots.$$

Hence, if we want to remove all the monomials of degree 3 in H we should look for a G_3 that solves the homo-logical equation $\{H_2, G_3\} = -H_3$.

Successive applications of generating functions G_n, homogeneous polynomials of degree n, $n = 4, 5, \ldots$ will be used to modify the part of order n of the current Hamiltonian. We also note that the application of the generating function G_n does not modify the homogeneous parts of the Hamiltonian, of degree less than n, obtained in previous steps.

The objective of our normal form is, in some way, to isolate the instability associated with the hyperbolic part of the Hamiltonian H. In the second order part, H_2, the instability is associated with the term $\lambda_0 q_3 p_3$, and for the linear approximations of the equations of motion that we obtain using H_2 instead of H, the center part corresponds to setting $q_3 = p_3 = 0$ which remains invariant for all time. If we want the trajectory to remain tangent to this space (this is, $q_3(t) = p_3(t) = 0$ for all $t > 0$) when considering also the nonlinear terms, we only need to require $\dot{q}_3(0) = \dot{p}_3(0) = 0$ due to the autonomous character of our Hamiltonian system.

Taking into account that the Hamiltonian equations associated with the couple q_3, p_3 have the form,

$$\dot{q}_3 = \frac{\partial H}{\partial p_3} = \lambda_0 q_3 + \sum_{n\geq 3} h_{ij}\, q_1^{i_1} q_2^{i_2} q_3^{i_3} p_1^{j_1} p_2^{j_2} p_3^{j_3-1},$$
$$\dot{p}_3 = -\frac{\partial H}{\partial q_3} = -\lambda_0 p_3 - \sum_{n\geq 3} h_{ij}\, q_1^{i_1} q_2^{i_2} q_3^{i_3-1} p_1^{j_1} p_2^{j_2} p_3^{j_3}$$

we get the condition $\dot{q}_3(0) = \dot{p}_3(0) = 0$ when $q_3(0) = p_3(0) = 0$ if the series expansion of the Hamiltonian has no monomials with $i_3 + j_3 = 1$. Of course we can have other choices, and in fact, in order to have high order expansions of the invariant manifolds of the libration point orbits it is convenient to remove all the monomials with $i_3 \neq j_3$. In any of these methods there are no problems of small divisors and the divergence of the process is very mild (see [15] and references therein for further information).

After all these changes of variables the Hamiltonian about the selected libration point has the form,

$$H(q,p) = \bar{H}_N(q,p) + R_N(q,p),$$

where $\bar{H}_N(q,p)$ is a polynomial of degree N where all the monomial have $i_3 = j_3$, and $R_N(q,p)$ is a remainder of order $N+1$, which is very small in a neighborhood of the libration point and it is neglected in further computations.

As a comment we can say that the reduction to the central manifold is similar to a Birkoff normal form process, although is "less aggressive" and the regions of practical convergence usually are bigger. It can be shown also that the Lindstedt-Poincaré expansions (6) are formally the solutions of the Hamiltonian equations corresponding to a Birkhoff normal form [5].

3 Applications to Classical Problems of Libration Point Mission Design

Invariant manifolds of libration point orbits can be applied with great success to many aspects of mission design. As classical problems we refer to the most common ones in mission analysis, like the transfer from the Earth to the libration orbit, the station keeping of the satellite or the transfer between libration point orbits.

As a general rule the insertion of the satellite in a stable manifold of an orbit provides a smooth way to reach libration point orbits without further maneuvers or just trajectory correction maneuvers. Using the Lindstedt-Poincaré expansions (6) a target torus is chosen by means of α_3 and α_4, the stable manifold taking $\alpha_1 = 0$ and a fixed $\alpha_2 \neq 0$. Finally the selected orbit in the manifold is selected choosing the appropriate phases. The procedure can also be adapted to include low thrust transfer arcs and to obtain trajectories from the Earth to a libration point orbit like the one represented in Fig. 1 (see [8]).

Station keeping methodologies using invariant manifolds are based on the fact that the deviation in position and velocity of the actual state of the satellite with respect to a nominal orbit, via a suitable change of coordinates, can be expressed as a linear combination of the Floquet modes with the objective to measure the projection of the deviation in the unstable, stable and central components. Then, suitable maneuvers are applied to cancel the unstable component [11, 14].

The correspondence between the usual position and velocity coordinates in the synodical frame and the normal form coordinates (q_i, p_i), or the Lindstedt Poincaré amplitudes and phases, is a tool which gives both qualitative and quantitative information of the state of the satellite in the neighborhood of the libration point regime. In this direction, better algorithms can be accommodated for station keeping as well as for transfer between near libration point orbits of more complex missions, such as formations of spacecraft, or to complement and assist the concept of low energy

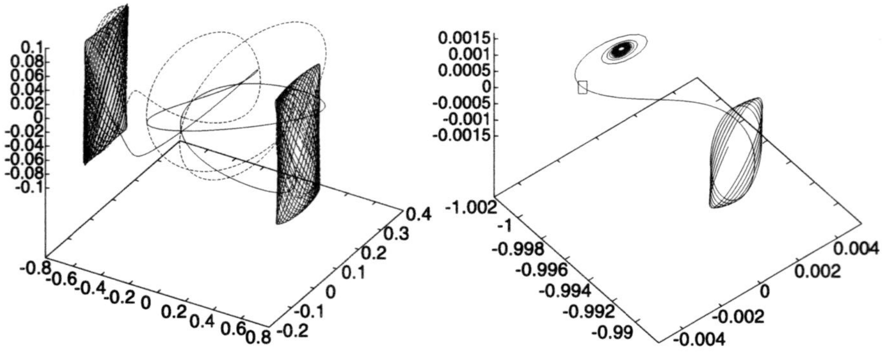

Fig. 1 (*Left*) Example of low thrust transfer from Earth to a libration point orbit. (*Right*) Heteroclinic trajectories between quasiperiodic Lissajous orbits about L_1 and L_2

transfers. In this sense families of heteroclinic orbits like the ones represented in Fig. 1 play the role of the skeleton of the dynamics however not free of complexity [9, 13, 16]. Orbits like this can be efficiently computed using Lindstedt-Poincaré, normal form, or a combination of both techniques.

In the area of low energy transfers is also very interesting the way that invariant manifolds of libration point orbits of different systems can be used to transfer between different regimes. For instance from Earth-Moon L_2 to Sun-Earth L_1 [1, 4, 18] or between the moons of Jupiter [17]. In these cases the complexity is increased because of the fact that the models are no longer autonomous.

There is a particular transfer between Lissajous orbits that can be used to efficiently display and use the dynamics about libration point orbits. Usually, a technical requirement for libration point satellites is to avoid an exclusion zone about the Sun when this is seen from Earth. For orbits about L_1 in the Sun–Earth system the exclusion zone is three degrees about the solar disk as seen from Earth. For orbits about L_2 in the Sun–Earth system sometimes the Earth half-shadow has to be avoided. In both cases, since Sun and Earth are located in the X axis, the exclusion zone is set as a disk in the YZ plane centered at the origin. A nice way to analyze this problem is using the action-angle variables used in the Lindstedt-Poincaré expansions of the Lissajous orbits (6). When the satellite is on a Lissajous orbit we have $\alpha_1 = \alpha_2 = 0$ and fixed values for α_3 and α_4. The trajectory is seen in the θ_1 and θ_2 variables moving uniformly in time since $\theta_1 = \omega t + \phi_1$ and $\theta_2 = \nu t + \phi_2$. Any trajectory in the torus, represented in the (θ_1, θ_2) plane, is seen as an initial point (ϕ_1, ϕ_2) which evolves in a straight line of slope and constant velocity ν/ω. The plane of the angular variables can be used to display many features of the dynamics, an not only in a local way. The simplest application is displayed in Fig. 2. See [3] for more details about this type of transfer trajectories.

The applications of the invariant manifolds extend also to astronomical problems, like to explain the arms formation in barred galaxies [21–23] or the mass transfer in the solar system among other areas of science [6, 19].

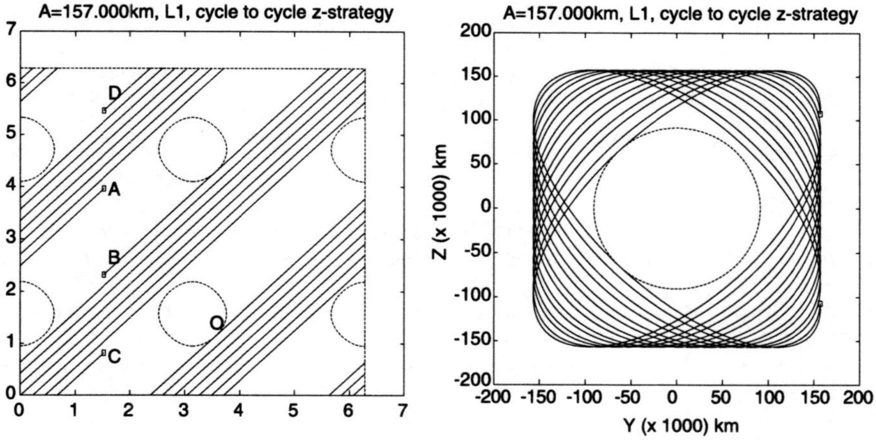

Fig. 2 (*Left*) The plane of angular variables where the exclusion zones are represented. Maneuvers are seen as jumps in the line trajectories and are marked with *a small box*. (*Right*) YZ projection of the libration trajectory in the CRTBP

Acknowledgements This work partially supported by the Spanish MCyT-FEDER Grant MTM2006-00478 and the Marie Curie Research Training Network Astronet Grant MCRTN-CT-2006-035151.

References

1. E.M. Alessi, G. Gómez, J.J. Masdemont, Leaving the Moon by means of invariant manifolds of libration point orbits. Commun. Nonlinear Sci. Numer. Simul. **14**, 4153–4167 (2009)
2. V.I. Arnold (ed.), *Dynamical Systems III*. Encyclopaedia of Mathematical Sciences, vol. 3 (Springer, Berlin, 1988)
3. E. Canalias, J. Cobos, J.J. Masdemont, Impulsive transfers between Lissajous libration point orbits. J. Astronaut. Sci. **51**, 361–390 (2003)
4. E. Canalias, J.J. Masdemont, Computing natural transfers between Sun–Earth and Earth–Moon Lissajous libration point orbits. Acta Astronaut. **63**, 238–248 (2008)
5. A. Delshams, J.J. Masdemont, P. Roldán, Computing the scattering map in the spatial Hill's problem. Discrete Contin. Dyn. Syst. Ser. B **10**, 455–483 (2008)
6. M. Deltnitz, O. Junge, M.W. Lo, J.E. Marsden, K. Padberg, R. Preix, S.D. Ross, B. Thiere, Transport of Mars-crossing asteroids from the quasi-Hilda region. Phys. Rev. Lett. **94**, 231102 (2005)
7. A. Deprit, Canonical transformations depending on a small parameter. Celest. Mech. **1**, 12–30 (1969)
8. P. Di Donato, J.J. Masdemont, P. Paglione, A.F. Prado, Low thrust transfers from the Earth to Halo orbits around the libration points of the Sun–Earth/Moon system, in *Proceedings of COBEM 2007* (2007), 9 pp
9. M. Gidea, J.J. Masdemont, Geometry of homoclinic connections in a planar circular restricted three-body problem. Int. J. Bifurc. Chaos **17**, 1151–1169 (2007)
10. A. Giorgilli, A. Delshams, E. Fontich, L. Galgani, C. Simó, Effective stability for a Hamiltonian system near an elliptic equilibrium point, with an application to the restricted three body problem. J. Differ. Equ. **77**, 167–198 (1989)

11. G. Gómez, K. Howell, J.J. Masdemont, C. Simó, Station keeping strategies for translunar libration point orbits. Adv. Astron. Sci. **99**, 949–967 (1998)
12. G. Gómez, A. Jorba, J.J. Masdemont, C. Simó, *Dynamics and Mission Design near Libration Points, vol. 3, Advanced Methods for Collinear Points* (World Scientific, Singapore, 2001), xvi+187
13. G. Gómez, W.S. Koon, M.W. Lo, J.E. Marsden, J.J. Masdemont, S.D. Ross, Connecting orbits and invariant manifolds in the spatial restricted three-body problem. Nonlinearity **17**, 1571–1606 (2004)
14. G. Gómez, J. Llibre, R. Martínez, C. Simó, *Dynamics and Mission Design near Libration Points, vol. 1, Fundamentals: The Case of Collinear Libration Points* (World Scientific, Singapore, 2001), xi+443
15. A. Jorba, J.J. Masdemont, Dynamics in the center manifold of the collinear points in the restricted three body problem. Physica D **132**, 189–213 (1999)
16. W.S. Koon, M.W. Lo, J.E. Marsden, S.D. Ross, Heteroclinic connections between periodic orbits and resonance transitions in celestial mechanics. Chaos **10**, 427–469 (2000)
17. W.S. Koon, M.W. Lo, J.E. Marsden, S.D. Ross, Constructing a low energy transfer between Jovian Moons. Contemp. Math. Am. Math. Soc. **292**, 129–145 (2001)
18. M.W. Lo, S.D. Ross, The lunar L_1 gateway: portal to the stars and beyond, in *AIAA Space 2001 Conference*. Albuquerque, New Mexico (2001)
19. J.E. Marsden, S.D. Ross, New methods in celestial mechanics and mission design. Bull. Am. Math. Soc. (New Ser.) **43**, 43–73 (2006)
20. J.J. Masdemont, High order expansions of invariant manifolds of libration point orbits with applications to mission design. Dyn. Syst. Int. J. **20**, 59–113 (2005)
21. M. Romero-Gómez, J.J. Masdemont, E. Athanassoula, C. García-Gómez, On the origin of rR_1 ring structures in barred galaxies. Astron. Astrophys. **453**, 39–45 (2006)
22. M. Romero-Gómez, J.J. Masdemont, E. Athanassoula, C. García-Gómez, The formation of spiral arms and rings in barred galaxies. Astron. Astrophys. **472**, 63–75 (2007)
23. M. Romero-Gómez, J.J. Masdemont, C. García-Gómez, E. Athanassoula, The role of the unstable equilibrium points in the transfer of matter in galactic potentials. Commun. Nonlinear Sci. Numer. Simul. **14**, 4123–4138 (2009).
24. V. Szebehely, *Theory of Orbits* (Academic Press, San Diego, 1967)

Solar Sail Orbits at the Earth-Moon Libration Points

Jules Simo and Colin R. McInnes

Abstract Solar sail technology offer new capabilities for the analysis and design of space missions. This new concept promises to be useful in overcoming the challenges of moving throughout the solar system. In this paper, novel families of highly non-Keplerian orbits for solar sail spacecraft at linear order are investigated in the Earth-Moon circular restricted three-body problem, where the third body is a solar sail. In particular, periodic orbits near the collinear libration points in the Earth-Moon system will be explored along with their applications. The dynamics are completely different from the Earth-Sun system in that the sun line direction constantly changes in the rotating frame but rotates once per synodic lunar month. Using an approximate, first order analytical solution to the nonlinear nonautonomous ordinary differential equations, periodic orbits can be constructed that are displaced above the plane of the restricted three-body system. This new family of orbits have the property of ensuring visibility of both the lunar far-side and the equatorial regions of the Earth, and can enable new ways of performing lunar telecommunications.

Keywords Periodic orbit · Solar sail · Circular restricted three-body problem

1 Introduction

Solar sailing technology appears as a promising form of advanced spacecraft propulsion, which can enable exciting new space-science mission concepts such as solar system exploration and deep space observation. Although solar sailing has been considered as a practical means of spacecraft propulsion only relatively recently,

J. Simo (✉) · C.R. McInnes
Department of Mechanical Engineering, University of Strathclyde, Glasgow, G1 1XJ, UK
e-mail: jules.simo@strath.ac.uk

C.R. McInnes
e-mail: colin.mcinnes@strath.ac.uk

J.A.T. Machado et al. (eds.), *Nonlinear Science and Complexity*,
DOI 10.1007/978-90-481-9884-9_19,

the fundamental ideas are by no means new (see McInnes [1] for a detailed description).

Solar sails can also be utilised for highly non-Keplerian orbits, such as closed orbits displaced high above the ecliptic plane (see Waters and McInnes [2]). Solar sails are especially suited for such non-Keplerian orbits, since they can apply a propulsive force continuously. This allows some exciting unique trajectories. In such trajectories, a sail will be used as communication satellites for high latitudes. For example, the orbital plane of the sail can be displaced above the orbital plane of the Earth, so that the sail can stay fixed above the Earth at some distance, if the orbital periods are equal. Orbits around the collinear points of the Earth-Moon system are of great interest because their unique positions are advantageous for several important applications in space mission design (see e.g. Szebehely [3] and Roy [4]). In the last decades several authors have tried to determine more accurate approximations (quasi-Halo orbits) of such equilibrium orbits. The orbits were first studied by Farquhar [5], Farquhar and Kamel [6], Breakwell and Brown [7], Howell [8]. Halo orbits near the collinear libration points in the Earth-Moon system are of great interest. If the orbit maintains visibility from Earth, a spacecraft on it can be used to assure communications between the equatorial regions of the Earth and the lunar far-side. Thus, the establishment of a bridge for radio communications is a crucial problem for incoming space missions, which plan to use the lunar far-side as a powerful observation point. McInnes [9] investigated a new family of displaced solar sail orbits near the Earth-Moon libration points. In Baoying and McInnes [10], the authors describe the new orbits which are associated with artificial lagrange points. These artificial equilibria have potential applications for future space physics and Earth observation missions. Most work has been done in the Sun-Earth system. In McInnes and Simmons [11], the authors investigate large new families of solar sail orbits, such as Sun-centered halo-type trajectories, with the sail executing a circular orbit of a chosen period above the ecliptic plane. In our study, we will demonstrate the possibility of such trajectories in the Earth-Moon system. Briefly, in the present study a new family of solar sail periodic orbits have been investigated in the Earth-Moon restricted three-body problem. The first-order approximation is introduced for the linearized system of equations. The Laplace transform is used to produce the first-order analytic solution of the out-of plane motion. We find families of periodic orbits above the ecliptic plane at linear order. It will be shown for example that, with a suitable sail attitude control program, a 3.5×10^3 km displaced, out-of-plane trajectory around the L_2 point may be executed with a sail accelerations of only $0.2\ \mathrm{mm\,s^{-2}}$ (see Fig. 2).

2 Solar Sail in the Earth-Moon Restricted Three-Body Problem

2.1 Qualitative Approach

In context of this work, we will assume that $m_1 = 1-\mu$ represents the larger primary (Earth), $m_2 = \mu$ the smaller primary (Moon) with $\mu = m_2/(m_1+m_2)$, and we will be concerned with the motion of the sail that has negligible mass (see Fig. 1).

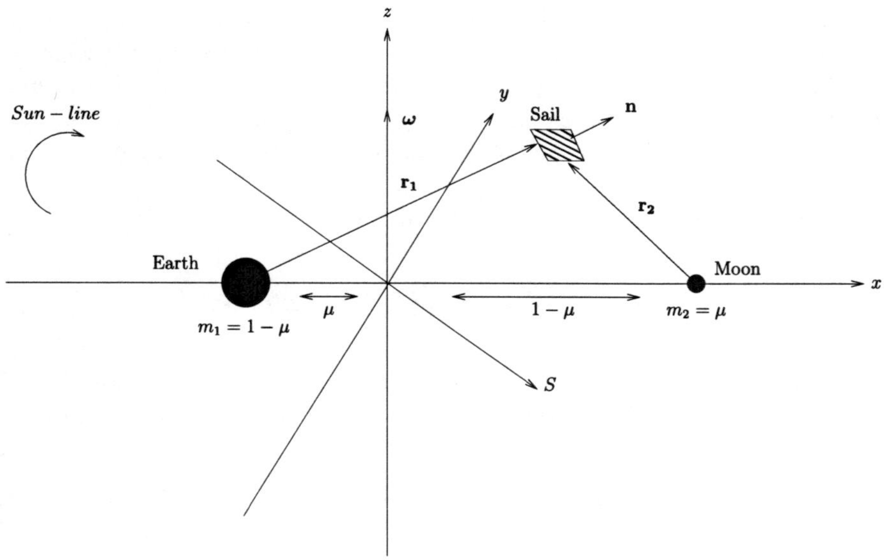

Fig. 1 Schematic geometry of the Earth-Moon restricted three-body problem

2.2 Equations of Motion in Presence of Solar Sail

The vector dynamical equation for the solar sail in a rotating frame of reference is described by

$$\frac{d^2\boldsymbol{r}}{dt^2} + 2\boldsymbol{\omega} \times \frac{d\boldsymbol{r}}{dt} + \nabla U(\boldsymbol{r}) = \boldsymbol{a}, \tag{1}$$

where $\boldsymbol{\omega} = \omega\hat{z}$ ($\hat{z}$ is a unit vector pointing in the direction of z) is the angular velocity vector of the rotating frame and $\boldsymbol{r}$ is the position vector of the solar sail relative to the center of mass of the two primaries. The three-body gravitational potential $U(\boldsymbol{r})$ and the solar radiation pressure acceleration $\boldsymbol{a}$ are defined by

$$U(\boldsymbol{r}) = -\left[\frac{1}{2}|\boldsymbol{\omega} \times \boldsymbol{r}|^2 + \frac{1-\mu}{r_1} + \frac{\mu}{r_2}\right], \qquad \boldsymbol{a} = a_0(\boldsymbol{S} \cdot \boldsymbol{n})^2\boldsymbol{n}, \tag{2}$$

where $\mu = 0.1215$ is the mass ratio for the Earth-Moon system, the sail position vectors w.r.t. the origin, m_1 and m_2 respectively, are defined as $\boldsymbol{r}_1 = [x + \mu, y, z\,]^T$ and $\boldsymbol{r}_2 = [x - (1-\mu), y, z]^T$, and a_0 is the magnitude of the solar radiation pressure force exerted on the sail. The unit normal to the sail $\boldsymbol{n}$ and the Sun line direction are given by

$$\boldsymbol{n} = [\cos(\gamma)\cos(\omega_\star t) \quad -\cos(\gamma)\sin(\omega_\star t) \quad \sin(\gamma)]^T, \tag{3}$$

$$\boldsymbol{S} = [\cos(\omega_\star t) \quad -\sin(\omega_\star t) \quad 0]^T, \tag{4}$$

where $\omega_\star = 0.923$ is the angular rate of the Sun line in the corotating frame in dimensionless synodic coordinate system. The dynamics of the sail in the neighborhood of the collinear libration points at $\boldsymbol{r}_L$ will be now investigated. Let a small displacement in $\boldsymbol{r}_L$ be $\delta\boldsymbol{r}$ such that $\boldsymbol{r} \to \boldsymbol{r}_L + \delta\boldsymbol{r}$. We will not consider the small annual changes in the inclination of the Sun line with respect to the plane of the system.

Also, the linear equations for the solar sail can be written as

$$\frac{d^2\delta\boldsymbol{r}}{dt^2} + 2\boldsymbol{\omega} \times \frac{d\delta\boldsymbol{r}}{dt} + \nabla U(\boldsymbol{r}_L + \delta\boldsymbol{r}) = \boldsymbol{a}(\boldsymbol{r}_L + \delta\boldsymbol{r}), \tag{5}$$

and retaining only the first-order term in $\delta\boldsymbol{r} = [\delta x, \delta y, \delta y]^T$ in a Taylor-series expansion, the gradient of the potential and the acceleration can be expressed as

$$\nabla U(\boldsymbol{r}_L + \delta\boldsymbol{r}) = \nabla U(\boldsymbol{r}_L) + \left.\frac{\partial \nabla U(\boldsymbol{r})}{\partial \boldsymbol{r}}\right|_{\boldsymbol{r}=\boldsymbol{r}_L} \delta\boldsymbol{r} + O(\delta\boldsymbol{r}^2), \tag{6}$$

$$\boldsymbol{a}(\boldsymbol{r}_L + \delta\boldsymbol{r}) = \boldsymbol{a}(\boldsymbol{r}_L) + \left.\frac{\partial \boldsymbol{a}(\boldsymbol{r})}{\partial \boldsymbol{r}}\right|_{\boldsymbol{r}=\boldsymbol{r}_L} \delta\boldsymbol{r} + O(\delta\boldsymbol{r}^2). \tag{7}$$

It is assumed $\nabla U(\boldsymbol{r}_L) = 0$, and the acceleration is constant with respect to the small displacement $\delta\boldsymbol{r}$, then

$$\left.\frac{\partial \boldsymbol{a}(\boldsymbol{r})}{\partial \boldsymbol{r}}\right|_{\boldsymbol{r}=\boldsymbol{r}_L} = 0. \tag{8}$$

The linear variational system associated with the collinear libration points at $\boldsymbol{r}_L$ can be determined through a Taylor series expansion by substituting (6) and (7) into (5)

$$\frac{d^2\delta\boldsymbol{r}}{dt^2} + 2\boldsymbol{\omega} \times \frac{d\delta\boldsymbol{r}}{dt} - K\delta\boldsymbol{r} = \boldsymbol{a}(\boldsymbol{r}_L), \tag{9}$$

where the matrix K is defined as

$$K = -\left[\left.\frac{\partial \nabla U(\boldsymbol{r})}{\partial \boldsymbol{r}}\right|_{\boldsymbol{r}=\boldsymbol{r}_L}\right]. \tag{10}$$

Using the matrix notation the linearized equation about the libration point (9) can be represented by the inhomogeneous linear system $\dot{\mathbf{X}} = A\mathbf{X} + \boldsymbol{b}(t)$, where the state vector $\mathbf{X} = (\delta\boldsymbol{r}, \delta\dot{\boldsymbol{r}})^T$, and $\boldsymbol{b}(t)$ is a 6×1 vector, which represents the solar sail acceleration.

The Jacobian matrix A has the general form

$$A = \begin{pmatrix} 0_3 & I_3 \\ K & \Omega \end{pmatrix}, \tag{11}$$

where I_3 is a identity matrix, and

$$\Omega = \begin{pmatrix} 0 & 2 & 0 \\ -2 & 0 & 0 \\ 0 & 0 & 0 \end{pmatrix}. \tag{12}$$

For convenience the sail attitude is fixed such that the sail normal vector $\boldsymbol{n}$, which is a unit vector that is perpendicular to the sail surface, points always along the direction of the Sun line with the following constraint $\boldsymbol{S} \cdot \boldsymbol{n} \geq 0$. Its direction is described by the pitch angle γ, which represents the sail attitude. This yields the linearized nondimensional equations of motion in component form of a solar sail near the collinear libration point

$$\ddot{\xi} - 2\dot{\eta} - U^o_{xx}\xi = a_0 \cos(\omega_\star t)\cos^3(\gamma), \tag{13}$$

$$\ddot{\eta} + 2\dot{\xi} - U^o_{yy}\eta = -a_0 \sin(\omega_\star t)\cos^3(\gamma), \tag{14}$$

$$\ddot{\zeta} - U^o_{zz}\zeta = a_0 \cos^2(\gamma)\sin(\gamma), \tag{15}$$

where U^o_{xx}, U^o_{yy}, and U^o_{zz} are the partial derivatives of the gravitational potential evaluated at the collinear libration points.

3 Solution of the Linearized Equations of Motion

The solution can be made to contain only oscillatory modes with the proper choice of initial conditions. Then, the solar sail will follow a periodic orbit about the libration point. Clearly, the out-of-plane motion (15) described by a driven harmonic oscillator is decoupled from the in-plane equations of motion (13)–(14). These conditions can be met by choosing a particular solution in the plane of the form (see Farquhar [12])

$$\xi(t) = \xi_0 \cos(\omega_\star t), \tag{16}$$

$$\eta(t) = \eta_0 \sin(\omega_\star t). \tag{17}$$

By inserting (16) and (17) in the differential equations, we obtain the linear system in ξ_0 and η_0,

$$\begin{cases} (U^o_{xx} - \omega_\star^2)\xi_0 - 2\omega_\star\eta_0 = a_0\cos^3(\gamma), \\ -2\omega_\star\xi_0 + (U^o_{yy} - \omega_\star^2)\eta_0 = -a_0\cos^3(\gamma). \end{cases} \tag{18}$$

Then the amplitudes ξ_0 and η_0 are given by

$$\xi_0 = a_0 \frac{(U^o_{yy} - \omega_\star^2 - 2\omega_\star)\cos^3(\gamma)}{(U^o_{xx} - \omega_\star^2)(U^o_{yy} - \omega_\star^2) - 4\omega_\star^2}, \tag{19}$$

$$\eta_0 = a_0 \frac{(-U^o_{xx} + \omega_\star^2 + 2\omega_\star)\cos^3(\gamma)}{(U^o_{xx} - \omega_\star^2)(U^o_{yy} - \omega_\star^2) - 4\omega_\star^2}, \tag{20}$$

and we have the equality

$$\frac{\xi_0}{\eta_0} = \frac{\omega_\star^2 + 2\omega_\star - U^o_{yy}}{-\omega_\star^2 - 2\omega_\star + U^o_{xx}}. \tag{21}$$

By applying the Laplace transform, the uncoupled out-of-plane ζ-motion defined by (15) can be solved. The transform version is obtained as

$$Z(s) = \frac{1}{s^2 - U^o_{zz}}\left(\dot{\xi}_0 + s\xi_0 + \frac{a_0\cos^2(\gamma)\sin(\gamma)}{s}\right). \tag{22}$$

Using Mathematica, we can find the inverse Laplace transform, which will be the general solution of the out-of-plane component

$$\begin{aligned}\zeta(t) &= \zeta_0\cos(\omega_\zeta t) + \dot{\zeta}_0|U^o_{zz}|^{-1/2}\sin(\omega_\zeta t)\\ &\quad + a_0\cos^2(\gamma)\sin(\gamma)|U^o_{zz}|^{-1}[U(t) - \cos(\omega_\zeta t)], \qquad (23)\\ &= U(t)a_0\cos^2(\gamma)\sin(\gamma)|U^o_{zz}|^{-1} + \dot{\zeta}_0|U^o_{zz}|^{-1/2}\sin(\omega_\zeta t)\\ &\quad + \cos(\omega_\zeta t)[\zeta_0 - a_0\cos^2(\gamma)\sin(\gamma)|U^o_{zz}|^{-1}], \qquad (24)\end{aligned}$$

where the nondimensional frequency is defined as $\omega_\zeta = |U^o_{zz}|^{1/2}$ and $U(t)$ is the unit step function. Specifically for the choice of the initial data $\dot{\zeta}_0 = 0$, (24) can be more conveniently expressed as

$$\begin{aligned}\zeta(t) &= U(t)a_0\cos^2(\gamma)\sin(\gamma)|U^o_{zz}|^{-1}\\ &\quad + \cos(\omega_\zeta t)[\zeta_0 - a_0\cos^2(\gamma)\sin(\gamma)|U^o_{zz}|^{-1}].\end{aligned} \tag{25}$$

The solution can be made to contain only the periodic oscillatory modes at an out-of-plane distance

$$\zeta_0 = a_0\cos^2(\gamma)\sin(\gamma)|U^o_{zz}|^{-1}. \tag{26}$$

Of course, this distance can be maximized by an optimal choice of the sail pitch angle determined by

$$\frac{d}{d\gamma^\star}\cos^2(\gamma^\star)\sin(\gamma^\star) = 0, \tag{27}$$

$$\gamma^\star = 35.264^\circ. \tag{28}$$

4 Numerical Integration of the Nonlinear Equations of Motion

This section is concerned with the computation of periodic orbits around the collinear libration points L_1 and L_2 in the Earth-Moon system. The linear approximation is only valid at small distances from the Lagrange points at linear order. Then, we should be able to find an out-of-plane halo-type motion around the Earth-Moon L_2 point. In order to model motion in the nonlinear case the linearized equations of motion are used to find a guess for the appropriate initial conditions to generate out-of-plane periodic (halo-type) trajectory around the collinear libration points. Thus, at the beginning the solutions given by analytical approximations are used as initial guess. These approximate analytical solutions are utilized in a numerical search to determine displaced periodic orbits in the full nonlinear model. However, the initial guess found is not good enough to close the orbit in the nonlinear system.

Our purpose, hereafter, is to apply a robust control approach, namely a time-delayed feedback control [13], which takes into account the solution found from the linearized dynamics to generate a periodic reference trajectory. For the nonlinear system

$$\dot{\boldsymbol{x}} = \boldsymbol{f}(\boldsymbol{x}, t), \qquad \boldsymbol{x}(t_0) = \boldsymbol{x}_0 \in \mathbb{R}^n. \tag{29}$$

We want to find a time-delayed feedback control ($\tau > 0$ is a delay-time)

$$\boldsymbol{u}(t) = -K(\mathbf{x}(t) - \boldsymbol{x}(t-\tau)), \tag{30}$$

to be added to $\boldsymbol{f}(\boldsymbol{x}, t)$ such that the controlled system orbit can track the target

$$\lim_{t\to\infty} \|\boldsymbol{x}(t) - \boldsymbol{x}(t-\tau)\| = 0. \tag{31}$$

Thus, the design problem is to determine the control gain matrix K to achieve the goal (31), such that

$$\dot{\boldsymbol{x}} = \boldsymbol{f}(\boldsymbol{x}, t) + K(\boldsymbol{x}(t) - \boldsymbol{x}(t-\tau)). \tag{32}$$

In order to obtain a one-month orbit, the delay-time will be $2\pi/\omega_\star$ and the matrix K is a scalar multiple of the identity matrix $I_{6\times 6}$ which is computed experimentally. The final trajectory that corresponds to the minimum feedback requirement will be used as a reference orbit (see Fig. 2).

It will be shown that the sail may execute an out-of-plane distance of 5×10^3 km with the semimajor and minor axes of 1.174×10^4 km ($\eta_0 = 3.051 \times 10^{-2}$) and 1.208×10^3 km ($\xi_0 = 3.140 \times 10^{-3}$) in the neighborhood of the cislunar libration point L_1. In the same way the trajectory around the L_2 point would be a narrow ellipse with semi-major and minor axes of 1.105×10^4 km ($\eta_0 = 2.876 \times 10^{-2}$) and 5.655×10^2 km ($\xi_0 = 1.471 \times 10^{-3}$) and a period of 28 days (synodic lunar month). Therefore the sail may be placed on such a trajectory by inserting it into a suitable elliptical trajectory about the L_2 point. The lunar far-side and the equatorial regions would be visible with only a sail acceleration of the order of 0.2 mm s^{-2}. This small performance solar sail could be used to demonstrate the use of the L_2 point for lunar far-side communications. At the Earth-Moon L_2 point, the same sail could be displaced into a modified equilibrium point 10^4 km following the trajectory with the

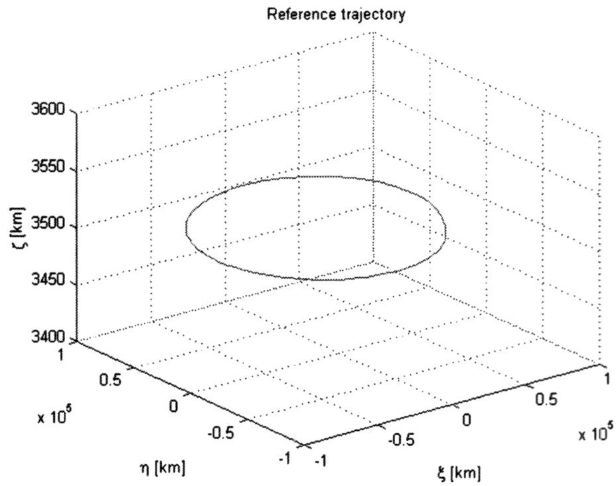

Fig. 2 Out-of-plane solar sail trajectory at the translunar libration point L_2

semimajor and minor axes of 3.116×10^4 ($\eta_0 = 8.21 \times 10^{-2}$) and 1.614×10^3 km ($\xi_0 = 4.197 \times 10^{-3}$).

Because of the instability of the collinear libration points, such orbits cannot be maintained without active control.

Future work will be focussed on linear control techniques to the problem of tracking and maintaining the solar sail on prescribed orbits. Also, we will use a linear feedback regulator (LQR) to track a periodic reference trajectory (Fig. 2) based on a time-delayed feedback mechanism.

5 Conclusion

In this study a new family of displaced solar sail orbits near the collinear libration points in the Earth-Moon system have been identified. The Laplace transform was used to give the general solution to the uncoupled out-of plane motion. It can be seen from this form of solution that once the sail is pitched from $\gamma = 0$ at $t = 0$, the motion of the sail is of the form of periodic oscillations at an out-of plane given distance. Also, by choosing this initial distance, the sail remains at this fixed distance. It was found that periodic orbits can be developed at linear order, that are displaced above the plane of the restricted three-body problem. These new families of highly non-Keplerian orbits that are unique to solar sails can enable new ways of performing space-science missions. Despite the fact that the accurate results by studying the linearized system may be found, it should be remembered that these solutions are only approximations to the real behaviour.

References

1. C.R. McInnes, *Solar Sailing: Technology, Dynamics and Mission Applications* (Springer Praxis, London, 1999)

2. T. Waters, C. McInnes, Periodic orbits above the ecliptic in the solar-sail restricted three-body problem. J. Guid. Control Dyn. **30**(3), 687–693 (2007)
3. V. Szebehely, *Theory of Orbits: The Restricted Problem of Three Bodies* (Academic Press, New York, 1967)
4. A.E. Roy, *Orbital Motion* (Institute of Phisics, Bristol, 2005)
5. R. Farquhar, The utilization of halo orbits in advanced lunar operations. NASA Technical Report (1971)
6. R.W. Farquhar, A.A. Kamel, Quasi-periodic orbits about the trans-lunar libration point. Celest. Mech. **7**, 458–473 (1973)
7. J.V. Breakwell, J.V. Brown, The 'halo' family of 3-dimensional periodic orbits in the earth-moon restricted 3-body problem. Celest. Mech. **20**, 389–404 (1979)
8. K.C. Howell, Three-dimensional, periodic, 'halo' orbits. Celest. Mech. **32**, 53–71 (1984)
9. C. McInnes, Solar sail trajectories at the lunar L_2 Lagrange point. J. Spacecraft Rocket **30**(6), 782–784 (1993)
10. H. Baoyin, C. McInnes, Solar sail halo orbits at the Sun-Earth artificial L_1 point. Celest. Mech. Dyn. Astron. **94**(2), 155–171 (2006)
11. C.R. McInnes, A.J.C. McDonald, J.F.C. Simmons, E.W. McDonald, Solar sail parking in restricted three-body systems. J. Guid. Control Dyn. **17**, 399–406 (1994)
12. R. Farquhar, The control and use of libration-point satellites, PhD Dissertation, Stanford University, 1968
13. K. Pyragus, Continuous control of chaos by self-controlling feedback. Phys. Lett. A **170**, 421–428 (1992)

Part IV
Mathematical Modeling of Nonlinear Structures in Bose-Einstein Condensates

Collisions of Discrete Breathers in Nonlinear Schrödinger and Klein–Gordon Lattices

J. Cuevas, A. Álvarez, F.R. Romero, and J.F.R. Archilla

Abstract Collisions between moving localized modes (moving breathers) in non-integrable lattices present a rich outcome. In this paper, some features of the interaction of moving breathers in Discrete Nonlinear Schrödinger and Klein–Gordon lattices, together with some plausible explanations, are exposed.

Keywords Discrete breathers · Moving breathers · Breather collisions · Klein-Gordon lattices · DNLS lattices

1 Introduction

Discrete breathers are localized modes that arise in nonlinear discrete lattices [1, 2]. Under certain conditions, these localized entities can move through the lattice and are denoted as moving breathers. Most of the models where discrete breathers exist are non-integrable, and, contrary to continuum nonlinear localized excitations (solitons) [3], collisions of discrete moving breathers exhibit a rich behaviour. One of the systems where discrete breathers have been extensively studied is the Discrete Nonlinear Schrödinger (DNLS) chain,

$$i\dot{u}_n + f(|u_n|^2)u_n + (u_{n+1} + u_{n-1} - 2u_n) = 0. \tag{1}$$

J. Cuevas (✉)
Grupo de Física No Lineal, Departamento de Física Aplicada I, EU Politécnica, Universidad de Sevilla, C/ Virgen de África, 7, 41011 Sevilla, Spain
e-mail: jcuevas@us.es

A. Álvarez · F.R. Romero
Grupo de Física No Lineal, Departamento de F.A.M.N. Facultad de Física, Universidad de Sevilla, Avda. Reina Mercedes s/n, 41012 Sevilla, Spain

J.F.R. Archilla
Grupo de Física No Lineal, Departamento de Física Aplicada I, ETSI Informática, Universidad de Sevilla, Avda. Reina Mercedes s/n, 41012 Sevilla, Spain

J.A.T. Machado et al. (eds.), *Nonlinear Science and Complexity*,
DOI 10.1007/978-90-481-9884-9_20,

Stationary solutions, oscillating with frequency ω_b, are given by $u_n(t) = \exp(i\omega_b t)v_n$, and can be calculated using methods based on the anti-continuous limit [4]. This equation has two conserved quantities: the Hamiltonian and the power (or norm). The latter is given by $P = \sum_n |u_n|^2$. Breathers can be put into movement by adding a momentum q so that the initial condition has the form $u_n(0) = v_n \exp(iqn)$.

Another system where discrete breathers appear is the Klein–Gordon chain,

$$\ddot{u}_n + V'(u_n) - C(u_{n+1} + u_{n-1} - 2u_n) = 0, \qquad (2)$$

where C is a coupling constant and $V(u)$ is the substrate potential. The method for calculating stationary breathers is similar to the used in the DNLS equation. Moving breathers are also generated by adding a momentum, a method analogous to the originally used in Ref. [5]. Both approaches consist in breaking the shift translational symmetry of the stationary breather.

The aim of this paper is to show some features of symmetric collisions between two breathers in DNLS and Klein–Gordon lattices, and pose some plausible explanations to the observed outcome.

2 DNLS Lattices

Symmetric collisions of discrete breathers in the DNLS equation with a cubic (or Kerr) nonlinearity, $f(|u|^2) = |u|^2$ in (1), were considered in Ref. [6]. The results of this paper show that, for small incoming velocities, breathers get trapped. For high velocities, the collision is quasi-elastic and breathers are refracted. This outcome is shown in Fig. 1 (left, 2 upmost panels). Both behaviours are separated by a critical value of the initial momentum, q_c. In that paper, inter-site and on-site collisions were considered. In the first case the value of q_c is an order of magnitude higher than the one of the second case.

Collisions in the DNLS equation with saturable nonlinearity, $f(|u|^2) = -\beta/(1 + |u|^2)$ in (1), were studied in Ref. [7] for $\beta = 2$, and for arbitrary values of β in

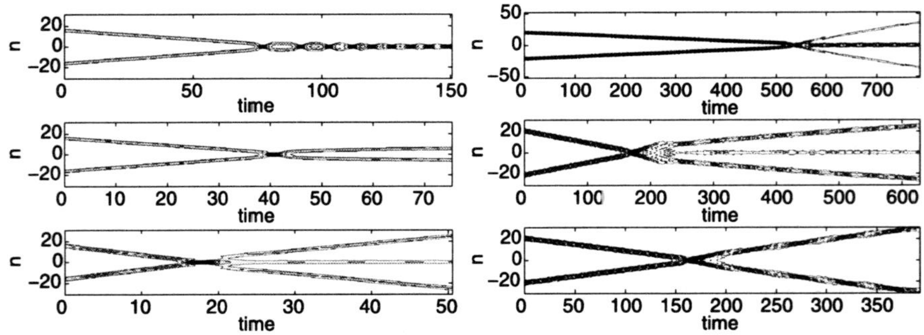

Fig. 1 Outcomes of discrete breathers collisions for saturable DNLS (*left*) and Klein–Gordon chains with $\omega_b = 0.8$ (*right*)

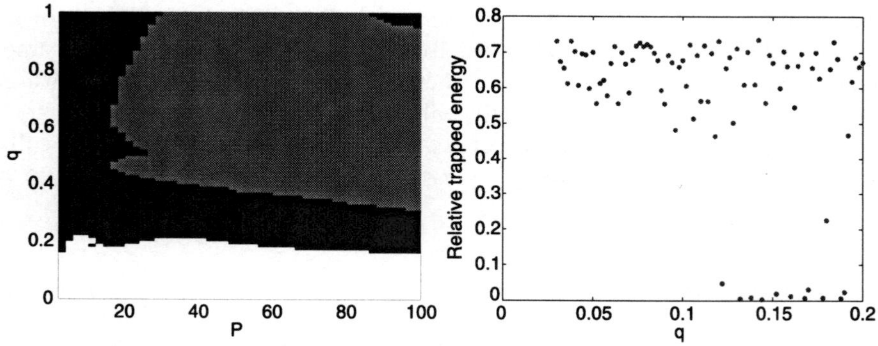

Fig. 2 (Color online) (*Left panel*) Different regimes observed for inter-site collisions in the saturable DNLS equation with $\beta = 2$. *Colours* represent the following: *white*—trapping; *black*—refraction; and *red*—breather creation. (*Right panel*) Relative trapped energy after a collision in the Klein–Gordon lattice with $C = 0.32$ and $\omega_b = 0.8$

Refs. [8–10]. This equation allows the existence of moving breathers with any arbitrarily high power, contrary to the cubic DNLS equation, where the power of the moving breathers is limited. In addition, three regimes are observed when moving breathers collide in this model, separated by two critical values, $q_{c1} < q_{c2}$. The order of magnitude of these critical values is the same for inter-site and on-site collisions. For $q < q_{c1}$, breathers get trapped; for $q \in (q_{c1}, q_{c2})$, breathers are refracted; and, for $q > q_{c2}$, breathers are refracted, while a large part of the energy remains trapped. This regime, also called *breather creation* takes place only if the power is high enough. Figure 1 (left) shows the above mentioned behaviours and Fig. 2 (left) depicts the parameter ranges for which these regimes are observed.

3 Klein–Gordon Lattices

Discrete breathers collisions in Klein–Gordon lattices have been considered in Ref. [11], where a Morse substrate potential, $V(u) = (\exp(-u) - 1)^2/2$, was chosen.[1] Two breather frequencies ω_b were chosen. For $\omega_b = 0.95$, which is close to the band of linear modes, the outcome is similar to the observed in the cubic DNLS equation, except for the fact that q_c has the same order of magnitude for both inter-site and on-site collisions. For $\omega_b = 0.8$, i.e. a frequency far from the linear modes band, the nonlinearity is high and a radically different outcome is observed. First of all, the collisions depend strongly on the phase, the outcome depending even on the initial distance between the incoming breathers. For this reason, it has no sense to distinguish between inter-site and on-site collisions. For small incoming velocities, partial trapping is observed. As Fig. 1 (right) shows, part of the energy is trapped

[1] Breather collisions in FPU lattices (i.e. Klein–Gordon lattices without substrate potential and nonlinear interaction potential) have been extensively studied in Ref. [12].

and another part is emitted as a pair of small-amplitude or high-amplitude moving breathers. For high velocities, the collision can be quasi-elastic although partial trapping is also observed. In consequence, there exists a critical value q_c below which no breather refraction (or equivalently, only partial trapping) is observed. Figure 2 (right) shows the relative trapped energy after the collision. For quasi-elastic collisions, this value drops to a value close to zero.

4 Interpretation

In this section, we provide some plausible explanations for the scenario of symmetric collisions observed in the previous sections. Some of the moving breathers features can be explained supposing that they behave as quasi-particles moving in a periodic potential, known as *Peierls-Nabarro potential*. In order to move a breather, some kinetic energy must be transferred to a stationary breather. The minimum value of this energy receives the name of Peierls–Nabarro barrier (PNB), and its value is the energy difference between an inter-site breather and an on-site breather, with the same power or action, for DNLS and Klein–Gordon breathers, respectively [13, 14]. Figure 3 (left) shows the PNB as a function of the power for the cubic and saturable DNLS. It can be observed that in the former, the PNB grows monotonically with the power, and, in consequence, there are no moving breathers for high values of the power. On the contrary, for the saturable DNLS, the PNB is bounded, and it is possible to find moving breathers for high powers. This is also related to the existence of non-radiating breathers (i.e. free of PNB) for the saturable DNLS [9, 10], and the non-existence of those entities in the cubic DNLS [16]. Figure 3 (right) shows the PNB as a function of the coupling constant C for Klein–Gordon lattices. It can be observed that the PNB is bounded, as in the saturable DNLS equation. The higher value of the PNB in the cubic DNLS equation is the reason for the large difference between the values of q_c when inter-site and on-site collisions are considered in that framework.

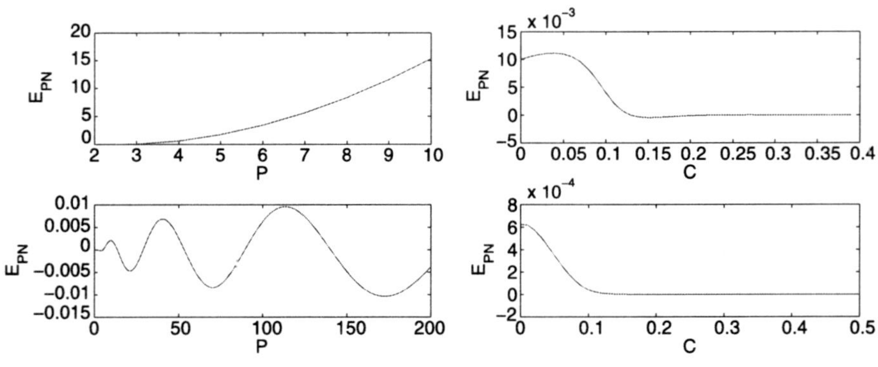

Fig. 3 (*Left panel*) Peierls–Nabarro barrier for saturable DNLS breathers with $\beta = 2$ (*top*) and cubic DNLS breathers (*bottom*). (*Right panel*) Peierls–Nabarro barrier for Klein–Gordon breathers with $\omega_b = 0.8$ (*top*) and $\omega_b = 0.95$ (*bottom*)

The scenario observed in the cubic DNLS equation was explained through a variational approach in Ref. [6]. Apart from this mathematical point of view, there is a rough physical explanation: when two breathers interact, they attract each other, creating a potential well. Thus, if their velocity is below the "escape" velocity of the potential well, they get trapped. If the velocity of the incoming breathers is above that critical value, they are refracted and no trapping is observed. The existence of the critical value q_c is explained by the increasing dependence of the incoming breathers velocity with q. This physical explanation is valid also for explaining the existence of q_{c1} in the saturable DNLS equation. In the Klein–Gordon equation for $\omega_b = 0.95$, which is close to the linear modes band, breathers can be approximated by envelope discrete breathers of the cubic DNLS equation [15]. Thus, the scenario should be similar to the observed in the cubic DNLS equation. This is not found for $\omega_b = 0.8$. Instead, no total trapping of the energy after the collision is observed, as two moving breathers escape from the potential well, additionally to the trapped one. Apart from this, there is a escape velocity which is not clearly defined. Contrary to the saturable DNLS breathers, in the Klein–Gordon case, the partial trapping can be observed for smaller velocities than those where reflection takes place.

The existence of partial trapping in Klein–Gordon lattices can be explained by means of an energetic balance. Incoming breathers can be described as quasiparticles with energy E_i, whereas the outgoing breathers energy is E_o; the trapped energy is denoted as U_{trap}. Then, neglecting the phonon radiation, it is fulfilled that $E_i = U_{\text{trap}}/2 + E_o$. For a localized trapped breather to be formed, its energy must be similar to that of a static breather. For a given value of C, the static breather energy possesses a maximum $\tilde{E}$, which corresponds to the minimum value of the frequency (i.e. to the resonance of the second harmonic of the breather frequency with the phonon band—see Fig. 4). Thus, $U_{\text{trap}} < \tilde{E}$, and, if $E_i > \tilde{E}/2$, there is an excess of energy that is emitted as two outgoing moving breathers. This is the case of breathers with $\omega_b = 0.8$. When $\omega_b = 0.95$, $E_i < \tilde{E}/2$, which explains the inex-

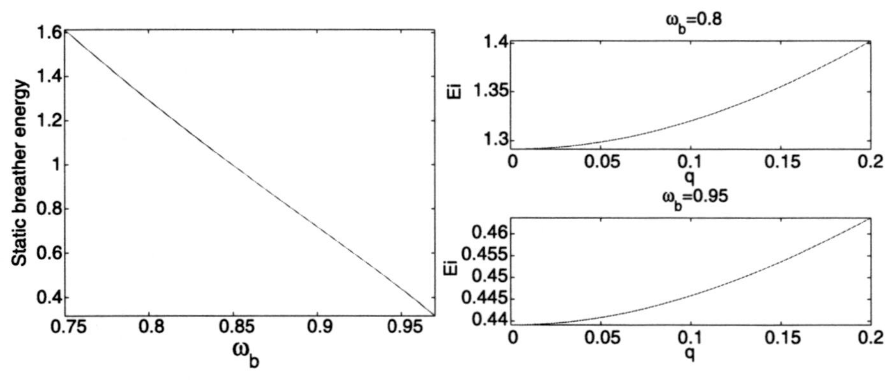

Fig. 4 (*Left panel*) Energy of static breathers in Klein–Gordon lattices with respect to ω_b for $C = 0.32$. The maximum value of the energies is $\tilde{E} = 1.5798$ and corresponds to $\omega_b = 0.7550$. (*Right panel*) Incoming moving breathers energies versus q for $\omega_b = 0.8$ and $\omega_b = 0.95$. Clearly, for $\omega_b = 0.95$, $E_i < \tilde{E}/2\ \forall q$

istence of outgoing breathers apart from the trapped one (see Fig. 4). This analysis cannot be done for the saturable DNLS as in this case the energy is not bounded [17].

References

1. S. Flach, C.R. Willis, Phys. Rep. **295**, 181 (1998)
2. P.G. Kevrekidis, K.O. Rasmussen, A.R. Bishop, Int. J. Mod. Phys. B **15**, 2833 (2001)
3. T. Dauxois, M. Peyrard, *Physics of Solitons* (Cambridge University Press, Cambridge, 2006)
4. J.L. Marín, S. Aubry, Nonlinearity **9**, 1501 (1996)
5. D. Chen, S. Aubry, G.P. Tsironis, Phys. Rev. Lett. **77**, 4776 (1996)
6. I.E. Papacharalampous, P.G. Kevrekidis, B.A. Malomed, D.J. Frantzeskakis, Phys. Rev. E **68**, 046604 (2003)
7. J. Cuevas, J.C. Eilbeck, Phys. Lett. A **358**, 15 (2006)
8. A. Maluckov, L. Hadzievski, M. Stepic, Eur. Phys. J. B **53**, 333 (2006)
9. T.R.O. Melvin, A.R. Champneys, P.G. Kevrekidis, J. Cuevas, Phys. Rev. Lett. **97**, 124101 (2006)
10. T.R.O. Melvin, A.R. Champneys, P.G. Kevrekidis, J. Cuevas, Physica D **237**, 551 (2008)
11. A. Álvarez, F.R. Romero, J. Cuevas, J.F.R. Archilla, Phys. Lett. A **372**, 1256 (2008)
12. Y. Doi, Phys. Rev. E **68**, 066608 (2003)
13. Yu.S. Kivshar, D.K. Campbell, Phys. Rev. E **48**, 3077 (1993)
14. J.A Sepulchre, in *Localization and Energy Transfer in Nonlinear Systems*, ed. by L. Vázquez, M.P. Zorzano, R.S. MacKay (World Scientific, Singapore, 2003)
15. O. Bang, M. Peyrard, Physica D **81**, 9 (1995)
16. O.F. Oxtoby, I.V. Barashenkov, Phys. Rev. E **76**, 036603 (2007)
17. J. Cuevas, J.C. Eilbeck, N.I. Karachalios, Discrete Contin. Dyn. Syst. **21**, 445 (2008)

Stability of BEC Systems in Nonlinear Optical Lattices

Lauro Tomio, F.K. Abdullaev, H.L.F. da Luz, and A. Gammal

Abstract The dynamics and stability of a Bose-Einstein Condensate, described by a two-dimensional nonlinear Schrödinger equation in a one-dimensional conservative plus dissipative nonlinear optical lattice, are investigated. In the case of focusing media (with attractive atomic systems), the collapse of the wave packet is arrested by the dissipative periodic nonlinearity. Confirmed by full numerical simulations, a stable soliton can exist in the defocusing media (repulsive case) with harmonic trap or linear periodic potential in one dimension (y-direction), with nonlinear optical lattice in the x-direction.

Keywords Two dimensional atomic condensate · Optical lattice · Localized states · Stability · Periodic dissipation

1 Introduction

We present some results on the stability of two-dimensional (2D) Bose-Einstein Condensate (BEC), with nonlinear periodic potential in one dimension (x-direction) and different configurations in the perpendicular y-direction: (a) harmonic trap or

L. Tomio (✉)
Instituto de Física Teórica, Universidade Estadual Paulista, Rua Pamplona, 145, 01405-900 São Paulo, Brazil
e-mail: tomio@ift.unesp.br

L. Tomio
Instituto de Física, Universidade Federal Fluminense, 24210-346 Niterói, RJ, Brazil

F.K. Abdullaev
Physical-Technical Institute, Uzbek Academy of Sciences, G. Mavlyanov str., 2-b, Tashkent-84, Uzbekistan

H.L.F. da Luz · A. Gammal
Instituto de Física, Universidade de São Paulo, 05315-970 São Paulo, Brazil

J.A.T. Machado et al. (eds.), *Nonlinear Science and Complexity*,
DOI 10.1007/978-90-481-9884-9_21,

no interaction [1]; and (b) linear periodic interaction. The influence of dissipative nonlinear optical lattice (NOL) on the dynamics and stability of solitons has been investigated in Ref. [1]. The role of dissipation can be crucial for the existence of solitons in multi dimensional NOLs, considering that homogeneous nonlinear dissipation can arrest collapse in the cubic focusing multi-dimensional nonlinear Schrödinger equation (NLSE) [2]. The dynamics of optical and matter wave solitons, under different management types for the system parameters, have been under intensive investigations [3, 4]. In atomic condensates periodic spatial management of nonlinearity can be realized by variation of the scattering length [5–12], as for example, by optically induced Feshbach resonances [13, 14]. Nonlinear optical lattices can be generated by two counter propagating laser beams [8–10]. However, the study of 1D nonlinear periodic potential in 2D NLSE shows that broad solitons are unstable. As reported in Ref. [15, 16], narrow solitons centered on the maximum of the lattice potential have a stability region so narrow that are physically unstable.

With optically induced Feshbach resonances it is also possible to generate mixture of conservative and dissipative NOL, as the imaginary part of the scattering length is non vanishing near the resonance. In view of that, we have studied a more general formalism with nonlinear dissipation. In the first part of the present contribution we consider conservative (a) attractive and (b) repulsive systems. The role of dissipation in the condensate stability is presented in the second part, before our conclusions.

Following Ref. [1], with dimensionless variables, the Gross-Pitaevskii (GP) equation for the wave function $u \equiv u(x, y, \tau)$ has the form

$$\begin{aligned} \mathrm{i}\frac{\partial u}{\partial \tau} &= -\frac{\partial^2 u}{\partial x^2} - \frac{\partial^2 u}{\partial y^2} - \{\gamma_0 + (\gamma_1 + \mathrm{i}\gamma_2)[1+\cos(\kappa x)]\}|u|^2 u \\ &\quad + V_0 \cos(k_2 y + \phi) u + \omega^2 y^2 u + \mathrm{i}\alpha_f u. \end{aligned} \tag{1}$$

γ_0 is related to the s-wave two-body scattering length a_s, with $\gamma_0 > 0$ ($\gamma_0 < 0$) for attractive (repulsive) condensates. The optical intensity is parametrized by $\gamma_1 (> 0)$; and dissipative effects, by γ_2. The wave-number parameters κ and k_2 adjust the lattice space in the x and y directions, respectively. The position of the minimum of the interaction $V_0 \cos(k_2 y + \phi)$ can be changed by the angle ϕ. For V_0 and ω, we consider cases with one of them equal zero and when both are zero. To extend the study of stability conditions in a few realistic cases, we have also studied the addition of a compression effect, by an adiabatic time variation of the scattering length background [5], by replacing the constant γ_0, as

$$\gamma_0 \to \gamma_0(\tau) = \gamma_0 \exp[2\alpha(\tau - \tau_c)\theta(\tau - \tau_c)]. \tag{2}$$

Compression effects can also be achieved by atoms feeding to condensate, that can be described by an additional term $\mathrm{i}\alpha_f u$ in the GP equation [17], as given in (1). Due to their similar role, we consider $\alpha_f = 0$ when $\alpha \neq 0$.

2 Conservative Systems ($\gamma_2 = 0$)

It is useful to describe shortly the solitons and their stability in the conservative case ($\gamma_2 = 0$). One dimensional conservative case has been considered by using a variational approach (VA) in Ref. [9, 10]. Using an exact approach, the 2D case with 1D nonlinear optical lattice was studied in Ref. [15, 16], where it is considered the case with attractive background nonlinearity ($\gamma_1 > 0$). Looking for perspective applications to BEC, we consider the 2D problem with 1D nonlinear optical lattice. Following Ref. [9, 10], we look for a VA solution of the form $u = v(x, y)\exp(-i\mu\tau)$, with the following Gaussian ansatz:

$$v(x, y) = A\exp\left(-[x^2/a_1^2 + y^2/a_2^2]\right), \qquad (3)$$

where A is the amplitude, and a_i ($i = 1, 2$) the corresponding widths in the x- and y-directions, respectively. The number of atoms is given by $N = \pi a_1 a_2 A^2$. From the averaged Lagrangian L and the corresponding Euler-Lagrange equations for the parameters, $\partial L/\partial N = 0$ and $\partial L/\partial a_{i=1,2} = 0$,

$$2\mu = \frac{1}{a_1^2} + \frac{1}{a_2^2} - \frac{N}{\pi a_1 a_2}(\tilde{\gamma}_0 + \gamma_1 e^{-\kappa^2 a_1^2/8}) + 2V_0\cos(\phi)e^{-k_2^2 a_2^2/4} + \omega^2 a_2^2,$$

$$N = \frac{4\pi a_2}{a_1}[\tilde{\gamma}_0 + \gamma_1 e^{-\kappa^2 a_1^2/8}(1 + \kappa^2 a_1^2/4)]^{-1}, \qquad (4)$$

$$\omega^2 a_2^4 + \frac{a_2^2}{a_1^2}\frac{[\tilde{\gamma}_0 + \gamma_1 e^{-\kappa^2 a_1^2/8} + (V_0/4)\cos(\phi)k_2^2 a_2 e^{-k_2^2 a_2^2/4}]}{[\tilde{\gamma}_0 + \gamma_1 e^{-\kappa^2 a_1^2/8}(1 + \kappa^2 a_1^2/4)]} - 1 = 0, \qquad (5)$$

where $\tilde{\gamma}_0 \equiv \gamma_0 + \gamma_1$, implying that attractive condensates are given by $\tilde{\gamma}_0 \equiv \gamma_1 + 1/2$, and repulsive ones by $\tilde{\gamma}_0 \equiv \gamma_1 - 1/2$.

2.1 Attractive Condensate ($\tilde{\gamma}_0 \equiv \gamma_1 + 1/2$)

The attractive case, corresponding to $\gamma_0 = 1/2$ and $\gamma_1 > 0$, has been investigated in Ref. [15, 16], for $\omega = 0$. In Fig. 1, we plot the corresponding results for the chemical potential μ as a function of N in the left-hand-side (lhs) frame and N as a function of a_1 in the right-hand-side (rhs) frame. Considering the Vakhitov-Kolokolov (VK) criterion [18] for the soliton stability, $d\mu/dN < 0$, we note that the solitons are unstable. This result is in agreement with the prediction given in Ref. [15, 16]. We note, from the VA results, that in the limit of large a_1 the system has a tendency to stabilize, indicating that with just a small trapping potential we can produce a stable region. This behavior can be verified by the VA results, as shown in Fig. 2 of Ref. [1]. The variational approach, besides an expected small quantitative shift, provides a good qualitative picture of the results when compared with full numerical predictions. If one is first concerned with the stability of the

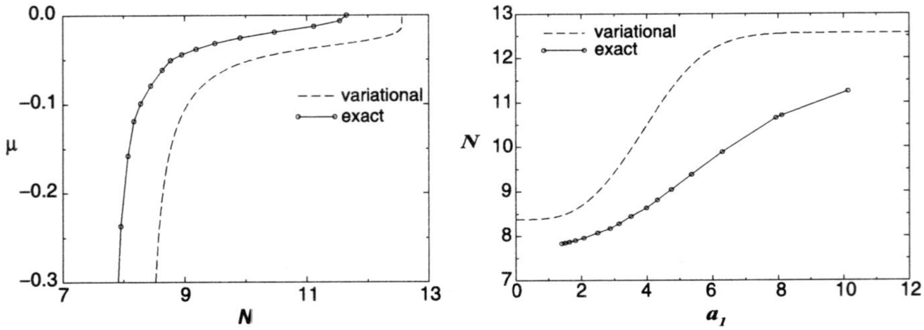

Fig. 1 Attractive case, with $\tilde{\gamma}_0 = 1$ and $\gamma_1 = 0.5$, presenting μ as a function of N (*lhs frame*) and N versus a_1 (*rhs frame*), using VA and full numerical calculations. The variational parameter for the width a_1 and $\sqrt{\langle x^2 \rangle}$ are related by $a_1 = \sqrt{2\langle x^2 \rangle}$

system (instead of the quantitative results of the observables), the VA provides a nice and reliable picture. We conclude that, for attractive condensates, the stability is not improved by increasing the strength of the lattice periodicity.

2.2 Repulsive Condensate ($\tilde{\gamma}_0 \equiv \gamma_1 - 1/2$)

For repulsive condensate, originally the particles have positive two-body scattering length, such that $\gamma_0 = -1/2$. So, given γ_1 (parameter of the spatial periodic variation of the scattering length), $\tilde{\gamma}_0 = \gamma_1 - 1/2$. If we also consider a negative background ($\tilde{\gamma}_0 < 0$), γ_1 is restricted to $0 < \gamma_1 < 1/2$. Some other limitations are applied, as the widths and N must be real and positive quantities. From the relation between a_2 and a_1, for $\omega = 0$:

$$e^{-\kappa^2 a_1^2/8} \geq 1/(2\gamma_1) - 1 \;\rightarrow\; a_{1,\max}^2 = -(8/\kappa^2)\ln[1/(2\gamma_1) - 1]. \tag{6}$$

This limit, $a_{1,\max}$, is necessary in order to have a_2 and N real and positive quantities for any values of ω. The cases with $\gamma_1 > 1/2$ are also allowed, without upper limit for a_1. However, such cases will correspond to positive background field, $\tilde{\gamma}_0 > 0$, which we have already considered.

In view of the above, the analytic VA limits are given by

$$a_2 \rightarrow \begin{cases} a_1, & \text{for } a_1 \ll 1; \\ 1/\sqrt{\omega}, & \text{for } a_1 = a_{1,\max}. \end{cases}$$

$$\mu \rightarrow \begin{cases} -1/a_1^2, & \text{for } a_1 \ll 1; \\ 1/(2a_1^2) + \omega, & \text{for } a_1 = a_{1,\max}. \end{cases}$$

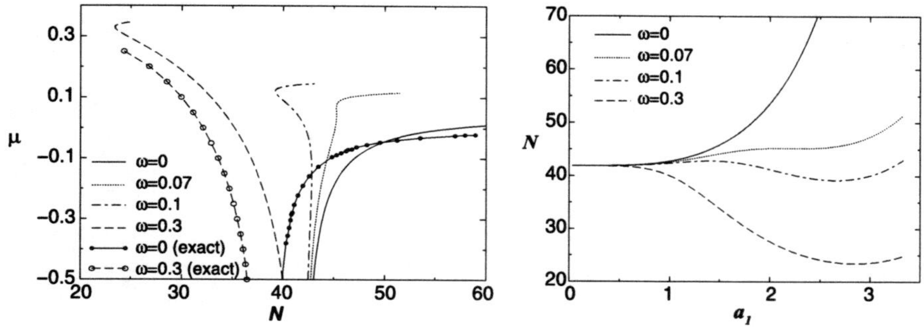

Fig. 2 Repulsive case, with $\tilde{\gamma}_0 = -0.1$ and $\gamma_1 = 0.4$, for μ versus N and N versus a_1. In both frames, we show the results using the VA, for $\omega = 0$, 0.07, 0.1 and 0.3. *In the lhs frame*, the exact PDE results are also shown in two cases: $\omega = 0$ (unstable) and $\omega = 0.3$ (stable). In this last case, near the region where the VA presents a small unstable branch ($22 < N < 25$). Exact numerical results are shown only for $N > 24$

$$N \to \begin{cases} 8\pi/(4\gamma_1 - 1), & \text{for } a_1 = 0; \\ 32\pi/[(1-2\gamma_1)\sqrt{\omega}\kappa^2 a_{1,\max}^3], & \text{for } a_1 = a_{1,\max}. \end{cases}$$

In Fig. 2, we plot N versus a_1 and the chemical potential μ versus N, for $\tilde{\gamma}_0 = -0.1$ and $\gamma_1 = 0.4$, considering VA and four values of ω (0, 0.07, 0.1, 0.3). In the case of $\omega = 0$, we also include results obtained from exact PDE calculations. Following the VK criterion for stability, $d\mu/dN < 0$, we notice that stable regions start to appear with $\omega \approx 0.1$. With $\omega > 0.3\kappa^2$ the unstable regions almost disappear. However, as one can observe in the rhs frame, the width a_1 is quite limited due to the condition (6). The observables μ and a_i depend on the wave parameter k of the spatial periodic variation of the atomic scattering length through some given scaling relations, as shown in Ref. [1]. Contrary to some discussions and conclusions of Ref. [15, 16], we note that specific values of the parameter k cannot affect the conclusions on stability. In such cases of conservative systems, the stability results from combined effects given by the parameters $\tilde{\gamma}_0$, γ_1 and ω. In conclusion, by taking $\omega = 0$, the optical lattice cannot stabilize the solutions, neither for repulsive nor for attractive condensates.

We also investigate the case with constant ω and different γ_1, in the repulsive case. From the results shown in Fig. 2, for $\gamma_1 = 0.4$, we found appropriate to consider $\omega = 0.07$, which has a marginal stability near $\mu \approx 0.05$. The results are given in Fig. 3, where we first observe that a larger γ_1 can help to allow the width a_1 to increase, within the limiting condition (6). However, the marginal stability remains for corresponding different values of the chemical potential. In order to keep the plots of Fig. 3 for different values of γ_1 in the same frames, we have normalize the number N such that it is equal to one when a_1 is zero. We determined the influence of the linear lattice to the dynamics and stability of a soliton in a 2D condensate, using full numerical (PDE) calculations and considering both nonlinear (x-direction) and linear (y-direction) lattices. Multidimensional solitons, in BEC with low dimensional linear optical lattices, are studied in Ref. [19]. This last one has the form

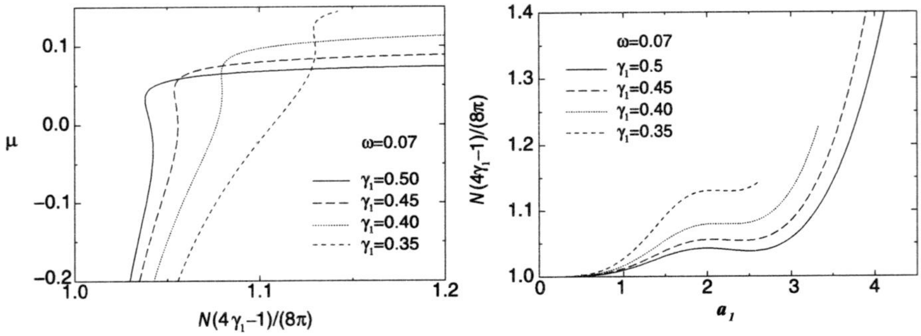

Fig. 3 Repulsive case—VA results with $\tilde{\gamma}_0 = -0.1$ and ω fixed to 0.07, considering $\gamma_1 = 0.35, 0.4,$ 0.45 and 0.5. *In the lhs frame* we have μ versus $N/N(a_1 = 0)$; and, *in the rhs frame*, $N/N(a_1 = 0)$ versus a_1

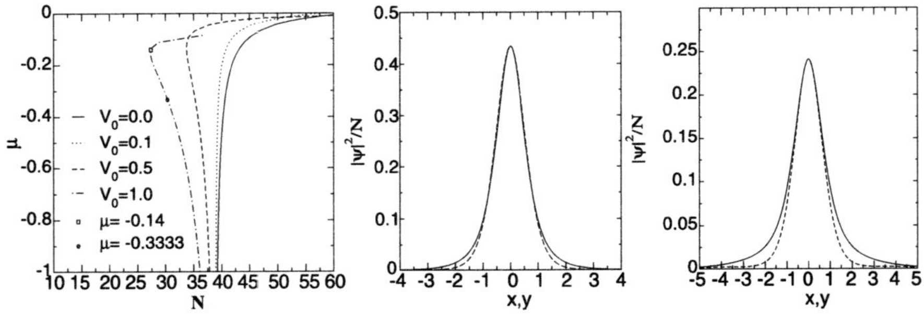

Fig. 4 *In the lhs frame*, we have μ versus N for a repulsive case with $\tilde{\gamma}_0 = -0.1$ and $\gamma_1 = 0.4$. Exact PDE results are shown for $V_0 = 0, 0.1, 0.5$ and 1.0. We can see stable branches appearing for V_0 =0.5 and 1.0. The initial soliton profiles, with cuts in the x (*solid*) and y (*dashed*) directions, for the stable (*center frame*) and unstable (*rhs frame*) regions are shown, respectively, for $\mu = -0.3333$ and $\mu = -0.14$, with $V_0 = 1.0$, $k_2 = 1.5$ and $\phi = \pi$. The μ points in terms of N are indicated *in the lhs frame*

$V(y) = V_0 \cos(k_2 y + \phi)$, where ϕ is an angle that only changes the position of the linear potential minimum and has no influence in the stabilization/dynamics. For now on we consider $\phi = \pi$, so that the soliton will be centered at origin. In Fig. 4 we have PDE results for μ versus N, with k_2=1.5 fixed. Following the VK criterion, we notice that stable regions start to appear near $V_0 \gtrsim 0.1$.

The initial soliton profiles for the two points ($\mu = -0.3333$ (star) and $\mu = -0.14$ (square)) are also shown in the center and rhs frames of Fig. 4. The first of them remains stable with the evolution in time. On the other hand, although the other soliton (square point) is quite close to the stable region, its wave function spreads to zero as the time goes to larger values (we consider τ up to $\sim$100).

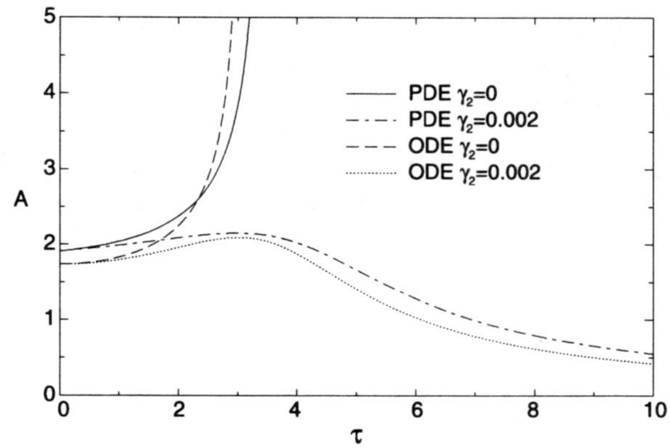

Fig. 5 Amplitude A as a function of τ, using VA and full numerical calculations. In both the cases, μ is fixed to the same value for $\tau = 0$, implying in a small shift of $A(0)$

3 Dissipative Systems

In the case of a homogeneous attractive condensate we can verify that the dissipation can arrest the collapse. Here we describe the influence of the dissipative lattice, with $\gamma_2 \neq 0$. To study the dynamics of localized states we apply the VA and full numerical simulations. The VA gives the system of five coupled ordinary differential equations for bright soliton parameters (see Ref. [1]). In Fig. 5 we present results of full numerical simulations for the evolution of matter-wave packets under combination of conservative and dissipative nonlinear optical lattices. As we can observe, the collapse is arrested by dissipative nonlinear optical lattices. The VA predictions are in good agreement with the full numerical results. It can be shown that the spreading out of the wave packet can be compensated by a proper variation in time (2) of the scattering length. Further, in Ref. [1], the role of compression effects have also being studied, in order to provide the stability conditions in realistic situations.

4 Conclusions

In resume, by considering 2D BEC, with conservative and dissipative nonlinear optical lattices in the x-direction, and with linear potentials (harmonic trap or optical lattice) in the y-direction, we have presented some investigations on the dynamics and stability of matter-wave solitons. In conclusion, for a conservative system, the nonlinear periodic lattice in the x-direction by itself (without a linear potential in the y-direction) cannot give stable solutions, satisfying the VK criterion [18]. We note that, for both attractive and repulsive systems, the general picture in respect to stability does not change when we keep ω fixed (zero or nonzero) and increase γ_1. Such periodic lattice in the x-direction cannot compensate the collapsing effect which results from the other dimension. Stable solutions can be obtained by controlling the soliton with a harmonic trap or a linear optical lattice in the y-direction. The stability in the last case, with linear optical lattice in the y-direction, was numerically verified in the repulsive case for certain region of parameters. When there is no potential in the y-direction, we also show that the collapsing effect can be arrested by a dissipative periodic nonlinearity.

Acknowledgements We thank M. Salerno for helpful discussions. For the partial support, F.Kh.A. is grateful to the Marie Curie IIF grant; and A.G., H.L.F. da Luz and L.T. thank FAPESP and CNPq (Brazil).

References

1. F.Kh. Abdullaev, A. Gammal, H.L.F. da Luz, L. Tomio, Phys. Rev. A **76**, 043611 (2007)
2. G. Fibich, SIAM. J. Appl. Math. **61**, 1680–1705 (2001)
3. F.Kh. Abdullaev, A. Gammal, A.M. Kamchatnov, L. Tomio, Int. J. Mod. Phys. B **19**, 3415–3473 (2005)
4. B.A. Malomed, *Soliton Management in Periodic Systems* (Springer, New York, 2006)
5. F.Kh. Abdullaev, M. Salerno, J. Phys. B **36**, 2851–2859 (2003)
6. F.Kh. Abdullaev, A. Gammal, L. Tomio, J. Phys. B **37**, 635–651 (2004)
7. G. Theocharis et al., Phys. Rev. A **72**, 033614 (2005)
8. F.Kh. Abdullaev, J. Garnier, Phys. Rev. A **72**, 061605(R) (2005)
9. H. Sakaguchi, B.A. Malomed, Phys. Rev. E **72**, 046610 (2005)
10. H. Sakaguchi, B.A. Malomed, Phys. Rev. E **73**, 026601 (2006)
11. J. Belmonte-Beitia, V.M. Perez-Garcia, V. Vekslerchik, Phys. Rev. Lett. **98**, 064102 (2007)
12. Y.V. Bludov, V.V. Konotop, Phys. Rev. A **74**, 043616 (2006)
13. P.O. Fedichev, Yu. Kagan, G.V. Schlyapnikov, J.T.M. Walraven, Phys. Rev. Lett. **77**, 2913–2916 (1996)
14. M. Theis et al., Phys. Rev. Lett. **93**, 123001 (2004)
15. G. Fibich, Y. Sivan, M.I. Weinstein, Physica D **217**, 31–57 (2006)
16. G. Fibich, Y. Sivan, M.I. Weinstein, Phys. Rev. Lett. **97**, 193902 (2006)
17. P.D. Drummond, K.V. Kheruntsyan, Phys. Rev. A **63**, 013605 (2001)
18. N.G. Vakhitov, A.A. Kolokolov, Radiophys. Quantum Electron. **16**, 783–789 (1973)
19. B.B. Baizakov, B.A. Malomed, M. Salerno, Phys. Rev. A **70**, 053613 (2004)

Nonlinear Schrödinger Equations with a Four-Well Potential in Two Dimensions: Bifurcations and Stability Analysis

C. Wang, G. Theocharis, P.G. Kevrekidis,
N. Whitaker, D.J. Frantzeskakis,
and B.A. Malomed

Abstract We report a full bifurcation diagram for trapped states in the two-dimensional (2D) nonlinear Schrödinger (NLS) equation with a symmetric four-well potential. Starting from the linear limit, we use a four-mode approximation to derive a system of ordinary differential equations, which makes it possible to trace the evolution of all trapped stationary modes, and thus to identify different branches of solutions bifurcating in the full NLS model. Their stability is examined within the framework of the linear stability analysis.

Keywords Nonlinear Schrödinger equations · Double-well potentials · Few-mode reduction · Linear stability analysis

1 Introduction

Over the last decade, the study of Bose-Einstein condensates (BECs) has been one of the focal points of experimental and theoretical investigations in atomic physics and nonlinear science [1]. Many of these studies have been devoted to macroscopic nonlinear structures that arise in BECs, following a similarity to nonlinear optics [2]. From the perspective of the nonlinear-wave theory, one of the appealing traits of such settings is the possibility to employ various external trapping potentials, such as harmonic, periodic, or combinations thereof [1, 2]. Hence, the existence, stability

C. Wang · G. Theocharis · P.G. Kevrekidis (✉) · N. Whitaker
Department of Mathematics and Statistics, University of Massachusetts,
Amherst, MA 01003-4515, USA
e-mail: kevrekid@math.umass.edu

D.J. Frantzeskakis
Department of Interdisciplinary Studies, Faculty of Engineering, Tel Aviv University,
Tel Aviv 69978, Israel

B.A. Malomed
Department of Physics, University of Athens, Panepistimiopolis, Zografos, Athens 157 84, Greece

J.A.T. Machado et al. (eds.), *Nonlinear Science and Complexity*,
DOI 10.1007/978-90-481-9884-9_22,

and evolution of nonlinear localized modes trapped in these potentials is a core subject of interest, especially within the context of the basic mean-field model, namely the Gross-Pitaevskii equation (GPE). The latter is actually a variant of the nonlinear Schrödinger (NLS) equation widely used in nonlinear optics [3], as well as in other branches of physics.

Among the potentials available in current BEC experiments, one that has drawn particular attention due to its relative simplicity and fundamental significance, is the double-well potential (DWP). A basic realization of DWP arises from the combination of a strong harmonic trap with a periodic, so-called, optical lattice (OL) potential [4]. Experiments with BECs loaded in DWPs have revealed a variety of fundamental phenomena, including Josephson oscillations for a relatively small number of atoms, or the macroscopic quantum self-trapping and asymmetric partition of atoms between the wells in condensates with a larger number of atoms [4]. On the theoretical side, DWPs have also stimulated studies of various topics, such as (*inter alia*) finite-mode reductions, analytical results for specially designed shapes of the potential, quantum depletion effects and a nonlinear DWP (alias pseudopotential) [5–15]. DWP settings have also been realized and studied in nonlinear optics, e.g., in self-guided twin-core laser beams in Kerr media [16] and optically-induced dual-core waveguiding structures in photorefractive crystals [17].

In the present work, our aim is to extend these considerations, which thus far were dealing with the one-dimensional (1D) geometry, to a two-dimensional (2D) setting. Although the trapping of quasi-2D BECs in harmonic traps combined with periodic OL potentials was studied previously [18, 19], here we use a few-mode reduction to deduce a discrete model for the setting based on a symmetric set of *four wells*. Using this discrete model, and starting from the linear limit, we analyze all possible trapped modes of the system in the case of repulsive interatomic interactions (i.e., for a self-defocusing cubic nonlinearity). Subsequently, we obtain the same modes from the full 2D GPE, and also report results of the linear stability analysis of these modes.

2 The Analytical Approach

Our starting point is the GPE in its usual rescaled form,

$$i\partial_t u = \hat{L}u + |u|^2 u - \mu u, \tag{1}$$

with operator $\hat{L} = -(1/2)\Delta + V(x, y)$ (Δ is the 2D Laplacian) and the potential $V(x, y) = (1/2)\Omega^2 r^2 + V_0[\cos(2kx) + \cos(2ky)]$, where $r^2 = x^2 + y^2$. The parameter μ represents the chemical potential (or the propagation constant in optics). We assume the following parameters of the potential: $\Omega = 0.21$, $V_0 = 0.5$ and $k = 0.3$, in which case the four smallest eigenvalues of operator $\hat{L}$ are $\omega_0 = 0.3585$, $\omega_1 = \omega_2 = 0.3658$ and $\omega_3 = 0.3731$.

To analyze the system, we resort to a natural four-mode reduction, based on eigenstates u_0 and $u_{1,2,3}$, i.e., the ground state and the first three excited states corresponding to ω_0 and $\omega_{1,2,3}$. Actually, it is more convenient to use a transformed basis, $(\Phi_0\ \Phi_1\ \Phi_2\ \Phi_3) = (u_0\ u_1\ u_2\ u_3)\mathbf{T}$, with an appropriate matrix $\mathbf{T}$. This basis is formed by an orthonormal set of functions Φ_j localized in each of the wells which constitute the four-well potential. Thus, we adopt the following decomposition of the full solution, $u(x,y,t) = \sum_{j=0}^{3} c_j(t)\Phi_j(x,y)$.

Substituting this expression into (1) and projecting onto the orthonormal basis $\{\Phi_0, \Phi_1, \Phi_2, \Phi_3\}$, we derive a system of four ODEs (ordinary differential equations) for complex amplitudes $c_j(t) \equiv \rho_j(t)e^{i\varphi_j(t)}$, $j = 0, 1, 2, 3$:

$$\begin{aligned} i\dot{c}_j &= \tilde{\omega}_j + A_j|c_j|^2 c_j + \sum_{k\neq j} B_{jk}\,(2|c_k|^2 c_j + c_k^2 c_j^*) \\ &\quad + \sum_{k\neq j} [D_{kj}|c_k|^2 c_k + D_{jk}\,(2|c_j|^2 c_k + c_j^2 c_k^*)] + \sum_{k\neq l\neq j\neq k} E_{kjl}\,(2|c_k|^2 c_l + c_k^2 c_l^*) \\ &\quad + \sum_{k\neq l\neq j\neq k} E_{jkl}\,(c_j^* c_k c_l + c_j c_k^* c_l + c_j c_k c_l^*) + G \sum_{\substack{k\neq l\neq m\neq k \\ k,l,m\neq j}} c_k^* c_l c_m, \end{aligned} \qquad (2)$$

with $k, l, m = 0, 1, 2, 3$, where $\tilde{\omega}_0 \equiv (1/4)[c_0(\gamma_0 - 4\mu) + c_1\gamma_1 + c_2\gamma_3 + c_3\gamma_2]$, $\tilde{\omega}_1 \equiv (1/4)[c_0\gamma_1 + c_1(\gamma_0 - 4\mu) + c_2\gamma_2 + c_3\gamma_3]$, $\tilde{\omega}_2 \equiv (1/4)[c_0\gamma_3 + c_1\gamma_2 + c_2(\gamma_0 - 4\mu) + c_3\gamma_1]$, $\tilde{\omega}_3 \equiv (1/4)[c_0\gamma_2 + c_1\gamma_3 + c_2\gamma_1 + c_3(\gamma_0 - 4\mu)]$, and $\gamma_0 \equiv \omega_0 + \omega_1 + \omega_2 + \omega_3$, $\gamma_1 \equiv \omega_0 + \omega_1 - \omega_2 - \omega_3$, $\gamma_2 \equiv \omega_0 - \omega_1 + \omega_2 - \omega_3$, $\gamma_3 \equiv \omega_0 - \omega_1 - \omega_2 + \omega_3$. Notice that $\gamma_1 = \gamma_2 = \omega_0 - \omega_3$ and $\gamma_3 \approx 0$, for the above-mentioned particular values of the eigenfrequencies.

These ODEs also involve a number of overlap integrals, *viz.* $A_n \equiv \iint \Phi_n^4 dxdy$, $n = 0, 1, 2, 3$; $B_{mn} \equiv \iint \Phi_m^2 \Phi_n^2 dxdy$, $m, n = 0, 1, 2, 3$, $m \neq n$; $D_{mn} \equiv \iint \Phi_m^3 \Phi_n dxdy$, $m, n = 0, 1, 2, 3$, $m \neq n$; $E_{lmn} \equiv \iint \Phi_l^2 \Phi_m \Phi_n dxdy$, $l, m, n = 0, 1, 2, 3$, with mutually different l, m, n; $G \equiv \iint \Phi_0\Phi_1\Phi_2\Phi_3 dxdy$. For our parameters (and, in fact, also for other cases corresponding to well-separated potential wells), other overlap integrals are much smaller than the A_n's. Then, neglecting these small overlap terms leads to an approximation of the tight-binding, alias discrete NLS type, which may actually be quite accurate, in comparison with numerical solutions of the underlying continuous NLS equation.

The resulting set of ODEs gives rise to both real and complex-valued solutions. Our analysis of the discrete model indicates that principal complex solutions in this setting correspond to discrete vortices with phase sets $\varphi_j = j(\pi/2)$ [20]. As this solution has been studied in detail elsewhere [18, 19, 21], we do not examine it further here. Instead, we focus on real solutions and a bifurcation diagram for them, based on the four algebraic equations derived from (2) by setting $\varphi_j = j\pi$. We will also compare this approximation with numerical solutions of the full GPE, and report results for the numerical linear-stability analysis for such solutions.

3 Numerical Results

The principal bifurcation diagram, shown in Fig. 1, displays the squared L^2 norm (physically representing the number of atoms in BEC, or the power in optics), N,

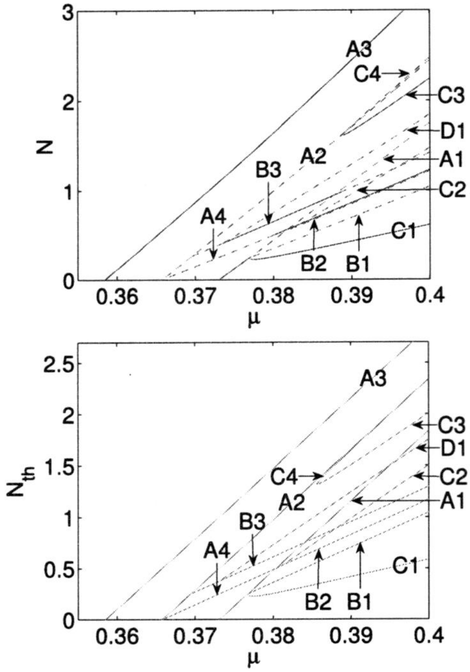

Fig. 1 Norm N of the numerically found solutions of (1) (*top*), and their counterparts predicted by the four-mode approximation (*bottom*), as a function of μ. The branches are explained in the text and their profiles and stability are detailed in subsequent figures. *Solid and dashed lines* denote stable and unstable solutions, respectively

as a function of chemical potential μ in (1). The version of the diagram shown in the bottom panel is obtained from the algebraic equations stemming from (2), and demonstrates very good agreement between the full numerical and semi-analytical results. Different branches labeled in the bifurcation diagram are displayed in detail in Fig. 2. We have also developed a symbolic representation in the form of 2×2 matrices, labeling different waveforms that arise in the diagram, as follows:

$$\mathrm{A1} \equiv \begin{pmatrix} -1 & 1 \\ 1 & -1 \end{pmatrix}, \quad \mathrm{A2} \equiv \begin{pmatrix} 1 & 1 \\ -1 & -1 \end{pmatrix}, \quad \mathrm{A3} \equiv \begin{pmatrix} 1 & 1 \\ 1 & 1 \end{pmatrix},$$

$$\mathrm{A4} \equiv \begin{pmatrix} 0 & 1 \\ -1 & 0 \end{pmatrix}, \quad \mathrm{B1} \equiv \begin{pmatrix} -1 & 1 \\ \varepsilon & -\varepsilon \end{pmatrix}, \quad \mathrm{B2} \equiv \begin{pmatrix} -\varepsilon & 1 \\ 1 & -\varepsilon \end{pmatrix},$$

$$\mathrm{B3} \equiv \begin{pmatrix} 1 & 1 \\ -\varepsilon & -\varepsilon \end{pmatrix}, \quad \mathrm{C1} \equiv \begin{pmatrix} 1 & -\varepsilon \\ -\varepsilon & \varepsilon \end{pmatrix}, \quad \mathrm{C2} \equiv \begin{pmatrix} -\varepsilon & 1 \\ 1 & -1+\varepsilon \end{pmatrix},$$

$$\mathrm{C3} \equiv \begin{pmatrix} -\varepsilon & 1 \\ 1 & 1+\varepsilon \end{pmatrix}, \quad \mathrm{C4} \equiv \begin{pmatrix} -1+\varepsilon & 1 \\ 1 & 1+\varepsilon \end{pmatrix}, \quad \text{and}$$

$$\mathrm{D1} \equiv \begin{pmatrix} 1+\varepsilon & 1 \\ -\varepsilon & -1+\varepsilon \end{pmatrix}.$$

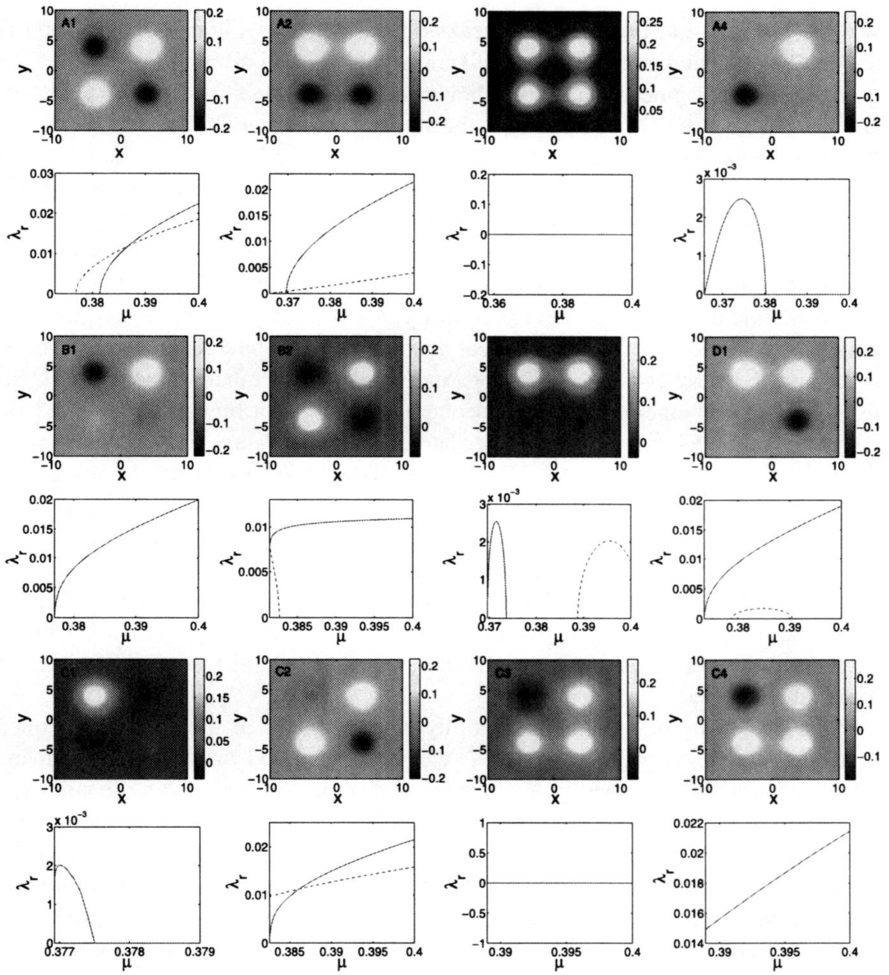

Fig. 2 *The first, third* and *fifth rows* show the wavefunctions of the bifurcation branches A1, A2, A3, A4, B1, B2, B3, D1, C1, C2, C3 and C4 at $\mu = 0.395$. *The second, fourth and sixth rows* show the real parts of unstable eigenvalues of the respective branches as a function of μ

In this representation, 1, -1 and 0 have the obvious meaning, by indicating that a particular well is or is not "populated", and its phase (0 and π in the cases of $+1$ and -1, respectively), when populated. The symbol ε, where $0 < \varepsilon \ll 1$, is used to denote either a small (but nonzero) population in one of the wells, or a symmetry-breaking effect (when some of the density peaks feature values $\pm 1 \pm \varepsilon$, as opposed to ± 1). The labeling of the branches has been chosen as follows: branches A1–A4 are the ones emerging directly from the linear limit; branches B1–B3 bifurcate from those of the A1–A4 types and feature two pairs of peaks with different amplitudes; branches C1–C4 have four different peaks which possess three different amplitudes, while branch D1 has all four of its peaks different. Clearly, branches B, C, and D are results of an elaborate sequence of symmetry-breaking bifurcations.

Examining the emergence of new branches and their stability in more detail, we conclude the following. Branch A3 is the stable ground state of the system. Branch A2 is immediately unstable, upon the departure from the linear limit, having a real eigenvalue pair. A supercritical pitchfork, leading to the emergence of branch B3, adds a second pair of real eigenvalues to the linearization around this branch. On the other hand, states A4 bifurcating from the same linear mode as A2, are unstable due to a quartet of eigenvalues near the linear limit, but become stabilized at higher values of chemical potential μ. On the contrary, A1 is stable near the linear limit, but becomes unstable due to a supercritical *and* a subcritical pitchfork, which involve, respectively, branches B1 and C1, apparently endowing branch A1 with *two* identical pairs of real eigenvalues. A subsequent bifurcation of branch B2 from A1 (at larger values of N) leads to yet another pair of real eigenvalues for the linearization around A1 (i.e., three pairs in total for N large enough). Although B2 remains unstable with two real pairs of eigenvalues, a subsequent bifurcation of C2 from it leads to a reduction of the number of such pairs to one, while the resulting branch C2 inherits the other two real pairs. Furthermore, C3 and C4 arise through a saddle-node bifurcation near $\mu = 0.389$, and one of them carries a real eigenvalue pair, while the other one does not (due to the nature of the bifurcation). Lastly, branch D1 bifurcates from B3 and always possesses a real eigenvalue pair, as well as potentially an eigenvalue quartet.

Note that many of the general stability characteristics of the above-mentioned branches, which are valid at *large N*, can be understood on the basis of a few simple principles developed in the context of discrete systems (from the so-called anti-continuum (AC) limit [20]) with the defocusing nonlinearity [22]. In particular, branches of states which feature a single-site shape in the AC limit should be stable, those based on a set of two nearest neighbors with opposite signs produce real eigenvalue pairs, while out-of-phase sets lead to complex eigenvalue quartets. Finally, sets of in-phase next-nearest-neighbors lead to real pairs, while sets of the same type, but with the out-of-phase arrangement, may produce quartets.

Naturally, these considerations do not apply to some of the asymmetric branches, such as C3 or D1, which cannot be examined in the AC setting. Nevertheless, this analysis provides a useful set of guidelines towards understanding most of the stability features of the fundamental branches.

4 Conclusions and Future Challenges

In this work, we have presented a systematic analysis of the emergence of a wide variety of branches in the two-dimensional GPE model including the self defocusing nonlinearity and four-well potential. Physically, this potential can be created as a combination of an isotropic parabolic trap with a 2D optical lattice. A four-mode decomposition has allowed us to identify different branches of solutions and bifurcations leading from four symmetric and antisymmetric linear modes of the system to asymmetric states, with the increase of the solution's norm. In addition to the existence of these nonlinear modes, we have investigated their linear stability, showing how pitchfork and saddle-node bifurcations are responsible for stability changes of the branches. We have also attempted to relate solutions for higher values of the norm to earlier findings stemming from the discrete NLS models.

It would be interesting to perform a similar analysis for the case of the self focusing nonlinearity. Furthermore, it would be interesting to investigate how these four-site configurations may be embedded into a larger potential pattern, with 9 or 16 wells, and examine whether the symmetry-breaking bifurcations considered above are sustained within the larger pattern. In this context, a relevant conjecture that would be worthwhile proving is that in an infinite periodic lattice formed by potential wells, the nonlinearity can support 2D solitons and localized vortices with various symmetries, but not confined asymmetric states [23].

References

1. L.P. Pitaevskii, S. Stringari, *Bose-Einstein Condensation* (Oxford University Press, Oxford, 2003)
2. P.G. Kevrekidis, D.J. Frantzeskakis, R. Carretero-González (eds.), *Emergent Nonlinear Phenomena in Bose-Einstein Condensates. Theory and Experiment* (Springer, Berlin, 2008)
3. Yu.S. Kivshar, G.P. Agrawal, *Optical Solitons: From Fibers to Photonic Crystals* (Academic Press, San Diego, 2003)
4. M. Albiez et al., Phys. Rev. Lett. **95**, 010402 (2005)
5. S. Raghavan et al., Phys. Rev. A **59**, 620 (1999)
6. S. Raghavan, A. Smerzi, V.M. Kenkre, Phys. Rev. A **60**, R1787 (1999)
7. E.A. Ostrovskaya et al., Phys. Rev. A **61**, 031601(R) (2000)
8. K.W. Mahmud, J.N. Kutz, W.P. Reinhardt, Phys. Rev. A **66**, 063607 (2002)
9. V.S. Shchesnovich, B.A. Malomed, R.A. Kraenkel, Physica D **188**, 213 (2004)
10. D. Ananikian, T. Bergeman, Phys. Rev. A **73**, 013604 (2006)
11. P. Ziń et al., Phys. Rev. A **73**, 022105 (2006)
12. T. Kapitula, P.G. Kevrekidis, Nonlinearity **18**, 2491 (2005)
13. G. Theocharis et al., Phys. Rev. E **74**, 056608 (2006)
14. D.R. Dounas-Frazer, L.D. Carr, arXiv:quant-ph/0610166
15. T. Mayteevarunyoo, B.A. Malomed, G. Dong, Phys. Rev. A **78**, 053601 (2008)
16. C. Cambournac et al., Phys. Rev. Lett. **89**, 083901 (2002)
17. P.G. Kevrekidis et al., Phys. Lett. A **340**, 275 (2005)
18. K.J.H. Law et al., Phys. Rev. A **77**, 053612 (2008)
19. K.J.H. Law et al., J. Phys. B **41**, 195303 (2008)
20. D.E. Pelinovsky, P.G. Kevrekidis, D.J. Frantzeskakis, Physica D **212**, 20 (2005)
21. T. Kapitula, P.G. Kevrekidis, D.J. Frantzeskakis, Chaos **18**, 023101 (2008)
22. P.G. Kevrekidis, H. Susanto, Z. Chen, Phys. Rev. E **74**, 066606 (2006)
23. R. Driben et al., Phys. Rev. E **76**, 066604 (2007)

Bose-Einstein Condensates and Multi-Component NLS Models on Symmetric Spaces of BD.I-Type. Expansions over Squared Solutions

V.S. Gerdjikov, D.J. Kaup, N.A. Kostov, and T.I. Valchev

Abstract A special class of multicomponent NLS equations, generalizing the vector NLS and related to the **BD.I**-type symmetric are shown to be integrable through the inverse scattering method (ISM). The corresponding fundamental analytic solutions are constructed thus reducing the inverse scattering problem to a Riemann-Hilbert problem. We introduce the minimal sets of scattering data $\mathfrak{T}$ which determines uniquely the scattering matrix and the potential Q of the Lax operator. The elements of $\mathfrak{T}$ can be viewed as the expansion coefficients of Q over the 'squared solutions' that are natural generalizations of the standard exponentials. Thus we demonstrate that the mapping $\mathfrak{T} \to Q$ is a generalized Fourier transform. Special attention is paid to two special representatives of this MNLS with three-component and five components which describe spinor ($F = 1$ and $F = 2$, respectively) Bose-Einstein condensates.

Keywords Multicomponent nonlinear Schrödinger equations · Inverse scattering method · Generalized Fourier transform

V.S. Gerdjikov (✉) · N.A. Kostov · T.I. Valchev
Institute for Nuclear Research and Nuclear Energy, Bulgarian Academy of Sciences, 72 Tsarigradsko chaussée, 1784 Sofia, Bulgaria
e-mail: gerjikov@inrne.bas.bg

N.A. Kostov
e-mail: nakostov@inrne.bas.bg

T.I. Valchev
e-mail: valtchev@inrne.bas.bg

D.J. Kaup
Department of Mathematics, University of Central Florida, Orlando, FL 32816-1364, USA
e-mail: kaup@mail.ucf.edu

J.A.T. Machado et al. (eds.), *Nonlinear Science and Complexity*,
DOI 10.1007/978-90-481-9884-9_23, © Springer Science+Business Media B.V. 2011

1 Introduction

Bose-Einstein condensates (BEC) with internal degrees of freedom, the so-called spinor BEC have attracted much attention experimentally and theoretically in recent years. Spinor BEC opens up a new paradigm, where the order parameter of condensates is described by a multicomponent vector. This situation can be realized by optically trapping cold atoms where all hyperfine states are liberated, while magnetic trapping freezes its freedom. So far ^{23}Na (the hyperfine state $F=1$), and ^{87}Rb ($F=2$) are extensively investigated, see [1–12] and the references therein. We consider BECs of alkali atoms with a hyperfine structure elongated in x direction and confined in the transverse directions y, z by purely optical means. Thus the assembly of atoms in the hyperfine state of spin F can be described by a normalized spinor wave vector with $2F+1$ components $\boldsymbol{\Phi}(x,t) = (\Phi_F(x,t), \Phi_{F-1}(x,t), \dots, \Phi_{-F}(x,t))^T$ whose components are labelled by the values of $m_F = F, \dots, 1, 0, -1, \dots, -F$.

The main tool for investigating BEC is the Gross-Pitaevski (GPE) equation. In the one-dimensional approximation the GPE system goes into a multicomponent nonlinear Schrödinger (MNLS) equation in 1D x-space [6]:

$$i\frac{\partial \boldsymbol{\Phi}}{\partial t} = \frac{\delta E_{\mathrm{GP}}[\boldsymbol{\Phi}]}{\delta \boldsymbol{\Phi}^*}, \tag{1}$$

where for $F=1$ the energy functional is given by:

$$E_{\mathrm{GP}} = \int dx \left\{ \frac{\hbar^2}{2m}|\partial_x \boldsymbol{\Phi}|^2 + \bar{c}\Big[|\Phi_1|^4 + |\Phi_{-1}|^4 + 2|\Phi_0|^2(|\Phi_1|^2 + |\Phi_{-1}|^2)\Big] \right.$$
$$\left. + (\bar{c}_0 - \bar{c}_2)|\Phi_1|^2|\Phi_{-1}|^2 + \frac{\bar{c}_0}{2}|\Phi_0|^4 + \bar{c}_2(\Phi_1^*\Phi_{-1}^*\Phi_0^2 + \Phi_0^{*2}\Phi_1\Phi_{-1}) \right\}. \tag{2}$$

The effective 1D couplings $\bar{c} = (\bar{c}_0 + \bar{c}_2)/2$, $\bar{c}_0 = c_0/2a_\perp^2$, $\bar{c}_2 = c_2/2a_\perp^2$, where $a_\perp$ is the size of the transverse ground state. In this expression, $c_0 = \pi\hbar^2(a_0 + 2a_2)/3m$, $c_2 = \pi\hbar^2(a_2 - a_0)/3m$, where a_f are the s-wave scattering lengths for the channel of total hyperfine spin f and m is the mass of the atom. We consider special (integrable) choice for the coupling constants $\bar{c}_0 = \bar{c}_2 \equiv -c < 0$. This situation corresponds to attractive mean-field interaction and ferromagnetic spin-exchange interaction. We will use dimensionless form: $\boldsymbol{\Phi} \to \{\Phi_1, \sqrt{2}\Phi_0, \Phi_{-1}\}^T$, where time and length are measured respectively in units of $\bar{t} = \hbar a_\perp/c$ and $\bar{x} = \hbar\sqrt{a_\perp/2mc}$.

For $F=2$ the energy functional is defined by[1–3, 5, 11]

$$E_{\mathrm{GP}}[\boldsymbol{\Phi}] = \int_{-\infty}^{\infty} dx \left(\frac{\hbar^2}{2m}|\partial_x \boldsymbol{\Phi}|^2 + \frac{\varepsilon c_0}{2} n^2 + \frac{c_2}{2}\mathbf{f}^2 + \frac{\varepsilon c_4}{2}|\Theta|^2 \right), \tag{3}$$

where $\varepsilon = \pm 1$. The number density n and the singlet-pair amplitude Θ are defined by [3, 5, 11]

$$n = (\boldsymbol{\Phi}, \boldsymbol{\Phi}^*) = \sum_{\alpha=-2}^{2} \Phi_\alpha \Phi_\alpha^*, \qquad \Theta = (\boldsymbol{\Phi}, s_0\boldsymbol{\Phi}) = 2\Phi_2\Phi_{-2} - 2\Phi_1\Phi_{-1} + \Phi_0^2.$$

The coupling constants c_i are real and can be expressed in terms of a transverse confinement radius and a linear combination of the s-wave scattering lengths of atoms [6, 7, 10] and **f** describe spin densities [11]. Choosing $c_2 = 0$, $c_4 = 1$ and $c_0 = -2$ we obtain an integrable version of the corresponding GPE equation by means of the inverse scattering transform method [6, 13].

The aim of present paper is to show that both GPE energy functionals (2), (3) correspond to integrable MNLS models [14] related to symmetric spaces [15] of **BD.I**-type: $\simeq SO(n+2)/SO(2)\times SO(n)$ with $n=3$ and $n=5$ respectively. In Sect. 2 we formulate the Lax representations and the explicit form of MNLS models, generalizing the vector NLS for any n. In Sect. 3 we construct the fundamental analytic solutions of the corresponding Lax operator L and reduce the inverse scattering problem (ISP) for L to a Riemann-Hilbert problem (RHP). We also introduce the minimal sets of scattering data $\mathfrak{T}_i$ each of which allow one to reconstruct both the scattering matrix $T(\lambda)$ and the corresponding potential $q(x,t)$. In Sect. 4 we explain that the ISM for this class of Lax operators can also be considered as a generalized Fourier transform. To this end we use the expansions of $q(x)$ and $\mathrm{ad}_J^{-1}\delta q$ over the 'squared solutions' of L. In Sect. 5 we briefly discuss the fundamental properties of these MNLS type equations.

2 MNLS Equations for BD.I. Series of Symmetric Spaces

MNLS equations for the **BD.I** series of symmetric spaces have a Lax representation $[L,M]=0$ as follows

$$L\psi(x,t,\lambda) \equiv i\partial_x\psi + (q(x,t) - \lambda J)\psi(x,t,\lambda) = 0. \tag{4}$$

$$M\psi(x,t,\lambda) \equiv i\partial_t\psi + (V_0(x,t) + \lambda V_1(x,t) - \lambda^2 J)\psi(x,t,\lambda) = 0, \tag{5}$$

$$V_1(x,t) = q(x,t), \qquad V_0(x,t) = i\,\mathrm{ad}_J^{-1}\frac{dq}{dx} + \frac{1}{2}\left[\mathrm{ad}_J^{-1}q, q(x,t)\right]. \tag{6}$$

We consider $n = 2r+1$ and choose $q(x,t) = \sum_{\alpha\in\Delta_1^+}(q_\alpha E_\alpha + p_\alpha E_{-\alpha})$ where the set of roots $\Delta_1^+ = \{e_1 - e_2, \dots, e_1 - e_r, e_r, e_1 + e_r, \dots, e_1 + e_2\}$. For the typical representation we have the matrix form:

$$q(x,t) = \begin{pmatrix} 0 & \mathbf{q}^T & 0 \\ \mathbf{p} & 0 & s_0\mathbf{q} \\ 0 & \mathbf{p}^T s_0 & 0 \end{pmatrix}, \qquad J = \mathrm{diag}(1, 0, \dots, 0, -1). \tag{7}$$

The n-component vectors $\mathbf{q} = (q_2, \dots, q_n)^T$ are formed by the coefficients q_α as follows: $q_k \equiv q_{e_1 - e_k}$, $q_{r+1} \equiv q_{e_1}$ and $q_{n+1-k} \equiv q_{e_1+e_k}$, $k = 1, \dots, r$; the vector $\mathbf{p} = (p_2, \dots, p_n)^T$ is formed analogously. The matrix $s_0 = S_0^{(n)}$ enters in the definition of $so(n)$, i.e. $X \in so(n)$, if $X + S_0^{(n)} X^T S_0^{(n)} = 0$, and for $n = 2r + 1$:

$$S_0^{(n)} = \sum_{s=1}^{n+1} (-1)^{s+1} E_{s,n+1-s}^{(n)}, \tag{8}$$

with this definition of orthogonality the Cartan subalgebra generators are represented by diagonal matrices. By $E_{sp}^{(n)}$ above we mean $n \times n$ matrix whose matrix elements are $(E_{sp}^{(n)})_{ij} = \delta_{si}\delta_{pj}$. In terms of these notations the generic MNLS type equations connected to **BD.I.** acquire the form

$$i\mathbf{q}_t + \mathbf{q}_{xx} + 2(\mathbf{q}, \mathbf{p})\mathbf{q} - (\mathbf{q}, s_0\mathbf{q}) s_0 \mathbf{p} = 0, \tag{9}$$

$$i\mathbf{p}_t - \mathbf{p}_{xx} - 2(\mathbf{q}, \mathbf{p})\mathbf{p} + (\mathbf{p}, s_0\mathbf{p}) s_0 \mathbf{q} = 0. \tag{10}$$

This equation allows the typical reduction $\mathbf{p} = \mathbf{q}^*$. The Hamiltonian for these MNLS equations (9)–(10) is given by

$$H_{\mathrm{MNLS}} = \int_{-\infty}^{\infty} dx \big((\partial_x \mathbf{q}, \partial_x \mathbf{q}^*) - (\mathbf{q}, \mathbf{q}^*)^2 + (\mathbf{q}, s_0\mathbf{q})(\mathbf{q}^*, s_0\mathbf{q}^*) \big). \tag{11}$$

For $r = 2$ we introduce the variables $\Phi_1 = q_2$, $\Phi_0 = q_3/\sqrt{2}$, $\Phi_{-1} = q_4$; for $r = 3$ we set $\Phi_2 = q_2$, $\Phi_1 = q_3$, $\Phi_0 = q_4$, $\Phi_{-1} = q_5$ and $\Phi_{-2} = q_6$. This reproduces the action functionals E_{GP} for $F = 1$ and $F = 2$.

3 The Inverse Scattering Problem

Solving the direct and the inverse scattering problem (ISP) for L uses the Jost solutions which are defined by, see [16] and the references therein

$$\lim_{x \to -\infty} \phi(x,t,\lambda) e^{i\lambda J x} = \mathbb{1}, \qquad \lim_{x \to \infty} \psi(x,t,\lambda) e^{i\lambda J x} = \mathbb{1} \tag{12}$$

and the scattering matrix $T(\lambda,t) \equiv \psi^{-1}\phi(x,t,\lambda)$. The choice of J and the fact that the Jost solutions and $T(\lambda,t)$ take values in the group $SO(n+2)$ means that we can use the following block-matrix structure of $T(\lambda,t)$

$$T = \begin{pmatrix} m_1^+ & -\mathbf{b}^{-T} & c_1^- \\ \mathbf{b}^+ & \mathbf{T}_{22} & -s_0\mathbf{B}^- \\ c_1^+ & \mathbf{B}^{+T} s_0 & m_1^- \end{pmatrix}, \qquad \hat{T} = \begin{pmatrix} m_1^- & \mathbf{b}^{-T} & c_1^- \\ -\mathbf{B}^+ & \hat{\mathbf{T}}_{22} & s_0\mathbf{b}^- \\ c_1^+ & -\mathbf{b}^{+T} s_0 & m_1^+ \end{pmatrix}, \tag{13}$$

where $\mathbf{b}^\pm(\lambda,t)$ and $\mathbf{B}^\pm(\lambda,t)$ are n-component vectors, $\mathbf{T}_{22}(\lambda)$ and $\mathbf{m}^\pm(\lambda)$ are $n \times n$ block matrices, and $m_1^\pm(\lambda)$, $c_1^\pm(\lambda)$ are scalars. Such parametrization is compatible with the generalized Gauss decompositions of $T(\lambda)$.

$$T(\lambda,t) = T_J^- D_J^+ \hat{S}_J^+ = T_J^+ D_J^- \hat{S}_J^-, \tag{14}$$

$$T_J^- = \begin{pmatrix} 1 & 0 & 0 \\ \boldsymbol{\rho}^+ & \mathbb{1} & 0 \\ c_1^{'',+} & \boldsymbol{\rho}^{+,T} s_0 & 1 \end{pmatrix}, \qquad S_J^+ = \begin{pmatrix} 1 & \boldsymbol{\tau}^{+,T} & c_1^{',-} \\ 0 & \mathbb{1} & s_0\boldsymbol{\tau}^+ \\ 0 & 0 & 1 \end{pmatrix}, \tag{15}$$

$$\begin{aligned} &S_J^-(\lambda) = \left(\hat{S}_J^+(\lambda^*)\right)^\dagger, \qquad T_J^+(\lambda) = \left(\hat{T}_J^-(\lambda^*)\right)^\dagger, \qquad D_J^-(\lambda) = \left(\hat{D}_J^+(\lambda^*)\right)^\dagger, \\ &c_1^{'',\pm} = \frac{1}{2}(\boldsymbol{\rho}^{\pm,T} s_0 \boldsymbol{\rho}^{\pm}), \qquad c_1^{',\pm} = \frac{1}{2}(\boldsymbol{\tau}^{\mp,T} s_0 \boldsymbol{\tau}^{\mp}) \end{aligned} \tag{16}$$

where the block-diagonal matrix $D_J^+ = \text{b-diag}\,(m_1^+, \mathbf{m}_2^+, 1/m_!^+)$.

$$\boldsymbol{\rho}^+ = \frac{\mathbf{b}^+}{m_1^+}, \qquad \boldsymbol{\tau}^+ = \frac{\mathbf{b}^-}{m_1^+}. \tag{17}$$

The matrix elements of $T(\lambda)$ satisfy a number of relations which ensure that both $T(\lambda)$ and its inverse $\hat{T}(\lambda)$ belong to $SO(n+2)$ and that $T(\lambda)\hat{T}(\lambda) = \mathbb{1}$. Some of them take the form:

$$\begin{aligned} &m_1^+ m_1^- + (\mathbf{b}^-, \mathbf{B}^+) + c_1^+ c_1^- = 1, \\ &\mathbf{b}^+ \mathbf{B}^{-T} + \mathbf{T}_{22} s_0 \mathbf{T}_{22}^T s_0 + s_0 \mathbf{B}^- \mathbf{b}^{+T} s_0 = \mathbb{1}. \end{aligned}$$

Important tools for reducing the ISP to a Riemann-Hilbert problem (RHP) are the fundamental analytic solution (FAS) $\chi^{\pm}(x,t,\lambda)$. Their construction is based on the generalized Gauss decomposition of $T(\lambda,t)$, see [17–19]:

$$\chi^{\pm}(x,t,\lambda) = \phi(x,t,\lambda) S_J^{\pm}(t,\lambda) = \psi(x,t,\lambda) T_J^{\mp}(t,\lambda) D_J^{\pm}(\lambda). \tag{18}$$

If $q(x,t)$ evolves according to (9)–(10) then the scattering matrix and its elements satisfy the following linear evolution equations

$$i\frac{d\boldsymbol{\rho}^+}{dt} + \lambda^2 \boldsymbol{\rho}^+(t,\lambda) = 0, \qquad i\frac{d\boldsymbol{\tau}^+}{dt} - \lambda^2 \boldsymbol{\tau}^+(t,\lambda) = 0, \qquad i\frac{dD+}{dt} = 0, \tag{19}$$

so the block-diagonal matrices $D^{\pm}(\lambda)$ can be considered as generating functionals of the integrals of motion. The fact that all $(2r-1)^2$ matrix elements of $\mathbf{m}_2^{\pm}(\lambda)$ for $\lambda \in \mathbb{C}_{\pm}$ generate integrals of motion reflect the superintegrability of the model and are due to the degeneracy of the dispersion law of (9)–(10). We remind that $D_J^{\pm}(\lambda)$ allow analytic extension for $\lambda \in \mathbb{C}_{\pm}$ and that their zeroes and poles determine the discrete eigenvalues of L.

The FAS for real λ are linearly related

$$\chi^+(x,t,\lambda) = \chi^-(x,t,\lambda) G_{0,J}(\lambda,t), \qquad G_{0,J}(\lambda,t) = \hat{S}_J^-(\lambda,t) S_J^+(\lambda,t) \tag{20}$$

and satisfy the normalization relation $\lim_{\lambda\to\infty}\chi^{\pm}(x,t,\lambda)e^{i\lambda Jx}=\mathbb{1}$. Then these FAS satisfy

$$\xi^{+}(x,t,\lambda)=\xi^{-}(x,t,\lambda)G_J(x,\lambda,t),\qquad G_J(x,\lambda,t)=e^{-i\lambda Jx}G^{-}_{0,J}(\lambda,t)e^{i\lambda Jx}. \tag{21}$$

Obviously the sewing function $G_j(x,\lambda,t)$ is uniquely determined by the Gauss factors $S_J^{\pm}(\lambda,t)$.

Lemma 1 *Let the potential $q(x,t)$ is such that the Lax operator L has no discrete eigenvalues. Then as minimal set of scattering data which determines uniquely the scattering matrix $T(\lambda,t)$ and the corresponding potential $q(x,t)$ one can consider either one of the sets $\mathfrak{T}_i$, $i=1,2$*

$$\mathfrak{T}_1\equiv\{\rho_\alpha^{+}(\lambda,t),\alpha\in\Delta_1^{+}\},\qquad \mathfrak{T}_2\equiv\{\tau_\alpha^{+}(\lambda,t),\alpha\in\Delta_1^{+}\},\qquad \lambda\in\mathbb{R}.$$

Given the solutions $\chi^{\pm}(x,t,\lambda)$ one recovers $q(x,t)$ via the formula

$$q(x,t)=\lim_{\lambda\to\infty}\lambda\left(J-\chi^{\pm}J\widehat{\chi}^{\pm}(x,t,\lambda)\right). \tag{22}$$

The main goal of the dressing method [17, 20–22] is, starting from a known solutions $\chi_0^{\pm}(x,t,\lambda)$ of $L_0(\lambda)$ with potential $q_{(0)}(x,t)$ to construct new singular solutions $\chi_1^{\pm}(x,t,\lambda)$ of L with a potential $q_{(1)}(x,t)$ with two (or more) additional singularities located at prescribed positions $\lambda_1^{\pm}$; the reduction $\mathbf{p}=\mathbf{q}^*$ ensures that $\lambda_1^{-}=(\lambda_1^{+})^*$. It is related to the regular one by a dressing factor $u(x,t,\lambda)$, for details see [13].

4 The Generalized Fourier Transforms for Non-regular J

It is known that the 'squared solutions' $\mathbf{e}_\alpha^{\pm}(x,\lambda)=\pi_{0J}(\chi^{\pm}E_\alpha\chi^{\pm}(x,t,\lambda))$, where $\pi_{0J}\cdot\equiv\operatorname{ad}_J^{-1}\operatorname{ad}_J\cdot$, form complete set of functions in the space of allowed potentials $q(x)$, see [16, 18, 19]. Skipping the details we write down the expansions of $q(x)$ and $\operatorname{ad}_J^{-1}\delta q(x)$ assuming L has no discrete spectrum:

$$q(x)=-\frac{i}{\pi}\int_{-\infty}^{\infty}d\lambda\sum_{\alpha\in\Delta_1^{+}}\left(\tau_\alpha^{+}(\lambda)\mathbf{e}_\alpha^{+}(x,\lambda)-\tau_\alpha^{-}(\lambda)\mathbf{e}_{-\alpha}^{-}(x,\lambda)\right), \tag{23}$$

$$\operatorname{ad}_J^{-1}\delta q(x)=\frac{i}{\pi}\int_{-\infty}^{\infty}d\lambda\sum_{\alpha\in\Delta_1^{+}}\left(\delta\tau_\alpha^{+}(\lambda)\mathbf{e}_\alpha^{+}(x,\lambda)+\delta\tau_\alpha^{-}(\lambda)\mathbf{e}_{-\alpha}^{-}(x,\lambda)\right). \tag{24}$$

These expansions can be viewed as a tool to establish an one-to-one correspondence between $q(x)$ (resp. $\operatorname{ad}_J^{-1}\delta q$ and each of the minimal sets of scattering data $\mathcal{T}_i$ (resp. $\delta\mathcal{T}_i$), $i=1,2$. To complete the analogy between the standard Fourier transform and the expansions over the 'squared solutions' we need the generating operators $\Lambda_{\pm}$:

$$\Lambda_{\pm}X(x)\equiv\operatorname{ad}_J^{-1}\left(i\frac{dX}{dx}+i\left[q(x),\int_{\pm\infty}^{x}dy\,[q(y),X(y)]\right]\right) \tag{25}$$

for which the 'squared solutions' are eigenfunctions:

$$(\Lambda_+ - \lambda)\mathbf{e}^+_{-\alpha}(x,\lambda) = 0, \qquad (\Lambda_+ - \lambda)\mathbf{e}^-_{\alpha}(x,\lambda) = 0, \tag{26}$$

$$(\Lambda_- - \lambda)\mathbf{e}^+_{\alpha}(x,\lambda) = 0, \qquad (\Lambda_- - \lambda)\mathbf{e}^-_{-\alpha}(x,\lambda) = 0. \tag{27}$$

5 Fundamental Properties of the MNLS Equations

The expansions (23), (24) and the explicit form of $\Lambda_\pm$ and (26)–(27) are basic for deriving the fundamental properties of all MNLS type equations related to the Lax operator L. Each of these NLEE is determined by its dispersion law which we choose to be of the form $F(\lambda) = f(\lambda)J$, where $f(\lambda)$ is polynomial in λ. The corresponding NLEE becomes:

$$i\,\mathrm{ad}_J^{-1} q_t + f(\Lambda_\pm)q(x,t) = 0. \tag{28}$$

Theorem 1 *The NLEE* (28) *are equivalent to*: (i) *the equations* (19) *and* (ii) *the following evolution equations for the generalized Gauss factors of* $T(\lambda)$:

$$i\frac{dS_J^+}{dt} + [F(\lambda), S_J^+] = 0, \qquad i\frac{dT_J^-}{dt} + [F(\lambda), T_J^-] = 0, \qquad \frac{dD_J^+}{dt} = 0. \tag{29}$$

The principal series of integrals is generated by the asymptotic expansion of $\ln m_1^+(\lambda) = \sum_{k=1}^{\infty} I_k \lambda^{-k}$. The first three integrals of motion:

$$I_1 = -\frac{i}{2}\int_{-\infty}^{\infty} dx\, \langle q(x), q(x)\rangle, \qquad I_2 = \frac{1}{2}\int_{-\infty}^{\infty} dx\, \langle q_x(x), \mathrm{ad}_J^{-1} q(x)\rangle. \tag{30}$$

Now iI_1 can be interpreted as the density of the particles, I_2 is the momentum and $I_3 = iH_{\mathrm{MNLS}}$. Indeed, the Hamiltonian equations of motion provided by $H_{(0)} = -iI_3$ with the Poisson brackets

$$\{q_k(y,t), p_j(x,t)\} = i\delta_{kj}\delta(x-y), \tag{31}$$

coincide with the MNLS equations (9)–(10). The above Poisson brackets are dual to the canonical symplectic form:

$$\Omega_0 = i\int_{-\infty}^{\infty} dx\, \mathrm{tr}\left(\delta\mathbf{p}(x) \underset{'}{\wedge} \delta\mathbf{q}(x)\right) = \frac{1}{2i}\left[\!\left[\mathrm{ad}_J^{-1}\delta q(x) \underset{'}{\wedge} \mathrm{ad}_J^{-1}\delta q(x)\right]\!\right],$$

where $\underset{'}{\wedge}$ means that taking the scalar or matrix product we exchange the usual product of the matrix elements by wedge-product.

The Hamiltonian formulation of (9)–(10) with Ω_0 and H_0 is just one member of the hierarchy of Hamiltonian formulations provided by:

$$\Omega_k = \frac{1}{i}\left[\!\left[\mathrm{ad}_J^{-1}\delta Q \underset{'}{\wedge} \Lambda^k \mathrm{ad}_J^{-1}\delta Q\right]\!\right], \qquad H_k = i^{k+3} I_{k+3} \tag{32}$$

where $\Lambda = \frac{1}{2}(\Lambda_+ + \Lambda_-)$. We can also calculate Ω_k in terms of the scattering data variations. Imposing the reduction $q(x) = q^\dagger(x)$ we get:

$$\Omega_k = \frac{1}{2\pi i}\int_{-\infty}^{\infty} d\lambda\, \lambda^k \left(\Omega_0^+(\lambda) - \Omega_0^-(\lambda)\right)$$
$$= \frac{1}{2\pi}\int_{-\infty}^{\infty} d\lambda\, \lambda^k \mathrm{Im}\left(m_1^+(\lambda)\left(\hat{\mathbf{m}}_2{}^+ \delta\boldsymbol{\rho}^+(\lambda) \wedge \delta\boldsymbol{\tau}^+(\lambda)\right)\right).$$

This allows one to prove that if we are able to cast Ω_0 in canonical form, then all Ω_k will also be cast in canonical form and will be pair-wise equivalent.

Acknowledgements One of us (V.S.G.) thanks the organizers of the Conference for their hospitality and for making his participation possible. This research has been supported in part by the National Science Foundation of the USA via grant DMS-0505566 and by the USA Air Force Office of Scientific Research.

References

1. T. Ohmi, K. Machida, J. Phys. Soc. Jpn. **67**, 1822 (1998)
2. T. Ho, Phys. Rev. Lett. **81**, 742 (1998)
3. C. Ciobanu, S. Yip, T. Ho, Phys. Rev. A **61**, 033607 (2000)
4. N. Klausen, J. Bohn, Ch. Greene, Phys. Rev. A **64**, 053602 (2001)
5. M. Ueda, M. Koashi, Phys. Rev. A **65**, 063602 (2002)
6. J. Ieda, T. Miyakawa, M. Wadati, Phys. Rev. Lett. **93**, 194102 (2004)
7. J. Ieda, T. Miyakawa, M. Wadati, J. Phys. Soc. Jpn. **73**, 2996 (2004)
8. T. Kuwamoto, K. Araki, T. Eno, T. Hirano, Phys. Rev. A **69**, 063604 (2004)
9. L. Li, Z. Li, B. Malomed, D. Mihalache, W. Liu, Phys. Rev. A **72**, 033611 (2005)
10. M. Uchiyama, J. Ieda, M. Wadati, J. Phys. Soc. Jpn. **75**, 064002 (2006)
11. M. Uchiyama, J. Ieda, M. Wadati, J. Phys. Soc. Jpn. **76**, 74005 (2007)
12. S. Uchino, T. Otsuka, M. Ueda, arXiv:0710.5210
13. V. Gerdjikov, N. Kostov, T. Valchev, Physica D (2008, in press). arXiv:0802.4398 [nlin.SI]
14. A. Fordy, P. Kulish, Commun. Math. Phys. **89**, 427–443 (1983)
15. S. Helgasson, *Differential Geometry, Lie Groups and Symmetric Spaces*. Graduate Studies in Mathematics, vol. 34 (AMS, Providence, 2001)
16. V. Gerdjikov, in *Geometry, Integrability and Quantization*, Softex, Sofia, ed. by Mladenov, I., Hirshfeld, A. (2005). nlin.SI/0604004
17. V. Zakharov, S. Manakov, S. Novikov, L. Pitaevskii, *Theory of Solitons. The Inverse Scattering Method* (Plenum Press (Consultant Bureau), New York, 1984)
18. V. Gerdjikov, Inverse Probl. **2**, 51–74 (1986)
19. V. Gerdjikov, Theor. Math. Phys. **92**, 374–386 (1992)
20. V. Gerdjikov, G. Grahovski, R. Ivanov, N. Kostov, Inverse Probl. **17**, 999–1015 (2001)
21. R. Ivanov, Nucl. Phys. B **694**, 509–524 (2004)
22. G. Grahovski, V. Gerdjikov, N. Kostov, V. Atanasov, in *Geometry, Integrability and Quantization VII*, Softex, Sofia, ed. by Mladenov, I., De Leon, M. (2006)

Part V
Mathematical Models in Engineering

Impulsive Boundary Layer Flow Past a Permeable Quadratically Stretching Sheet

V. Kumaran, A. Vanav Kumar, and J. Sarat Chandra Babu

Abstract In this paper boundary layer flow over a permeable sheet due to an impulsive quadratic stretching and linear cross flow is considered. The governing equations are solved numerically using an implicit finite difference method of Crank-Nicholson type. The steady state solutions are compared with available results in the literature. Profiles of velocity, skin friction are presented for various values of the parameters governing the stretching and cross flow.

Keywords Boundary layer flow · Quadratically stretching · Linear cross flow

1 Introduction

Boundary layer flow past a stretching sheet has been studied by many researchers due to its applications in extrusion processes. Crane [1] obtained a closed form solution for boundary layer flow past a linearly stretching sheet. Gupta and Gupta [2] extended analytic solution of Crane [1] by including suction and injection. Flow past a nonlinearly stretching sheet has also attracted many researchers. Few among them are Vajravelu [4], Rafael Cortell [6], Kumaran and Ramanaiah [3] and Liao [5]. In this paper the unsteady boundary layer flow over a stretching sheet subjected to a impulsive quadratic stretching and linear cross flow is analyzed. This unsteady problem is solved numerically using finite difference scheme. The steady solution is compared with the exact solution of Kumaran and Ramanaiah [3] for a particular case.

V. Kumaran (✉) · A.V. Kumar
Department of Mathematics, National Institute of Technology, Tiruchirappalli 620015, India
e-mail: kumaran@nitt.edu

J.S.C. Babu
Department of Chemical Engineering, National Institute of Technology, Tiruchirappalli 620015, India
e-mail: sarat@nitt.edu

J.A.T. Machado et al. (eds.), *Nonlinear Science and Complexity*,
DOI 10.1007/978-90-481-9884-9_24,

2 Governing Equations

Consider a two-dimensional, laminar flow of a viscous incompressible fluid past a permeable elastic sheet. The coordinates x' is measured along the horizontal direction and y' is measured along the vertical direction. u',v' are the horizontal and vertical velocity components respectively. For $t' \leq 0$, the fluid and the sheet is maintained at rest for all $x' \geq 0$, $y' \geq 0$. At $t' > 0$, the sheet is suddenly stretched quadratically in x' and a linear cross flow is applied across the sheet. The governing equations of the boundary layer fluid flow reduce to,

$$\frac{\partial u'}{\partial x'} + \frac{\partial v'}{\partial y'} = 0, \tag{1}$$

$$\frac{\partial u'}{\partial t'} + u'\frac{\partial u'}{\partial x'} + v'\frac{\partial u'}{\partial y'} = \nu\frac{\partial^2 u'}{\partial y'^2} \tag{2}$$

subjected to the initial and boundary conditions:

$$t' \leq 0: \quad u' = 0, \qquad v' = 0, \quad \forall x', y' \geq 0, \tag{3}$$

$$t' > 0: \quad u' = \beta x' + \alpha x'^2, \qquad v' = v_c + \delta x' \quad \text{on } y' = 0, x' \geq 0,$$
$$u' \to 0 \quad \text{as} y' \to \infty, x' \geq 0 \tag{4}$$

where t' is the time, β, α, v_c, δ are constants and ν is the kinematic viscosity of the fluid. Using the following dimensionless variables

$$x' = x\sqrt{\frac{\nu}{\beta}}, \qquad y' = y\sqrt{\frac{\nu}{\beta}}, \qquad t' = \frac{t}{\beta}, \qquad u' = u\sqrt{\nu\beta}, \qquad v' = v\sqrt{\nu\beta} \tag{5}$$

and defining the parameters x_c, b, $s(>0)$, a as

$$x_c = \frac{s - \frac{1}{s}}{2b}, \qquad b = \frac{\delta}{2\beta}, \qquad \frac{1}{s} - s = \frac{v_c}{\sqrt{\beta\nu}}, \qquad a = \left(\alpha - \frac{\delta s}{2}\right)\frac{1}{\beta}\sqrt{\frac{\nu}{\beta}}, \tag{6}$$

(1)–(4) become

$$\frac{\partial u}{\partial x} + \frac{\partial v}{\partial y} = 0, \tag{7}$$

$$\frac{\partial u}{\partial t} + u\frac{\partial u}{\partial x} + v\frac{\partial u}{\partial y} = \frac{\partial^2 u}{\partial y^2} \tag{8}$$

with initial and boundary conditions,

$$t \leq 0: \quad u = 0, \qquad v = 0, \quad \text{for } x, y \geq 0, \tag{9}$$

$$t > 0: \quad u = x + (sb + a)x^2, \qquad v = -s + \frac{1}{s} + 2bx \quad \text{on } y = 0, x \geq 0, \tag{10}$$

$$u \to 0 \quad \text{as } y \to \infty, x \geq 0. \tag{11}$$

The skin friction at the sheet in dimensionless form is given by

$$\tau_x = \frac{\tau'_{x'}}{\mu\beta} = -\left.\frac{\partial u}{\partial y}\right|_{y=0}. \tag{12}$$

3 Results and Discussion

Computations are carried out only for $a = 0$. Also when $a = 0$, the exact steady state expressions of the horizontal velocity, vertical velocity and skin friction are given by Kumaran and Ramanaiah [3],

$$u = (x + sbx^2)e^{-sy}, \qquad v = \left(\frac{1}{s} + 2bx\right)e^{-sy} - s, \qquad \tau_x = s(x + sbx^2). \tag{13}$$

The computation for unsteady case have been carried out using the numerical technique given by Muthucumaraswamy and Ganesan [7]. Truncation error is $O(\Delta y^2 + \Delta t^2 + \Delta x)$. The computation domain for x, y, t are 0 (0.002) 1, 0 (0.0125) 7.5 and 0 (0.01) steady time, respectively. Good agreement is found between the computed values and the exact values given by (13). The graphs of u, v plotted are for $x = 1$. Steady skin friction profiles are shown in Fig. 3.

3.1 Case (i): Linear Injection ($s \leq 1$ and $b \geq 0$)

In this case (Serial no. (1) to (8) in Table 1), it is seen that u, v increases with b for a particular value of s (Figs. 1(a)–1(d)). Also, it is seen that there is an increase in u and decrease in v for an increase in the parameter value s. The time required to

Table 1 Steady state values of time for case (i) and case (ii)

Serial No.	s	b	Steady time	Serial No.	s	b	Steady time
(1)	5/8	0	4.85	(9)	1	−39/80	3.43
(2)		15/112	4.87	(10)		−7/24	3.46
(3)		7/24	4.87	(11)		−15/112	3.48
(4)		39/80	4.89	(12)		0	3.49
(5)	1	0	3.49	(13)	8/5	−39/80	1.77
(6)		15/112	3.53	(14)		−7/24	2.07
(7)		7/24	3.56	(15)		−15/112	2.18
(8)		39/80	3.65	(16)		0	2.23

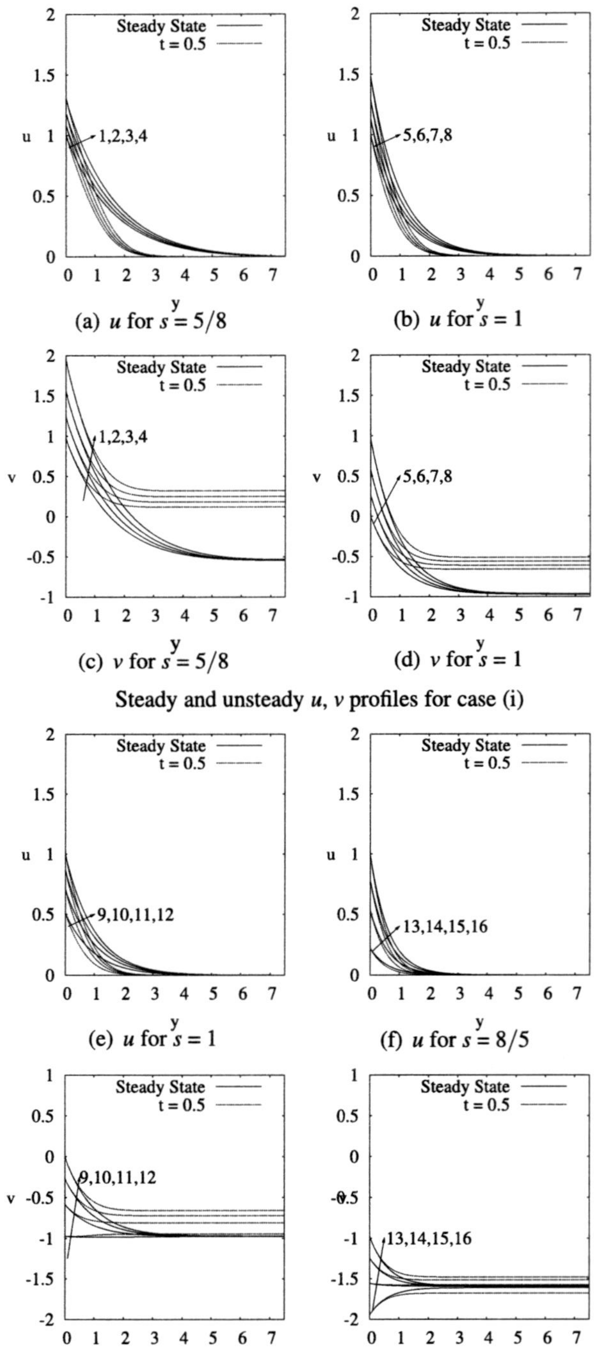

Fig. 1 Profiles of u, v for the cases (i) and (ii)

(a) u for $s = 5/8$ (b) u for $s = 1$

(c) v for $s = 5/8$ (d) v for $s = 1$

Steady and unsteady u, v profiles for case (i)

(e) u for $s = 1$ (f) u for $s = 8/5$

(g) v for $s = 1$ (h) v for $s = 5/8$

Steady and unsteady u, v profiles for case (ii)

attain the steady state increases with b for a particular s. Also, for an increase in the value of s the steady state time decreases. When b increases from 0 to 39/80 their is an increase of around 0.82% in steady state time for $s = 5/8$ compared to around 4.58% for $s = 1$.

3.2 Case (ii): Linear Suction ($s \geq 1$ and $b \leq 0$)

In this case (Serial no. (9) to (16) in Table 1) it is seen that u, v increases with b for a particular value of s (Figs. 1(e)–1(h)). When b increases from $-39/80$ to 0 there is an increase of around 1% in the steady state time for $s = 1$ and when b increases from $-39/80$ to 0 their is an increase of around 25% in the steady state time for $s = 8/5$.

3.3 Case (iii): Linear Suction for $x < x_c$ and Linear Injection for $x > x_c$ ($s > 1$ and $b > 0$)

In this case, from Figs. 2(a)–2(d) (Serial no. (17) to (22)) similar features are observed as in case (i) and (ii) for u and v. Here, when b increases from 15/112 to 39/80 there is an increase of 6% in the steady state time when $s = 8/7$ and when $s = 8/5$ there is an increase of 29% in the steady state time. Also, when s increases the steady state time decreases. Table 2 reveals that x_c decreases when b increases and x_c increases when s increases. Also increase in x_c has resulted in an decrease of steady state time.

3.4 Case(iv): Linear Injection for $x < x_c$ and Linear Suction for $x > x_c$ ($s < 1$ and $b < 0$)

In this case (Serial no. (23) to (28)), similar features are observed in general as in the above 3 cases (Figs. 2(e)–2(h) and the Table 2). The steady state time decreases by

Table 2 Steady state values of time for case (iii) and case (iv)

Serial No.	s	b	Steady time	x_c	Serial No.	s	b	Steady time	x_c
(17)	8/7	15/112	3.16	1	(23)	5/8	−39/80	4.92	1
(18)		7/24	3.19	0.459183673	(24)		−7/24	4.89	1.671428571
(19)		39/80	3.36	0.274725274	(25)		−15/112	4.88	3.64
(20)	8/5	15/112	2.29	3.64	(26)	7/8	−39/80	3.87	0.274725274
(21)		7/24	2.43	1.671428571	(27)		−7/24	3.87	0.459183673
(22)		39/80	2.97	1	(28)		−15/112	3.87	1

Fig. 2 Profiles of u, v for the cases (iii) and (iv)

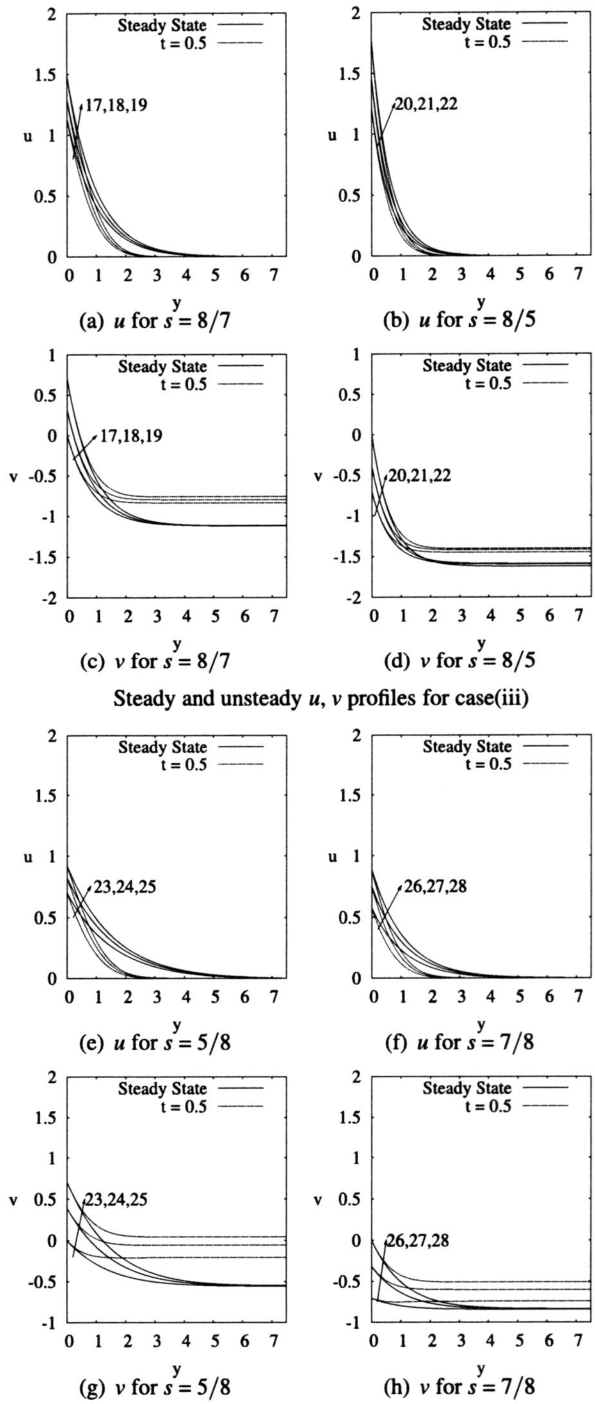

(a) u for $s = 8/7$

(b) u for $s = 8/5$

(c) v for $s = 8/7$

(d) v for $s = 8/5$

Steady and unsteady u, v profiles for case(iii)

(e) u for $s = 5/8$

(f) u for $s = 7/8$

(g) v for $s = 5/8$

(h) v for $s = 7/8$

Steady and unsteady u, v profiles for case (iv)

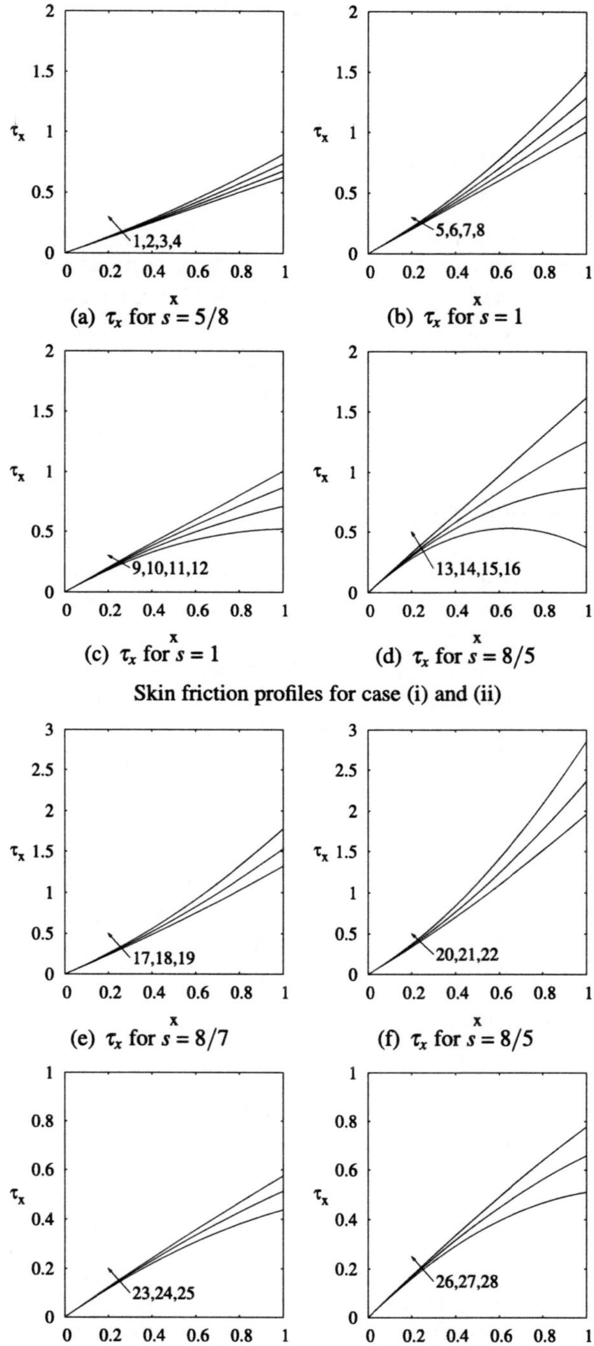

Fig. 3 Profiles of the skin friction

(a) τ_x for $s = 5/8$ (b) τ_x for $s = 1$

(c) τ_x for $s = 1$ (d) τ_x for $s = 8/5$

Skin friction profiles for case (i) and (ii)

(e) τ_x for $s = 8/7$ (f) τ_x for $s = 8/5$

(g) τ_x for $s = 5/8$ (h) τ_x for $s = 7/8$

Skin friction profiles for case (iii) and (iv)

0.8% when b increases from $-39/80$ to $-15/112$ for $s = 5/8$ and remains constant for $s = 7/8$.

3.5 Note on the Value of s

For $s = 5/8$, 1, 8/7, 8/5 when $b = 15/112$ the values of s increases by 56% and the steady state time decreases by 52%. Similarly, if we consider the values of s to be 8/5, 1, 7/8 and 5/8 when $b = -39/80$, s increases by 56% and the steady state time decreases by 64%. If we examine the value of b, it decreases by 364% but the steady state time increases by almost 3%. This shows that the parameter s has a much greater influence on the flow of the fluid compared to b.

4 Conclusions

In this paper an analysis of boundary layer flow past a stretching sheet due to a sudden quadratic stretching and linear cross flow governed by the stretching parameter b and suction/injection parameter s is carried out. It is found that an increase in values of s and b leads to an decrease in the steady state time. The horizontal velocity and the skin friction increase with increase in the value of both s and b. The vertical velocity decreases with an increase in the value of s whereas vertical velocity increases with increasing b. It is observed that there is a greater effect of the parameter s compared to the parameter b on the steady state time.

Acknowledgements The authors gratefully acknowledge the support rendered by the Director, National Institute of Technology, Tiruchirappalli, India for this research, under TEQIP grant. The authors also thank the anonymous referees for their useful comments.

References

1. L. Crane, Flow past a stretching plate. Z. Angew. Math. Phys. **21**, 645–647 (1970)
2. P.S. Gupta, A.S. Gupta, Heat and Mass transfer on a stretching sheet with suction or blowing. Can. J. Chem. Eng. **55**, 744–746 (1977)
3. V. Kumaran, G. Ramanaiah, A note on the flow over a stretching sheet. Acta Mech. **116**, 229–233 (1996)
4. K. Vajravelu, Vicious flow over a non-linear stretching sheet. Appl. Math. Comput. **124**, 281–288 (2001)
5. S.J. Liao, An analytic solution of unsteady boundary layer flows caused by an impulsively stretching plate. Commun. Nonlinear Sci. Numer. Simul. **11**, 326–339 (2006)
6. R. Cortell, Viscous flow and heat transfer over a nonlinearly stretching sheet. Appl. Math. Comput. **184**(2), 864–873 (2007)
7. R. Muthucumaraswamy, P. Ganesan, Flow past an impulsively started vertical plate with constant heat flux and mass transfer. Comput. Methods Appl. Mech. Eng. **187**, 79–90 (2000)

Complete Dynamic Modeling of a Stewart Platform Using the Generalized Momentum Approach

António Mendes Lopes and E.J. Solteiro Pires

Abstract Dynamic modeling is of great importance regarding computer simulation and advanced control of parallel manipulators.

Dynamic modeling of parallel manipulators presents an inherent complexity, mainly due to system closed-loop structure and kinematic constraints.

In this paper an approach based on the manipulator generalized momentum is explored and applied to the dynamic modeling of a Stewart platform. The generalized momentum is used to compute the kinetic component of the generalized force acting on each manipulator rigid body. Analytic expressions for the rigid bodies' inertia and Coriolis and centripetal terms matrices are obtained, which can be added, as they are expressed in the same frame. Gravitational part of the generalized force is obtained using the manipulator potential energy.

Keywords Dynamic model · Parallel manipulator · Robotics · Generalized momentum

1 Introduction

The dynamic model of a parallel manipulator operated in free space can be mathematically represented, in the Cartesian space, by a system of nonlinear differential equations that may be written in matrix form as:

$$\mathbf{I}(\mathbf{x}).\ddot{\mathbf{x}} + \mathbf{V}(\mathbf{x}, \dot{\mathbf{x}}).\dot{\mathbf{x}} + \mathbf{G}(\mathbf{x}) = \mathbf{f}, \tag{1}$$

A.M. Lopes (✉)
Unidade de Integração de Sistemas e Processos Automatizados, Faculdade de Engenharia, Universidade do Porto, Rua Dr. Roberto Frias, 4200-465 Porto, Portugal
e-mail: aml@fe.up.pt

E.J.S. Pires
Centro de Investigação e de Tecnologias Agro-Ambientais e Biológicas, Dep. de Engenharias, Universidade de Trás-os-Montes e Alto Douro, Quinta de Prados, 5000-911 Vila Real, Portugal
e-mail: epires@utad.pt

J.A.T. Machado et al. (eds.), *Nonlinear Science and Complexity*,
DOI 10.1007/978-90-481-9884-9_25,

$\mathbf{I}(\mathbf{x})$ being the inertia matrix, $\mathbf{V}(\mathbf{x}, \dot{\mathbf{x}})$ the Coriolis and centripetal terms matrix, $\mathbf{G}(\mathbf{x})$ a vector of gravitational generalized forces, $\mathbf{x}$ the generalized position of the moving platform (end-effector) and $\mathbf{f}$ the controlled generalized force applied on the end-effector. Thus,

$$\mathbf{f} = \mathbf{J}^T(\mathbf{x}).\tau, \tag{2}$$

where τ is the generalized force developed by the actuators and $\mathbf{J}(\mathbf{x})$ is a Jacobian matrix.

The dynamic model of parallel manipulators is usually developed using the Newton-Euler or the Lagrange methods. Do and Yang [1] and, Reboulet and Berthomieu [2] use the Newton-Euler method on the dynamic modeling of a Stewart platform. Ji [3] presents a study on the influence of leg inertia on the dynamic model of a Stewart platform. Dasgupta and Mruthyunjaya [4] used the Newton-Euler approach to develop a closed-form dynamic model of the Stewart platform. This method was also used by Khalil and Ibrahim [5], Riebe and Ulbrich [6], and Guo and Li [7], among others.

On the other hand, Nguyen and Pooran [8] use the Lagrange method to model a Stewart platform, modeling the legs as point masses. Lebret et al. [9] follow an approach similar to the one used by Nguyen and Pooran [8]. Lagrange's method was also used by Gregório and Parenti-Castelli [10] and Caccavale et al. [11], for example.

Unfortunately the dynamic models obtained from these classical approaches usually present high computational loads. Therefore, alternative methods have been searched, namely the ones based on the principle of virtual work [12, 13], screw theory [14], and the generalized momentum approach [15].

In this paper the authors present a new approach to the dynamic modeling of a six degrees-of-freedom (dof) Stewart platform: the use of the generalized momentum concept.

2 Stewart Platform Kinematic Structure

A Stewart platform comprises a fixed platform (base) and a moving (payload) platform, linked together by six independent, identical, open kinematic chains (Fig. 1). Each chain (leg) comprises a cylinder and a piston (or spindle) that are connected together by a prismatic joint, l_i. The upper end of each leg is connected to the moving platform by a spherical joint whereas the lower end is connected to the fixed base by a universal joint. Points B_i and P_i are the connecting points to the base and moving platforms, respectively. They are located at the vertices of two semi-regular hexagons.

For kinematic modeling purposes, two frames, {P} and {B}, are attached to the moving and base platforms, respectively. The generalized position of frame {P} relative to frame {B} may be represented by the vector:

$${}^B\mathbf{x}_{P|B|E} = [x_P \quad y_P \quad z_P \quad \psi_P \quad \theta_P \quad \varphi_P]^T = \left[{}^B\mathbf{x}^T_{P(pos)|B} \quad {}^B\mathbf{x}^T_{P(o)|E}\right]^T, \tag{3}$$

Fig. 1 Stewart platform kinematic structure

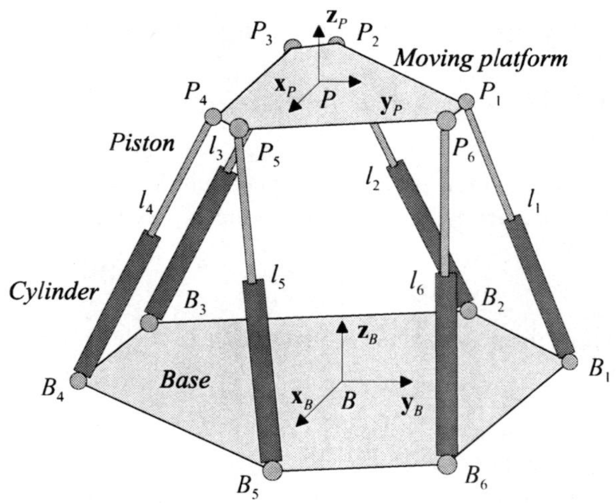

where ${}^B\mathbf{x}_{P(pos)|_B} = [x_p\ y_p\ z_p]^T$ is the position of the origin of frame {P} relative to frame {B}, and ${}^B\mathbf{x}_{P(o)|_E} = [\psi_p\ \theta_p\ \varphi_p]^T$ defines an Euler angles system representing orientation of frame {P} relative to {B}.

The velocity kinematics is represented by the Euler angles Jacobian matrix, $\mathbf{J}_E$, or the kinematic Jacobian, $\mathbf{J}_C$. These Jacobian's relate the velocities of the active joints (actuators) to the generalized velocity of the moving platform:

$$\dot{\mathbf{l}} = \mathbf{J}_E.{}^B\dot{\mathbf{x}}_{P|_{B|E}} = \mathbf{J}_E.\begin{bmatrix} {}^B\dot{\mathbf{x}}_{P(pos)|_B} \\ {}^B\dot{\mathbf{x}}_{P(o)|_E} \end{bmatrix}, \tag{4}$$

$$\dot{\mathbf{l}} = \mathbf{J}_C.{}^B\dot{\mathbf{x}}_{P|_B} = \mathbf{J}_C.\begin{bmatrix} {}^B\dot{\mathbf{x}}_{P(pos)|_B} \\ {}^B\boldsymbol{\omega}_{P|_B} \end{bmatrix}, \tag{5}$$

where

$$\dot{\mathbf{l}} = \begin{bmatrix} \dot{l}_1 & \dot{l}_2 & \ldots & \dot{l}_6 \end{bmatrix}^T, \tag{6}$$

$${}^B\boldsymbol{\omega}_{P|_B} = \mathbf{J}_A.{}^B\dot{\mathbf{x}}_{P(o)|_E}, \tag{7}$$

$$\mathbf{J}_A = \begin{bmatrix} 0 & -S\psi_P & C\theta_P C\psi_P \\ 0 & C\psi_P & C\theta_P S\psi_P \\ 1 & 0 & -S\theta_P \end{bmatrix}. \tag{8}$$

Vectors ${}^B\dot{\mathbf{x}}_{P(pos)|_B} \equiv {}^B\mathbf{v}_{P|_B}$ and ${}^B\boldsymbol{\omega}_{P|_B}$ represent the linear and angular velocity of the moving platform relative to {B}, and ${}^B\dot{\mathbf{x}}_{P(o)|_E}$ represents the Euler angles time derivative.

3 Dynamic Modeling Using the Generalized Momentum

3.1 Moving Platform Modeling

The linear momentum of the manipulator moving platform, written in frame {B}, may be obtained from the following expression:

$$\mathbf{Q}_{P|B} = m_p.{}^B\mathbf{v}_{P|B} = \mathbf{I}_{P(tra)}.{}^B\mathbf{v}_{P|B}, \tag{9}$$

$\mathbf{I}_{P(tra)}$ is the translational inertia matrix of the moving platform,

$$\mathbf{I}_{P(tra)} = \mathrm{diag}([m_P \quad m_P \quad m_P]), \tag{10}$$

m_P being its mass.

The angular momentum, also written in frame {B}, is:

$$\mathbf{H}_{P|B} = \mathbf{I}_{P(rot)|B}.{}^B\boldsymbol{\omega}_{P|B}, \tag{11}$$

$\mathbf{I}_{P(rot)|B}$ represents the rotational inertia matrix of the moving platform, expressed in the base frame {B}. The rotational inertia matrix of the moving platform, expressed in frame {P}, may be written as:

$$\mathbf{I}_{P(rot)|P} = \mathrm{diag}([I_{P_{xx}} \quad I_{P_{yy}} \quad I_{P_{zz}}]). \tag{12}$$

This inertia matrix can be written in frame {B} using the following transformation:

$$\mathbf{I}_{P(rot)|B} = {}^B\mathbf{R}_P.\mathbf{I}_{P(rot)|P}.{}^B\mathbf{R}_P^T. \tag{13}$$

The generalized momentum of the moving platform, expressed in frame {B}, can be obtained from the simultaneous use of (9) and (11):

$$\mathbf{q}_{P|B} = \begin{bmatrix} \mathbf{I}_{P(tra)} & \mathbf{0} \\ \mathbf{0} & \mathbf{I}_{P(rot)|B} \end{bmatrix} . \begin{bmatrix} {}^B\mathbf{v}_{P|B} \\ {}^B\boldsymbol{\omega}_{P|B} \end{bmatrix}. \tag{14}$$

The combination of (7) and (11) results into:

$$\mathbf{H}_{P|B} = \mathbf{I}_{P(rot)|B}.\mathbf{J}_A.{}^B\dot{\mathbf{x}}_{P(o)|E}. \tag{15}$$

Accordingly, (14) may be rewritten as:

$$\mathbf{q}_{P|B} = \begin{bmatrix} \mathbf{I}_{P(tra)} & \mathbf{0} \\ \mathbf{0} & \mathbf{I}_{P(rot)|B} \end{bmatrix} . \begin{bmatrix} \Im & 0 \\ 0 & \mathbf{J}_A \end{bmatrix} . \begin{bmatrix} {}^B\mathbf{v}_{P|B} \\ {}^B\dot{\mathbf{x}}_{P(o)|E} \end{bmatrix}, \tag{16}$$

$$\mathbf{q}_{P|B} = \mathbf{I}_{P|B}.\mathbf{T}.{}^B\dot{\mathbf{x}}_{P|B|E}, \tag{17}$$

$\mathbf{T}$ being a matrix transformation defined by:

$$\mathbf{T} = \mathrm{diag}([\Im \quad \mathbf{J}_A]) \tag{18}$$

The time derivative of (17) results into:

$$^{P}\mathbf{f}_{P(kin)|B} = \dot{\mathbf{q}}_{P|B} = \frac{\mathrm{d}}{\mathrm{d}t}\left(\mathbf{I}_{P|B}.\mathbf{T}\right).^{B}\dot{\mathbf{x}}_{P|B|E} + \mathbf{I}_{P|B}.\mathbf{T}.^{B}\ddot{\mathbf{x}}_{P|B|E}, \tag{19}$$

$^{P}\mathbf{f}_{P(kin)|B}$ is the kinetic component of the generalized force acting on {P}, due to the moving platform motion, expressed in frame {B}. The corresponding actuating forces, $\tau_{P(kin)}$, may be computed from the following relation:

$$\tau_{P(kin)|B} = \mathbf{J}_C^{-T}.^{P}\mathbf{f}_{P(kin)|B}. \tag{20}$$

where

$$^{P}\mathbf{f}_{P(kin)|B} = \left[{}^{P}\mathbf{F}^{T}_{P(kin)|B} \quad {}^{P}\mathbf{M}^{T}_{P(kin)|B}\right]^{T}. \tag{21}$$

Vector $^{P}\mathbf{F}_{P(kin)|B}$ represents the force vector acting on the centre of mass of the moving platform, and $^{P}\mathbf{M}_{P(kin)|B}$ represents the moment vector acting on the moving platform, expressed in the base frame, {B}.

From (19) it can be concluded that two matrices playing the roles of the inertia matrix and the Coriolis and centripetal terms matrix are:

$$\mathbf{I}_{P|B}.\mathbf{T}, \tag{22}$$

$$\frac{\mathrm{d}}{\mathrm{d}t}\left(\mathbf{I}_{P|B}.\mathbf{T}\right). \tag{23}$$

3.2 Cylinder Modeling

If the centre of mass of each cylinder is located at a constant distance, b_C, from the cylinder to base platform connecting point, B_i (Fig. 2), then its position relative to frame {B} is:

$$^{B}\mathbf{p}_{C_i|B} = b_C.\hat{\mathbf{l}}_i + \mathbf{b}_i, \tag{24}$$

where

$$\hat{\mathbf{l}}_i = \frac{\mathbf{l}_i}{\|\mathbf{l}_i\|} = \frac{\mathbf{l}_i}{l_i}, \tag{25}$$

$$\mathbf{l}_i = {}^{B}\mathbf{x}_{P(pos)|B} + {}^{P}\mathbf{p}_{i|B} - \mathbf{b}_i. \tag{26}$$

The linear velocity of the cylinder centre of mass, $^{B}\dot{\mathbf{p}}_{C_i|B}$, relative to {B} and expressed in the same frame, may be computed as:

$$^{B}\dot{\mathbf{p}}_{C_i|B} = {}^{B}\boldsymbol{\omega}_{l_i|B} \times b_C.\hat{\mathbf{l}}_i, \tag{27}$$

where $^{B}\boldsymbol{\omega}_{l_i|B}$ represents the leg angular velocity, which can be found from:

$$^{B}\boldsymbol{\omega}_{l_i|B} \times \mathbf{l}_i = {}^{B}\mathbf{v}_{P|B} + {}^{B}\boldsymbol{\omega}_{P|B} \times {}^{P}\mathbf{p}_{i|B}. \tag{28}$$

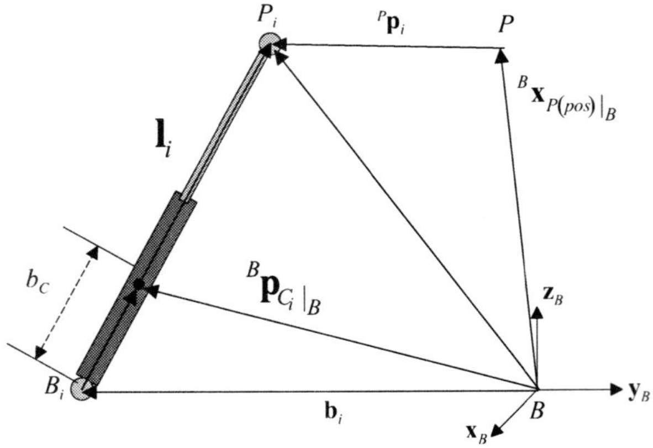

Fig. 2 Position of the centre of mass of the cylinder i

As the leg (both the cylinder and piston) cannot rotate along its own axis, the angular velocity along $\hat{\mathbf{l}}_i$ is always zero, and vectors $\mathbf{l}_i$ and ${}^{B}\boldsymbol{\omega}_{l_i|B}$ are always perpendicular. This enables (28) to be rewritten as:

$$ {}^{B}\boldsymbol{\omega}_{l_i|B} = \frac{1}{\mathbf{l}_i^T.\mathbf{l}_i}.\left[\mathbf{l}_i \times \left({}^{B}\mathbf{v}_{P|B} + {}^{B}\boldsymbol{\omega}_{P|B} \times {}^{P}\mathbf{p}_{i|B}\right)\right] \tag{29} $$

or,

$$ {}^{B}\boldsymbol{\omega}_{l_i|B} = \mathbf{J}_{D_i}.\begin{bmatrix} {}^{B}\mathbf{v}_{P|B} \\ {}^{B}\boldsymbol{\omega}_{P|B} \end{bmatrix}, \tag{30} $$

where Jacobian $\mathbf{J}_{D_i}$ is given by:

$$ \mathbf{J}_{D_i} = \begin{bmatrix} \tilde{\bar{\mathbf{l}}}_i & \tilde{\bar{\mathbf{l}}}_i.{}^{P}\tilde{\mathbf{p}}_{i|B}^T \end{bmatrix} \tag{31} $$

$$ \bar{\mathbf{l}}_i = \frac{\mathbf{l}_i}{\mathbf{l}_i^T.\mathbf{l}_i} \tag{32} $$

and, for a given vector $\mathbf{a} = [a_x\ a_y\ a_z]^T$,

$$ \tilde{\mathbf{a}} = \begin{bmatrix} 0 & -a_z & a_y \\ a_z & 0 & -a_x \\ -a_y & a_x & 0 \end{bmatrix}. \tag{33} $$

On the other hand, (27) can be rewritten as:

$$ {}^{B}\dot{\mathbf{p}}_{C_i|B} = \mathbf{J}_{B_i}.\begin{bmatrix} {}^{B}\mathbf{v}_{P|B} \\ {}^{B}\boldsymbol{\omega}_{P|B} \end{bmatrix}, \tag{34} $$

where $\mathbf{J}_{B_i}$ the Jacobian is given by:

$$ \mathbf{J}_{B_i} = \begin{bmatrix} b_C.\hat{\mathbf{l}}_i^T.\tilde{\bar{\mathbf{l}}}_i & b_C.\hat{\mathbf{l}}_i^T.\tilde{\bar{\mathbf{l}}}_i.{}^{P}\tilde{\mathbf{p}}_{i|B}^T \end{bmatrix}. \tag{35} $$

The linear momentum of each cylinder, $\mathbf{Q}_{C_i|B}$, can be represented in frame {B} as (36) where m_C is the cylinder mass.

$$\mathbf{Q}_{C_i|B} = m_c.{}^B\dot{\mathbf{p}}_{C_i|B}. \tag{36}$$

Introducing Jacobian $\mathbf{J}_{B_i}$ and matrix transformation $\mathbf{T}$ in the previous equation results into:

$$\mathbf{Q}_{C_i|B} = m_c.\mathbf{J}_{B_i}.\mathbf{T}.{}^B\dot{\mathbf{x}}_{P|B|E}. \tag{37}$$

The kinetic component of the force applied to the cylinder, due to its translation, and expressed in {B} can be obtained from the time derivative of (37):

$${}^{C_i}\mathbf{f}_{C_i(kin)(tra)|B} = \dot{\mathbf{Q}}_{C_i|B} = m_c.\frac{\mathrm{d}}{\mathrm{d}t}\left(\mathbf{J}_{B_i}.\mathbf{T}\right).{}^B\dot{\mathbf{x}}_{P|B|E} + m_c.\mathbf{J}_{B_i}.\mathbf{T}.{}^B\ddot{\mathbf{x}}_{P|B|E}. \tag{38}$$

When (38) is multiplied by $\mathbf{J}_{B_i}^T$, the kinetic component of the force applied to {P} due to each cylinder translation is obtained in frame {B}:

$$\begin{aligned}{}^P\mathbf{f}_{C_i(kin)(tra)|B} &= \mathbf{J}_{B_i}^T.{}^{C_i}\mathbf{f}_{C_i(kin)(tra)|B}\\ &= m_c.\mathbf{J}_{B_i}^T.\frac{\mathrm{d}}{\mathrm{d}t}\left(\mathbf{J}_{B_i}.\mathbf{T}\right).{}^B\dot{\mathbf{x}}_{P|B|E} + m_c.\mathbf{J}_{B_i}^T.\mathbf{J}_{B_i}.\mathbf{T}.{}^B\ddot{\mathbf{x}}_{P|B|E}.\end{aligned} \tag{39}$$

The inertia matrix and the Coriolis and centripetal terms matrix of the translating cylinder being:

$$m_c.\mathbf{J}_{B_i}^T.\mathbf{J}_{B_i}.\mathbf{T}, \tag{40}$$

$$m_c.\mathbf{J}_{B_i}^T.\frac{\mathrm{d}}{\mathrm{d}t}\left(\mathbf{J}_{B_i}.\mathbf{T}\right). \tag{41}$$

These matrices represent the inertia matrix and the Coriolis and centripetal terms matrix of a virtual moving platform that is equivalent to each translating cylinder.

On the other hand, the angular momentum of each cylinder can be represented in frame {B} by:

$$\mathbf{H}_{C_i|B} = \mathbf{I}_{C_i(rot)|B}.{}^B\boldsymbol{\omega}_{l_i|B}. \tag{42}$$

It is convenient to express the inertia matrix of the rotating cylinder in a frame fixed to the cylinder itself, $\{C_i\} \equiv \{\mathbf{x}_{C_i}, \mathbf{y}_{C_i}, \mathbf{z}_{C_i}\}$. So,

$$\mathbf{I}_{C_i(rot)|B} = {}^B\mathbf{R}_{C_i}.\mathbf{I}_{C_i(rot)|C_i}.{}^B\mathbf{R}_{C_i}^T, \tag{43}$$

where ${}^B\mathbf{R}_{C_i}$ is the orientation matrix of each cylinder frame, $\{C_i\}$, relative to base frame, {B}.

Cylinder frames were chosen in the following way: axis $\mathbf{x}_{C_i}$ coincides with the leg axis and points towards P_i, axis $\mathbf{y}_{C_i}$ is perpendicular to $\mathbf{x}_{C_i}$ and always parallel to the base plane, this condition being possible given the existence of a universal joint at B_i, that negates any rotation along its own axis; axis $\mathbf{z}_{C_i}$ completes the referential following the right hand rule, and its projection along axis $\mathbf{z}_B$ is always positive. Thus, matrix ${}^B\mathbf{R}_{C_i}$ becomes:

$$ {}^B\mathbf{R}_{C_i} = \begin{bmatrix} \mathbf{x}_{C_i} & \mathbf{y}_{C_i} & \mathbf{z}_{C_i} \end{bmatrix}, \tag{44} $$

where

$$ \mathbf{x}_{C_i} = \hat{\mathbf{l}}_i, \tag{45} $$

$$ \mathbf{y}_{C_i} = \begin{bmatrix} -\frac{l_{iy}}{\sqrt{l_{ix}^2+l_{iy}^2}} & \frac{l_{ix}}{\sqrt{l_{ix}^2+l_{iy}^2}} & 0 \end{bmatrix}^T, \tag{46} $$

$$ \mathbf{z}_{C_i} = \mathbf{x}_{C_i} \times \mathbf{y}_{C_i}. \tag{47} $$

So, the inertia matrices of the cylinders can be written as (48) where $I_{C_{xx}}$, $I_{C_{yy}}$ and $I_{C_{zz}}$ are the cylinders moments of inertia expressed in its own frame.

$$ \mathbf{I}_{C_i(rot)|C_i} = \mathrm{diag}\left(\begin{bmatrix} I_{C_{xx}} & I_{C_{yy}} & I_{C_{zz}} \end{bmatrix}\right). \tag{48} $$

Introducing Jacobian $\mathbf{J}_{D_i}$ and matrix transformation $\mathbf{T}$ in (42) results into:

$$ \mathbf{H}_{C_i|B} = \mathbf{I}_{C_i(rot)|B}.\mathbf{J}_{D_i}.\mathbf{T}.{}^B\dot{\mathbf{x}}_{P|B|E}. \tag{49} $$

The kinetic component of the generalized force applied to the cylinder, due to its rotation and expressed in {B} can be obtained from the time derivative of (49):

$$ \begin{aligned} {}^{C_i}\mathbf{f}_{C_i(kin)(rot)|B} &= \dot{\mathbf{H}}_{C_i|B} \\ &= \frac{\mathrm{d}}{\mathrm{d}t}\left(\mathbf{I}_{C_i(rot)|B}.\mathbf{J}_{D_i}.\mathbf{T}\right).{}^B\dot{\mathbf{x}}_{P|B|E} + \mathbf{I}_{C_i(rot)|B}.\mathbf{J}_{D_i}.\mathbf{T}.{}^B\ddot{\mathbf{x}}_{P|B|E}. \end{aligned} \tag{50} $$

When (50) is pre-multiplied by $\mathbf{J}_{D_i}^T$ the kinetic component of the generalized force applied to {P} due to each cylinder rotation is obtained in frame {B}:

$$ \begin{aligned} {}^P\mathbf{f}_{C_i(kin)(rot)|B} &= \mathbf{J}_{D_i}^T.{}^{C_i}\mathbf{f}_{C_i(kin)(rot)|B} \\ &= \mathbf{J}_{D_i}^T.\frac{\mathrm{d}}{\mathrm{d}t}\left(\mathbf{I}_{C_i(rot)|B}.\mathbf{J}_{D_i}.\mathbf{T}\right).{}^B\dot{\mathbf{x}}_{P|B|E} \\ &\quad + \mathbf{J}_{D_i}^T.\mathbf{I}_{C_i(rot)|B}.\mathbf{J}_{D_i}.\mathbf{T}.{}^B\ddot{\mathbf{x}}_{P|B|E}. \end{aligned} \tag{51} $$

The inertia matrix and the Coriolis and centripetal terms matrix of the rotating cylinder may be written as:

$$ \mathbf{J}_{D_i}^T.\mathbf{I}_{C_i(rot)|B}.\mathbf{J}_{D_i}.\mathbf{T}, \tag{52} $$

$$ \mathbf{J}_{D_i}^T.\frac{\mathrm{d}}{\mathrm{d}t}\left(\mathbf{I}_{C_i(rot)|B}.\mathbf{J}_{D_i}.\mathbf{T}\right). \tag{53} $$

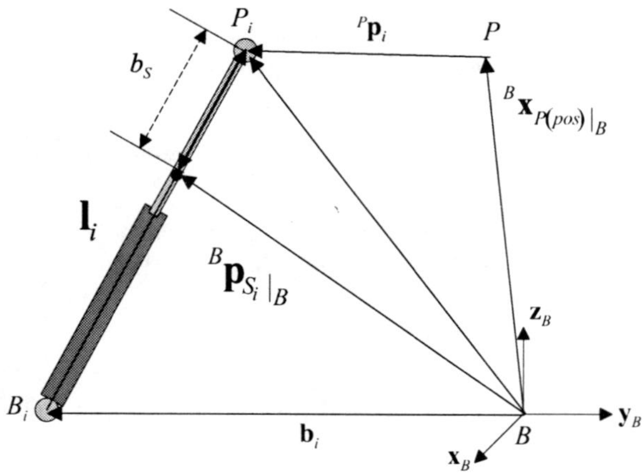

Fig. 3 Position of the centre of mass of the piston *i*

3.3 Piston Modeling

If the centre of mass of each piston is located at a constant distance, b_S, from the piston to moving platform connecting point, P_i (Fig. 3), then its position relative to frame {B} is:

$$^B\mathbf{p}_{S_i|B} = -b_S.\hat{\mathbf{l}}_i + ^B\mathbf{p}_{i|B} + ^B\mathbf{x}_{P(pos)|B}. \tag{54}$$

The linear velocity of the piston centre of mass, $^B\dot{\mathbf{p}}_{S_i|B}$, relative to {B} and expressed in the same frame, may be computed as:

$$^B\dot{\mathbf{p}}_{S_i|B} = \dot{\mathbf{l}}_i + ^B\boldsymbol{\omega}_{l_i|B} \times \left(-b_S.\hat{\mathbf{l}}_i\right), \tag{55}$$

$$^B\dot{\mathbf{p}}_{S_i|B} = \mathbf{J}_{G_i}\cdot\begin{bmatrix} ^B\mathbf{v}_{P|B} \\ ^B\boldsymbol{\omega}_{P|B} \end{bmatrix}, \tag{56}$$

where the Jacobian $\mathbf{J}_{G_i}$ is given by:

$$\mathbf{J}_{G_i} = \left[\mathbf{I} - b_S.\tilde{\hat{\mathbf{l}}}_i^T.\tilde{\hat{\mathbf{l}}}_i \left(\mathbf{I} - b_S.\tilde{\hat{\mathbf{l}}}_i^T.\tilde{\hat{\mathbf{l}}}_i\right).^P\tilde{\mathbf{p}}_{i|B}^T\right]. \tag{57}$$

The linear momentum of each piston, $\mathbf{Q}_{S_i|B}$, can be represented in frame {B} as:

$$\mathbf{Q}_{S_i|B} = m_S.^B\dot{\mathbf{p}}_{S_i|B}, \tag{58}$$

where m_S is the piston mass.

Introducing Jacobian $\mathbf{J}_{G_i}$ and matrix transformation $\mathbf{T}$ in the previous equation results into:

$$\mathbf{Q}_{S_i|B} = m_S.\mathbf{J}_{G_i}.\mathbf{T}.^B\dot{\mathbf{x}}_{P|B|E}. \tag{59}$$

Kinetic component of the force applied to the piston, due to its translation, expressed in {B}, is obtained by the time derivative of (59):

$$^{S_i}\mathbf{f}_{S_i(kin)(tra)|B} = \dot{\mathbf{Q}}_{S_i|B} = m_S.\frac{\mathrm{d}}{\mathrm{d}t}\left(\mathbf{J}_{B_i}.\mathbf{T}\right).^{B}\dot{\mathbf{x}}_{P|B|E} + m_S.\mathbf{J}_{B_i}.\mathbf{T}.^{B}\ddot{\mathbf{x}}_{P|B|E}. \tag{60}$$

When (60) is multiplied by $\mathbf{J}_{G_i}^T$, the kinetic component of the force applied to {P} due to each piston translation is obtained in frame {B}:

$$\begin{aligned} ^{P}\mathbf{f}_{S_i(kin)(tra)|B} &= \mathbf{J}_{G_i}^T.^{S_i}\mathbf{f}_{S_i(kin)(tra)|B} \\ &= m_S.\mathbf{J}_{G_i}^T\frac{\mathrm{d}}{\mathrm{d}t}\left(\mathbf{J}_{G_i}.\mathbf{T}\right).^{B}\dot{\mathbf{x}}_{P|B|E} + m_S.\mathbf{J}_{G_i}^T.\mathbf{J}_{G_i}.\mathbf{T}.^{B}\ddot{\mathbf{x}}_{P|B|E}. \end{aligned} \tag{61}$$

The inertia matrix and the Coriolis and centripetal terms matrix of the translating piston are:

$$m_S.\mathbf{J}_{G_i}^T.\mathbf{J}_{G_i}.\mathbf{T}, \tag{62}$$

$$m_S.\mathbf{J}_{G_i}^T\frac{\mathrm{d}}{\mathrm{d}t}\left(\mathbf{J}_{G_i}.\mathbf{T}\right). \tag{63}$$

On the other hand, the angular momentum of each piston can be represented in frame {B} by (65) and (65), where $^{B}\mathbf{R}_{S_i}$ is the orientation matrix of each piston frame, $\{S_i\}$, relative to the base frame, {B}.

$$\mathbf{H}_{S_i|B} = \mathbf{I}_{S_i(rot)|B}.^{B}\boldsymbol{\omega}_{l_i|B}, \tag{64}$$

$$\mathbf{I}_{S_i(rot)|B} = ^{B}\mathbf{R}_{S_i}.\mathbf{I}_{S_i(rot)|S_i}.^{B}\mathbf{R}_{S_i}^T. \tag{65}$$

As the relative motion between cylinder and piston is a pure translation, $\{S_i\}$ can be chosen parallel to $\{C_i\}$ and, therefore, $^{B}\mathbf{R}_{S_i} = ^{B}\mathbf{R}_{C_i}$.

So, the inertia matrices of the pistons can be written as (66) where $I_{S_{xx}}$, $I_{S_{yy}}$ and $I_{S_{zz}}$ are the pistons moments of inertia expressed in its own frame.

$$\mathbf{I}_{S_i(rot)|S_i} = \mathrm{diag}\left(\left[\, I_{S_{xx}} \quad I_{S_{yy}} \quad I_{S_{zz}} \,\right]\right). \tag{66}$$

Introducing Jacobian $\mathbf{J}_{D_i}$ and matrix transformation $\mathbf{T}$ in (64) results into:

$$\mathbf{H}_{S_i|B} = \mathbf{I}_{S_i(rot)|B}.\mathbf{J}_{D_i}.\mathbf{T}.^{B}\dot{\mathbf{x}}_{P|B|E}. \tag{67}$$

The kinetic component of the generalized force applied to the piston, due to its rotation and expressed in {B} can be obtained from the time derivative of (67):

$$\begin{aligned} ^{S_i}\mathbf{f}_{S_i(kin)(rot)|B} &= \dot{\mathbf{H}}_{S_i|B} \\ &= \frac{\mathrm{d}}{\mathrm{d}t}\left(\mathbf{I}_{S_i(rot)|B}.\mathbf{J}_{D_i}.\mathbf{T}\right).^{B}\dot{\mathbf{x}}_{P|B|E} + \mathbf{I}_{S_i(rot)|B}.\mathbf{J}_{D_i}.\mathbf{T}.^{B}\ddot{\mathbf{x}}_{P|B|E}. \end{aligned} \tag{68}$$

Pre-multiplied by $\mathbf{J}_{D_i}^T$, the kinetic component of the generalized force applied to {P} due to each piston rotation is obtained in frame {B}:

$$\begin{aligned} {}^P\mathbf{f}_{S_i(kin)(rot)|B} &= \mathbf{J}_{D_i}^T . {}^{S_i}\mathbf{f}_{S_i(kin)(rot)|B} \\ &= \mathbf{J}_{D_i}^T . \frac{\mathrm{d}}{\mathrm{d}t}\left(\mathbf{I}_{S_i(rot)|B} . \mathbf{J}_{D_i} . \mathbf{T}\right) . {}^B\dot{\mathbf{x}}_{P|B|E} \\ &\quad + \mathbf{J}_{D_i}^T . \mathbf{I}_{S_i(rot)|B} . \mathbf{J}_{D_i} . \mathbf{T} . {}^B\ddot{\mathbf{x}}_{P|B|E} . \end{aligned} \tag{69}$$

The inertia matrix and the Coriolis and centripetal terms matrix of the rotating piston will be:

$$\mathbf{J}_{D_i}^T . \mathbf{I}_{S_i(rot)|B} . \mathbf{J}_{D_i} . \mathbf{T}, \tag{70}$$

$$\mathbf{J}_{D_i}^T . \frac{\mathrm{d}}{\mathrm{d}t}\left(\mathbf{I}_{S_i(rot)|B} . \mathbf{J}_{D_i} . \mathbf{T}\right). \tag{71}$$

It should be noted that rigid bodies' inertia and Coriolis and centripetal terms matrices can be added, as they are expressed in the same frame.

3.4 Dynamic Model Gravitational Components

Given a general frame $\{x, y, z\}$, with $\mathbf{z} \equiv -\hat{\mathbf{g}}$, the potential energy of a rigid body is given by (72) where m_c is the body mass, g is the modulus of the gravitational acceleration and z_c the distance, along $\mathbf{z}$, from the frame origin to the body centre of mass.

$$P_c = m_c . g . z_c . \tag{72}$$

The gravitational components of the generalized forces acting on {P} can be obtained from the potential energy of the bodies that compose the system:

$${}^P\mathbf{f}_{P(gra)|B|E} = \frac{\partial P_p({}^B\mathbf{x}_{P|B|E})}{\partial {}^B\mathbf{x}_{P|B|E}}, \tag{73}$$

$${}^P\mathbf{f}_{C_i(gra)|B|E} = \frac{\partial P_{C_i}({}^B\mathbf{x}_{P|B|E})}{\partial {}^B\mathbf{x}_{P|B|E}}, \tag{74}$$

$${}^P\mathbf{f}_{S_i(gra)|B|E} = \frac{\partial P_{S_i}({}^B\mathbf{x}_{P|B|E})}{\partial {}^B\mathbf{x}_{P|B|E}} \tag{75}$$

The three vectors ${}^P\mathbf{f}_{P(gra)|B|E}$, ${}^P\mathbf{f}_{C_i(gra)|B|E}$ and ${}^P\mathbf{f}_{S_i(gra)|B|E}$ represent the gravitational components of the generalized forces acting on {P}, expressed using the Euler angles system, due to the moving platform, each cylinder and each piston. Therefore, to be added to the kinetic force components, these vectors must be transformed, to be expressed in frame {B}. This may be done pre-multiplying the gravitational components force vectors by the matrix:

$$\mathrm{diag}\left(\left[\Im \quad \mathbf{J}_A^{-T}\right]\right). \tag{76}$$

4 Conclusions

In this paper an approach based on the manipulator generalized momentum was explored and applied to the dynamic modeling of a Stewart platform. Analytic expressions for the rigid bodies' inertia and Coriolis and centripetal terms matrices are obtained, which can be added, as they are expressed in the same frame. Having these matrices, the kinetic component of the generalized force acting on the moving platform may be easily computed. This component can be added to the gravitational part of the generalized force, which is obtained through the manipulator potential energy.

References

1. W. Do, D. Yang, Inverse dynamic analysis and simulation of a platform type of robot. J. Robot. Syst. **5**, 209–227 (1988)
2. C. Reboulet, T. Berthcmieu, Dynamic models of a six degree of freedom parallel manipulators, in *IEEE Int. Conf. on Robotics and Automation* (1991), pp. 1153–1157
3. Z. Ji, Dynamics decomposition for Stewart platforms. ASME J. Mech. Des. **116**, 67–69 (1994)
4. B. Dasgupta, T. Mruthyunjaya, A Newton-Euler formulation for the inverse dynamics of the Stewart platform manipulator. Mech. Mach. Theory **34**, 711–725 (1998)
5. W. Khalil, O. Ibrahim, General solution for the dynamic modelling of parallel robots. J. Intell. Robot Syst. **49**, 19–37 (2007)
6. S. Riebe, H. Ulbrich, Modelling and online computation of the dynamics of a parallel kinematic with six degrees-of-freedom. Arch. Appl. Mech. **72**, 817–829 (2003)
7. H. Guo, H. Li, Dynamic analysis and simulation of a six degree of freedom Stewart platform manipulator. J. Mech. Eng. Sci. **220**, 61–72 (2006). *Proceedings of the Institution of Mechanical Engineers, Part C*
8. C. Nguyen, F. Pooran, Dynamic analysis of a 6 dof CKCM robot end-effector for dual-arm telerobot systems. Robot. Auton. Syst. **5**, 377–394 (1989)
9. G. Lebret, F. Lewis, Dynamic analysis and control of a Stewart platform manipulator. J. Robot. Syst. **10**, 629–655 (1993)
10. R.D. Gregório, V. Parenti-Castelli, Dynamics of a class of parallel wrists. J. Mech. Des. **126**, 436–441 (2004)
11. F. Caccavale, B. Siciliano, L. Villani, The tricept robot: dynamics and impedance control. IEEE/ASME Trans. Mech. **8**, 263–268 (2003)
12. S. Staicu, X.J. Liu, J. Wang, Inverse dynamics of the half parallel manipulator with revolute actuators. Nonlinear Dyn. **50**, 1–12 (2007)
13. L.W. Tsai, Solving the inverse dynamics of a Stewart-Gough manipulator by the principle of virtual work. J. Mech. Des. **122**, 3–9 (2003)
14. J. Gallardo, J. Rico, A. Frisoli, D. Checcacci, M. Bergamasco, Dynamics of parallel manipulators by means of screw theory. Mech. Mach. Theory **38**, 1113–1131 (2003)
15. A. Lopes, A computational efficient approach to the dynamic modeling of 6-dof parallel manipulators, in *Proc. of the ENOC'08* (2008)

Numerical Solution of a PDE System with Non-Linear Steady State Conditions that Translates the Air Stripping Pollutants Removal

Ana C. Meira Castro, J. Matos, and A. Gavina

Abstract This work deals with the numerical simulation of air stripping process for the pre-treatment of groundwater used in human consumption. The model established in steady state presents an exponential solution that is used, together with the Tau Method, to get a spectral approach of the solution of the system of partial differential equations associated to the model in transient state.

Keywords Tau method · Partial differential equations · Air stripping · Volatile organic compounds

1 Background

The air stripping process in packed columns is a physical process traditionally used in the groundwater volatile organic compounds (VOCs) removal [1–4]. This operation, that is carried out without any chemical reaction, has as main characteristic

A.C. Meira Castro · J. Matos · A. Gavina
LEMA—Laboratório de Engenharia Matemática, Porto, Portugal

A.C. Meira Castro (✉) · J. Matos · A. Gavina
ISEP—Instituto Superior de Engenharia do Porto, Porto, Portugal
e-mail: amc@isep.ipp.pt

J. Matos
e-mail: jma@isep.ipp.pt

A. Gavina
e-mail: alg@isep.ipp.pt

A.C. Meira Castro
CIGAR—Centro de Investigação em Geo-Ambiente e Recursos, Porto, Portugal

J. Matos
CMUP—Centro de Matemática da Universidade do Porto, Porto, Portugal

J.A.T. Machado et al. (eds.), *Nonlinear Science and Complexity*,
DOI 10.1007/978-90-481-9884-9_26,

the fact that operates with counter-current phases. Thus, through a pump group, the groundwater is caught from the soil to be introduced at the top of the column as drops, which constitutes a discontinuous phase, as far as the drops are able to flow through the packing material at the same time a compressor introduces, in counter-current from the base of the column, clean air as a continuous phase. In this air stripping operation, the packing material is used to supply the area for contact between the gas and the liquid needed for the contaminant mass transfer. This type of technology operates under level values for pressure and temperature generally near the typical ones from the common environment, is ideal for pollutant concentration levels under 200 mg/l and offers a level of removal often higher than 90% [5–8].

2 The Differential Model

Castro [9] presents a mathematical model that translates the space-time dynamics of the air stripping process in a packed column. In this model it is considered that exists only one space dimension, that the variation in time is limitless, that the mass transfer is based on the "Two Films Theory" [10, 11], that the air used in the VOCs removal is pure, that the flows are constant in all column. This model also considers that the system works under constant temperature and pressure values and in uniform conditions [12, 13].

Considering the following referential, see Fig. 1, in which the origin of the space is the base of the column, for the velocities u_L and u_G, corresponding VOC mass concentrations, x_{in} and y_{in}.

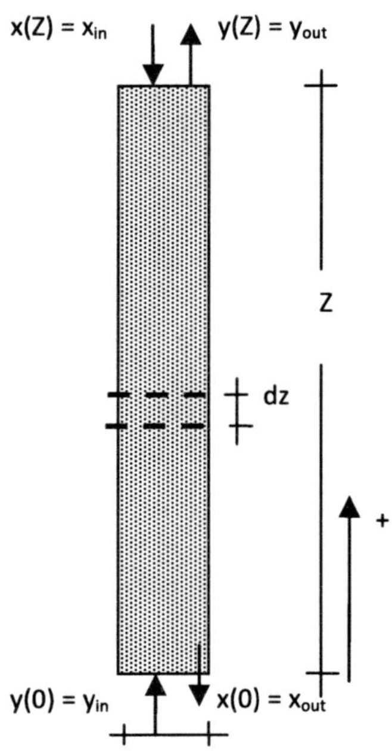

Fig. 1 The dynamic model referential

Considering that γ represents the volumetric relation of debits, K the global mass transfer coefficient, H the inverse of the dimensionless value for the Henry's constant for the VOC and ϵ the porosity of the packing material, the global dynamic system that translates the air stripping process is represented by the following system of equations [14–20]

$$\begin{cases} \dfrac{\epsilon}{1+\gamma}\dfrac{\partial x}{\partial t} + \dfrac{\epsilon\gamma}{1+\gamma}\dfrac{\partial y}{\partial t} = u_L\dfrac{\partial x}{\partial z} - u_G\dfrac{\partial y}{\partial z}, \\ \dfrac{\epsilon}{1+\gamma}\dfrac{\partial x}{\partial t} = u_L\dfrac{\partial x}{\partial z} - K(x - Hy), \end{cases} \quad 0 < z < Z,\ t > 0. \tag{1}$$

2.1 The Boundary Conditions

In this model, the characterization of the dynamic state of the air stripping operation implies the consideration of Dirichlet boundary conditions that can be translated with the input data of the concentrations of the liquid and gaseous phases, at the column entrance, i.e.

$$\begin{cases} x(t, Z) = x_{in}(t), \\ y(t, 0) = 0, \end{cases} \quad t \geq 0. \tag{2}$$

This problem is exactly determined and can be solved when the values of the boundary conditions and the disturbances at the entrance are specified.

2.2 The Steady State

The representation of the steady state can be achieved from the consideration, in the global model, that the time derivatives are all null, which implies that the VOC concentration gradients, in the gaseous phase and the liquid phase, can be given by

$$\begin{cases} \dfrac{\partial}{\partial z}x(0, z) = \dfrac{K}{u_L}[x(0, z) - Hy(0, z)], \\ \dfrac{\partial}{\partial z}y(0, z) = \dfrac{K}{u_G}[x(0, z) - Hy(0, z)], \end{cases} \quad 0 < z < Z. \tag{3}$$

The analytic solution for the steady state equations can be found deriving the first equation of (3) with respect to z

$$\frac{\partial^2}{\partial z^2}x(0, z) = \frac{K}{u_L}\left(\frac{\partial}{\partial z}x(0, z) - H\frac{\partial}{\partial z}y(0, z)\right) \tag{4}$$

and, since we know that $\frac{\partial}{\partial z}y(0, z) = \frac{u_L}{u_G}\frac{\partial}{\partial z}x(0, z)$, then (4) can be written as a second-order linear homogeneous ordinary differential equation (ODE) with constant coefficients

$$\frac{\partial^2}{\partial z^2}x(0,z) - D\frac{\partial}{\partial z}x(0,z) = 0 \tag{5}$$

with $D = K(\frac{1}{u_L} - \frac{H}{u_G}) \neq 0$, in order to guarantee that the process occurs. The general solution for this equation is given by

$$x(0,z) = m_{11} + m_{12}e^{Dz}. \tag{6}$$

In the same way, we have the general solution for $y(0,z)$

$$y(0,z) = m_{21} + m_{22}e^{Dz}. \tag{7}$$

According to the boundary conditions, for $t = 0$, we have

$$\begin{cases} x(0,Z) = x_{in}(0), \\ y(0,0) = 0. \end{cases} \tag{8}$$

If we substitute (6) and (7) in (8) we have the solution for the system of differential equations in steady state

$$\begin{cases} x_S(z) \equiv x(0,z) = M\left(\dfrac{u_G}{u_L}e^{Dz} - H\right), & 0 < z < Z \\ y_S(z) \equiv y(0,z) = M(e^{Dz} - 1), \end{cases} \tag{9}$$

with

$$M = X_0\left(\frac{u_G}{u_L}e^{DZ} - H\right)^{-1}$$

and $X_0 = x_{in}(0)$.

3 The Transient State

Defining $A = \varepsilon/(1+\gamma)$, (1) of the model in transient state can be written as

$$\begin{cases} A\dfrac{\partial x}{\partial t} = u_L\dfrac{\partial x}{\partial z} - K(x - Hy), \\ A\gamma\dfrac{\partial y}{\partial t} = -u_G\dfrac{\partial y}{\partial z} + K(x - Hy), \end{cases} \tag{10}$$

with $x = x(t,z)$ and $y = y(t,z)$ for $0 < z < Z,\ t > 0$.

In order to approximate the solution of (10), subject to initial conditions (2), the functions x and y can be represented by

$$\begin{cases} x(t,z) = tx_S(z) + x_T(t,z), \\ y(t,z) = ty_S(z) + y_T(t,z), \end{cases} \tag{11}$$

where x_S and y_S are the solutions (9) for the steady state, and x_T and y_T are the initial solutions for the transient state. Replacing this expressions in (10) and using (3), we have for x_T and y_T the following differential equations

$$\begin{cases} A\dfrac{\partial x_T}{\partial t} - u_L\dfrac{\partial x_T}{\partial z} + K(x_T - Hy_T) = -Ax_S, \\ A\gamma\dfrac{\partial y_T}{\partial t} + u_G\dfrac{\partial y_T}{\partial z} - K(x_T - Hy_T) = -A\gamma y_S. \end{cases} \tag{12}$$

And, for the initial conditions,

$$\begin{cases} x_T(t, Z) = x_{in}(t) - tX_0, \\ y_T(t, 0) = 0. \end{cases} \tag{13}$$

This representation allows to approach the solution for the system of differential equations, maintaining the exact form of the solutions for the steady state and reducing the error to the initial solutions of the transient state.

4 The Lanczos's Tau Method

Given the differential problem

$$\begin{cases} Dy = f, \\ D_j y = f_j, \quad j = 1 : \nu, \end{cases} \tag{14}$$

where D represents a linear differential operator with polynomial coefficients, f and f_j are polynomials and D_j represents the ν initial or boundary conditions associated to the problem, the Tau method [21, 22] can be formulated by solving the perturbed problem

$$\begin{cases} Dy_n = f + \tau_n, \\ D_j y_n = f_j, \quad j = 1 : \nu, \end{cases}$$

where τ_n is a perturbation term that allows the perturbed differential problem to have a unique polynomial solution y_n.

Following the implementation of the Tau method proposed in Matos et al. [23], based in Ortiz's operational approach [24] and in the formulation of the Tau method presented in [25, 26] we begin by chose m and n the degrees of the polynomial

$$\begin{cases} x_{m,n}(t, z) = \displaystyle\sum_{i,j=0}^{m,n} x_{i,j} t^i z^j, \\ y_{m,n}(t, z) = \displaystyle\sum_{i,j=0}^{m,n} y_{i,j} t^i z^j, \end{cases} \tag{15}$$

and we proceed by determining the coefficients $x_{i,j}$ and $y_{i,j}$ imposing that $x_{m,n}(t, z)$ and $y_{m,n}(t, z)$ satisfies the initial and the boundary conditions of the problem and satisfies the perturbed differential equation.

4.1 Initial Conditions and Differential Equations

In the case of the initial conditions (13), with

$$x_{in}(t) = \sum_{i=0}^{m} X_i t^i$$

being a, at least approximated, polynomial representation of the liquid phase initial VOC concentration, in the top of the column $x_{in}(t) = x(t, Z)$ then

$$\begin{cases} \sum_{j=0}^{n} x_{0,j} Z^j = X_0, \\ \sum_{j=0}^{n} x_{1,j} Z^j = X_1 - X_0, \\ \sum_{j=0}^{n} x_{i,j} Z^j = X_i, \quad i = 2:m. \end{cases}$$

The initial condition for the gas phase initial VOC concentration, in the bottom of the column, is given by $y_{m,n}(t, 0) = 0$. Then, we get

$$y_{i,0} = 0, \quad i = 0:m.$$

The determination of the $2(m+1)(n+1)$ unknowns $x_{i,j}$ and $y_{i,j}$ needs more $2(m+1)n$ equations that come from the differential equations imposed to $x_{m,n}$ and to $y_{m,n}$ as far as possible [25, 26], in the sense of the respective polynomial degrees.

In that case, the non polynomial character of the right hand side of the differential equations, introduced by the exponential solution of the steady state problem, introduces another approximation problem. To have a polynomial approximation of the steady state problem solution, we proceed by use the Taylor approximation of the exponential function

$$e^{Dz} \approx \sum_{j=0}^{n} \frac{D^j}{j!} z^j,$$

to get

$$\begin{cases} x_S(z) \approx M\left(\frac{u_G}{u_L} - H\right) + M\frac{u_G}{u_L} \sum_{j=1}^{n} \frac{D^j}{j!} z^j, \\ y_S(z) \approx M \sum_{j=1}^{n} \frac{D^j}{j!} z^j. \end{cases}$$

Differentiating (15) we get

$$\begin{cases} \dfrac{\partial}{\partial t} x_{m,n}(t,z) = \displaystyle\sum_{i=0}^{m-1}\sum_{j=0}^{n}(i+1)x_{i+1,j}t^i z^j, \\ \dfrac{\partial}{\partial z} x_{m,n}(t,z) = \displaystyle\sum_{i=0}^{m}\sum_{j=0}^{n-1}(j+1)x_{i,j+1}t^i z^j, \end{cases} \tag{16}$$

for the partial derivatives of the polynomial $x_{m,n}(t,z)$ and analog expressions for the partial derivatives of $y_{m,n}(t,z)$. Introducing those formulas in (12) we get

$$A\sum_{i=0}^{m-1}\sum_{j=0}^{n}(i+1)x_{i+1,j}t^i z^j - u_L\sum_{i=0}^{m}\sum_{j=0}^{n-1}(j+1)x_{i,j+1}t^i z^j$$
$$+ K\sum_{i=0}^{m}\sum_{j=0}^{n}(x_{i,j} - Hy_{i,j})t^i z^j = -AM\left(-H + \frac{u_G}{u_L}\sum_{j=0}^{n}\frac{D^j}{j!}z^j\right)$$

and

$$A\gamma\sum_{i=0}^{m-1}\sum_{j=0}^{n}(i+1)y_{i+1,j}t^i z^j + u_G\sum_{i=0}^{m}\sum_{j=0}^{n-1}(j+1)y_{i,j+1}t^i z^j$$
$$- K\sum_{i=0}^{m}\sum_{j=0}^{n}(x_{i,j} - Hy_{i,j})t^i z^j = -A\gamma M\sum_{j=1}^{n}\frac{D^j}{j!}z^j$$

and, identifying term by term the corresponding coefficients, this results in a algebraic over-determinated system of $2(m+1)(n+1)$ linear equations.

4.2 Algebraic Equations

By applying the Tau method we choose suitable [23, 27] $2(m+1)n$ equations defining a regular linear system of algebraic equations whose unique solution $x_{i,j}$ and $x_{i,j}$ give us the polynomial approximation of the exact solution.

With $x_{in}(t) = X_0$ constant, the solution $x(t,z)$ and $y(t,z)$ of the problem coincides with $x(0,z)$ e $y(0,z)$, the solution of the steady state (9).

For the more general situation where $x_{in}(t)$ depends on t, we have supposed that $x_{in}(t) = \sum_{i=0}^{m} X_i t^i$ at least by approximation. That is, if the function $x_{in}(t)$ is not a $\leq m$ degree polynomial, it is replaced by a polynomial approximation.

In the following test, see Fig. 2, we consider a periodic function for $x_{in}(t)$. In the case, we made

$$x_{in}(t) = 100(1+\sin(t)/5) = 100\left(1 + \frac{1}{5}\sum_{i\geq 0}\frac{(-1)^i}{(2i+1)!}t^{2i+1}\right).$$

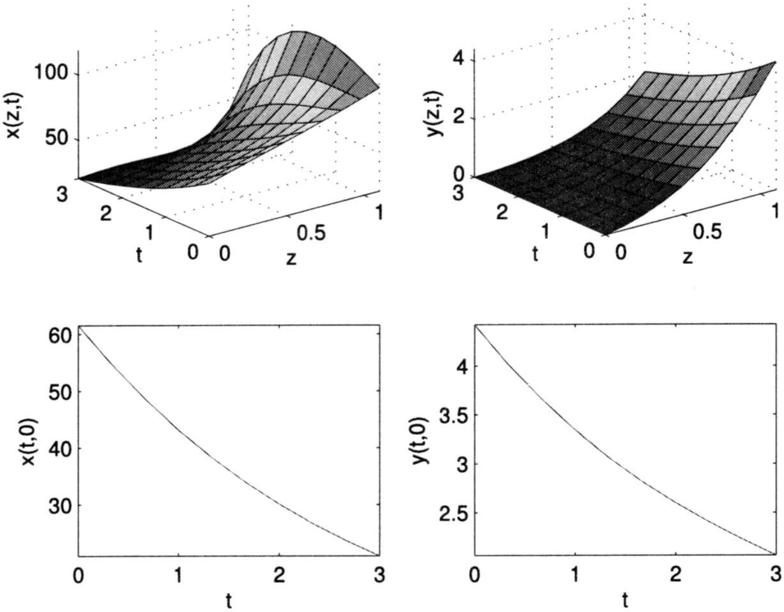

Fig. 2 Variation of the contaminant concentration in the water $x(t,z)$ and in the air $y(t,z)$ with $m=n=3$

5 Conclusions and Future Work

The Tau method was tested for several polynomial degrees and for the case study presented in [9].

The numerical results that were found revealed to be coherent with the experimentally achieved data [9], which suggests that the model is consistent under a numerical, physical and mathematical approach and that the use of the Tau method allows an approach, with reasonable precision, to the exact solution of the model.

As the numerical results achieved show some increasing instability with the time, it can be suggested that the use of orthogonal polynomials will be able to allow the increment of the precision of the results in an larger time interval.

References

1. E. Bouwer, M.C. Kavanaugh, Coping with groundwater contamination. J. Water Polut. Control Fed. **60**, 1415–1427 (1988)
2. A.O. Bradley, Remove VOCs from wastewater by air stripping. Chem. Eng. Progr. **60**, 89–93 (1992)
3. W.V. Chang, Stripping of TCE from water at low air/water ratios. Environ. Progr. **14**, 111–114 (1995)
4. D. Hand, Design and evaluation of an air stripping tower for removing VOCs from groundwater. J. Am. Water Works Assoc., 87–97 (1986)

5. J.G. Kunesh, Recent developments in packed columns. Can. J. Chem. Eng. **65**, 907–913 (1987)
6. A. Mersmann, A. Deixler, Packed columns. Ger. Chem. Eng., 265–276 (1986)
7. R.W. Rousseau, Analyzing chemical absorbers and strippers. Chem. Eng. **June**, 91–95 (1985)
8. W. Yang, Removal of trihalomethanes from water by air stripping. Water Treat. **June**, 267–282 (1991)
9. A.C.M. Castro, Projecto e simulação dinâmica utilizando um sistema de parâmetros distribuidos, PhD Thesis, Faculdade de Engenharia da Universidade do Porto, Portugal, 2004
10. J.D. Seader, E.J. Henley, *Separation Process Principles* (Wiley, New York, 1998)
11. J.L. Schnoor, *Environmental Modelling—Fate and Transport of Pollutants in Water, Air and Soil* (Wiley, New York, 1996)
12. I.A. Furzer, Liquid dispersion in packed columns I. Chem. Eng. Sci. **39**, 70–76 (1984)
13. I.A. Furzer, Liquid dispersion in packed columns II. Chem. Eng. Sci. **39**, 77–80 (1984)
14. E. James, W. Byers, Limitations and practical use of mass transfer model for predicting air stripper performance. Environ. Progr. **12**, 61–66 (1993)
15. W. Ball, M.C. Kavanaugh, Mass transfer of volatile organic compounds in packed tower aeration. J. Water Polut. Control Fed. **56**, 127–136 (1984)
16. P. Lamarche, R. Droste, Air stripping mass transfer correlations for volatile organics. J. Am. Water Works Assoc., 78–89 (1989)
17. M. Leva, Reconsider packed-tower pressure-drop correlations. Chem. Eng. Progr. **January**, 65–72 (1992)
18. K. Onda, Mass transfer coefficients between gas and liquid phases in packed columns. J. Chem. Eng. Jpn. **1**, 56–62 (1968)
19. C. Yaws, Henry's law constants for organic compounds in water. Chem. Eng. **November**, 179–185 (1991)
20. V. Stanek, Transients of the hydrodynamics of counter-current packed-bed columns. Chem. Eng. Sci. **45**, 449–455 (1990)
21. C. Lanczos, Trigonometric interpolation of empirical and analytical functions. J. Math. Phys. **17**, 123–199 (1938)
22. C. Lanczos, *Applied Analysis* (Pitman, London, 1957)
23. J. Matos, M.J. Rodrigues, P.B. Vasconcelos, New implementation of the Tau method for PDEs. J. Comput. Appl. Math. **164–165**, 555–567 (2004)
24. E.L. Ortiz, The Tau method. SIAM J. Numer. Anal. Optim. **12**, 480–492 (1969)
25. M.R. Silva, Numerical treatment of differential equations with the Tau method. J. Comput. Appl. Math. **20**, 167–173 (1987)
26. M.R. Silva, M.J. Rodrigues, A simple alternative principle for rational τ-method approximation, in *Nonlinear Numerical Methods and Rational Approximation*, ed. by A. Cuyt (Reidel, Dordrecht, 1988), pp. 427–434
27. M. Hosseini Ali Abadi, E.L. Ortiz, Numerical treatment of moving and free boundary value problems with the Tau method. Comput. Math. Appl. **35**, 53–61 (1998)

Three Behavioural Scenarios for Contingent Claims Valuation in Incomplete Markets

L. Boukas, D. Pinheiro, A.A. Pinto, S.Z. Xanthopoulos, and A.N. Yannacopoulos

Abstract We describe three different but related scenarios for determination of asset prices in an incomplete market: one scenario uses a market game approach whereas the other two are based on risk sharing or regret minimizing considerations.

Keywords Incomplete markets · Market games · Risk sharing · Regret

1 Introduction

Realistic markets are incomplete, due to either the existence of frictions or to the nonexistence of the necessary assets needed to achieve the complete replication of any contingent claim by linear combinations of available (traded) assets. Incomplete

L. Boukas (✉)
Department of Information & Communication Systems Engineering, University of the Aegean, Samos, Greece
e-mail: lboukas@aegean.gr

D. Pinheiro
CEMAPRE, ISEG, Universidade Técnica de Lisboa, Lisboa, Portugal
e-mail: dpinheiro@iseg.utl.pt

A.A. Pinto
Departamento de Matemática, Universidade do Minho, Braga, Portugal
e-mail: aapinto@math.uminho.pt

S.Z. Xanthopoulos
Department of Statistics and Actuarial-Financial Mathematics, University of the Aegean, Samos, Greece
e-mail: sxantho@aegean.gr

A.N. Yannacopoulos
Department of Statistics, Athens University of Economics and Business, Athens, Greece
e-mail: ayannaco@aueb.gr

J.A.T. Machado et al. (eds.), *Nonlinear Science and Complexity*,
DOI 10.1007/978-90-481-9884-9_27,

markets is a very interesting field of economic theory and finance which through seminal studies (see [1–3]) has led to important results that have helped the community to reach a deeper understanding of the function of financial markets.

While there is a rich literature on this field, the majority of these works focuses on the determination of bounds on the prices that are consistent with general equilibrium considerations. It is well known that in an incomplete market set up there is no longer a unique pricing kernel (martingale measure) and the existence of more than one pricing kernel may at best point out a whole band of prices that do not allow for arbitrage opportunities. The determination of a single price, at which an asset will eventually be traded in this market, out of this whole band, is still an open problem. There exists an extensive and very interesting literature on the subject, focusing on the determination of the upper and lower hedging prices (see [2]) as well as a number of suggestions on the price selected by the market (e.g. Kuhlback–Leibner or related entropy minimization criteria [4]) which lead to interesting implications, some of which are testable from real market data, however, a complete theory of price selection in incomplete markets is still missing.

The aim of the papers [5, 6] is to contribute to this literature by providing types of scenarios on price selection in incomplete markets. The scenarios are based on behavioural considerations and lead to interesting results. We deliberately use a simple one period discrete model in order to reveal and clarify the concepts and ideas behind price formation rather than linger on laborious technical details that would be of interest to a more specialized audience.

2 Three Scenarios for Price Selection

We consider a one period market model with N tradable assets $a_1, \dots, a_N$ and K states of the world $\omega_1, \dots, \omega_K$. One of the assets is a riskless asset with return rate r and without loss of generality we may assume that this asset is a_1. Let $\mathbf{p} = (p_1, \dots, p_N)$ denote the price vector of the N assets at time $t = 0$, i.e. for each $i = 1, \dots, N$ we denote by p_i the price at time $t = 0$ of the asset a_i. Let also $\mathbf{D} = (d_{ij})_{\substack{i=1,\dots,K \\ j=1,\dots,N}}$ denote the $K \times N$ matrix of the assets payoffs at time $t = 1$, that is, d_{ij} denotes the value of the asset a_j at time $t = 1$ when the prevailing state of the world is ω_i. In the sequel we may use the notation $d_j(\omega_i)$ instead of d_{ij} and we may write d_j to indicate the j-th column vector of $\mathbf{D}$ without these causing any confusion. We also assume that the value of the riskless asset at time $t = 1$ is equal to 1 no matter what the prevailing state of the world is, i.e. $d_1(\omega_i) = 1$ for all $i = 1, \dots, K$ (clearly, $p_1 = 1/(1+r)$). Moreover, without loss of generality, we assume that none of the assets $a_1, \dots, a_N$ is redundant, that is, none of the columns of the payoff matrix $\mathbf{D}$ can be expressed as a linear combination of the remaining columns of the matrix. Furthermore we assume that $N < K$, that is, we face an incomplete markets situation where there are more states of the world than assets and thus the existing assets are not enough to reproduce every possible contingent claim at time $t = 1$.

2.1 Allocation of Wealth and Indifference Pricing

Assume that an agent's preferences are described by a smooth utility function $U(\cdot)$ such that $U' > 0$ (strict non-satiation) and $U'' < 0$ (strict risk aversion). These two properties guarantee the strict quasiconvexity of the utility function (see. e.g. [7]).

The agent desires to allocate her initial wealth in the market in a way that her expected utility is maximized. Suppose that the agent will allocate her initial wealth W_0 to the N assets $a_1, \ldots, a_N$ in proportions $\boldsymbol{\pi} = (\pi_1, \ldots, \pi_N)$, i.e. proportion π_i of the initial wealth will be placed in asset a_i and $\pi_1 + \cdots + \pi_N = 1$. The final wealth of this portfolio at time $t = 1$ is the random variable W_1 given by

$$W_1 = \left[W_0 \sum_{j=1,\ldots,N} \frac{\pi_j}{p_j} d_j(\omega_i) \right]_{i=1,\ldots,K}.$$

The agent will choose $\boldsymbol{\pi}$ so as to maximize $E[U(W_1)]$ where the expectation is taken under a measure $Q = (q_1, \ldots, q_K)$, reflecting the beliefs of the agent about the probabilities of occurrence of the future states of the world. So q_i denotes the perceived by the agent probability that the state ω_i will occur.

It is well known that under the assumptions made here this problem has a solution. Let $\boldsymbol{\pi}_*$ denote the solution to this maximization problem and let U_* denote the resulting maximized expected utility. Clearly both $\boldsymbol{\pi}_*$ and U_* are functions of the probability measure Q and of the initial wealth W_0.

Assume now that the seller has initial wealth W_0^S and preferences described by a utility function $U_S(\cdot)$ and that she has issued a contingent claim with payoff to the buyer, at time $t = 1$, given by the vector $d_{N+1} = (d_{N+1}(\omega_i))_{i=1,\ldots,K}$. If the seller had not sold the contingent claim, she would have allocated proportions $\boldsymbol{\pi}^S = (\pi_1^S, \ldots, \pi_N^S)$ of her wealth W_0^S to the N assets, according to the previous discussion, in order to achieve maximum expected utility $U_*^S(W_0^S; Q^S)$ under her perceived probability measure Q^S about the future states of the world.

Assume now that the seller decides to issue the contingent claim and receives the price p_{N+1}^S at time $t = 0$. The seller desires now to invest her initial wealth $W_0^S + p_{N+1}^S$ among the N assets $a_1, \ldots, a_N$ in proportions $\bar{\boldsymbol{\pi}}^S = (\bar{\pi}_1^S, \ldots, \bar{\pi}_N^S)$, i.e. proportion $\bar{\pi}_i^S$ of the initial wealth will be placed in asset a_i and $\bar{\pi}_1^S + \cdots + \bar{\pi}_N^S = 1$. The final wealth of this portfolio at time $t = 1$ is the random variable $\bar{W}_1^S$ given by

$$\bar{W}_1^S = \left[(W_0^S + p_{N+1}^S) \sum_{j=1,\ldots,N} \frac{\bar{\pi}_j^S}{p_j} d_j(\omega_i) - d_{N+1}(\omega_i) \right]_{i=1,\ldots,K}.$$

The seller will choose $\bar{\boldsymbol{\pi}}^S$ so as to maximize $E[U^S(\bar{W}_1^S)]$, where the expectation is taken under the measure $Q^S = (q_1^S, \ldots, q_K^S)$, reflecting the beliefs of the seller about the probabilities of occurrence of the future states of the world. So q_i^S denotes the perceived by the seller probability that the state ω_i will occur.

Let $\bar{\boldsymbol{\pi}}_*^S$ denote the solution to this maximization problem and let $\bar{U}_*^S$ denote the resulting maximized expected utility. Clearly both $\bar{\boldsymbol{\pi}}_*^S$ and $\bar{U}_*^S$ are functions of the probability measure Q^S, of the initial wealth $W_0^S + p_{N+1}^S$ and of the random payoff of the liability undertaken by the writer of the contingent claim d_{N+1}. The seller of the contingent claim will choose a price for this contract so that these two decisions leave her indifferent, i.e. she will choose a price p_{N+1}^S as the solution of the equation

$$U_*^S(W_0^S; Q^S) = \bar{U}_*^S(W_0^S + p_{N+1}^S, d_{N+1}; Q^S).$$

The solution of this equation will provide the seller's indifference or reservation price of the contingent claim.

Similarly, assume that the buyer has initial wealth W_0^B and preferences described by a utility function $U_B(\cdot)$ and that she decided to buy the contingent claim with payoff to the buyer, at time $t = 1$, given by the vector $d_{N+1} = (d_{N+1}(\omega_i))_{i=1,\dots,K}$. As in the case of the seller, the buyer will choose her price so that she will stay indifferent between buying or not buying the contingent claim contract. The price p_{N+1}^B will be given as the solution of the algebraic equation

$$U_*^B(W_0^B; Q^B) = \bar{U}_*^B(W_0^B - p_{N+1}^B, d_{N+1}; Q^B).$$

Assuming that such a solution exists, it is called the indifference or reservation price for the buyer of the contingent claim contract. In general, when markets are incomplete, the reservation price of a contingent claim for the seller and the reservation price of the same contingent claim for the buyer do not coincide.

2.2 Market Games Approach

The first scenario is a market game where the buyer and the seller bargain on the price of the derivative and choose the bargaining strategy that minimizes maximum regret. Given their initial valuations, this mechanism offers a unique bargaining strategy that will lead to at most one unique price (depending on their initial valuations).

Let P_B denote the value of the contingent claim to the buyer and similarly, let P_S denote the value of the contingent claim to the seller. We assume that the support of the buyer's prior as to the distribution of P_B is $[\alpha, \beta]$, equal to the support of the seller's prior as to the distribution of P_S. Let $\tilde{P}_B = \mathcal{P}_B(\tilde{Q}^B)$ be the intended bid price of the buyer and $\tilde{P}_S = \mathcal{P}_S(\tilde{Q}^S)$ be the intended ask price of the seller. Trade occurs if and only if $\tilde{P}_B \geq \tilde{P}_S$. If trade occurs, the price is $P = \frac{\tilde{P}_B+\tilde{P}_S}{2}$. Thus the buyer's profit is $\Pi_B = P_B - \frac{\tilde{P}_B+\tilde{P}_S}{2}$ if $\tilde{P}_B \geq \tilde{P}_S$, otherwise it is zero. On the other hand the sellers profit is $\Pi_S = \frac{\tilde{P}_B+\tilde{P}_S}{2} - P_S$ if $\tilde{P}_B \geq \tilde{P}_S$, otherwise it is zero. Let $\Pi_B^* = \max_{\tilde{P}_B} \Pi_B$ be the maximum profit of the buyer (i.e. the best the buyer could have done, had he known the sellers ask) and let $R_B = \Pi_B^* - \Pi_B$ be the maximum regret of the buyer and similarly for the seller.

Proposition 1 *There exist functions $\tilde{P}_B(P_B, \alpha, \beta)$ and $\tilde{P}_S(P_S, \alpha, \beta)$ minimizing, respectively, the maximum regret R_B and R_S of the buyer and seller.*

2.3 The Risk Sharing Approach

The second scenario which leads to a unique price for the asset is based on the concept of risk sharing price for the asset. In this scenario we assume that each of the agents has firm beliefs about the future prices of the world but deliberately undertakes some risk so that the transaction will be made possible. The unique price of the asset is defined by the solution of an optimization problem, in which the risk undertaken by each agent is chosen so that a convex combination of the risks undertaken by the agents is minimized, under the constraint that the transaction is made possible, i.e. under the constraint that the buyer's price is greater or equal than the seller's price.

Let $U_*^A(W^A; Q^A)$ be the maximum expected utility of an agent A that has not undertaken any position on the contingent claim, where W^A is the initial wealth of the agent and Q^A is the probability measure reflecting the beliefs of the agent about the future states of the world. Let us now assume that the seller issues the contingent claim at an initial price P_S while simultaneously adopting a position in the underlying market so that she maximizes her expected utility. In this case the maximum expected utility achieved is equal to $\bar{U}_*^S(W^S + P_S, d_{N+1}; Q^S)$. If the seller decides to undertake risk ϵ_S then the price P_S corresponding to this risk position will be given by the solution of the equation

$$U_*^S(W^S; Q^S) - \bar{U}_*^S(W^S + P_S, d_{N+1}; Q^S) = \epsilon_S.$$

We will denote the solution of this algebraic equation by $P_S(\epsilon_S)$. This depends on the risk undertaken as well as on the beliefs on the future states of the world. Note that $P_S(0)$ is the indifference price for the seller.

Let us now assume that the buyer buys the contingent claim at an initial price P_B while simultaneously adopting a position in the underlying market so that she maximizes her expected utility. The maximum utility is then $\bar{U}_*^B(W^B - P_B, d_{N+1}; Q^B)$. If the buyer decides to undertake risk ϵ_B then the price P_B corresponding to this risk position will be given by the solution of the equation

$$U_*^B(W^B; Q^B) - \bar{U}_*^B(W^B - P_B, d_{N+1}; Q^B) = \epsilon_B.$$

We will denote the solution of this algebraic equation by $P_B(\epsilon_B)$. This also depends on the risk undertaken as well as on the beliefs on the future states of the world. Note that $P_B(0)$ is the indifference price for the buyer.

In the next lemma we summarize some properties of the functions $P_S(\epsilon_S)$ and $P_B(\epsilon_B)$.

Lemma 1 *Assume that the seller and the buyer make their decisions with expected utility functions $U_S(W) = E[u_S(W)]$ and $U_B(W) = E[u_B(W)]$, respectively. Furthermore, assume that $u_i' > 0$ and $u_i'' < 0$ for $i = S, B$. Then,*

(i) *P_S is strictly decreasing and strictly quasiconcave in ϵ_S.*
(ii) *P_B is strictly increasing and strictly quasiconvex in ϵ_B.*
(iii) *The function $P_S(\epsilon_S) - P_B(\epsilon_B)$ is strictly quasiconcave.*

We may define the total risk undertaken by both agents by the convex combination

$$R(\epsilon_S, \epsilon_B) := \lambda\epsilon_S + (1-\lambda)\epsilon_B.$$

If $\lambda \in (0,1)$ this corresponds to sharing the total risk undertaken by the agents in the proportion $\lambda/(1-\lambda)$.

We then define the price of the contingent claim by the solution of the following optimization problem

$$\min_{(\epsilon_S,\epsilon_B)} \lambda\epsilon_S + (1-\lambda)\epsilon_B$$
$$\text{subject to the constraint} \quad P_S(\epsilon_S) \leq P_B(\epsilon_B). \tag{1}$$

If this problem has a unique solution $(\epsilon_S^*, \epsilon_B^*)$ this would lead to a unique price $P_S(\epsilon_S^*) = P_B(\epsilon_B^*)$. The following result shows that this optimization problem admits a unique solution and sheds some light on its properties.

Theorem 1 *Under the conditions of Lemma* 1 *the following statements hold for the optimization problem* (1):

(i) *Problem* (1) *has a unique solution which will be denoted as* $(\epsilon_B^*(\lambda), \epsilon_S^*(\lambda))$.
(ii) *The function* $\epsilon_B^*(\lambda)$ *is nonincreasing in* $\lambda \in (0,1)$ *whereas the function* $\epsilon_S^*(\lambda)$ *is nondecreasing in* $\lambda \in (0,1)$.
(iii) *On the optimal risk bearing allocation* $P_S(\epsilon_S) = P_B(\epsilon_B)$.

2.4 Optimal Choice of the Agents Market Price of Risk

The third scenario models the situation where the two agents do not have firm beliefs about the future states of the world but are willing to update their beliefs as part of the bargaining procedure. Their quoted prices thus do not entail any risk but there is some potential loss, which we call regret. The potential loss for agent 1 comes about from not being able to persuade agent 2 to accept her original belief (that would lead to the best possible price for her) and similarly for agent 2. A unique price is then chosen by the solution of an optimization problem in which the beliefs are chosen so that the convex combination of the regrets of the two agents is minimized under the constraint that the transaction eventually takes place.

The seller of the asset will of course wish to obtain at least the price that corresponds to her beliefs about the future states of the world $Q^S = \overline{Q}$. By changing her belief to some new Q^S the seller compromises to give away the potential "extra" profit corresponding to $\overline{Q} - Q^S$. On the other hand the buyer of the asset would wish to obtain at most the price that corresponds to her beliefs about the future states of the world $Q^B = \underline{Q}$. By changing her belief to some new Q^B the buyer compromises to give away the potential "extra" profit corresponding to $Q^B - \underline{Q}$. The quantity $\lambda(\overline{Q} - Q^S) + (1-\lambda)(Q^B - \underline{Q})$ can be considered as the total potential loss of the two agents, where the parameter λ gives us information on the way that this "loss" is divided between the two agents.

We may consider now the case where the agents update their beliefs so that the total potential loss is minimized. This would correspond to choosing Q^S, Q^B so that

$$\min_{Q^S,Q^B} \lambda d_S(\overline{Q} - Q^S) + (1-\lambda) d_B(Q^B - \underline{Q})$$

$$\text{subject to the constraint} \quad P_S(Q^S) \leq P_B(Q^B) \tag{2}$$

which would in turn lead to a price for the asset. In the above problem, d_B and d_S are distance functions in the unit simplex Δ^K quantifying the buyer and seller's regret respectively.

Lemma 2 *The price functions $P_S(Q^S)$, $P_B(Q^B)$ satisfy the following conditions:*

(i) *The price function $P_B(Q^B)$ is a continuous function of $Q^B \in \Delta^K$ and achieves a minimum value $\underline{P}_B$ and a maximum value $\overline{P}_B$.*
(ii) *The price function $P_S(Q^S)$ is a continuous function of $Q^S \in \Delta^K$ and achieves a minimum value $\underline{P}_S$ and a maximum value $\overline{P}_S$.*
(iii) *If $[\underline{P}_B, \overline{P}_B] \cap [\underline{P}_S, \overline{P}_S] \neq \emptyset$, there exist $Q^B, Q^S \in \Delta^K$ such that $P_B(Q^B) \geq P_S(Q^S)$.*

Using the above lemma, we may prove the existence of a price that minimizes the total regret of the buyer and the seller.

Theorem 2 *Assume further that the functions $P_B(Q^B)$ and $P_S(Q^S)$ are strictly convex. Then, the following statements hold:*

(i) *There exists a unique solution to the minimization problem* (2).
(ii) *The unique choice of the agents beliefs Q^S and Q^B corresponds to a unique price $P_B = P_S$.*

3 Conclusion

It is well known that in an incomplete markets setting, if equivalent martingale measures exist, they are not unique. Therefore, this leads to more than one possible prices, all of which are consistent with the absence of arbitrage arguments. Other criteria will therefore be needed in order to select the price at which a particular asset is traded in an incomplete market.

We propose three different, but ultimately related, scenarios for the price selection in incomplete markets. The first approach is a market game approach, the second is a risk sharing approach, whereas the third is one in which the agents update their beliefs about the possible prices of the states of the world, in a way which is consistent with the minimization of total regret.

Acknowledgements We thank the Calouste Gulbenkian Foundation, PRODYN-ESF, POCTI, and POSI by FCT and Ministério da Ciência, Tecnologia e Ensino Superior, Centro de Matemática da Universidade do Minho, CEMAPRE, and Centro de Matemática da Universidade do Porto for their financial support.

S. Xanthopoulos would like to acknowledge that this project is co-funded by the European Social Fund and National Resources—(EPEAEK-II) PYTHAGORAS.

D. Pinheiro would also like to acknowledge the financial support from "Programa Gulbenkian de Estímulo à Investigação 2006" and FCT—Fundação para a Ciência e Tecnologia grant with reference SFRH / BPD / 27151 / 2006.

References

1. J. Geanakoplos, H. Polemarchakis, Existence, Regularity and Constrained Suboptimality of Competitive Allocations When the Asset Market Is Incomplete, in *Essays in Honour of K. Arrow, vol. III*, ed. by W. Heller, D. Starrett (Cambridge University Press, Cambridge, 1986)
2. I. Karatzas, S. Shreve, *Methods of Mathematical Finance* (Springer, New York, 1998)
3. M. Magill, W. Shafer, Incomplete Markets, in *Handbook of Mathematical Economics, vol. IV*, ed. by W. Hildenbrand, H. Sonnenschein (North-Holland, Amsterdam, 1991). Chap. 30
4. D. Hobson, A survey of mathematical finance. Proc. R. Soc. Lond. A **460**, 3369–3401 (2004)
5. S.Z. Xanthopoulos, A.N. Yannacopoulos, Scenarios for price determination in incomplete markets. Int. J. Theor. Appl. Finance **11**, 415–445 (2008)
6. L. Boukas, D. Pinheiro, A.A. Pinto, S.Z. Xanthopoulos, A.N. Yannacopoulos, Behavioural and dynamical scenarios for contingent claims valuation in incomplete markets. J. Differ. Equ. Appl. (2009, to appear)
7. A. Mas-Colell, M. Whinston, J.R. Green, *Microeconomic Theory* (Oxford University Press, New York, 1995)

Undesired Oscillations in Pneumatic Systems

João Falcão Carneiro
and Fernando Gomes de Almeida

Abstract Automatic positioning devices are worldwide used in tasks like handling or assembly, making them key components of modern manufacturing systems. Pneumatic solutions are usually less expensive than their electrical counterparts, are more reliable and require less maintenance. However, the complex nonlinear nature and high model order of pneumatic systems lead to a very difficult control task. This paper illustrates these difficulties by presenting a study where several linear controllers are experimentally tested. In fact, despite their simplicity, these controllers can lead to undesired oscillations in the system output. The causes of these oscillations are described and justified in detail.

Keywords Pneumatic systems control · Limit cycles · Describing functions

1 Introduction

Servopneumatic systems have a high power-to-weight ratio and low maintenance cost. However, when medium to high accuracy positioning tasks are needed, electrical solutions are typically chosen. The main reason for this is the complexity in controlling servopneumatic systems. Air compressibility, seal friction and non-linear behaviour of servovalves make classical control theory unable to provide the same good results as it does with the more "well behaved" electrical devices. As a consequence, pneumatic solutions are discarded in most industrial applications where fine motion control is needed. This setback is even more relevant in applications where pneumatics may constitute the only solution. For instance, Kagawa

J. Falcão Carneiro (✉) · F. Gomes de Almeida
IDMEC, Faculdade de Engenharia, Universidade do Porto, Rua Doutor Roberto Frias, s/n, 4200-465 Porto, Portugal
e-mail: jpbrfc@fe.up.pt

F. Gomes de Almeida
e-mail: fga@fe.up.pt

J.A.T. Machado et al. (eds.), *Nonlinear Science and Complexity*,
DOI 10.1007/978-90-481-9884-9_28, © Springer Science+Business Media B.V. 2011

et al. [1] describe an application where the fast and fine positioning of semiconductor wafers must be obtained in an environment without electromagnetic fields or excessive heat generation. Pneumatic systems provide a natural solution to these demands.

These arguments have been motivating the scientific community to enhance the capabilities of pneumatic systems. In terms of modelling tasks, new pneumatic servovalve models were introduced in [2–4], detailed studies on the models of the actuator chambers were presented in [5, 6] and a comparison between two friction models was presented in [7]. In terms of control tasks, some major advances have been achieved in recent years. For instance, by using advanced nonlinear controllers, a positioning accuracy of 5 μm has been achieved in [8, 9]. Research in control is mainly directed towards nonlinear controllers since common linear ones (like proportional or PID) may lead to a poor performance or even to unwanted oscillations. Nevertheless, one cannot find in literature a work focused on the possible causes of oscillations in pneumatic systems. This work tries to fill that gap.

It is well known that systems with static friction may exhibit limit cycles when the controller includes an integral term. For instance, in [10–12] this result was obtained for a second order system without stiffness forces, based on an algebraic analysis of the system model. In this study the same result is obtained for a third order system in which stiffness forces play a major role. A different approach is used, based on a describing function analysis retrieved from classical literature on hydraulic servosystems. Experimental results validate the theoretical analysis.

In order to reduce the possibility of limit cycles one might be tempted to use low friction actuators. However, this solution may lead to another type of undesired oscillations that can appear even with a simple proportional controller. The justification of this second type of oscillations has only been recently provided [13] and will be fully presented and justified with experimental data.

This paper is organized as follows. Section 2 presents the experimental setup and the system nonlinear and linear models. Section 3 is devoted to friction generated limit cycles and to the tools used to predict them. It ends with a comparison between the foreseen oscillations and the experimental results. The discrepancies found in these comparisons are justified in Section 4. Finally, Section 5 resumes the major conclusions drawn from this work.

2 Pneumatic System

2.1 Experimental Setup

The system under study is represented in Fig. 1 and includes a low friction, double acting, asymmetric Asco-Joucomatic actuator that moves a load with mass M. The power modulation is achieved by two industrial Festo servovalves and the system is instrumented with pressure and position transducers. The main parameters of this system are resumed in Table 1.

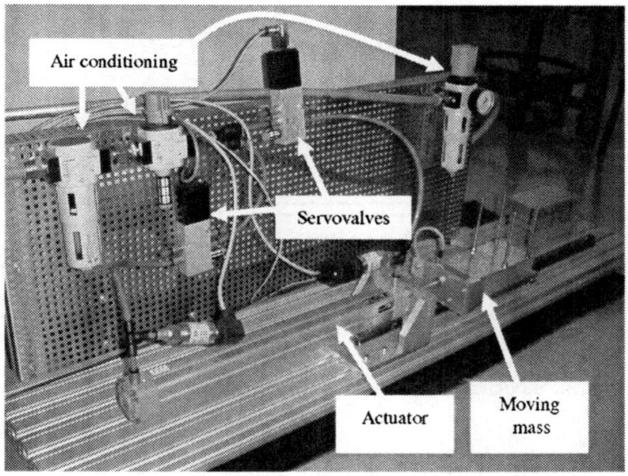

Fig. 1 Pneumatic servosystem

Table 1 System parameters

Variable	Description	Value
L	Actuator stroke	0.4 m
A_A	Chamber A area	8.04×10^{-4} m^2
A_B	Chamber B area	6.91×10^{-4} m^2
A_h	Rod area	1.13×10^{-4} m^2
k_0	Actuator thermal conductance	0.25 W K^{-1}
M	Moving mass	2.7 kg

2.2 Nonlinear Model

Consider the pneumatic system diagram in Fig. 2, where F_{ext} represents all external and friction forces acting on the piston and $x, \dot{x}, \ddot{x}$ the position, velocity and acceleration of the load. $P_{A,B}$ and $T_{A,B}$ represent absolute pressure and temperature of chambers A and B, respectively.

The analysis of the mathematical model of this system reveals three main blocks (see Fig. 3): the servovalves, the actuator chambers and the motion model. In Fig. 3 $u_{A,B}$ and $\dot{m}_{A,B}$ are the command input and mass flow into chambers A and B, respectively.

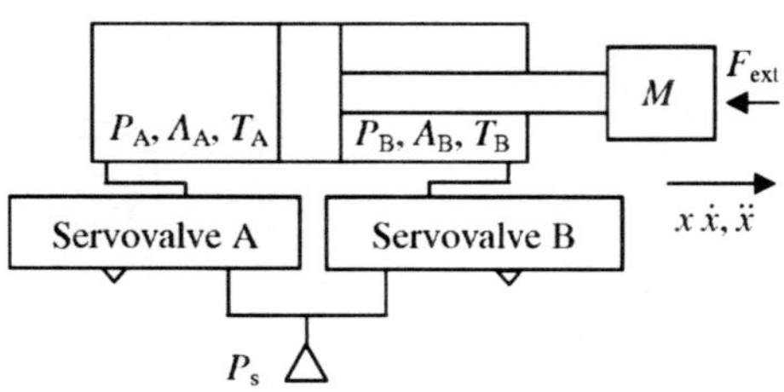

Fig. 2 Pneumatic servosystem components

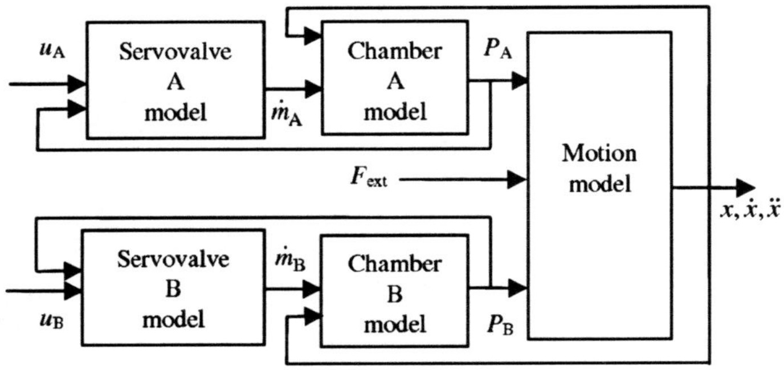

Fig. 3 Block models of a pneumatic system controlled by two servovalves

2.2.1 Pneumatic Chamber Model

The temperature dynamics in each chamber is typically neglected and a polytropic evolution of temperature is usually considered [5, 14–20]. Under these assumptions, the pressure dynamics of each chamber may be generically given by

$$\dot{P} = f_1(P, x, \dot{x}, T_{\text{amb}}) + f_2(P, x, T_{\text{amb}}).\dot{m} \tag{1}$$

where functions f_1 and f_2 are dependent on the particular polytropic process adopted and T_{amb} is the ambient temperature.

2.2.2 Servovalve Model

Consider Fig. 4 that represents each servovalve used in this work. The working orifice is connected to the cylinder, port S to the pressure source and port R to atmosphere. $x_v(u)$ is the spool position.

The mass flow crossing the working orifice in Fig. 4 can be determined by subtracting the mass flow crossing restriction R_2 ($\dot{m}_2$) from the mass flow crossing restriction R_1 ($\dot{m}_1$):

$$\dot{m} = \dot{m}_1 - \dot{m}_2. \tag{2}$$

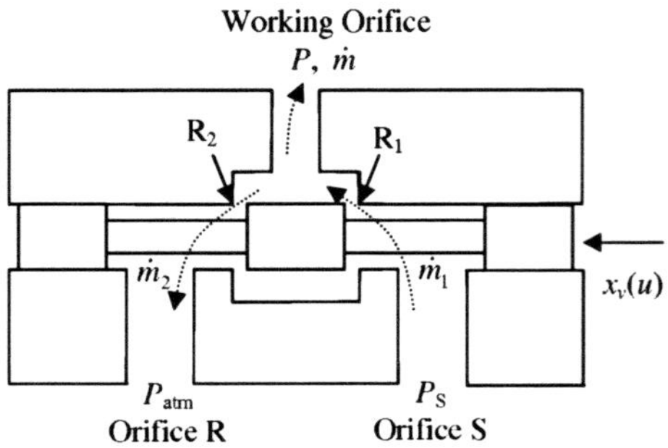

Fig. 4 Schematic representation of a 3 orifice servovalve

According to ISO 6358, the flow in a restriction is given by (3) and (4), where P_{ui} is the pressure upstream of the restriction, P_{di} the downstream pressure, $C_i(u)$ is the sonic conductance of the restriction and $b_i(u)$ the experimental critical pressure ratio.

$$\dot{m}_i = C_i(u) P_{ui} \rho_0 \sqrt{\frac{293.15}{T_{ui}}} Y_1, \tag{3}$$

where

$$Y_1 = \begin{cases} 1 & \text{if } P_{di}/P_{ui} \leq b_i(u), \\ \sqrt{1 - (\frac{P_{di}/P_{ui} - b_i(u)}{1 - b_i(u)})^2} & \text{if } P_{di}/P_{ui} > b_i(u). \end{cases} \tag{4}$$

Neglecting the cylinder chamber temperature fluctuations and applying (3) and (4) to both restrictions in (2) leads to the model for each servovalve [3, 4]:

$$\dot{m} = \rho_0 \sqrt{\frac{293.15}{T_{amb}}} \left(C_1(u) P_S Y_1 \left(\frac{P}{P_S} \right) - C_2(u) P Y_1 \left(\frac{P_{atm}}{P} \right) \right). \tag{5}$$

2.2.3 Mechanical Model

In this work it is assumed that no external force is applied on the actuator. The motion dynamics can be obtained by applying Newton's second law:

$$M\ddot{x} = F_i - k_a \dot{x} - F_{atr}. \tag{6}$$

In (6) F_i is the available pneumatic force defined by:

$$F_i = P_A A_A - P_B A_B - P_{atm} A_h. \tag{7}$$

The total friction force is divided into two parcels: $k_a \dot{x}$ represents the viscous friction and F_{atr} the nonlinear static force defined by (8) and represented in Fig. 5. The complete system model is therefore a fourth order one given by (1) applied to each chamber, by (5) applied to each servovalve and by (6).

$$F_{atr} = \begin{cases} F_{cb}\,\mathrm{sgn}(\dot{x}) & \text{if } \dot{x} \neq 0, \\ F_i & \text{if } \dot{x} = 0 \text{ and } |F_i| \leq F_s, \\ F_s & \text{if } \dot{x} = 0 \text{ and } |F_i| > F_s. \end{cases} \tag{8}$$

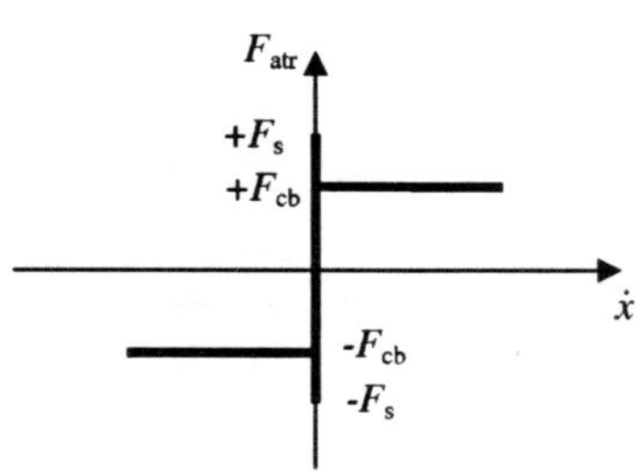

Fig. 5 Static friction model

2.3 Linearised Model

Classical linear models in servopneumatics are developed assuming that control actions are symmetric ($\delta u_A = -\delta u_B = \delta u$). Furthermore, in order to reduce the model order, the pressure time constants of the actuator chambers τ_A and τ_B are supposed to be the same and equal to their harmonic mean τ_m [5, 6]:

$$\frac{1}{\tau_m} = \frac{1}{2}\left(\frac{1}{\tau_A} + \frac{1}{\tau_B}\right) \tag{9}$$

with

$$\tau_{A,B} = -\frac{1}{\left.\frac{\partial \dot{P}_{A,B}}{\partial P_{A,B}}\right|_0} = -\frac{1}{\left(\frac{\gamma R T_{A,B}}{V_{A,B}} G_{PA,PB}\right)\Big|_0 - \frac{(\gamma-1)k_0}{V_{A0,B0}}\frac{(n-1)}{n}\frac{T_{A0,B0}}{P_{A0,B0}}} \tag{10}$$

and

$$G_{PA,PB} = \partial \dot{m}_{A,B} / \partial P_{A,B}. \tag{11}$$

In (10) n is the index of the polytropic evolution of temperature, k_0 is the actuator thermal conductance, γ is the ratio of specific heats for air ($\gamma = 1.4$) and R is the constant of air as a perfect gas.

Using these assumptions and the equilibrium values derived for the system presented in Sect. 2.1 (see Table 2) when $P_s = 7$ bar, it is possible to write the third order linear model presented in (12), where u is the control action provided by the controller, $G_{uA,uB}$ are the flow gain values of each valve defined by (13) and $\psi_{A,B}$ are constants defined by (14).

$$\begin{aligned}\frac{d^3x}{dt^3} = \frac{1}{M}\Bigg[&-\left(\frac{k_a}{\tau_m} - A_A\psi_A + A_B\psi_B\right)\dot{x} - \left(\frac{M}{\tau_m} + k_a\right)\ddot{x} \\ &+ \left(\frac{A_A\gamma R T_{A0}}{V_{A0}} G_{uA}\Big|_0 + \frac{A_B\gamma R T_{B0}}{V_{B0}} G_{uB}\Big|_0\right)u - \dot{F}_{atr} - \frac{1}{\tau_m}F_{atr}\Bigg],\end{aligned} \tag{12}$$

$$G_{uA,uB} = \partial \dot{m}_{A,B} / \partial u_{A,B}, \tag{13}$$

$$\begin{aligned}\psi_A &= -\gamma P_{A0} A_A V_{A0}, \\ \psi_B &= \gamma P_{B0} A_B V_{B0}.\end{aligned} \tag{14}$$

Table 2 Equilibrium values of the linearised model

x (m)	$\dot{x}$ (m/s)	$P_{A,B}$ (bar)	$T_{A,B}$ (K)	$u_{A,B}$ (V)
$x_0 = 0$	$\dot{x}_0 = 0$	$P_{A0} = 4.97$	$T_{A0} = 293.15$	$u_{A0} = 4.8$
		$P_{B0} = 5.61$	$T_{B0} = 293.15$	$u_{B0} = 5.1$

3 Friction Generated Oscillations

3.1 Describing Function Analysis

Single input describing functions (SIDF) are a useful tool to predict limit cycles in systems that can be modeled by a nonlinear element $w = f(e)$ in series with a linear one $G(j\omega)$—see Fig. 6. The SIDF $G_d(Am, \omega)$ is a simplified model of the nonlinear element that is in general dependent on the amplitude Am and on the frequency ω of its input. It enables the use of linear analysis techniques, namely the extended Nyquist criterion [21] that predicts a limit cycle when the characteristic equation of Fig. 6 is zero:

$$G(j\omega)H(j\omega) = -\frac{1}{G_d(Am, \omega)}. \tag{15}$$

It is possible to find several recent studies in the literature that exclude SIDF as a valid tool to predict friction caused limit cycles [10–12]. However, the SIDF considered in those studies are memoryless and cannot correctly represent stiction since stiction implies a dynamic behavior. In this work a dynamic SIDF is retrieved from classical servohydraulic literature [22] to explain friction caused limit cycles in pneumatic systems.

Consider the system represented in Fig. 7, consisting of a mass M_a pushed against a spring of stiffness K by a force F_{i1}. The equation of motion for this system is given by:

$$F_{i1} = M_a\ddot{y} + B\dot{y} + Ky + F_{atr}. \tag{16}$$

During a limit cycle in a pneumatic servosystem, the velocity and acceleration of the system have low values. Given the high air compressibility, it is therefore acceptable to assume that the friction nonlinearity acts predominantly on stiffness forces. This is highlighted in the block diagram of Fig. 8.

The mapping between the input ($F_{ak} = F_{atr} + Ky$) and output (Ky) of the nonlinearity G_d is represented in Fig. 9. This relation can be obtained with some physical reasoning. At rest, $F_{i1} = F_{ak}$ since the inertial and viscous forces are null. Starting

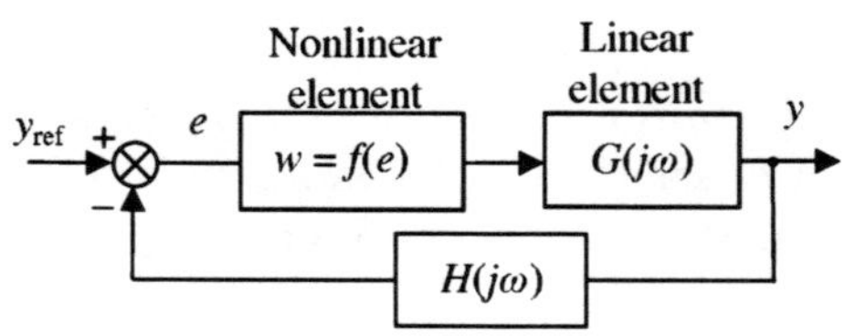

Fig. 6 Non-linear system

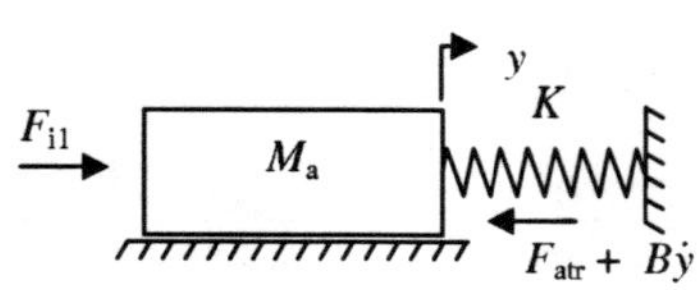

Fig. 7 Friction acting on spring

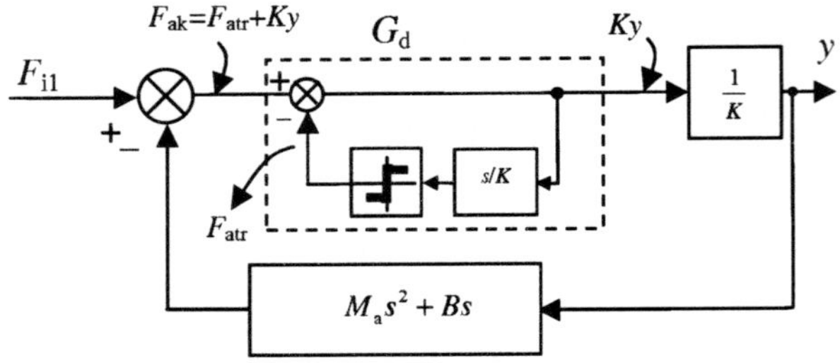

Fig. 8 Block diagram representing (16)

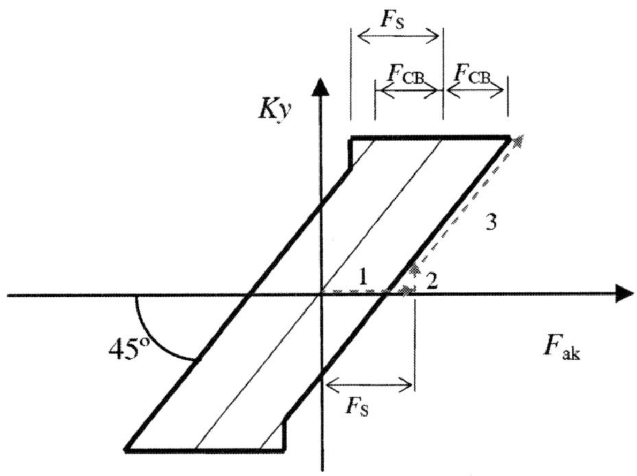

Fig. 9 Relation between F_{ak} and Ky

from this situation, an increase in F_{i1} will only lead to an increase in y after F_s is surpassed (path 1). If F_{i1} continues to rise, the discontinuity in F_{atr} due to the difference between F_s and F_{cb} will lead to a discontinuity in y (path 2). The mass starts to move and since F_{atr} is constant and equal to F_{cb}, the relation between F_{ak} and Ky is linear with a 45° slope (path 3). When the motion is reversed, the process restarts in the opposite direction. Notice that during a limit cycle only the full bold line is present; the dotted one was only used for illustration purposes. Furthermore, the discontinuity in y (path 2) is only possible if the inertial and viscous forces are negligible, i.e., if:

$$F_{ak} \gg M_a \ddot{y} + B\dot{y}. \tag{17}$$

As previously explained, this is an acceptable assumption during a limit cycle in a pneumatic servosystem. When F_{ak} is a sinusoidal signal, the output of G_d is only dependent on its amplitude and is given by [22]:

$$G_d(Am) = G_d = \left[\left(\frac{c_1}{Am}\right)^2 + \left(\frac{c_2}{Am}\right)^2\right]^{1/2} \angle(\tan^{-1}(c_1/c_2)) \tag{18}$$

with coefficients c_1 and c_2 given by:

$$c_1 = \frac{Am}{\pi}\left[4\left(\frac{F_{cb}}{Am}\right)^2\left(1+\frac{F_s}{F_{cb}}\right) - 4\frac{F_{cb}}{Am} - \left(\frac{F_{cb}}{Am}\right)^2\left(1+\frac{F_s}{F_{cb}}\right)^2\right],$$

$$c_2 = \frac{Am}{\pi}\left[\frac{\pi}{2} + \sin^{-1}\left(1 - \frac{F_{\text{cb}}}{Am}\left(1 + \frac{F_{\text{s}}}{F_{\text{cb}}}\right)\right)\right.$$
$$\left. + \left(1 - \frac{F_{\text{cb}}}{Am}\left(3 - \frac{F_{\text{s}}}{F_{\text{cb}}}\right)\right)\left(\frac{2F_{\text{cb}}}{Am}\left(1 + \frac{F_{\text{s}}}{F_{\text{cb}}}\right) - \left(\frac{F_{\text{cb}}}{Am}\right)^2\left(1 + \frac{F_{\text{s}}}{F_{\text{cb}}}\right)^2\right)^{1/2}\right].$$

In order to apply (18) to the pneumatic system analysis it is necessary to reorganize model (12) so that elastic and nonlinear friction forces are highlighted. Replacing the output of the nonlinear system, y, with the piston position, x, and applying the Laplace transform to the model of (12), it is possible to obtain (19), where all the force contributions are detailed.

$$\underbrace{Mk_{\text{lin}}\omega_{\text{n}}^2 U(s)}_{\text{mass flow terms}} - \Big(\underbrace{MY(s)s^2}_{\text{inertial terms}} + \underbrace{k_{\text{a}}Y(s)s}_{\text{viscous terms}}\Big)\left(s + \frac{1}{\tau_{\text{m}}}\right)$$
$$= \underbrace{KY(s)}_{\text{stiffness terms}} .s + F_{\text{atr}}\left(s + \frac{1}{\tau_{\text{m}}}\right) \tag{19}$$

with

$$k_{\text{lin}}\omega_{\text{n}}^2 = \frac{A_{\text{A}}\gamma RT_{\text{A0}}}{MV_{\text{A0}}}G_{\text{uA}}|_0 + \frac{A_{\text{B}}\gamma RT_{\text{B0}}}{MV_{\text{B0}}}G_{\text{uB}}|_0 \tag{20}$$

and K, the air chamber stiffness, given by:

$$K = -A_{\text{A}}\psi_{\text{A}} + A_{\text{B}}\psi_{\text{B}}. \tag{21}$$

Notice that in many practical applications the influence of the term $1/\tau_{\text{m}}$ is small when compared to the frequency of the limit cycle. In this situation,

$$F_{\text{atr}}\left(s + \frac{1}{\tau_{\text{m}}}\right) \approx F_{\text{atr}}s. \tag{22}$$

Furthermore, during a limit cycle, the mean value of $\text{F}_{\text{atr}}/\tau_{\text{m}}$ is zero, so it is acceptable that the relation between F_{ak} and Ky presented in Fig. 9 is not significantly affected by this simplification. In fact, since the term $\text{F}_{\text{atr}}/\tau_{\text{m}}$ is constant during piston movement, it does not affect the dynamics of velocity but only its value. With this assumption, the block diagram of the system model (19) including the non linearity G_{d} can be represented as in Fig. 10.

Consider now that the system represented in Fig. 10 is controlled by a controller C with direct and feedback branches transfer functions C_G and C_H respectively—see Fig. 11. The closed loop transfer function of this system is given by:

$$\frac{Y(s)}{Y_{\text{ref}}(s)} = \frac{Mk_{\text{lin}}\omega_{\text{n}}^2 C_G}{s(Ms^2 + (\frac{M}{\tau_{\text{m}}} + k_{\text{a}})s + \frac{K}{G_{\text{d}}} + \frac{k_{\text{a}}}{\tau_{\text{m}}}) + Mk_{\text{lin}}\omega_{\text{n}}^2 C_G C_H}. \tag{23}$$

After manipulating the block diagram of Fig. 11 to achieve a structure similar to the one in Fig. 6, it is possible to obtain the equation that allows the detection of limit cycles:

$$\frac{M\tau_{\text{m}}s^3 + (M + k_{\text{a}}\tau_{\text{m}})s^2 + k_{\text{a}}s + C_G C_H Mk_{\text{lin}}\omega_{\text{n}}^2\tau_{\text{m}}}{K\tau_{\text{m}}s} = -\frac{1}{G_{\text{d}}}. \tag{24}$$

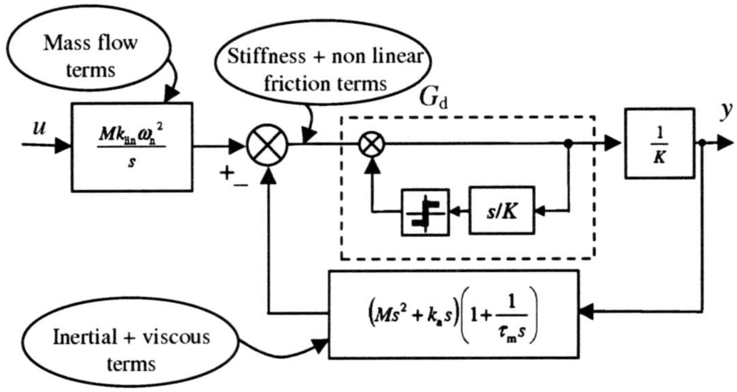

Fig. 10 Block diagram representing (19) and G_d

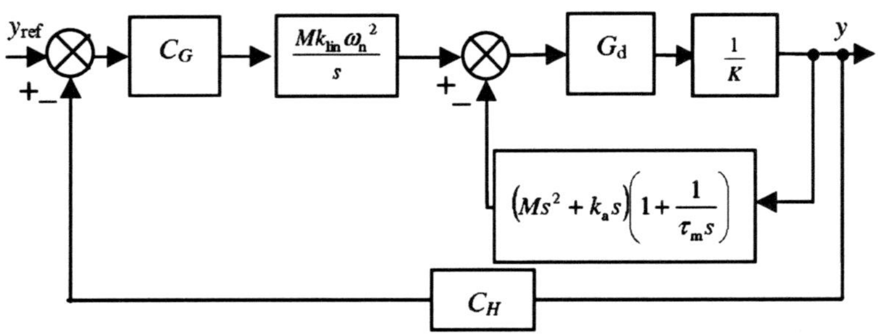

Fig. 11 The system of Fig. 10 controlled by C

3.2 Limit Cycle Prediction

This section is devoted to the prediction of the system oscillations with several linear controllers: proportional (P), proportional plus integral and derivative actions (PID) and state feedback with (ISF) and without (SF) integral action. Table 3 resumes the transfer functions C_G and C_H for each of these controllers. Further details on how to obtain these transfer functions can be found for example in [23].

By substitution of the transfer functions presented in Table 3 in (24), the following general expression can be obtained:

$$\frac{a_5 s^5 + a_4 s^4 + a_3 s^3 + a_2 s^2 + a_1 s + a_0}{b_3 s^3 + b_2 s^2 + b_1 s + b_0} = -\frac{1}{G_d}. \tag{25}$$

The coefficients of (25) depend on the system parameters and on the specific controller. Its actual values do not need to be known since the analysis can be made with the following assumptions: (i) the controller and system parameters are strictly positive and (ii) the controller parameters insure a stable closed loop system when static friction is absent. If these conditions are met, the coefficients of (25) satisfy Table 4. Based on the data presented in Table 4 it is possible to detect limit cycles by representing both sides of (25) in the complex plane—see Fig. 12.

Table 3 Linear controllers transfer functions

Controller	Transfer functions
P	$C_G = k_p$ $C_H = 1$
PID	$G_c = k_{PID}\frac{(T_d/N+T_d)s^2+(1+T_d/T_{int}/N)s+1/T_{int}}{(T_d/N)s^2+s}$ $C_H = 1$
SF	$C_G = k_1$ $C_H = 1 + \frac{k_2}{k_1}s + \frac{k_3}{k_1}s^2$
ISF	$C_G = \frac{k_i}{s}$ $C_H = 1 + \frac{k_{11}}{k_i}s + \frac{k_{22}}{k_i}s^2 + \frac{k_{33}}{k_i}s^3$

Table 4 Equation (25) coefficients for different controllers

	PID	ISF	P	SF
a_0	$\neq 0$	$\neq 0$	$\neq 0$	$\neq 0$
a_1	$\neq 0$	$\neq 0$	$\neq 0$	$\neq 0$
a_2	$\neq 0$	$\neq 0$	$\neq 0$	$\neq 0$
a_3	$\neq 0$	$\neq 0$	$\neq 0$	$\neq 0$
a_4	$\neq 0$	$\neq 0$	0	0
a_5	$\neq 0$	0	0	0
b_0	0	0	0	0
b_1	0	0	$\neq 0$	$\neq 0$
b_2	$\neq 0$	$\neq 0$	0	0
b_3	$\neq 0$	0	0	0

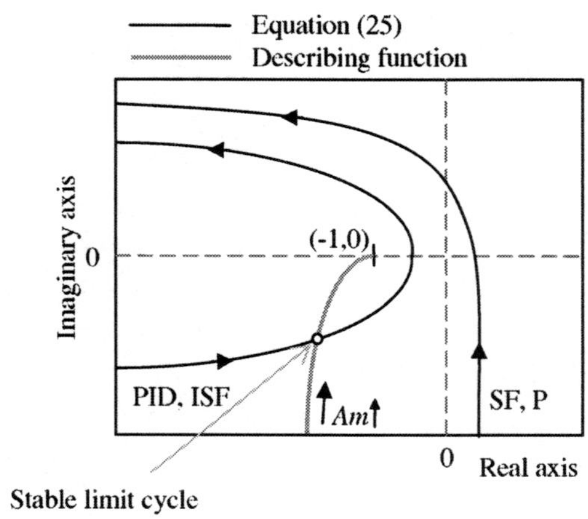

Fig. 12 Limit cycle prediction in the complex plane

It is possible to see that a (stable) limit cycle occurs when the PID and the ISF controllers are used. This limit cycle, represented in Fig. 12 by a small circumference, takes place with every possible combination of parameters satisfying (i) and (ii) above since the plot in the complex plane always starts at $-180°$, ends at $180°$ and does not encircle point $(-1,0)$. By the same line of argument it can be seen that a friction induced limit cycle cannot occur in the P and SF controllers: the plot always starts at $-90°$, tends towards $180°$ and does not encircle point $(-1,0)$.

3.3 Experimental Results

The four controllers were experimentally tested with the system presented in Section 2.1. The P and PID controllers were adjusted in two steps: first the Ziegler-Nicols rules [23] were used and then a fine tuning was experimentally performed. The state feedback controllers were adjusted using the pole placement technique. The system response to a step reference is presented in Figs. 13 and 14.

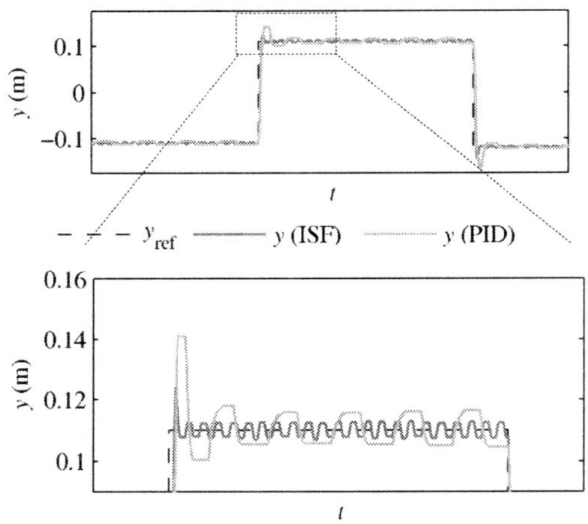

Fig. 13 Step response (PID and ISF)

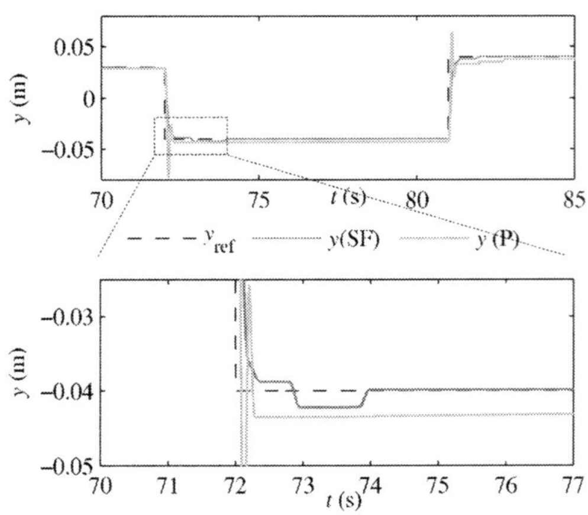

Fig. 14 Step response (P and SF)

The (stable) limit cycles predicted in last section for the PID and ISF controllers can be clearly seen in Fig. 13. Figure 14 presents the results obtained for the SF and P controllers. The zoom in this figure highlights, for the SF controller, a strange phenomenon: after stopping at $x \approx -0.039$ m ($t \approx 72.1$ s) with an error of about 2 mm, the piston suddenly moves away from the target position to $x \approx -0.042$ m; it remains there for approximately 1s and then it moves again, this time approaching the reference position. Notice that this phenomenon does not occur, for this particular example, in the proportional controller. The justification for these events will be presented in Section 4.

4 Pressure Dynamics Generated Oscillations

The oscillations appearing in Fig. 14 are, in a sense, surprising, since they occur even though the control action is constant. A complete justification for this occurrence,

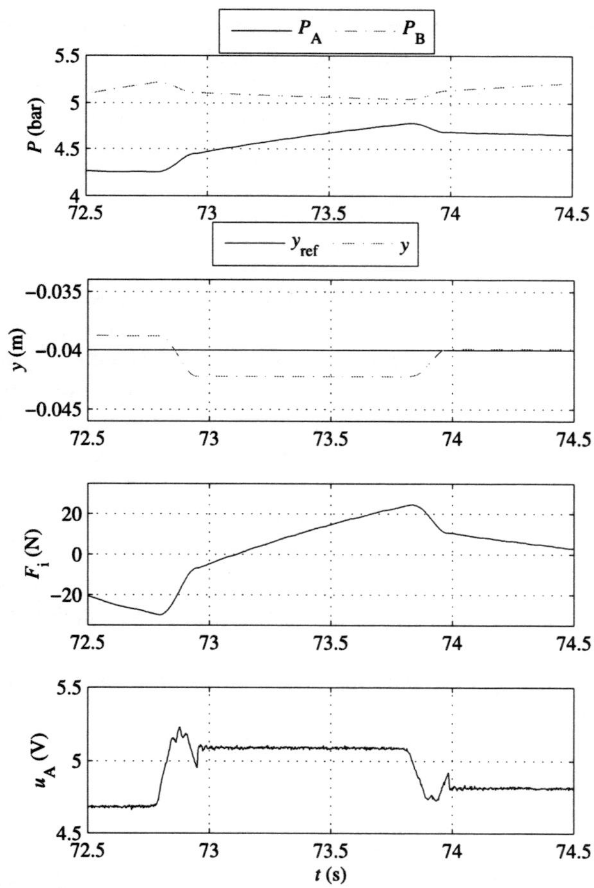

Fig. 15 Evolution of pressure, position, available force and control action during a piston sticking and restarting phenomenon

named sticking and restarting phenomenon (SRP), was recently presented in [13]. Consider Fig. 15 that presents the evolution of pressure, available force (7), control action and piston position during the experiment previously presented in Fig. 14 for the SF controller.

The SRP phenomenon occurs twice, in time instants ca. 72.8 s and 73.8 s. Consider the first of these two SRP: from 72.5 s until 72.8 s the control action is constant since there is mechanical equilibrium (y, $\dot{y}$ and $\ddot{y}$ are zero). However, during this time interval the pressures inside chambers A and B are changing towards their equilibrium values in the valve pressure gain curves [3], leading to an increase in the available force F_{i}. When F_{i} becomes higher than the static friction force ($t \approx 72.8$ s), the piston moves. The frequency of this phenomenon is highly dependent on two factors: the static friction force value and the asymmetry of the cylinder.

Notice that this analysis can be generalized to any controller whose control action is constant during mechanical equilibrium and is therefore extendable to the proportional controller. Notice also that this is an extremely difficult phenomenon to predict and therefore to avoid. This can be illustrated with the example of Fig. 14: the SRP occurred twice for the state feedback controller and only once for the proportional one ($t \approx 82$ s). Finally, it must be emphasized that there are some theoretical results on the prediction of SRP [13]. However, their practical application is hindered by the deep knowledge of the model of the system that is required.

5 Conclusions

This paper has presented in detail the main causes of unwanted oscillations in pneumatic systems when controlled with typical linear controllers. This presentation was supported by experimental data retrieved from a servopneumatic system.

Static friction force, when combined with a PID or ISF controller, causes unwanted oscillations in servopneumatic systems. This paper has presented a justification for this fact based on a describing function retrieved from classical hydraulic literature.

A strategy to cope with friction induced limit cycles may be the use of a low friction cylinder. However, the decrease of friction forces may reveal another type of undesired oscillations: the sticking and restarting phenomenon. This phenomenon can only be avoided with controllers that observe, direct or indirectly, the pressure dynamics.

All these aspects highlight the complexity of servopneumatic systems and justify the need of advanced nonlinear control strategies.

References

1. T. Kagawa, L. Tokashiki, T. Fujita, Accurate positioning of a pneumatic servosystem with air bearings, in *Proc. of the Bath Workshop on Power Transmission and Motion Control* (2000), pp. 257–268

2. D. Thomasset, S. Scavarda, S. Sesmat, M. Belgharbi, Analytical model of the flow stage of a pneumatic servo-distributor for simulation and nonlinear control, in *Proc. of the Sixth Scandinavian International Conference on Fluid Power* (1999), pp. 848–860
3. J.F. Carneiro, F.G. de Almeida, Modeling pneumatic servovalves using neural networks, in *Proc. of the 2006 IEEE Conference on Computer Aided Control Systems Design* (2006), pp. 790–795
4. J.F. Carneiro, F.G. de Almeida, Pneumatic servovalve models using artificial neural networks, in *Proc. of the Bath Symposium on Power Transmission and Motion Control* (2006), pp. 195–208
5. J.F. Carneiro, F.G. de Almeida, Reduced order thermodynamic models for servopneumatic actuator chambers. J. Syst. Control Eng. **220**(4), 301–314 (2006). *Proc. Inst. Mech. Eng., Part I*
6. J.F. Carneiro, F.G. de Almeida, Heat transfer evaluation on industrial pneumatic cylinders. J. Syst. Control Eng. **221**(1), 119–128 (2007). *Proc. Inst. Mech. Eng., Part I*
7. J.F. Carneiro, F.G. de Almeida, Comparação entre dois modelos do atrito num sistema servopneumático, in *Proc. 8° Congresso Iberoamericano de Engenharia Mecânica* (2007)
8. F. Xiang, J. Wikander, Block-oriented approximate feedback linearization for control of pneumatic actuator system. Control Eng. Pract. **12**(4), 387–399 (2004)
9. J.F. Carneiro, Modelação e controlo de actuadores pneumáticos utilizando redes neuronais artificiais, PhD Thesis, University of Porto, 2007
10. B. Armstrong-Hélouvry, B. Amin, PID control in the presence of static friction: exact and describing function analysis, in *Proc. Proceedings of the American Control Conference* (1994), pp. 597–601
11. B. Armstrong-Hélouvry, B. Amin, PID control in the presence of static friction: a comparison of algebraic and describing function analysis. Automatica **32**(5), 679–692 (1996)
12. H. Olsson, K. Astrom, Friction generated limit cycles. IEEE Trans. Control Syst. Technol. **9**(4), 629–636 (2001)
13. X. Brun, S. Sesmat, D. Thomasset, S. Scavarda, Study of "Sticking and Restarting Phenomenon" in electropneumatic positioning systems. ASME J. Dyn. Syst. Meas. Control **127**(1), 173–184 (2005)
14. S. Pandian, Y. Hayakawa, Y. Kanazawa, Y. Kamoyama, S. Kawamura, Practical design of a sliding mode controller for pneumatic actuators. ASME J. Dyn. Syst. Meas. Control **119**(4), 666–674 (1997)
15. S. Drakunov, G.D. Hanchin, W.C. Su, U. Ozguner, Nonlinear control of a rodless pneumatic servoactuator, or sliding modes versus Coulomb friction. Automatica **33**(7), 1401–1408 (1997)
16. S. Pandian, F. Takemura, Y. Hayakawa, S. Kawamura, Pressure observer-controller design for pneumatic actuators. IEEE/ASME Trans. Mechatron. **7**(4), 490–499 (2002)
17. E. Richard, De la commande lineaire et non lineaire en position des systems electropneumatiques, PhD Thesis, Institut National des Sciences Appliqués de Lyon, 1990
18. E. Richard, S. Scavarda, Comparison between linear and nonlinear control of an electropneumatic servodrive. ASME J. Dyn. Syst. Meas. Control **118**(2), 245–252 (1996)
19. E. Richer, Y. Hurmuzlu, A high performance pneumatic force actuator system: Part I—nonlinear mathematical model. ASME J. Dyn. Syst. Meas. Control **122**(3), 416–425 (2000)
20. E. Richer, Y. Hurmuzlu, A high performance pneumatic force actuator system: Part II—nonlinear controller design. ASME J. Dyn. Syst. Meas. Control **122**(3), 426–434 (2000)
21. J.J. Slotine, W. Li, *Applied Nonlinear Control* (Prentice-Hall, New York, 1991)
22. H. Merritt, *Hydraulic Control Systems* (Wiley, New York, 1967)
23. K. Ogata, *Modern Control Engineering* (Prentice-Hall, New York, 2001)

A Study of Correlation and Entropy for Multiple Time Series

José A.O. Matos, Sílvio M.A. Gama, Heather J. Ruskin, Adel Al Sharkasi, and Martin Crane

Abstract In this work we study multiple related (multivariate) time series from worldwide markets. We search for signs of coherence and/or synchronization using the main index as representative of the whole market. In order to better understand the relations between the time series we use two different techniques, entropy and variance-covariance matrices. We apply each procedure in a time dependent way to better understand the underlying dynamics of the system. We found that both methods show that world markets, regardless of their maturity status (mature or emergent), are behaving more and more alike over the last years. The simultaneous use of correlation and entropy to study multivariate time series is a promising approach in the sense that they capture different aspects of the collective system dynamics.

Keywords Econophysics · Long term memory processes · Entropy · Variance-covariance matrix · Multivariate time series

J.A.O. Matos (✉) · S.M.A. Gama
Centro de Matemática da Universidade do Porto, Edifício dos Departamentos de Matemática da FCUP, Rua do Campo Alegre 687, 4169-007 Porto, Portugal
e-mail: jamatos@fep.up.pt

J.A.O. Matos
Grupo de Matemática e Informática, Faculdade de Economia da Universidade do Porto, Rua Roberto Frias, 4200-464 Porto, Portugal

S.M.A. Gama
Departamento de Matemática Aplicada, Faculdade de Ciências da Universidade do Porto, Rua do Campo Alegre 687, 4169-007 Porto, Portugal

H.J. Ruskin · M. Crane
School of Computing, Dublin City University, Dublin 9, Ireland

A.A. Sharkasi
Department of Statistics, Faculty of Science, Garyounis University, Benghazi, Libya

J.A.T. Machado et al. (eds.), *Nonlinear Science and Complexity*,
DOI 10.1007/978-90-481-9884-9_29,

1 Introduction

1.1 Goals

The main goal of this study is the analysis of stock exchange world indices searching for signs of coherence and/or synchronization across the set of studied markets.

We have expanded the scope of previous work on the PSI-20 (Portuguese Standard Index), since results there [17] seemed to provide a basis for a wider ranging study of coherence and entropy.

With that purpose we applied econophysics techniques related to measures of "disorder"/complexity (entropy). As a measure of coherence among a selected set of markets we have studied the eigenvalues of the correlation matrices for two different set of markets [8], exploring the dichotomy represented by emerging and mature markets and proposing a more refined classification. The indices are used to represent or characterise the respective market.

The classification of markets into mature or emergent is not a simple issue. The International Finance Corporation (IFC) uses income per capita and market capitalisation relative to Gross National Product (GNP) for classifying equity markets. If either (i) a market resides in a low or middle-income economy, or (ii) the ratio of the investable market capitalisation to GNP is low, then the IFC classifies the market as emerging, otherwise the classification is mature.

The data used in this study was taken daily for a set of worldwide market indices. As is usual in this kind of analysis [1] we base our results on the study of log returns $\eta_i = \log \frac{x_i}{x_{i-1}}$, where η_i is the log return at time step i.

2 Entropy

The Shannon entropy for blocks of size m for an alphabet of k symbols is [9]

$$\widetilde{H}(m) = -\sum_{j=0}^{k^m-1} p_j \log p_j. \tag{1}$$

The entropy of the source is then

$$\widetilde{h} = \lim_{m\to\infty} \frac{\widetilde{H}(m)}{m}. \tag{2}$$

This definition is attractive for several reasons: it is easy to calculate and it is well defined for a source of symbol strings. In the particular case of returns, if we choose a symmetrical partition we know that half of the symbols represent losses and half of the symbols represent gains. If the sequence is predictable, we have the same losses and gains sequences repeated everytime, i.e. the entropy will be lower; if however all sequences are equally probable the uncertainty will be higher and so it will be the entropy. Entropy is thus a good measure of uncertainty and much work has been done to relate predictability and entropy [12, 13]. In general for series, such as those in finance, high levels of complexity apply and the nature of disorder or uncertainty is captured by measures of coherence (or lack of this—entropy). Where entropy measures can be used to characterise the complexity, predictability becomes more understandable.

This particular method has limitations, the entropy depends on the choice of encoding and it is not a unique characteristic for the underlying continuous time series. Also since the number of possible states grows exponentially with m, it becomes difficult in practical terms, after a short number of sequences, to find a sequence that repeats itself. This entropy is not invariant under smooth coordinate changes, both in time and encoding. This is a strong handicap for its direct usage into financial time series study.

We have applied the Shannon entropy for blocks of size 5 and an alphabet of 50 symbols, to a set of markets previously studied. We should recall that using blocks of size 5 corresponds to a week in trading time. Notice also that we have only considered trading days, as for our previous analyses, so we ignore any holidays or days where the market was closed.

It should be noted that results are robust to the choice of the total number of bins (the size of our alphabet). That is, we have repeated the analysis with a different choice of the number of partitions yielding similar results.

In order to enhance the time dependence of results we have evaluated the entropy of the set for periods of 100 trading days (roughly corresponding to half a year). The primary motivation of the analysis as a whole is thus to investigate how the entropy evolves in time.

The results displayed in Fig. 1 show improved coherence (i.e. reduced entropy) after 1997 as compared with previous periods for all markets. Higher entropy implies less predictability, in general, although the nature of shocks qualifies this statement to some extent. The notable feature of this graphic is that both mature and developing markets are affected similarly which suggests that global behaviour patterns are becoming more coherent or linked because of the progressive globalisation of markets. This is in stark contrast to the situation from 1982 to 1997 where, despite common features (coincident highs and lows), entropy levels are very different. This is in line with the findings of [14] where we found the Hurst exponent for different markets to be decreasing with time.

3 Covariance

In the previous section we have used the block entropy applied to several markets. The analysis of co-movements suggested a multivariate analysis. This method shares with block entropy applied in the previous Section the emphasis on time dependency.

The time dependent covariance matrix (see [5]) studies the multivariate case (several random variables at once).

This method shares with the entropy analysis of last section the time dependent results that allows to evaluate the time evolution of the set.

The covariance matrix with variable weights at time T, over an horizon M, $\sigma^T(M)$, is given by:

$$\sigma_{ij}^T(M) = \frac{\sum_{s=0}^{M} W_s r_{i,T-s} r_{j,T-s}}{\sum_{s=0}^{M} W_s}. \tag{3}$$

Where $r_{i,t}$ is the value of return r_i at time t, and W_s is the weight given for the covariance at delay s (time $T-s$).

The weight vector, $\mathbf{W}$, has decreasing components since we give higher weights to closer times for moments closer to the time we are analysing. One example traditionally used and the same that is used in this work is $W_i = R^i$, with $0 < R < 1$. Then we have $\sum_{s=0}^{T} W_{T-s} = \frac{R^T}{1-R^T}$, and W_i corresponds to a geometric series. Typical values (see [5]) are $R = 0.9$ and $T = 20$.

According to the findings of [2–4, 6, 7, 10, 15] the correlation (or covariance) matrices of financial time series, apart from a few large eigenvalues and their corresponding eigenvectors, appear to contain such a large amount of noise that their structure can essentially be regard as random.

Such as in [3, 8] we will consider the three larger eigenvalues and its respective eigenvectors as carrying meaningful information.

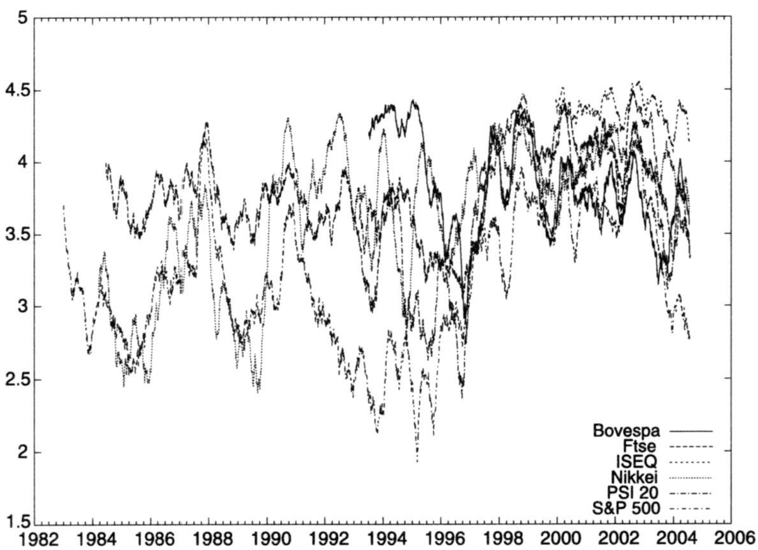

Fig. 1 Weekly entropy for various market indexes

In the multivariate signal processing problem, one key issue might be when instabilities occur in signal patterns and how we might determine if the fluctuations are damped, remain at low level, or combine in some way as to cause a major event,

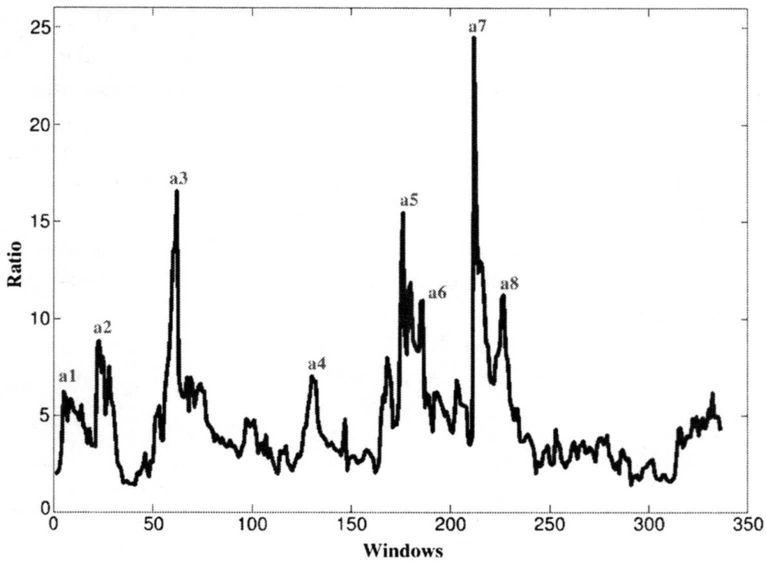

Fig. 2 Evolution of for $\frac{\lambda_1}{\lambda_3}$ emerging markets

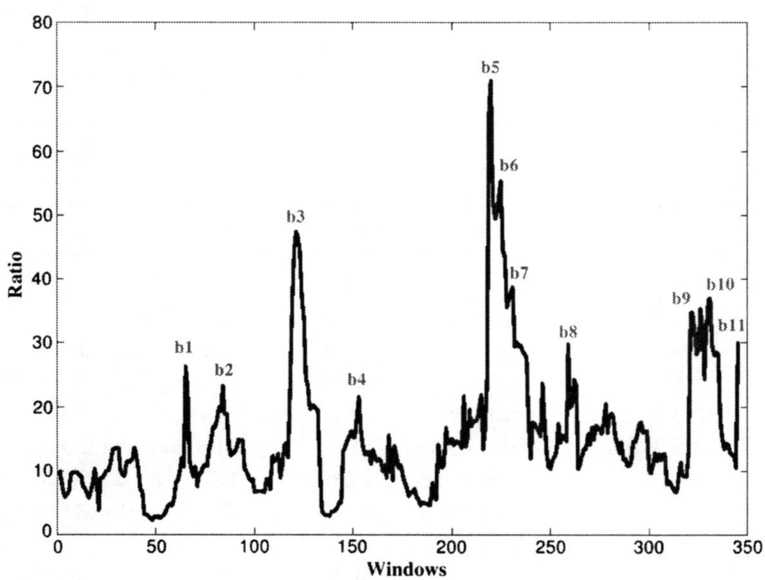

Fig. 3 Evolution of $\frac{\lambda_1}{\lambda_3}$ for mature markets

e.g. a market crash. Crashes are also interesting since the market dynamics changes during the event, see [11, 16].

The work explored here was developed, by the author and collaborators, in [8]. We use the covariance matrix to study the coherence of various set of markets, with different degrees of maturity (for this study we have considered the traditional distinction between mature and emerging markets as the initial point). We are interested in the time dependency of the (three) most significant eigenvalues of the covariance matrix, since as seen above, those are the only eigenvalues which carry meaningful information.

We have used the typical value of parameters, $R = 0.9$ and an horizon of 20 trading days (for details see [5]). In line with the analysis of the previous section, weekly periods have been used to estimate the returns.

In Fig. 2, we represent the ratio between the first and the third most important eigenvalues ($\frac{\lambda_1}{\lambda_3}$) for a given set of emerging markets. The same analysis applies for mature markets, see Fig. 3.

Again, interest lies in the fact that spikes in Figs. 2 and 3 correlate with real events, as summarised in Tables 1 and 2, respectively.

Table 1 Table of events (emerging)

Mark	Window No	Last week included	Events
*a*1	5	first week of 7/1997	Asian Crash
*a*2	23	second week of 11/1997	Asian Crash
*a*3	62	fourth week of 8/1998	Global Crash
*a*4	130	second week of 1/2000	
*a*5	176	second week of 12/2000	Effects of DotCom Crash
*a*6	186	second week of 3/2001	
*a*7	212	second week of 9/2001	September 11th Crash
*a*8	227	fourth week of 1/2002	

Table 2 Table of events (mature)

Mark	Window No	Last week included	Events
*b*1	65	first week of 9/1998	Global Crash
*b*2	84	fourth week of 12/1998	Global Crash
*b*3	121	third week of 10/1999	Last October in the 20th Century
*b*4	153	second week of 6/2000	DotCom Crash
*b*5	220	second week of 9/2001	September 11th Crash
*b*6	225	first week of 11/2001	Effects of 9/11 Crash
*b*7	231	second week of 12/2001	Effects of 9/11 Crash
*b*8	259	first week of 5/2002	The Stock Market Downturn
*b*9	322	first week of 10/2003	
*b*10	331	first week of 12/2003	General Threat Level Raised
*b*11	345	third week of 3/2004	Madrid Bomb

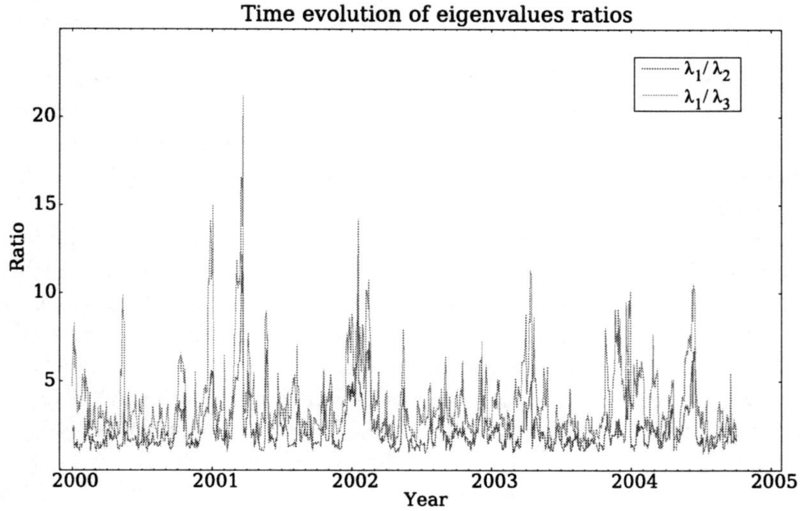

Fig. 4 Evolution of eigenvalue ratios for emergent markets (daily data)

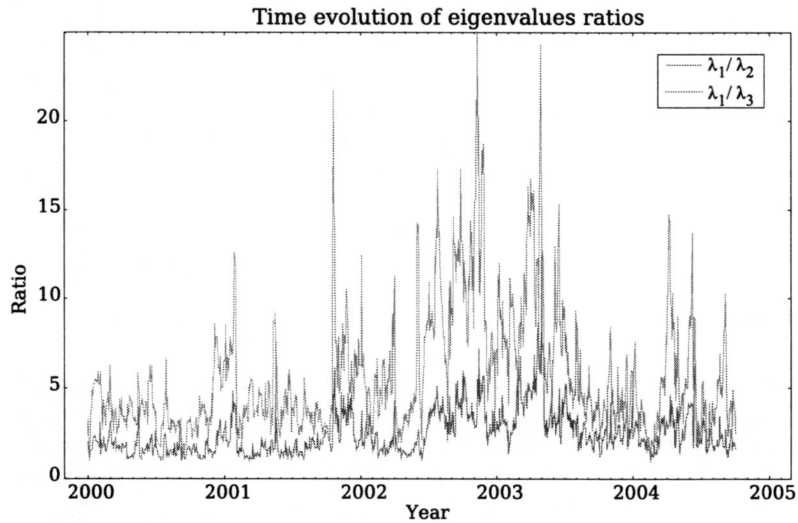

Fig. 5 Evolution of eigenvalue ratios for mature markets (daily data)

We have considered the evolution of the major eigenvalues assuming weekly data. Applying the same analysis for daily data we get the results displayed in Figs. 4 and 5. This analysis highlights the role of the data granularity, the coarse grained approach, in the results. We have a better resolution on the events and the results are qualitatively the same.

3.1 Data

We have considered, in this study, the major and most active markets worldwide from America (North and South), Asia, Africa, Europe and Oceania. All the data on the respective market indices are public and came from Yahoo Finance (finance.yahoo.com). We have considered the daily closure as the value for the day, to obviate any time zone difficulties.

The choice of the markets used in this study was determined by the aim to study major markets across the world in an effort to ensure that tests and conclusions could be as general as possible. Despite the breadth of the markets studied, results for a selection only are presented here for illustration. Hence from the results we have divided the markets according to mature: AEX General (Netherlands); Dow Jones (U.S.); CAC 40 (France); FTSE 100 (United Kingdom); DAX (Germany); S&P 500 Index (U.S.); Nasdaq (U.S.); Seoul Composite (South Korea); Nikkei 225 (Japan); NYSE Composite Index (United States) and Stockholm General (Sweden). The list of hybrid markets is smaller: All Ordinaries (Australia); Bovespa (Brazil); S&P TSX Composite (Canada); NZSE 10 (New Zealand); Madrid General (Spain) and Swiss Market (Switzerland).

All the other markets from our study behave as emergent: ATX (Austria); BEL-20 (Belgium); BSE 30 (India); CMA (Egypt); All Share (Sri Lanka); Hang Seng (Hong Kong); IPSA (Chile); ISEC Small Cap (Ireland); ISEC Small Cap Techno (Ireland); Irish SE Index (Ireland); Jakarta Composite (Indonesia); KFX (Denmark); KLSE Composite (Malaysia); Karachi 100 (Pakistan); MerVal (Argentina); MIB-Tel (Italy); IPC (Mexico); OSE All Share (Norway); PSE Composite (Philippines); PSI 20 (Portugal); PX50 (Czech Republic); Shanghai Composite (China); Straits Times (Singapore); TA-100 (Israel); Taiwan Weighted (Taiwan) and ISE National-100 (Turkey).

4 Conclusions

We have focused on aspects of time dependence, explored by several econophysics techniques, applied to markets, categorised as emerging or mature and subject to diverse levels of disorder or volatility in their financial series. The outcome shows clear synchronisation of world markets, observed in the weekly entropy of individual markets or groups. The results show that world markets tend to influence each other and reduce individual market levels of disorder (i.e. reduced entropy) demonstrating a clear synchronism of responses which is more or less robust depending on the nature of the market. The entropy measure here is considered over a week, a fairly long time in terms of market behaviour, but the results obtained for daily results show the same qualitatively behaviour.

Despite evidence that stability is linked to this synchronisation and low energy or equilibrium state, it is evident that shocks upset the balance and disorder increases with very high entropy levels in some instances. These occurrences correspond usually to crashes in markets, as it can be seen associating the events in Tables 1 and 2 with their corresponding spikes in Figs. 2, 3, 4 and 5. Nevertheless it is a characteristic of the more mature markets that this period of increased entropy is relatively short, with smaller recovery times. See both in Figs. 2 and 4 how it takes almost two months for emerging markets to reflect 9/11 effects while for mature markets (Figs. 3 and 5) this effect is instantaneous.

This distinction is not always clearcut, however and under different conditions markets may exhibit more than one type of behaviour (see in Tables 1 and 2 where for certain peaks we were not able to associate any known event).

Particularly interesting is that, despite differences in behaviour between emerging and mature markets, we find worldwide, that entropy measures in recent years are convergent. Thus, in general, markets appear to move ever-more-rapidly towards mature behaviour (reflected e.g. in improved time to recovery after a significant episode). A plausible explanation for this phenomenon is, clearly, the progressive globalization of financial markets.

Acknowledgements One of authors (JAOM) would like to thank ESF (European Science Foundation) for COST action P10-STSM 00421, that made possible a visit to Dublin City University where part of this work was initiated.

References

1. R.N. Mantegna, H.E. Stanley, *An Introduction to Econophysics* (Cambridge University Press, Cambridge, 2000)
2. L. Laloux, P. Cizeau, J.P. Bouchaud, M. Potters, Noise dressing of financial correlation matrices. Phys. Rev. Lett. **83**, 1467–1470 (1999)
3. D. Wilcox, T. Gebbie, On the analysis of cross-correlations in South African market data. Physica A **344**, 294–298 (2004)
4. S. Sharifi, M. Crane, A. Shamaie, H.J. Ruskin, Random matrix theory for portfolio optimization: a stability approach. Physica A **335**, 629–643 (2004)
5. R. Litterman, K. Winkelmann, Estimating Covariance Matrices, in *Goldman-Sachs Risk Management Series*, ed. by R.A. Krieger (Goldman, Sachs and Co., Munich, 1998)
6. V. Plerou, P. Gopikrishnan, B. Rosenow, L.A.N. Amaral, H.E. Stanley, Universal and non-universal properties of cross correlations in financial time series. Phys. Rev. Lett. **83**, 1471–1474 (1999)
7. S. Gallucio, J.P. Bouchaud, M. Potters, Rational decisions, random matrices and sping glasses. Physica A **259**, 449–456 (1998)
8. A. Sharkasi, M. Crane, H.J. Ruskin, J.A.O. Matos, The reaction of stock markets to crashes and events: a comparison study between emerging and mature markets using wavelet transforms. Physica A **368**, 511–521 (2006)
9. C. Shannon, A mathematical theory of communication. Bell Syst. Techn. J. **27**, 379–423 (1948)
10. V. Plerou, P. Gopikrishnan, B. Rosenow, Collective behaviour of stock price movement: a random matrix theory approach. Physica A **299**, 175–180 (2001)
11. R. Vilela Mendes, T. Araújo, F. Louçã, Reconstructing an economic space from a market metric. Physica A **323**, 635–650 (2003)
12. G. Boffetta, M. Cencini, M. Falconi, A. Vulpiani, Predictability: a way to characterize complexity. Phys. Rep. **356**, 367–474 (2002)
13. G.A. Darbellay, D. Wuertz, The entropy as a tool for analysing statistical dependences in financial time series. Physica A **287**, 429–439 (2000)

14. J.A.O. Matos, S.M.A. Gama, A. Sharkasi, H.J. Ruskin, M. Crane, Temporal and scale DFA applied to stock markets (2008). doi:10.1016/j.physa.2008.01.060
15. L. Laloux, P. Cizeau, M. Potters, Random matrix theory and financial correlations. Int. J. Theor. Appl. Finance **3**(3), 391–397 (2000)
16. T. Araújo, F. Louçã, Complex behavior of stock markets: process of synchronization and desynchronization during crises, Perspectives on Econophysics, Universidade de Évora, Portugal, 2006
17. J.A.O. Matos, S.M.A. Gama, H.J. Ruskin, J.A.M.S. Duarte, An econophysics approach to the Portuguese Stock Index—PSI-20. Physica A **342**, 665–676 (2004)

Characterization and Parameterization of the Singular Manifold of a Simple 6–6 Stewart Platform

Tiago Charters and Pedro Freitas

Abstract This paper presents a study of the singular manifold of the six-degree-of-freedom parallel manipulator commonly known as the Stewart platform. We consider a platform with base vertices in a circle and for which the bottom and top plates are related by a rotation and a contraction. It is shown that in this case the platform is always in a singular configuration and that the singular manifold can be parameterized by a scalar parameter.

Keywords Stewart platform · Dynamics · Singular solutions

1 Introduction

The Stewart platform is a parallel manipulator with six degrees of freedom [1]. We will use the (standard) variables $x, y, z, pitch, roll$ and yaw, where x, y and z are the coordinates of the centre of the top platform, and $pitch, roll$ and yaw denote the Euler angles defining the inclination of this platform with respect to the bottom platform, see Fig. 1.

T. Charters
Department of Mechanical Engineering, Instituto Superior de Engenharia de Lisboa (IPL), Rua Conselheiro Emidio Navarro, 1, 1959-007 Lisboa, Portugal

T. Charters (✉)
Centro de Física Teórica e Computacional, University of Lisbon Complexo Interdisciplinar, Av. Prof. Gama Pinto 2, 1649-003 Lisboa, Portugal
e-mail: tca@cii.fc.ul.pt

P. Freitas
Department of Mathematics, Faculdade de Motricidade Humana (TU Lisbon) Group of Mathematical Physics of the University of Lisbon Complexo Interdisciplinar, Av. Prof. Gama Pinto 2, 1649-003 Lisboa, Portugal
e-mail: freitas@cii.fc.ul.pt

J.A.T. Machado et al. (eds.), *Nonlinear Science and Complexity*, 255
DOI 10.1007/978-90-481-9884-9_30,

Fig. 1 The Stewart platform

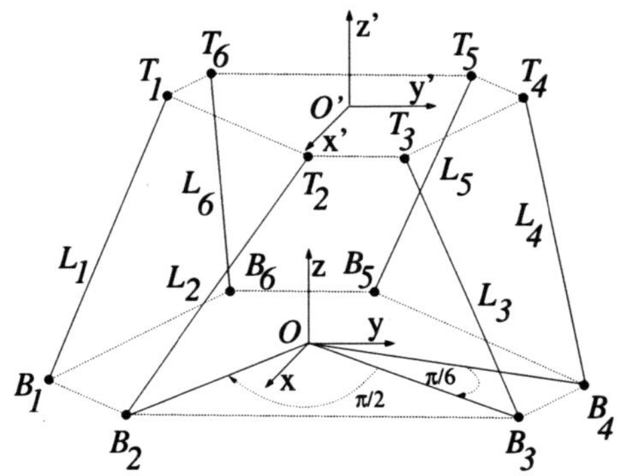

The aim of this paper is to study the singular manifold which is defined by the physical configurations for which it will not be possible to determine the position of the platform uniquely by fixing the lengths of the legs. This is a well-known problem affecting parallel manipulators [1]. The configuration considered here was motivated by a problem posed at the 60th European Study Group with Industry in 2007, and a first approach to these problems was presented in [2].

The solution to the forward kinematics problem may be divided in a natural way into a singular and a non-singular. In the non-singular case we recall the work [3] of Ji and Wu and show that there are 8 possible isolated singular solutions corresponding to the same legs lengths. In the singular case we extend the previous analysis and show how to obtain, for a given set of length legs, a set of singular solutions all of which may be parameterized by a scalar parameter. These solutions form a continuous curve in position space and in rotation space in which the platform moves without changing the values of the leg lengths. This fully characterizes the singular manifold and shows that the platform is, in this case, completely singular.

Spatial rotations in three dimensions can be parameterized using both Euler angles (ϕ, θ, ψ) and unit quaternions $\mathbf{q} = (q_0, q_1, q_2, q_3)$, $\|\mathbf{q}\| = 1$ [3, 4]. A unit quaternion may be described as a vector in $\mathbb{R}^4$

$$\mathbf{q} = (q_0, q_1, q_2, q_3), \tag{1}$$

$$\mathbf{q}^T \mathbf{q} = q_0^2 + q_1^2 + q_2^2 + q_3^2 = 1, \tag{2}$$

and the rotation matrix is then given by

$$R = \begin{pmatrix} 2q_0^2 - 1 + 2q_1^2 & 2q_1q_2 - 2q_0q_3 & 2q_0q_2 + 2q_1q_3 \\ 2q_1q_2 + 2q_0q_3 & 2q_0^2 - 1 + 2q_2^2 & 2q_2q_3 - 2q_0q_1 \\ 2q_1q_3 - 2q_0q_2 & 2q_0q_1 + 2q_2q_3 & 2q_0^2 - 1 + 2q_3^2 \end{pmatrix}. \tag{3}$$

Consider the Stewart platform shown in Fig. 1. As shown there, the two coordinate systems O and O' are fixed to the base and the mobile platforms. The platform geometry can be described by vectors $\mathbf{L}_i$, $i = 1, 2, \ldots, 6$, defined by $\mathbf{L}_i = \mathbf{P} + \mathbf{T}_i - \mathbf{B}_i$, $i = 1, 2, \ldots, 6$, where $\mathbf{B}_i$ and $\mathbf{T}_i$ are the base and top vertices' coordinates, respectively, and $\mathbf{P}$ is the center point of the top plate. We assume that these points are related by

$$\mathbf{T}_i = \mu A \mathbf{B}_i, \quad i = 1, 2, \ldots, 6, \tag{4}$$

where A is a 3×3 orthogonal matrix ($A^T A = I$, I being the 3×3 identity matrix) and $\mu \in]0, 1[$ is called the rescaling factor. The coordinates of the base vertices are given by

$$\mathbf{B}_i = (x_i, y_i, 0), \quad i = 1, 2, \ldots, 6. \tag{5}$$

Given the position $\mathbf{P} = (x, y, z)$ and the transformation matrix R between the two coordinate systems, the leg vectors may be written as

$$\mathbf{L}_i = \mathbf{T}_i - \mathbf{B}_i + \mathbf{P}, \tag{6}$$

$$= (\mu R A - I)\mathbf{B}_i + \mathbf{P}, \quad i = 1, 2, \ldots, 6. \tag{7}$$

So the length for each i-leg is given by

$$\mathbf{L}_i{}^T \mathbf{L}_i = ((\mu R A - I)\mathbf{B}_i + \mathbf{P})^T ((\mu R A - I)\mathbf{B}_i + \mathbf{P}) \tag{8}$$

Given $\mathbf{q}$, A and $\mathbf{P}$ the leg lengths are given by

$$L_i = \sqrt{((\mu R A - I)\mathbf{B}_i + \mathbf{P})^T ((\mu R A - I)\mathbf{B}_i + \mathbf{P})}. \tag{9}$$

2 Forward Kinematics

In the forward kinematics the six leg lengths L_i, $i = 1, 2, \ldots, 6$, are given, while R and $\mathbf{P}$ are unknown. Let $\mathbf{e}_x = (1, 0, 0)$, $\mathbf{e}_y = (0, 1, 0)$, $\mathbf{e}_z = (0, 0, 1)$ and expand (8), then one gets,

$$\begin{aligned} L_i^2 = {} & \mathbf{P}^T \mathbf{P} + \mathbf{B}_i{}^T \big((\mu(RA)^T - I)(\mu R A - I)\big)\mathbf{B}_i \\ & + 2\mathbf{B}_i{}^T (\mu(RA)^T - I)\mathbf{P}, \end{aligned} \tag{10}$$

or

$$\begin{aligned} L_i^2 = {} & \mathbf{P}^T \mathbf{P} + 2x_i \big(\mathbf{e}_x{}^T (\mu(RA)^T \mathbf{P} - \mathbf{P})\big) + 2y_i \big(\mathbf{e}_y{}^T (\mu(RA)^T \mathbf{P} - \mathbf{P})\big) \\ & - 2\mu\big[x_i^2 (\mathbf{e}_x{}^T RA\mathbf{e}_x) + x_i y_i (\mathbf{e}_x{}^T RA\mathbf{e}_y + \mathbf{e}_y{}^T RA\mathbf{e}_x) \\ & + y_i^2 (\mu\mathbf{e}_y{}^T RA\mathbf{e}_y)\big] + (1 + \mu^2)(x_i^2 + y_i^2). \end{aligned} \tag{11}$$

Define $\mathbf{w} = (w_1, w_2, w_3, w_4, w_5, w_6)$ as

$$w_1 = \mathbf{P}^T \mathbf{P}, \tag{12}$$

$$w_2 = 2\mu\mathbf{e}_x{}^T ((RA)^T \mathbf{P} - \mathbf{P}), \tag{13}$$

$$w_3 = 2\mu \mathbf{e}_y{}^T((RA)^T\mathbf{P} - \mathbf{P}), \tag{14}$$

$$w_4 = -2\mu \mathbf{e}_x{}^T RA\mathbf{e}_x, \tag{15}$$

$$w_5 = -2\mu\left(\mathbf{e}_x{}^T RA\mathbf{e}_y + \mathbf{e}_y{}^T RA\mathbf{e}_x\right), \tag{16}$$

$$w_6 = -2\mu \mathbf{e}_y{}^T RA\mathbf{e}_y, \tag{17}$$

and $\mathbf{d} = (d_1, d_2, d_3, d_4, d_5, d_6)$, where

$$d_i = L_i^2 - (1+\mu^2)(x_i^2 + y_i^2), \quad i = 1, 2, \ldots, 6. \tag{18}$$

Then relation (11) can be written as a linear system with the form

$$Q\mathbf{w} = \mathbf{d}, \tag{19}$$

where the matrix Q is given by

$$Q = \begin{pmatrix} 1 & x_1 & y_1 & x_1^2 & x_1y_1 & y_1^2 \\ 1 & x_2 & y_2 & x_2^2 & x_2y_2 & y_2^2 \\ 1 & x_3 & y_3 & x_3^2 & x_3y_3 & y_3^2 \\ 1 & x_4 & y_4 & x_4^2 & x_4y_4 & y_4^2 \\ 1 & x_5 & y_5 & x_5^2 & x_5y_5 & y_5^2 \\ 1 & x_6 & y_6 & x_6^2 & x_6y_6 & y_6^2 \end{pmatrix}. \tag{20}$$

Note that if the base points are all different and belong to a conic section then $\det Q = 0$. The matrix given by (20) corresponds to the well known Braikenridge-Maclaurin construction [5].

In the next sections we will show that one can obtain the rotation matrix R and the position $\mathbf{P}$ in terms of the solution $\mathbf{w} = (w_1, w_2, \ldots, w_6)$ of the linear system given by (19). The solution to the forward kinematics problem naturally divides into two cases, namely, a non-singular case where $\det Q \neq 0$ and a singular case where $\det Q = 0$.

In the singular case, we obtain for a given set of length legs, $L_1, L_2, \ldots, L_6$, a singular solution parameterized by a scalar parameter. These solutions are curves in position space and in rotation space in which the platform moves without changing the values of the leg lengths.

2.1 Non-singular Case

In the case where the six base vertices are not on a conic section, one gets $\det Q \neq 0$, and so the solution of (19), $\mathbf{w} = (w_1, w_2, w_3, w_4, w_5)$, can be obtained from

$$\mathbf{w} = Q^{-1}d. \tag{21}$$

The first three equations (12), (13) and (14) determines the rotation parameters, namely, $\mathbf{q}$, and the last three (15), (16) and (17) the position $\mathbf{P} = (x, y, z)$.

To determine the rotation parameters consider the equations

$$w_4 = -2\mu(2{q_1}^2 + 2{q_0}^2 - 1), \tag{22}$$

$$w_5 = -8\mu q_1 q_2, \tag{23}$$

$$w_6 = -2\mu(2{q_2}^2 + 2{q_0}^2 - 1), \tag{24}$$

which are obtained from (15), (16) and (17), respectively. Eliminating q_0, one gets,

$${q_1}^2 - {q_2}^2 = -(w_4 - w_6)/(4\mu), \tag{25}$$

$$q_1 q_2 = -w_5/(8\mu). \tag{26}$$

Let

$$\alpha = \frac{w_4 - w_6}{4\mu}, \qquad \beta = -\frac{w_5}{8\mu}. \tag{27}$$

Then the above equations can be written as

$$q_1^4 + \alpha q_1^2 - \beta^2 = 0, \tag{28}$$

$$q_2^4 - \alpha q_2^2 - \beta^2 = 0. \tag{29}$$

So,

$$q_1^2 = \frac{-\alpha + \gamma}{2}, \tag{30}$$

$$q_2^2 = \frac{\alpha + \gamma}{2}, \tag{31}$$

where

$$\gamma = \sqrt{\alpha^2 + 4\beta^2}. \tag{32}$$

Substituting yields

$$q_3^2 = \frac{1}{2} + \frac{w_4}{4\mu} - \frac{\alpha + \gamma}{2}, \tag{33}$$

$$q_0^2 = \frac{1}{2} - \frac{w_4}{4\mu} + \frac{\alpha - \gamma}{2}. \tag{34}$$

Assuming $q_0 \geq 0$ and that (33) and (31) have two roots each, then, q_1 is determined by (23). Consequently, we have a total of four different quaternions. These are

$$\mathbf{s_1} = (\bar{q}_0, \bar{q}_1, \bar{q}_2, \bar{q}_3), \tag{35}$$

$$\mathbf{s_2} = (\bar{q}_0, \bar{q}_1, \bar{q}_2, -\bar{q}_3), \tag{36}$$

$$\mathbf{s_3} = (\bar{q}_0, -\bar{q}_1, -\bar{q}_2, \bar{q}_3), \tag{37}$$

$$\mathbf{s_4} = (\bar{q}_0, -\bar{q}_1, -\bar{q}_2, -\bar{q}_3), \tag{38}$$

where $(\bar{q}_0, \bar{q}_1, \bar{q}_2, \bar{q}_3)$ are the roots.

To determine the position, consider the equations

$$\mathbf{u}^T = 2\mu {\mathbf{e}_x}^T ((RA)^T - I), \tag{39}$$

$$\mathbf{v}^T = 2\mu {\mathbf{e}_y}^T ((RA)^T - I). \tag{40}$$

Thus

$$\mathbf{P}^T\mathbf{P} = w_1, \tag{41}$$

$$\mathbf{u}^T\mathbf{P} = w_2, \tag{42}$$

$$\mathbf{v}^T\mathbf{P} = w_3. \tag{43}$$

Obviously (42) and (43) represent two planes and their intersection is a line with equation given by

$$\mathbf{P} = \mathbf{r}_0 + t\mathbf{r}_1, \tag{44}$$

where t is the parameter of the line. The vectors $\mathbf{r}_0$ and $\mathbf{r}_1$ are given by

$$\mathbf{r}_0 = \frac{(\mathbf{v}^T\mathbf{v})w_2 - (\mathbf{u}^T\mathbf{v})w_3}{(\mathbf{u}^T\mathbf{u})(\mathbf{v}^T\mathbf{v}) - (\mathbf{u}^T\mathbf{v})^2}\mathbf{u} - \frac{-(\mathbf{u}^T\mathbf{v})w_2 + (\mathbf{u}^T\mathbf{u})w_3}{(\mathbf{u}^T\mathbf{u})(\mathbf{v}^T\mathbf{v}) - (\mathbf{u}^T\mathbf{v})^2}\mathbf{v}, \tag{45}$$

$$\mathbf{r}_1 = \frac{\mathbf{u} \times \mathbf{v}}{\|\mathbf{u} \times \mathbf{v}\|}. \tag{46}$$

The line (44) intersects the sphere (41) at two points $P_\pm$ given by

$$P_\pm = \mathbf{r}_0 \pm t^*\mathbf{r}_1, \tag{47}$$

where

$$t^* = \sqrt{w_1 - \mathbf{r}_0^T\mathbf{r}_0}. \tag{48}$$

Note that in order to $P_\pm$ exist one should have

$$w_1 \geq \mathbf{r}_0^T\mathbf{r}_0. \tag{49}$$

So, both R and $\mathbf{P}$ are found, and totally they have eight possible different solutions for a given set of leg lengths.

2.2 Singular Case

In this case, we assume that all points belong to a circle $x_i^2 + y_i^2 = 1$ (we can assume $r = 1$ without loss of generality), $i = 1, 2, \ldots, 6$. In this case the matrix

$$Q = \begin{pmatrix} 1 & x_1 & y_1 & x_1^2 & x_1 y_1 & 1 - x_1^2 \\ 1 & x_2 & y_2 & x_2^2 & x_2 y_2 & 1 - x_2^2 \\ 1 & x_3 & y_3 & x_3^2 & x_3 y_3 & 1 - x_3^2 \\ 1 & x_4 & y_4 & x_4^2 & x_4 y_4 & 1 - x_4^2 \\ 1 & x_5 & y_5 & x_5^2 & x_5 y_5 & 1 - x_5^2 \\ 1 & x_6 & y_6 & x_6^2 & x_6 y_6 & 1 - x_6^2 \end{pmatrix} \tag{50}$$

is singular, that is, $\det Q = 0$ and in fact, if all points are different and belong to a conic section the rank of Q is five (corresponding to the Braikenridge-Maclaurin construction). This will be the case if $x_i^2 + y_i^2 = 1$, $i = 1, 2, \ldots, 6$, and $(x_i, y_i) \neq (x_j, y_j)$ for $i \neq j$, $i, j = 1, 2, \ldots, 6$ [5].

This fact enables us to explicitly compute the LU factorization of the matrix Q in terms of the coordinate of the vertices of the base (x_i, y_i), $i = 1, 2, \ldots, 6$. These expressions are to big to be shown here but a script for the maxima computer algebra system [6] is available upon request to the author.

So the linear system $Q\mathbf{w} = d$ can be put into the form

$$U\mathbf{w} = L^{-1}d, \tag{51}$$

where $\det L = 1$ and U is a matrix with rank 5. The solution of (51) is given in terms of a solution $(w_2, w_3, w_4, w_5, w_6)$ which depends on the value of w_1, which we take to be a free parameter. Notice that any other quantity could be used for this purpose, although expression (41) suggests that w_1 is the good choice. So the expressions given by (30), (31), (33) and (34) can be used to determine the values of the quaternion $\mathbf{q}$, the rotation matrix, and the point $\mathbf{P}$ as a function of the free parameter w_1.

3 Conclusions

The singular manifold of a Stewart platform is defined as the set of physical configurations for which it will not be possible to determine the position of the platform uniquely by fixing the lengths of the legs. By considering a simple Stewart platform, for which the base vertices are in a circle (although the result also holds for any conic section) and the bottom and top plates are related by a rotation and a contraction, it was shown that the platform is always in a singular configuration, i.e., will always have a singular set of solutions is parametrized by a scalar parameter. It was also shown how to characterize the singular manifold in this case and how it can be parameterize by a scalar parameter.

References

1. J.P. Merlet, *Parallel Robots* (Springer, Berlin, 2006)
2. T. Charters, R. Enguiça, P. Freitas, The 60th European Study Group with Industry Report on the Stewart platform problem. Math.-Ind. Case Stud. J. **1**, 66–80 (2009)
3. P. Ji, H. Wu, A closed-form kinematics solutions for the $6-6^{p}$ Stewart platform. IEEE Trans. Robot. Autom. **17**(4), 522–526 (2001)
4. J. Diebel, *Representing attitude: Euler angles, quaternions, and rotation vectors*. http://ai.stanford.edu/~diebel/attitude/attitude.pdf
5. H.S.M. Coxeter, *Projective Geometry*, 2nd edn. (Springer, New York, 1987)
6. http://maxima.sourceforge.net

Part VI
Fractional Calculus Applications

Some Advances on Image Processing by Means of Fractional Calculus

E. Cuesta

Abstract Fractional calculus and fractional time integration was introduced by the author in problems related with image processing. In particular in the mentioned work the author uses fractional integral equations and the backward Euler convolution quadrature method for the time discretization. However, a naive implementation of this procedure shows some drawbacks in the field of image processing and filtering as, for example, the high computational cost and the homogeneous treatment of non-homogeneous images (in the sense of images with high gradient variations). In the present work a local treatment is proposed as well as more efficient numerical methods leading us to improve the filtering and the efficiency of the implementation. Practical illustrations will be provided.

Keywords Fractional calculus · Adaptive quadratures · Image filtering

1 Introduction

The applications of the fractional calculus is attracting an increasing interest in several and different fields of science and engineering (see Prüss [9]). In particular, a pioneer work in the application of the fractional calculus to image denoising and processing (see Cuesta and Codés [10]) has promoted further and interesting works on this line (see e.g. Duits et al. [7] and Didas et al. [6]).

In Cuesta and Codés [10] the starting point turns out to be the two-dimensional heat equation

$$\begin{cases} \partial_t u(t,\mathbf{x}) = \Delta u(t,\mathbf{x}), & (t,\mathbf{x}) \in [0,T] \times \Omega, \\ u(0,\mathbf{x}) = u_0(\mathbf{x}), \end{cases} \tag{1}$$

E. Cuesta (✉)
Department of Applied Mathematic, E.U.P., University of Valladolid C, Francisco Mendizabal 1, 47014 Valladolid, Spain
e-mail: eduardo@mat.uva.es

J.A.T. Machado et al. (eds.), *Nonlinear Science and Complexity*, 265
DOI 10.1007/978-90-481-9884-9_31, © Springer Science+Business Media B.V. 2011

where Ω is a square domain in $\mathbb{R}^2$, $\Delta = \partial^2/\partial_x^2 + \partial^2/\partial_y^2$ with a suitable set of boundary conditions (typically homogeneous Newman conditions), and ∂_t stands for time derivative. In this framework the initial data $u_0(\mathbf{x})$ is understood as the original image which evolves in time as the analytic solution $u = u(t, \mathbf{x})$ of (1).

The smoothing properties of (1) are very well known and play a crucial role in the understanding of the heat transfer, however in the framework of the image processing, smoothing is partially undesirable. To be more precise, image denoising procedures are usually assumed to be edges preserving, i.e. while in some parts of the original image some smoothing is wanted, in some others (e.g. edges) this effect should be neglected. To overcome this drawback non-linear models have been proposed in literature as for example, the Perona-Malik based models (see Aubert and Kornprobst [3] and references therein, and more recently Bartels and Prohl [4], and Amann [5]). In these models the smoothing effect is handled by a non-linear term, frequently depending on the variation of the gradient. However regarding some mathematical facts of the mentioned non-linear problem as, for example, the well-possessedness of the problem, the authors are obliged to consider *perturbations* of the model initially proposed yielding results not so good as expected.

In Cuesta and Codés [10] we proposed a new approach to image denoising by means of partial differential equations with fractional time derivatives which reads, in an abstract format as,

$$\begin{cases} \partial_t^\alpha u(t, \mathbf{x}) = \Delta u(t, \mathbf{x}), & (t, \mathbf{x}) \in [0, T] \times \Omega, \\ u(0, \mathbf{x}) = u_0(\mathbf{x}). \end{cases} \tag{2}$$

In (2) ∂_t^α stands for the Riemann-Liouville fractional time derivative of order α, $1 < \alpha < 2$, and the boundary conditions are going to be homogeneous Newmann boundary conditions. Since the problem (2) interpolates a diffusion problem (for $\alpha = 1$ we have the heat equation) and a conservative problem (for $\alpha = 2$ we have the wave equation with zero initial velocity), the solution of (2) will satisfy intermediate properties. In fact the diffusion in (2) is handled by the viscosity parameter α so that the maximal diffusion is reached for $\alpha = 1$ and there is no diffusion at all for $\alpha = 2$. Let us notice that this approach allows us to control diffusion avoiding tricky non-linear terms and drawbacks as the well-possessedness mentioned above.

Integrating in both sides of the equation in (2) we have an equivalent formulation of (2) as an integral equation which in an abstract setting can be written as

$$u(t) = u_0 + \int_0^t \frac{(t-s)^{\alpha-1}}{\Gamma(\alpha)} \Delta u(s)\, \mathrm{d}s, \quad 0 \le t \le T, \tag{3}$$

where $u(t) = u(t, \mathbf{x})$ and $u_0 = u_0(\mathbf{x})$. Let us recall that the integral

$$\int_0^t \frac{(t-s)^{\beta-1}}{\Gamma(\beta)} g(s)\, \mathrm{d}s,$$

stands for the fractional integral of g of order β, $1 < \beta < 2$, in the sense of Riemann-Liouville.

Besides the theoretical properties of the solutions of (3), numerical methods and its implementation play an important role in the reliability of these models. In fact, the memory of the processes modeled by means of convolution equations as (3), reflects in time discretization. In particular, the *runtime* and the memory required for these algorithms are larger than for classical methods applied to ordinary differential equations. In this paper we show an efficient time discretization and different implementations looking for the reliability of these models. Actually several ideas concerning the fractional calculus and image denoising are putting into practice at present by Cuesta et al. but without a clear explanation of the results obtained yet. Hopefully, new and interesting results will be reached in forthcoming works.

On the other hand, different approaches lying in the framework of the partial differential equations and fractional calculus has been proposed in literature. Let us mention, e.g. Bai [8] where an anisotropic model with fractional space derivatives is considered, or Mathieu et al. [1] whose authors analyze an edge detection procedure based on fractional calculus.

The paper organizes as follows. In Sect. 2 we show the main result concerning the time discretization of (3). Section 3 is devoted to present several numerical illustrations and Sect. 4 we show some conclusions on this work. Finally we include the references cited throughout the paper.

2 Time Discretization

In this section we focus on the analysis of the time discretization of (3). Since the space discretization of the Laplacian in (3) is not an issue in this paper, we will consider a classical second order finite difference scheme in an uniform $M \times M$ mesh grid Ω_h with step length h. Thus let us denote Δ_h the discrete two-dimensional Laplacian with homogeneous Newmann boundary conditions, and let us consider the semi-discretization of (3)

$$\mathbf{u}(t) = \mathbf{u}_0 + \int_0^t \frac{(t-s)^{\alpha-1}}{\Gamma(\alpha)} \Delta_h \mathbf{u}(s)\, \mathrm{d}s, \quad 0 \leq t \leq T, \tag{4}$$

where $\mathbf{u}(t) = u(t, \mathbf{x})$ and the initial data $\mathbf{u}_0 = u_0(\mathbf{x})$, for $0 \leq t \leq T, \mathbf{x} \in \Omega_h$, are vector-valued functions.

In Cuesta and Codés [10], (4) was discretized by means of the convolution quadrature method based on the backward Euler method, thus with step-size $\tau > 0$. Since the computational cost is a serious drawback for these kind of discretizations, many attempts to get faster and more efficient algorithms can be found lately in literature. One of these attempts concerns to the adaptive quadratures (see Cuesta [2]). In this way, and since the inner structure of the convolution quadrature methods does not allow variable time step-size formulations, our choice to discretize (4) will be a classical adaptive quadrature to approximate the convolution integral, in particular we choose the one based on the right rectangle rule as in Cuesta [2]. In fact, given $0 = t_0 < t_1 < t_2 < \cdots < t_N = T$ and $\tau_j = t_j - t_{j-1}$, the numerical scheme we consider reads

$$\mathbf{u}_n = \mathbf{u}_0 + \sum_{j=1}^{n-1} \tau_j \frac{(t_n - t_j)^{\alpha-1}}{\Gamma(\alpha)} \Delta_h \mathbf{u}_j, \quad n \geq 1, \tag{5}$$

where $\mathbf{u}_j$ stand for the approximation to $\mathbf{u}(t_j)$, for $j = 0, 1, 2, \dots, N$. Let us observe that the computational cost to calculate $\mathbf{u}_N$ is dominated by the cost to evaluate the quadrature which turns out to be $O(N^2)$ (computational cost due to the space discretization is not taken into account in this discussion). Method (5) combined with a suitable step-size setting allows us to reduce the number of steps N keeping the error under control. This scheme has been studied in Cuesta [2] as well as the mentioned step-size setting strategy. In particular, in Cuesta [2] we propose a dynamical time step-size setting strategy based on an a posteriori error analysis which allows to the user to compute the numerical solution keeping the step-by-step error under a given tolerance.

Despite a precise analysis of the convergence of (5) can be found in Cuesta [2], for the sake of commodity of reader we show below the main result of the mentioned paper applied to (5).

Theorem 1 *Let* $\{\tau_1, \tau_2, \dots, \tau_N\}$ *be a step-size setting,* u_0 *belonging to the domain of the two-dimensional Laplacian with homogeneous Newmann boundary conditions* $D(\Delta)$. *Then there exists* $C > 0$ *independent of the data such that*

$$\|\mathbf{u}(t_n) - \mathbf{u}_n\| \le C \sum_{j=1}^{n-1} h_j^2 \|\Delta u_0\| (t_n - t_j)^{\alpha-2}, \quad 1 \le n \le N.$$

Let us note that the error is measured in the norm $\|\cdot\|$ of the functional space where $D(\Delta)$ lives, typically $L^p(\Omega)$. Moreover, let us clarify that the error bound in Theorem 1 requires a slight modification of (5) which is precisely described in Cuesta [2] and which does not change significantly that scheme.

In Cuesta [2] can be also found numerical illustrations showing the efficiency of the method (5) combined with the mentioned step-size setting strategy when applying to equations as (4), also when a complementary term appears in the right hand size of the equation (i.e. for non homogeneous equation).

3 Numerical Experiments

Since the efficiency of methods of type (5) has been showed in Cuesta [2] in this paper we focus on the reliability of the fractional equations when applying to image filtering.

Some facts on the implementation deserve to be commented and highlighted. In particular, a naive implementation of the numerical scheme we propose lies in fixing $1 < \alpha < 2$ then applying (5) to the whole image (i.e. to $\mathbf{u}_0$). The computational cost is then dominated by $O(N^2 \cdot M^2)$ (named approach AP1 in the next).

However, since he smoothing is handled by α, splitting the original image (e.g. into rectangular sub-images) and applying (5) with different values of α on each sub-image is allowed (named approach AP2 in the next). In such a case, α should be chosen close to 1 if the sub-image is flat (seen as a three-dimensional surface), i.e.

if the variation of gradient is small (smoothing). On the other hand, the parameter α should be close to 2 for the sub-images having edges, i.e. where the variation of gradient is large (edges preserving). Intermediates values of α apply to intermediate situations.

Different and sophisticated procedures can automatically provide the values of α for each sub-image. In the experiments below we proceed by means of a naive argument: we take $\alpha = 1.01$ and $\alpha = 1.99$ for the sub-images with minimum and maximum mean gradient respectively (we avoid the singular cases $\alpha = 1, 2$), and we linearly assign the values of α for the intermediate mean gradients.

Let us point out that the splitting of the image carries out a significant computational cost reduction, in fact if $M = M_1 + M_2 + \cdots + M_k$, then the computational cost is $O(N^2 \cdot (M_1^2 + M_2^2 + \cdots + M_k^2))$, thus much less than $O(N^2 \cdot M^2)$. Therefore, among the filtering improvements due to the splitting we get a notable computational improvement.

To illustrate the features of our approach we show below several practical and simple examples where the splitting into rectangular areas has been *hand-done*, merely by observing the edges of the image. The next experiments has been carried out with $\alpha = 1.5$ for the whole image (central figure in each), final time $T = 0.03$, and a tolerance $TOL = 10^{-2}$ for the time step setting strategy which is reached for $N = 3$ (in all cases below).

In Fig. 1 we show a gray-scale picture (leftmost) and we can observe that the approach (AP1) yields a damaged image (center) because of the uniform smoothing. However, the approach (AP2) (rightmost) turns out to be better because edges are preserved and in the *flat* areas the *pixelating* is blurred. The final image yielded by (AP2) can be improved by considering smaller tolerances, i.e. greater N for the time discretization.

Similar results can be observed with color images, in particular RGB images as the ones considered in Figs. 2 and 3. Now (AP1) and (AP2) apply on the three color-levels separately and, as in Fig. 1, the approach (AP1) blurs severely edges and the approach (AP2) preserves edges while smoothing the rest of the images (see rightmost images).

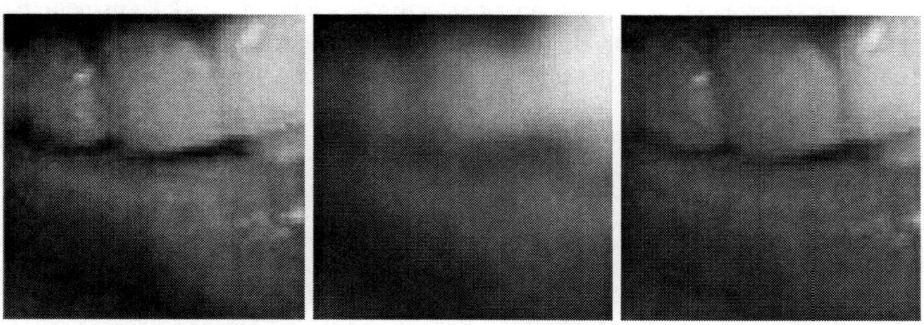

Fig. 1 Gray-scale image: Original (*leftmost*), approach AP1 (*center*), and local approach (*rightmost*)

Fig. 2 RGB image: Original (*leftmost*), approach AP1 (*center*), and local approach (*rightmost*)

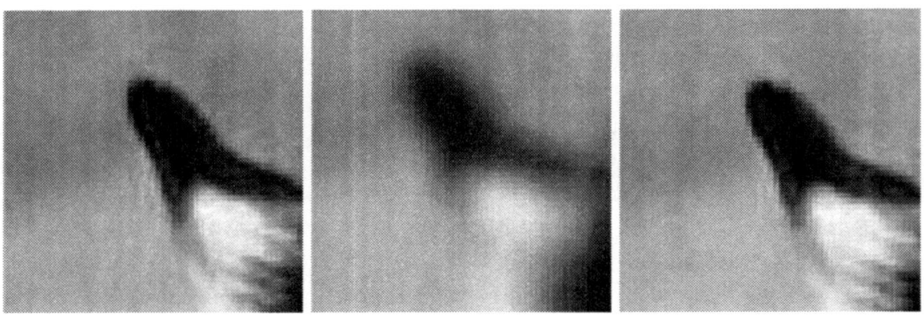

Fig. 3 RGB: Original (*leftmost*), approach AP1 (*center*), and local approach (*rightmost*)

4 Conclusions

The way of the fractional calculus in the framework of the image processing is now open more than ever.

Concerning the experiments shown in Sect. 3 it easily understandable that other parameter setting (α, *tolerance*, splitting, ...) can notably improve the results, and might be, a precise analysis of the improvements in terms, e.g. of confusion matrices or the Kappa index, could be carried out.

On the other hand, in view of the subjects discussed in this work let us note that several challenges are still pending, for example a suitable dynamical splitting image strategy to apply efficiently the approach (AP2), including, why not, non rectangular sub-images.

References

1. B. Mathieu, P. Melchior, A. Oustaloup, C. Ceyral, Fractional differentiation for edge detection. Signal Process. **83**, 2421–2432 (2002)
2. E. Cuesta, Adaptive discretizations in time for convolution equations in Banach spaces with a posteriori error control (2008, submitted for publication)

3. G. Aubert, P. Kornprobst, *Mathematical Problems in Image Pocessing*. Applied Mathematical Sciences, vol. 147 (Springer, New York, 2002)
4. S. Bartels, A. Prohl, Stable discretization of a scalar and constrained vectorial Perona–Malik equation (2008, submitted for publication)
5. H. Amann, Time-delayed Perona–Malik type problems. Acta Math. Univ. Comenianae **LXXVI**, 15–38 (2007)
6. S. Didas, B. Burgeth, A. Imiya, J. Weickert, Regularity and scale–space properties of fractional high order linear filtering. Scale Spaces PDE Methods Comput. Vis., 13–25 (2005)
7. R. Duits, M. Felsberg, L. Florack, B. Platel, α scale spaces on a bounded domain, in *Proc. 4th Int. Conf. Scale Spaces* (2003), pp. 494–510
8. J. Bai, X. Feng, Fractional-order anisotropic diffusion for image processing. IEEE Trans. Image Process. **16**(10), 2492–2502 (2007)
9. J. Prüss, *Evolutionary Integral Equations and Applications* (Birkhäuser, Basel, 1993)
10. E. Cuesta, J.F. Codes, Image processing by means of linear integro-differential equation, in *Proc. 3rd IASTED Int. Conf. Visualization, Imaging and Image Processing*, vol. 12 (2003), pp. 1579–1590

Application of Genetic Algorithms in the Design of an Electrical Potential of Fractional Order

Isabel S. Jesus, J.A. Tenreiro Machado, and Ramiro S. Barbosa

Abstract Fractional calculus (FC) is currently being applied in many areas of science and technology. In fact, this mathematical concept helps the researches to have a deeper insight about several phenomena that integer order models overlook. Genetic algorithms (GA) are an important tool to solve optimization problems that occur in engineering. This methodology applies the concepts that describe biological evolution to obtain optimal solution in many different applications. In this line of thought, in this work we use the FC and the GA concepts to implement the electrical fractional order potential. The performance of the GA scheme, and the convergence of the resulting approximation, are analyzed. The results are analyzed for different number of charges and several fractional orders.

Keywords Fractional order · Electrical potential · Genetic algorithms

1 Introduction

A new look of several phenomena present in electrical systems [1], induced an approach based in the fractional calculus (FC) viewpoint. Some authors [2, 3] verified that well-known expressions for the electrical potential are related through integer-order integral and derivatives and have proposed its generalization based on the concept of fractional-order poles. Nevertheless, the mathematical generalization towards FC lacks a comprehensive method for its practical implementation.

I.S. Jesus (✉) · J.A.T. Machado · R.S. Barbosa
Dept. of Electrotechnical Engineering, Institute of Engineering of Porto, Porto 4200-072, Portugal
e-mail: isj@isep.ipp.pt

J.A.T. Machado
e-mail: jtm@isep.ipp.pt

R.S. Barbosa
e-mail: rsb@isep.ipp.pt

J.A.T. Machado et al. (eds.), *Nonlinear Science and Complexity*,
DOI 10.1007/978-90-481-9884-9_32,

This article addresses the synthesis of fractional-order multipoles. In Sect. 2 we recall the classical expressions for the static electric potential and we analyze them in the perspective of FC. Based on this re-evaluation we develop a GA scheme for implementing fractional-order electrical potential approximations. Finally, in Sect. 3 we outline the main conclusions.

2 Integer and Fractional Electrical Potential

For a homogeneous, linear and isotropic media, the electric potential φ at a point P produced by a single charge (1a), a dipole (1b), a quadrupole (1c), an infinite straight filament carrying a charge λ per unit length (2a), two opposite charged filaments (2b), and a planar surface with charge density σ (3), are given by [4]:

$$\varphi = \frac{q}{4\pi\varepsilon_0}\frac{1}{r} + C \tag{1a}$$

$$\varphi = \frac{ql\cos\theta}{4\pi\varepsilon_0}\frac{1}{r^2} + C, \quad r \gg l \tag{1b}$$

$$\varphi = \frac{ql^2(3\cos^2\theta - 1)}{4\pi\varepsilon_0}\frac{1}{r^3} + C, \quad r \gg l \tag{1c}$$

$$\varphi = -\frac{\lambda}{2\pi\varepsilon_0}\ln r + C, \tag{2a}$$

$$\varphi = \frac{\lambda l\cos\theta}{2\pi\varepsilon_0}\frac{1}{r} + C, \quad r \gg l \tag{2b}$$

$$\varphi = -\frac{\sigma}{2\varepsilon_0}r + C, \tag{3}$$

where $C \in \Re$, ε_0 represents the permittivity, q the electric charge, r the radial distance and θ the corresponding angle with the axis.

Analyzing expressions (1a)–(3) we verify the relationship $\varphi \sim r^{-3}, r^{-2}, r^{-1}$, $\ln r$, r, corresponds to the application of integer-order derivatives and integrals.

The integer-order differential nature of the potential expressions (1a)–(3) motivated several authors [3] to propose its generalization in a FC perspective. Therefore, a fractional multipole produces at point P a potential $\varphi \sim r^{\alpha}, \alpha \in \Re$. Nevertheless, besides the abstract manipulation of mathematical expressions, the truth is that there is no practical method, and physical interpretation, for establishing the fractional potential [2, 3, 5–7].

Inspired by the integer-order recursive approximation of fractional-order transfer functions [11, 12], we adopt a genetic algorithm (GA) [8–10] for implementing a fractional order potential. Similarly to what occur with transfer function, the electrical integer-order potential has a *global* nature and fractional-order potentials can have only a *local* nature, that is, possible to capture only in a restricted region. This observation leads to an implementation approach conceptually similar to the one described in [5, 11–14] that is, to an approximation scheme based on a recursive placement of integer-order functions.

In this line of thought, we develop a one-dimensional GA that places n charges at the positions x_i and determines the corresponding values q_i. Our goal is to compare the approximate potential φ_{app} given by:

$$\varphi_{\text{app}} = \sum_{i=1}^{n} \frac{q_i}{4\pi\varepsilon_0|x - x_i|} \tag{4}$$

that mimics the desired reference potential $\varphi_{\text{ref}} = kx^{\alpha}$ in a given interval $x_{\text{mim}} < x < x_{\text{max}}$.

It is important to refer that a reliable execution and analysis of a GA usually requires a large number of simulations to provide that stochastic effects have been properly considered. Therefore, in this study the experiments consist on executing the GA several times, in order to generate a combination of positions and charges that lead to an electrical potential with fractional slope similar to the desire reference potential. In the first case of study, the values of GA parameters are: population number $\text{P} = 40$, crossover $\text{C}(\%) = 85.0\%$, mutation $\text{M}(\%) = 1.0\%$ and an elitist strategy $\text{ES}(\%) = 10.0\%$. The chromosome has $2n$ genes: the first n genes correspond to the charges and the last n genes indicate their positions. The gene codifications adopts a Gray Code with a string length of $l = 16$ bits. The optimization fitness function corresponds to the minimization of the index:

$$J = \sum_{k=1}^{m} \left(\ln \left| \frac{\varphi_{\text{app}}}{\varphi_{\text{ref}}} \right| \right)^2, \qquad \min_i(J), \quad i = 0, 1, \ldots, n-1 \tag{5}$$

where m is the number of sampling points along the interval $x_{\text{mim}} < x < x_{\text{max}}$. We establish a maximum number of iterations $I_{\text{Max}} = 100$ and a stopping scheme when $J < 10^{-10}$ for the best individual (i.e., solution) of the GA population.

Figure 1(a) shows a pre-defined number of $n = 5$ charge approximation and $\varphi_{\text{ref}} = 1.0x^{-1.5}$, $0.2 < x < 0.8$, leading to $\{q_1, q_2, q_3, q_4, q_5\} = \{0.737, 0.846, -0.777, 0.382, -0.225\}$ [C] (with scale factor $\times(4\pi\varepsilon_0)^{-1}$), located at $\{x_1, x_2, x_3, x_4, x_5\} = \{-0.06, 0.092, 0.147, -0.106, 0.117\}$ [m], respectively. In this case, the GA needs $I = 51$ iterations to satisfy the adopted fitness function stopping threshold.

The results show a good fit between φ_{ref} and φ_{app} and we verify that it is possible to find more than one 'good' solution (Fig. 1(b)). Nevertheless, for a given application, a superior precision may be required and, in that case, a larger number of charges must be used. In this line of thought, we study the performance of this method for different number of charges, namely from $n = 1$ up to $n = 10$ charges, and we compare the necessary number of GA iterations when the number of charges increases. In order to analyze the precision of this distribution of charges, we study the require number of iterations I and the computational time T when the number of charges varies from $n = \{1, \ldots, 10\}$.

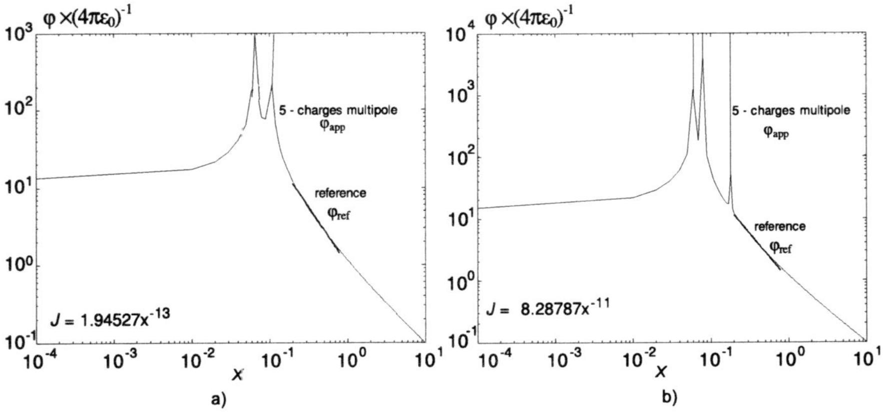

Fig. 1 Comparison of the electric potential φ_{app} and φ_{ref} versus the position x for $\varphi_{\text{ref}} = 1.0x^{-1.5}$ [volt], $0.2 < x < 0.8$ [m], and a $n = 5$ charge approximation, in both cases

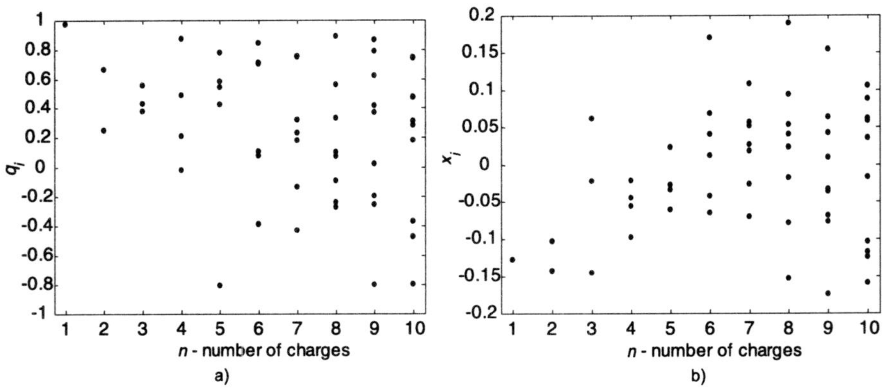

Fig. 2 Values of (**a**) charges q_i and the (**b**) corresponding positions x_i versus n, for a distribution of charges with $n = \{1, \ldots, 10\}$, $\varphi_{\text{ref}} = 1.0x^{-1.5}$ [volt], $0.2 < x < 0.8$ [m]

Figure 2 shows the values of the charges q_i and the corresponding positions x_i, for $n = \{1, \ldots, 10\}$. We verify that the value of the charge and the location pattern *versus* the number of charges is not clear.

Figure 3(a) depict the minimum, average and maximum of the number of required GA iterations I versus n. This chart reveals clearly that the required number of iterations increases with n.

We can also evaluate the GA computational time T for different number of charges. Therefore, we test the GA scheme for identical parameters and fitness function J (5). Figure 4(b) illustrates the corresponding minimum, average and maximum of T versus n. We verify that we get a smaller approximation error J but a larger computational time T for larger values of n.

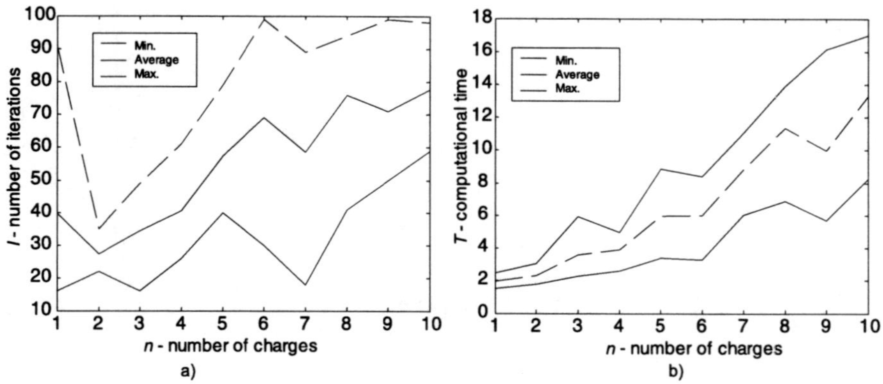

Fig. 3 Performance of the GA scheme versus the number charges $n = \{1, \ldots, 10\}$ for $\varphi_{\text{ref}} = 1.0x^{-1.5}$ [volt], $0.2 < x < 0.8$ [m], (**a**) number of required iterations I, (**b**) computational time T

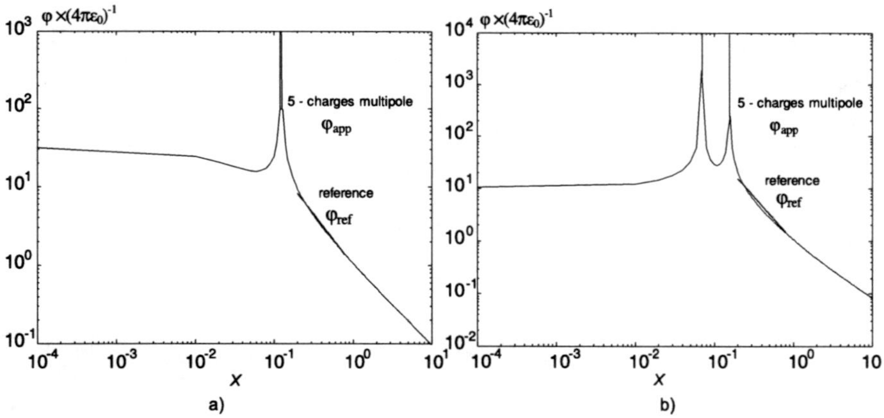

Fig. 4 Comparison of the electrical potential φ_{app} and φ_{ref} versus the position x for (**a**) $\varphi_{\text{ref}} = 1.0x^{-1.3}$ [volt] and (**b**) $\varphi_{\text{ref}} = 1.0x^{-1.7}$ [volt], $0.2 < x < 0.8$ [m] and a $n = 5$ charge approximation

With the proposed method it is also possible to have a reference potential with other slope values α. Figure 5 shows a five charge approximation for $0.2 < x < 0.8$ and $\varphi_{\text{ref}} = 1.0x^{-1.3}$, leading to $\{q_1, q_2, q_3, q_4, q_5\} = \{0.471, 0.464, 0.578, -0.371, -0.173\}$ [C] (with scale factor $\times(4\pi\varepsilon_0)^{-1}$), located at $\{x_1, x_2, x_3, x_4, x_5\} = \{-0.125, 0.029, 0.037, 0.132, 0.152\}$ [m] and for $\varphi_{\text{ref}} = 1.0x^{-1.7}$, $0.2 < x < 0.8$, leading to $\{q_1, q_2, q_3, q_4, q_5\} = \{0.753, 0.535, 0.429, -0.218, -0.681\}$ [C] (with scale factor $\times(4\pi\varepsilon_0)^{-1}$), located at $\{x_1, x_2, x_3, x_4, x_5\} = \{-0.157, -0.070, 0.171, 0.188, 0.200\}$ [m], respectively.

The charges are also function of the slope α and, therefore, we apply the GA with identical parameters, for $0.2 < x < 0.8$ [m] while varying α. Figure 5 depicts

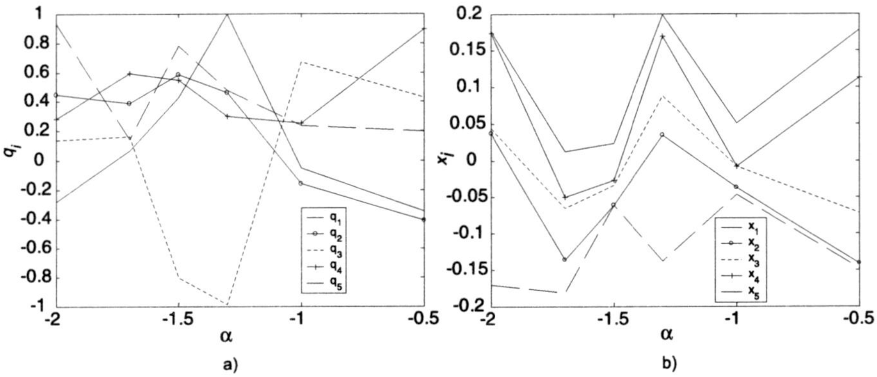

Fig. 5 Values of (**a**) charges q_i and the (**b**) corresponding positions x_i versus α, for a $n = 5$ distribution of charges and for $0.2 < x < 0.8$ [m]

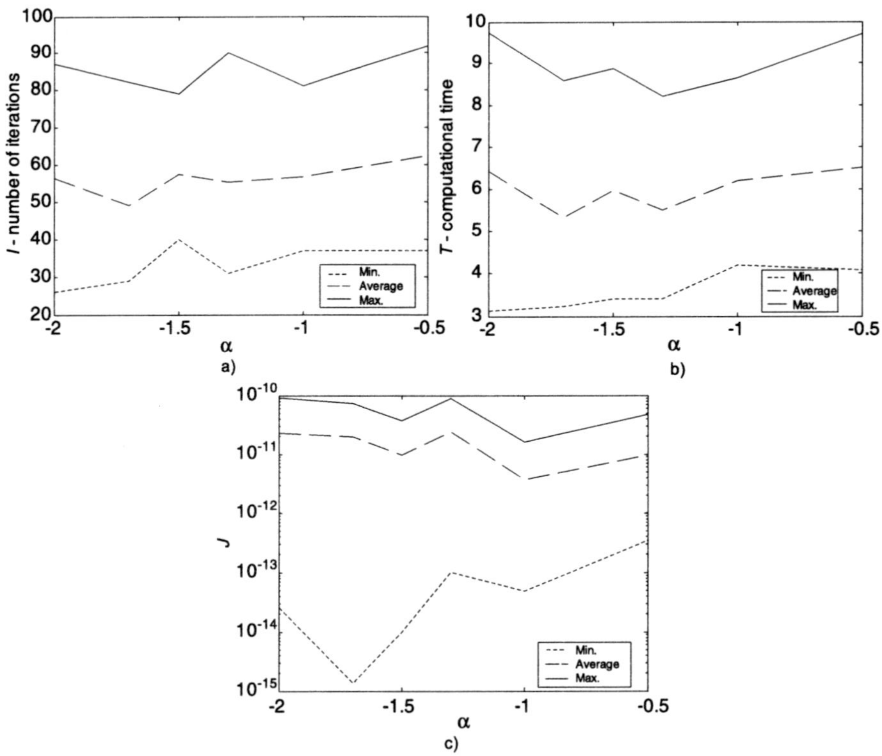

Fig. 6 Performance of the GA scheme versus α for $0.2 < x < 0.8$ [m], (**a**) number of required iterations I, (**b**) computational time T, (**c**) error J for $n = 5$

q_i and x_i versus α, without revealing again any clear pattern. On the other hand, the number of iterations I, the GA computational time T and the error J versus α, reveal a smooth evolution. Figure 6 illustrates the corresponding minimum, average and maximum of I, T and J as function of α.

In conclusion, the fit between φ_{app} and φ_{ref} is adequate and there is no obvious pattern for the charge distribution as n increases.

This lack of 'order' is due to the large number of possible solutions. Therefore, the GA has a high freedom, choosing solutions that are almost not correlated. However, we believe that further study imposing more strict restrictions may lead to the emergence of a comprehensive scheme.

3 Conclusions

This paper addressed the problem of implementing a fractional-order electric potential through a genetic algorithm. The results reveal the necessity of a larger number of iterations when the number of charges increases. The GA reveals a good compromise between the accuracy and computational time. The GA approach constitutes a step towards the development of a simple design technique and, consequently, several of its aspects must be further evaluated.

References

1. N. Engheta, IEEE Trans. Antennas Propag. **44**(4), 554–566 (1996)
2. J.T. Machado, I. Jesus, A. Galhano, A.W. Malpica, F. Silva, J.K. Tar, Fractional order dynamics in classical electromagnetic phenomena, in *Proc. of Fifth EUROMECH Nonlinear Dynamics Conference (ENOC'05)*, Eindhoven (2005), pp. 1322–1326
3. J.T. Machado, I. Jesus, A. Galhano, A fractional calculus perspective in electromagnetics, in *Proc. of Int. Design Engineering Technical Conf. & Computers and Information in Engineering Conference—5th Int. Conf. on Multibody Systems, Nonlinear Dynamics and Control (ASME'05)*, USA (2005)
4. L. Bessonov, *Applied Electricity for Engineers* (MIT Press, Moscow, 1968)
5. I.S. Jesus, J.A.T. Machado, J.B. Cunha, Application of genetic algorithms to the implementation of fractional electromagnetic potentials, in *Proc. of The Fifth International Conference on Engineering Computational Technology (ECT'06)*, Spain (2006)
6. J.T. Machado, I. Jesus, A. Galhano, J.B. Cunha, Signal processing. Fract. Calc. Appl. Signals Syst. **86**(10), 2637–2644 (2006), Special Issue (EURASIP/Elsevier)
7. J.T. Machado, I. Jesus, A. Galhano, Electric fractional order potential, in *Proc. of XII International Symposium on Electromagnetics Fields in Mechatronics, Electrical and Electronic Engineering (ISEF'05)*, Spain (2005)
8. D.E. Goldberg, *Genetic Algorithms in Search Optimization and Machine Learning* (Addison-Wesley, Reading, 1989)
9. Z. Michalewicz, *Genetic Algorithms + Data Structures = Evolution Programs* (Springer, Berlin, 1996)
10. M. Mitchell, *An Introduction to Genetic Algorithms* (MIT Press, Cambridge, 1998)
11. K.B. Oldham, J. Spanier, *The Fractional Calculus: Theory and Application of Differentiation and Integration to Arbitrary Order* (Academic Press, San Diego, 1974)

12. A. Oustaloup, *La Commande CRONE: Commande Robuste d'Ordre Non Entier* (Hermes, Paris, 1991)
13. I.S. Jesus, J.T. Machado, Fractional calculus and applied analysis. Int. J. Theory Appl. **3**(11), 237–248 (2008)
14. I.S. Jesus, J.T. Machado, S.B. Ramiro, Implementing an electrical fractional potential through a genetic algorithm, in *2nd Conference on Nonlinear Science and Complexity (NSC'08)*, Porto, Portugal

Mellin Transform for Fractional Differential Equations with Variable Potential

M. Klimek and D. Dziembowski

Abstract Fractional differential equations with the t^{β} potential and fractional derivatives of order α are solved in a finite time interval applying the Mellin transform. The solution is obtained in the form of the Meijer G-function series. The results are compared with the solutions derived by Kilbas and Saigo using the fixed point theorem. The analytical formulas for some higher Meijer G-functions representing them as power functions are proved. The Mellin transform yields also solutions for the nonhomogeneous equations. As examples the equations with $\beta = 0$ and $\beta = -\alpha/2$ are studied.

Keywords Riemann-Liouville derivative · Caputo derivative · Fractional differential equations · Mellin transform · Meijer G-function

1 Introduction

Fractional differential and integral equations arise in mathematical modelling in various fields of mechanics, physics, engineering, bioengineering and finance (compare [1–7] and the references therein). Hence, the procedures of solving such equations and the investigation of the properties of their solutions are an important part of fractional calculus. Extensive review of the results derived using fixed point theorems, integral transforms and properties of special functions can be found in [8–10].

The present paper is devoted to the application of the Mellin transform method in solving equations with left-sided derivatives and variable potential. We consider

M. Klimek (✉) · D. Dziembowski
Institute of Mathematics, Czestochowa University of Technology, ul. Dabrowskiego 73, 42-200 Czestochowa, Poland
e-mail: klimek@im.pcz.pl

D. Dziembowski
e-mail: im@im.pcz.pl

J.A.T. Machado et al. (eds.), *Nonlinear Science and Complexity*,
DOI 10.1007/978-90-481-9884-9_33,

here the homogeneous and nonhomogeneous case. Such equations were also studied by Kilbas and Saigo [8, 11, 12] using the fixed point theorem.

The Mellin transform yields an intermediate difference equation in the complex halfplane. We solve it explicitly and obtain after the inverse Mellin transform analytical solutions in finite time interval $[0, b]$ obeying the respective initial conditions. The obtained results and the Mellin transform method can be easily extended to equations in which the left-sided operators are replaced by right-sided ones and consequently to equations containing both: left- and right-sided derivatives.

The paper is organized as follows: in Sect. 2 we recall the definitions and properties of fractional operators and of the Mellin integral transform. In Sect. 3 a fractional differential equation with a left-sided Riemann-Liouville derivative and potential term—t^{β} is solved. The results for an analogous equation containing the Caputo derivative are enclosed in Sect. 4. In both cases the obtained solutions are compared with the results by Kilbas and Saigo [8, 11, 12] and analytical formulas for higher Meijer G-functions are derived. Next, in Sect. 5 the class of equations is extended to the nonhomogeneous case and the analytical solutions are again the result of the application the Mellin transform.

2 Fractional Operators

We recall some definitions of fractional operators and their properties. The left-sided fractional Riemann-Liouville integral of real, positive order $\alpha \in R_+$ is defined as follows [8, 13]:

$$I^{\alpha}_{0+} f(t) := \frac{1}{\Gamma(\alpha)} \int_0^t \frac{f(u)du}{(t-u)^{1-\alpha}} \quad t > 0. \tag{1}$$

Using this fractional integral, the left-sided Riemann-Liouville derivative is constructed. For real order $\alpha \in (n-1, n)$ it looks as follows (we have denoted the classical derivative as $D := \frac{d}{dt}$) [8, 13]:

$$D^{\alpha}_{0+} f(t) := D^n I^{n-\alpha}_{0+} f(t). \tag{2}$$

Having the Riemann-Liouville derivative we can define the left-sided Caputo derivative:

$${}^{c}D^{\alpha}_{0+} f(t) := D^{\alpha}_{0+}\left[f(t) - \sum_{k=0}^{n-1} \frac{D^k f(0)}{k!} t^k \right]. \tag{3}$$

The useful property of fractional operators are their composition rules, which we shall apply in the procedure of solving certain fractional differential equations. For any function $f \in L_1(R_+)$ we have the corresponding relations [8, 13] for the Riemann-Liouville and Caputo derivatives:

$$D^{\alpha}_{0+} I^{\alpha}_{0+} f(t) = f(t), \qquad {}^{c}D^{\alpha}_{0+} I^{\alpha}_{0+} f(t) = f(t) \tag{4}$$

valid almost everywhere in time interval $[0, b]$. When we assume that function $f \in C[0, b]$, then relations (4) hold at any point of this interval.

2.1 The Mellin Transform and Its Properties

We propose to apply one of the integral transforms as the method of solving certain fractional equations with a variable potential. Namely, we shall use the Mellin transform which looks as follows for sufficiently good functions [8, 14]:

$$\mathcal{M}[f](s) := \int_0^\infty t^{s-1} f(t)dt. \tag{5}$$

Similarly to the Laplace transform, the Mellin transform also has its convolution defined by the formula:

$$f * g(t) := \int_0^\infty f(u) g\left(\frac{t}{u}\right)\frac{du}{u}. \tag{6}$$

When the Mellin transform acts on a Mellin convolution of two functions, the result is the multiplication of the corresponding transforms of both functions:

$$\mathcal{M}[f * g](s) = \mathcal{M}[f](s) \cdot \mathcal{M}[g](s). \tag{7}$$

We shall also apply the following shifting property of the Mellin transform:

$$\mathcal{M}[t^\beta f(t)](s) = \mathcal{M}[f(t)](s + \beta). \tag{8}$$

Finally, we quote a Lemma describing the Mellin transform for fractional integral I^α_{0+} from [8, 15].

Lemma 2.1 *Let $s \in C$ and $\int_0^\infty | t^{s+\alpha-1} f(t)|dt < \infty$. The following formula holds for* $\mathrm{Re}(s) < 1 - \alpha$:

$$\mathcal{M}[I^\alpha_{0+} f](s) = \frac{\Gamma(1 - \alpha - s)}{\Gamma(1 - s)}\mathcal{M}[f](s + \alpha). \tag{9}$$

3 Fractional Linear Equation with Riemann-Liouville Derivative and t^β-Potential

We shall solve a fractional equation involving the variable t^β-coefficient in finite time interval $[0, b]$:

$$(D^\alpha_{0+} - \lambda t^\beta) f(t) = 0 \quad t \in [0, b]. \tag{10}$$

Let us extend the equation and solution f to R_+ using the new version of the solution as the function defined for positive, real t by the formula:

$$f_0(t) := f(t)\Delta H(t) \quad t \in R_+,$$

where we have denoted the difference of the Heaviside's functions as $\Delta H(t) = H(t) - H(t-b)$. We assume now $t^\beta f_0 \in L_1(R_+)$ and $t^\beta f_0 \in C[0,b]$, then we apply composition rule (4) to obtain the equivalent integral form of equation (10):

$$(1 - \lambda I^\alpha_{0+} t^\beta) f_0(t) = f^{st}_\alpha(t), \tag{11}$$

where the stationary functions of the D^α_{0+} derivative in interval $[0,b]$ are known [13] as the linear combinations of certain power functions (here $c_k \in R$ are arbitrary constant coefficients):

$$f^{st}_\alpha(t) = \sum_{k=0}^{n} c_k t^{\alpha-k} \Delta H(t). \tag{12}$$

We solve (10), (11) using the Mellin transform. After the transformation (11) becomes the following difference equation:

$$(1 - \lambda g(s) T_{\alpha+\beta}) \mathcal{M}[f_0](s) = \mathcal{M}[f^{st}_\alpha](s) \tag{13}$$

valid in complex halfplane $\mathrm{Re}(s+\alpha) < 1$ with function g given as

$$g(s) = \frac{\Gamma(1-\alpha-s)}{\Gamma(1-s)}$$

and the translation operator acting on the functions of the complex variable as follows: $T_{\alpha+\beta}\, v(s) := v(s+\alpha+\beta)$. The above difference equation is solved by the series:

$$\mathcal{M}[f_0](s) = \sum_{m=0}^{\infty} \lambda^m (g(s) T_{\alpha+\beta})^m \mathcal{M}[f^{st}_\alpha](s) \tag{14}$$

absolutely convergent in the complex halfplane $\mathrm{Re}(s+\alpha) < 1$ and the convergence is uniform with respect to imaginary part $\mathrm{Im}(s)$.

We shall evaluate solution f_0 using the inverse Mellin transform. Let us note that the following equality is valid for the products of the g-functions:

$$\prod_{l=0}^{m-1} g(s + l(\alpha+\beta)) = \mathcal{G}^{0,m}_{m,m}\left[\begin{matrix}\mathbf{a}_m \\ \mathbf{b}_m\end{matrix}\,\middle|\, s\right], \tag{15}$$

where $\mathcal{G}$ denotes the Mellin transform of the respective Meijer G-function defined by the vectors (see the properties of these functions in [8, 15]):

$$\mathbf{a}_m = (\alpha+\beta)\mathbf{j}_m + \alpha \mathbf{e}_m, \qquad \mathbf{b}_m = (\alpha+\beta)\mathbf{j}_m \tag{16}$$

with component vectors $\mathbf{j}_m$ and $\mathbf{e}_m$ given by the formulas:

$$\mathbf{j}_m := [0, 1, \dots, m-1] \in R^m, \qquad \mathbf{e}_m := [1, \dots, 1] \in R^m.$$

In the general case, the Meijer G-functions are defined as a subset of the Fox H-functions given by the following inverse Mellin transform:

$$G^{m,n}_{p,q}\left[\begin{matrix} (a_1, \dots, a_p) \\ (b_1, \dots, b_q) \end{matrix} \middle| z \right] := \frac{1}{2\pi i} \int_{\mathcal{C}} \mathcal{G}^{m,n}_{p,q}(s) z^{-s} ds, \tag{17}$$

where the components of the vectors are complex numbers $a_l, b_j \in C$ for $j = 1, \dots, q, l = 1, \dots, p$ as well as variable $z \in C$. Kernel $\mathcal{G}^{m,n}_{p,q}(s)$ is described by the formula:

$$\mathcal{G}^{m,n}_{p,q}\left[\begin{matrix} (a_1, \dots, a_p) \\ (b_1, \dots, b_q) \end{matrix} \middle| s \right] := \frac{\prod_{j=1}^{m} \Gamma(b_j + s) \prod_{l=1}^{n} \Gamma(1 - a_l - s)}{\prod_{l=n+1}^{p} \Gamma(a_l + s) \prod_{j=m+1}^{q} \Gamma(1 - b_j - s)}.$$

As our transformed equation and solution are valid in the vertical halfplane we should invert it using the vertical contour of type $\mathcal{L}_{i\gamma\infty}$. Let us check the parameters defining the existence of the inverse Mellin transforms for kernels $\mathcal{G}^{0,m}_{m,m}$. We apply here Theorems 1.1 and 3.3 from [15] (see also Theorem 1.6 in [8]) for $m \geq 2$:

$$\Delta = 0, \qquad a^* = 0, \qquad \mu = -m\alpha. \tag{18}$$

The solution is correctly defined in interval $[0, b]$ when the following condition is fulfilled:

$$\Delta\gamma + \text{Re}(\mu) < -1 \quad \Longrightarrow \quad \alpha > \frac{1}{2}. \tag{19}$$

Hence, in our further investigations we shall assume fractional order $\alpha > \frac{1}{2}$. The case of order $\alpha \in (0, \frac{1}{2}]$ will be studied separately in a subsequent paper.

Now for $\alpha > \frac{1}{2}$ solution f_0 in finite time interval $[0, b]$ is given as a series of Meijer G-functions:

$$f_0(t) = \sum_{m=0}^{\infty} \lambda^m G^{0,m}_{m,m}\left[\begin{matrix} \mathbf{a}_m \\ \mathbf{b}_m \end{matrix} \middle| t \right] * t^{m(\alpha+\beta)} f^{st}_{\alpha}(t). \tag{20}$$

We observe that each term of the series contains the Mellin convolution. In order to calculate this convolution explicitly we follow the theorem on the integration of Fox H-functions (Theorem 2.7 from [15]) and reformulate it for Meijer G-functions.

When the condition for the components of vector $\mathbf{a}$

$$\max_{1 \leq i \leq n}[\text{Re}(a_i) - 1] + \text{Re}(\omega) + 1 < 0 \tag{21}$$

is fulfilled, then the following integration formula for Meijer G-functions holds:

$$I^1_{-} u^{\omega} G^{m,n}_{p,q}\left[\begin{matrix} \mathbf{a} \\ \mathbf{b} \end{matrix} \middle| u \right](x) = x^{\omega+1} G^{m+1,n}_{p+1,q+1}\left[\begin{matrix} [\mathbf{a}, -\omega] \\ [-\omega - 1, \mathbf{b}] \end{matrix} \middle| x \right]. \tag{22}$$

It is easy to check that each vector $\mathbf{a}_m$ fulfills the above requirement for $\omega = -m(\alpha + \beta) - \alpha + k - 1$ provided $\{\alpha\} + \beta > 0$ ($\{\alpha\}$ is the fractional part of number α). The Mellin convolutions in formula (20) produce a Meijer G-function for each $m \in N$ and $k = 1, \ldots, n$:

$$G^{0,m}_{m,m}\left[\begin{array}{c}\mathbf{a}_m\\ \mathbf{b}_m\end{array}\middle| t\right] * t^{m(\alpha+\beta)+\alpha-k}\Delta H(t) = b^{m(\alpha+\beta)+\alpha-k} G^{1,m}_{m+1,m+1}\left[\begin{array}{c}\mathbf{A}_{m,k}\\ \mathbf{B}_{m,k}\end{array}\middle| \frac{t}{b}\right] \quad (23)$$

with new defining vectors $\mathbf{A}_{m,k}$ and $\mathbf{B}_{m,k}$:

$$\begin{aligned}\mathbf{A}_{m,k} &:= [\mathbf{a}_m, m(\alpha+\beta) + \alpha - k + 1] \in R^{m+1},\\ \mathbf{B}_{m,k} &:= [m(\alpha+\beta) + \alpha - k, \mathbf{b}_m] \in R^{m+1}.\end{aligned} \quad (24)$$

The above result is valid when the respective Meijer G-function on the right-hand side exists, that means that the poles of the Gamma functions in the numerator of its Mellin transform do not coincide (compare for example condition (1.1.6) from [15] or (1.12.5) from [8]). This general condition looks as follows:

$$b_j + l \neq a_i - k - 1, \quad i = 1, \ldots, n, \ j = 1, \ldots, m, \ k, l \in N_0. \quad (25)$$

In our case this requirement yields the condition:

$$\{\alpha\} + \beta > 0,$$

where $\{\alpha\}$ is the fractional part of number α. When this condition is obeyed we have an explicit analytical expression for each component series given by the component of the stationary function—$t^{\alpha-k}\Delta H(t)$ for $k = 1, \ldots, n$:

$$f_0^k(t) = b^{\alpha-k}\sum_{m=0}^{\infty}(\lambda b^{\alpha+\beta})^m G^{1,m}_{m+1,m+1}\left[\begin{array}{c}\mathbf{A}_{m,k}\\ \mathbf{B}_{m,k}\end{array}\middle| \frac{t}{b}\right]. \quad (26)$$

We can establish the initial conditions for solution f_0 investigating the asymptotics of each component series f_0^k (we apply here Theorem 1.11 from [15], see also formula (1.12.21) in [8]). Namely, we observe that close to $t = 0$ the Meijer G-functions can be represented as follows:

$$G^{1,m}_{m+1,m+1}\left[\begin{array}{c}\mathbf{A}_{m,k}\\ \mathbf{B}_{m,k}\end{array}\middle| \frac{t}{b}\right] = h^*_{1,m,k} t^{m(\alpha+\beta)+\alpha-k} + o\left[\left(\frac{t}{b}\right)^{m(\alpha+\beta)+\alpha-k}\right] \quad (27)$$

with the coefficient:

$$h^*_{1,m,k} = \lambda^m \frac{\prod_{l=1}^{m}\Gamma(1 + (\alpha+\beta)(m-l+1) - k)}{\prod_{l=1}^{m}\Gamma(1 + (\alpha+\beta)(m-l+1) - k + \alpha)}.$$

This asymptotical representation yields for components f_0^k the set of initial conditions for $j = 1, \ldots, n$:

$$D^{\alpha-j}_{0+} f_0^k(t)\,|_{t=0} = \Gamma(\alpha - j + 1)\delta_{jk}. \quad (28)$$

We summarize the above considerations and results in the following proposition.

Proposition 3.1 *Let $\alpha \in (n-1, n)$ and $\alpha > \frac{1}{2}$, $\{\alpha\} + \beta > 0$. Then the equation:*

$$[D^{\alpha}_{0+} - \lambda t^{\beta}] f_0(t) = 0, \quad t \in [0, b]$$

has the solution in the $C^{\alpha}_{n-\alpha}[0, b]$ space given by the formula:

$$f_0(t) = \sum_{k=1}^{n} \frac{c_k}{\Gamma(\alpha - k + 1)} f_0^k(t), \tag{29}$$

where components f_0^k look as follows:

$$f_0^k(t) = b^{\alpha-k} \sum_{m=0}^{\infty} (\lambda b^{\alpha+\beta})^m G^{1,m}_{m+1,m+1}\left[\begin{matrix}\mathbf{A}_{m,k} \\ \mathbf{B}_{m,k}\end{matrix} \,\middle|\, \frac{t}{b}\right].$$

This solution fulfills the initial conditions:

$$D^{\alpha-j}_{0+} f_0(t) \,|_{t=0} = c_j, \quad j = 1, \ldots, n. \tag{30}$$

Remark 3.2 Let us note that when $\beta \geq 0$ then the assumptions of Theorem 4.2 from [8] are obeyed, thus, we can identify our solution with the function given in interval $[0, b]$ by the formula:

$$f_0(t) = \sum_{k=1}^{n} \frac{c_k t^{\alpha-k}}{\Gamma(\alpha - k + 1)} \sum_{m=0}^{\infty} (\lambda t^{\alpha+\beta})^m \prod_{l=1}^{m} \frac{\Gamma(l(\alpha+\beta) - k + 1)}{\Gamma(l(\alpha+\beta) + \alpha - k + 1)}.$$

This remark leads to the following analytical formula for the Meijer G-functions in interval $[0, b]$ for $m \in N$ and $k = 1, \ldots, n$:

$$G^{1,m}_{m+1,m+1}\left[\begin{matrix}\mathbf{A}_{m,k} \\ \mathbf{B}_{m,k}\end{matrix} \,\middle|\, \frac{t}{b}\right] = \left(\frac{t}{b}\right)^{m(\alpha+\beta)+\alpha-k} \prod_{l=1}^{m} \frac{\Gamma(b_{l,m} + (\alpha+\beta) - k + 1)}{\Gamma(a_{l,m} + (\alpha+\beta) - k + 1)}, \tag{31}$$

where $a_{l,m}, b_{l,m}$ are the components of vectors $\mathbf{a}_m$ and $\mathbf{b}_m$ from (16), (24).

3.1 Example: Solution for Case $\beta = 0$

Let us now assume $\beta = 0$ and let us apply the Mellin transform to this simplest version of (10). The translation operator is T_α and the product of the g functions in the transformed equation looks as follows:

$$\prod_{l=0}^{m-1} g(s + l\alpha) = \frac{\Gamma(1 - m\alpha - s)}{\Gamma(1 - s)}. \tag{32}$$

We use the inverse Mellin transform and obtain:

$$\mathcal{M}^{-1}\left[\frac{\Gamma(1-m\alpha-s)}{\Gamma(1-s)}\right]=\frac{1}{\Gamma(m\alpha)}H(t-1)(t-1)^{m\alpha-1}. \tag{33}$$

The Mellin convolution with the $t^{\alpha-k}\Delta H(t)$ component of the stationary function is then the power function (here $m\in N, k=1,\dots,n$):

$$H(t-1)(t-1)^{m\alpha-1}*t^{m\alpha+\alpha-k}\Delta H(t)=t^{m\alpha+\alpha-k}\frac{\Gamma(\alpha-k+1)}{\Gamma(m\alpha+\alpha-k+1)}. \tag{34}$$

The above convolutions lead to the closed form of the f_0^k component which is simply one of the generalized Mittag-Leffler functions [8]:

$$f_0^k(t)=\Gamma(\alpha-k+1)t^{\alpha-k}E_{\alpha,\alpha-k+1}(\lambda t^\alpha). \tag{35}$$

3.2 Example: Solution for Case $\beta=-\alpha/2$

Let us now consider case $\beta=-\frac{\alpha}{2}$. Then the assumptions of Proposition 3.1 yield condition $\{\alpha\}-\frac{\alpha}{2}>0$, which requires order $\alpha\in(0,1)$. For such values of parameters α,β, the following reduction property is valid (compare Property 2.2 from [15] and (1.12.43–44) from [8]) for $m\geq 2$:

$$G^{1,m}_{m+1,m+1}\left[\begin{matrix}\mathbf{A}_{m,1}\\ \mathbf{B}_{m,1}\end{matrix}\,\middle|\,\frac{t}{b}\right]=G^{1,2}_{3,3}\left[\begin{matrix}[\frac{\alpha}{2}m,\frac{\alpha}{2}(m+1),\frac{\alpha}{2}(m+2)]\\ [\frac{\alpha}{2}(m+2)-1,0,\frac{\alpha}{2}]\end{matrix}\,\middle|\,\frac{t}{b}\right]. \tag{36}$$

Concluding, we obtain solution f_0^1 from Proposition 3.1 in the following simple form:

$$\begin{aligned}b^{1-\alpha}f_0^1(t)&=G^{1,0}_{1,1}\left[\begin{matrix}[\alpha]\\ [\alpha-1]\end{matrix}\,\middle|\,\frac{t}{b}\right]+\lambda b^{\alpha/2}G^{1,1}_{2,2}\left[\begin{matrix}[\alpha,\alpha]\\ [\alpha-1,0]\end{matrix}\,\middle|\,\frac{t}{b}\right]\\ &\quad+\sum_{m=2}^{\infty}(\lambda b^{\alpha/2})^m G^{1,2}_{3,3}\left[\begin{matrix}[\frac{\alpha}{2}m,\frac{\alpha}{2}(m+1),\frac{\alpha}{2}(m+2)]\\ [\frac{\alpha}{2}(m+2)-1,0,\frac{\alpha}{2}]\end{matrix}\,\middle|\,\frac{t}{b}\right].\end{aligned} \tag{37}$$

4 Fractional Linear Equation with Caputo Derivative and t^β-Potential

In the previous section we described in detail the procedure of solving (10) by means of the Mellin transform. We notice that all the calculations can be repeated for an equation, in which the Riemann-Liouville derivative is replaced by the Caputo derivative:

$$({}^cD^\alpha_{0+}-\lambda t^\beta)f_0(t)=0\quad t\in R_+. \tag{38}$$

The stationary function in this case is an arbitrary polynomial of degree $n-1$ when order $\alpha \in (n-1, n)$:

$$f_\alpha^{st}(t) = \sum_{k=0}^{n-1} c_k t^k \Delta H(t)$$

and $c_k \in R$ are arbitrary constant coefficients. For component $t^k \Delta H(t)$ of the stationary function we get component series f_0^k of solution f_0:

$$f_0^k(t) = \sum_{m=0}^{\infty} \lambda^m G_{m,m}^{0,m} \left[\begin{matrix} \mathbf{a}_m \\ \mathbf{b}_m \end{matrix} \middle| t \right] * t^{m(\alpha+\beta)+k} \Delta H(t). \tag{39}$$

After the evaluation of the Mellin convolutions and the investigation of their asymptotic properties at $t = 0$ we arrive at results analogous to those obtained for (10) with the Riemann-Liouville derivative. We conclude that the following proposition is valid.

Proposition 4.1 *Let $\alpha \in (n-1, n)$ and $\alpha > \frac{1}{2}$, $\beta > -1$. Then the equation*:

$$[{}^c D_{0+}^\alpha - \lambda t^\beta] f_0(t) = 0 \quad t \in [0, b]$$

has the solution in the $C_\gamma^{\alpha, n-1}[0, b]$ space given by the formula:

$$f_0(t) = \sum_{k=0}^{n-1} \frac{c_k}{k!} f_0^k(t), \tag{40}$$

where components f_0^k look as follows:

$$f_0^k(t) = b^k \sum_{m=0}^{\infty} (\lambda b^{\alpha+\beta})^m G_{m+1,m+1}^{1,m} \left[\begin{matrix} \mathbf{A}'_{m,k} \\ \mathbf{B}'_{m,k} \end{matrix} \middle| \frac{t}{b} \right]$$

with vectors $\mathbf{A}'_{m,k}$ and $\mathbf{B}'_{m,k}$ given by the formulas:

$$\mathbf{A}'_{m,k} := [\mathbf{a}_m, m(\alpha+\beta)+k+1], \qquad \mathbf{B}'_{m,k} := [m(\alpha+\beta)+k, \mathbf{b}_m].$$

This solution fulfills the initial conditions:

$$D^j f_0(t) \mid_{t=0} = c_j \quad j = 0, \ldots, n-1. \tag{41}$$

Remark 4.2 Let us note that when $\beta \geq 0$, then the assumptions of Theorem 4.4 from [8] are obeyed thus we can identify our solution with the function given in interval $[0, b]$ as follows

$$f_0(t) = \sum_{k=0}^{n-1} \frac{c_k t^k}{k!} \sum_{m=0}^{\infty} (\lambda t^{\alpha+\beta})^m \prod_{l=1}^{m} \frac{\Gamma(l(\alpha+\beta)-\alpha+k+1)}{\Gamma(l(\alpha+\beta)+k+1)}.$$

This remark leads to the following analytical formula for the Meijer G-functions in finite interval $[0, b]$ when $m \in N$ and $k = 1, \ldots, n$:

$$G^{1,m}_{m+1,m+1}\left[\begin{matrix}\mathbf{A}'_{m,k} \\ \mathbf{B}'_{m,k}\end{matrix} \middle| \frac{t}{b}\right] = \left(\frac{t}{b}\right)^{m(\alpha+\beta)+k} \prod_{l=1}^{m} \frac{\Gamma(b_{l,m}+\beta+k+1)}{\Gamma(a_{l,m}+\beta+k+1)}, \tag{42}$$

where $a_{l,m}$ and $b_{l,m}$ are the corresponding components of vectors $\mathbf{a}_m$ and $\mathbf{b}_m$ given in formula (16).

5 Nonhomogeneous Fractional Equations with t^β-Potential

We now extend the results from Propositions 3.1 and 4.1 to the case of inhomogeneous equation in finite time interval $[0, b]$:

$$(D^{\alpha}_{0+} - \lambda t^{\beta}) f^{non}(t) = g(t) \quad t \in [0, b], \tag{43}$$

$$({}^{c}D^{\alpha}_{0+} - \lambda t^{\beta}) f^{non}(t) = g(t) \quad t \in [0, b]. \tag{44}$$

According to our previous investigations it is clear that for $g_0 := g\Delta H$ fulfilling the assumptions: $g_0 \in C[0, b]$ and $g_0 \in L_1(R_+)$, we can work with the equivalent integral versions of the above equations:

$$(1 - \lambda I^{\alpha}_{0+} t^{\beta}) f_0(t) = f^{st}_{\alpha}(t) + I^{\alpha}_{0+} g_0(t) \quad t \in R_+, \tag{45}$$

where stationary function f^{st}_{α} is taken respectively for the Riemann-Liouville derivative or for the Caputo derivative.

Using the Mellin transform and its inverse we obtain the following solutions of the above inhomogeneous equations:

$$f^{non}_0(t) = f_0(t) + \sum_{m=0}^{\infty} \lambda^m G^{0,m}_{m,m}\left[\begin{matrix}\mathbf{a}_m \\ \mathbf{b}_m\end{matrix} \middle| t\right] * t^{m(\alpha+\beta)} I^{\alpha}_{0+} g_0(t), \tag{46}$$

where the solution of the homogeneous part—f_0 is described in formula (29) for the Riemann-Liouville derivative and in formula (40) for the Caputo derivative.

6 Final Remarks

In the paper we studied applications of the Mellin transform to certain classes of fractional differential equations with variable coefficients.

When order α and parameter β are restricted by additional conditions, then the obtained analytical solutions coincide with those derived by Kilbas and Saigo [8, 11, 12] via the fixed point theorem. This fact leads to certain analytical formulas for Meijer G-functions. It appears that in some cases they can be identified with power functions.

The results enclosed in the present paper include nonhomogeneous equations. Close inspection of the procedure of solving the discussed equations implies that it can be extended to equations of the same type but containing right-sided fractional derivatives. Preliminary results are discussed in paper [16]. Let us point out that the Mellin transform method and the Banach theorem on a fixed point were also effectively applied to equations containing both types of fractional derivatives, namely the left- and the right-sided ones in [17, 18]. The mixing of fractional derivatives is a characteristic feature of fractional mechanics. First it was shown in the paper by Riewe [19, 20] then in subsequent work by Klimek [21, 22], Agrawal [23–25], Baleanu and his collaborators [26–28] and lately by Cresson [29]. The described procedure of solving equations with variable coefficients will be extended to fractional variational equations with variable coefficients in our further work.

References

1. O.P. Agrawal, J.A.T. Machado, J. Sabatier (eds.), *Fractional Derivatives and Their Application: Nonlinear Dynamics*, vol. 38 (Springer, Berlin, 2004)
2. R. Herrmann, J. Phys. G, Nucl. Phys. **34**, 607–625 (2007)
3. R. Hilfer (ed.), *Applications of Fractional Calclus in Physics* (Singapore, World Scientific, 2000)
4. R.L. Magin, *Fractional Calculus in Bioengineering* (Begell House, Redding, 2006)
5. R. Metzler, J. Klafter, J. Phys. A **37**, R161–R208 (2004)
6. J. Sabatier, O.P. Agrawal, J.A.T. Machado (eds.), *Advances in Fractional Calculus. Theoretical Developments and Applications in Physics and Engineering* (Springer, Berlin, 2007)
7. B.J. West, M. Bologna, P. Grigolini, *Physics of Fractional Operators* (Springer, Berlin, 2003)
8. A.A. Kilbas, H.M. Srivastava, J.J. Trujillo, *Theory and Applications of Fractional Differential Equations* (Elsevier, Amsterdam, 2006)
9. K.S. Miller, B. Ross, *An Introduction to the Fractional Calculus and Fractional Differential Equations* (Wiley, New York, 1993)
10. I. Podlubny, *Fractional Differential Equations* (Academic Press, San Diego, 1999)
11. A.A. Kilbas, M. Saigo, Differ. Uravn. **33**, 195–204 (2007)
12. M. Saigo, A.A. Kilbas, Integral Transform. Spec. Funct. **7**, 97–112 (1998)
13. S.G. Samko, A.A. Kilbas, O.I. Marichev, *Fractional Integrals and Derivatives* (Gordon & Breach, Amsterdam, 1993)
14. H.-J. Glaeske, A.P. Prudnikov, K.A. Skórnik, *Operational Calculus and Related Topics* (Chapman & Hall/CRC, Boca Raton, 2007)
15. A.A. Kilbas, M. Saigo, *H-Transforms. Theory and Applications* (Chapman & Hall/ CRC, Boca Raton, 2004)
16. M. Klimek, D. Dziembowski, Sci. Res. Inst. Math. Comput. Sci. **2**(7), 31–41 (2008)
17. M. Klimek, in *XXVI Workshop on Geometric Methods in Physics, Bialowieza 2007*, ed. by P. Kielanowski, A. Odzijewicz, M. Schlichenmaier, T. Voronov. AIP Conference Proceedings, vol. 956 (AIP, New York, 2007), pp. 73–78
18. M. Klimek, J. Eur. Syst. Autom. **42**, 653–664 (2008)
19. F. Riewe, Phys. Rev. E **53**, 1890–1899 (1996)
20. F. Riewe, Phys. Rev. E **55**, 3581–3592 (1997)
21. M. Klimek, Czech. J. Phys. **51**, 1348–1354 (2001)
22. M. Klimek, Czech. J. Phys. **52**, 1247–1253 (2002)

23. O.P. Agrawal, J. Math. Anal. Appl. **272**, 368–379 (2002)
24. O.P. Agrawal, J. Phys. A **39**, 10375–10384 (2006)
25. O.P. Agrawal, J. Phys. A **40**, 5469–5476 (2007)
26. D. Baleanu, T. Avkar, Nuovo Cimento **119**, 73–79 (2004)
27. D. Baleanu, S.I. Muslih, Czech. J. Phys. **55**, 633–642 (2005)
28. D. Baleanu, Signal Process. **86**, 2632–2636 (2006)
29. J. Cresson, J. Math. Phys. **48**, 033504 (2007)

Phase Plane Characteristics of Marginally Stable Fractional Order Systems

**Narges Nazari, Mohammad Haeri,
and Mohammad Saleh Tavazoei**

Abstract When an integer order linear time invariant system possesses unrepeated pure imaginary poles it can generate oscillatory response which is represented by invariant closed contours in the phase plane. In linear time invariant fractional order systems with the same property, due to their special characteristics, this behavior will be more complicated and the contours would not be invariant. In this paper we will investigate the behavior of fractional order systems under such conditions.

Keywords LTI fractional order system · Phase plane · Marginal stability · Oscillation

1 Introduction

Fractional calculus has attracted attention of researchers from different fields in the recent years. While it was developed by mathematicians few hundred years ago, efforts on its usage in practical applications have been made only recently.

Some actual systems are fractional in nature, so it is more effective to model them by means of fractional order than integer order systems. Applications including modeling of damping behavior of viscoelastic materials, cell diffusion processes and transmission of signals through strong magnetic fields are some samples [1–3]. Studies have shown that for fractional order systems, a fractional order controller

N. Nazari · M. Haeri (✉) · M.S. Tavazoei
Advanced Control System Lab., Electrical Engineering Department,
Sharif University of Technology, Tehran, Iran
e-mail: haeri@sina.sharif.edu

N. Nazari
e-mail: Nazari@ee.sharif.edu

M.S. Tavazoei
e-mail: tavazoei@sharif.edu

J.A.T. Machado et al. (eds.), *Nonlinear Science and Complexity*,
DOI 10.1007/978-90-481-9884-9_34,

can provide better performances than an integer order one. Also fractional order controllers lead to more robust control performance [4].

For an effective use of fractional calculus in modeling a dynamical system or designing a controller, it is necessary to analyze the dynamical behavior that can be appeared due to mathematics of this representation. Fractional order representations possess long memory characteristics that make the system behave in more complicated manner. Fractional order differential equations and their solution have been studied widely in literature [5–7]. Some researchers have tried to find closed form solutions for these equations [8, 9]. Oscillations in fractional order systems have also been investigated in many papers such as [10–12].

In this paper we concern the concept of marginal stability and phase plane oscillations in a special case of linear fractional order time invariant systems. Also we have obtained a closed form solution for these systems. The paper is organized as follows; Sect. 2 includes analysis of marginally stable systems of integer and fractional orders. In Sect. 3 the simulation results which confirm the previous analysis are presented. The paper is concluded in Sect. 4.

2 Analysis of the Phase Plane

In this section we analyze the phase plane characteristics of marginally stable systems in two subsections, in Sect. 2.1, integer order systems are discussed while in Sect. 2.2 fractional order systems are analyzed.

2.1 Analysis of Phase Plane in Marginally Stable LTI Integer Order Systems

It is well known that an integer order LTI system is asymptotically stable if all roots of its characteristic polynomial have negative real parts. So the margin of stability is the imaginary axis. Now consider the following integer order system which has two poles on imaginary axis and therefore is marginally stable:

$$\frac{dx}{dt} = Ax, \quad A = \begin{bmatrix} 0 & 1 \\ -1 & 0 \end{bmatrix}. \tag{1}$$

For the system defined above one can see circular contours of different radius (depends on the initial conditions) in the phase plane as shown in Fig. 1. If we choose an initial condition on one of the contours, the trajectory stays on the same contour at all future times. This means that the contours are invariant. In general, the contours might have elliptical shape for marginally stable integer order systems having 2 states. Solutions of system (1) are determined as follows where $x(0)$ represents the initial condition.

$$x(t) = e^{At}x(0). \tag{2}$$

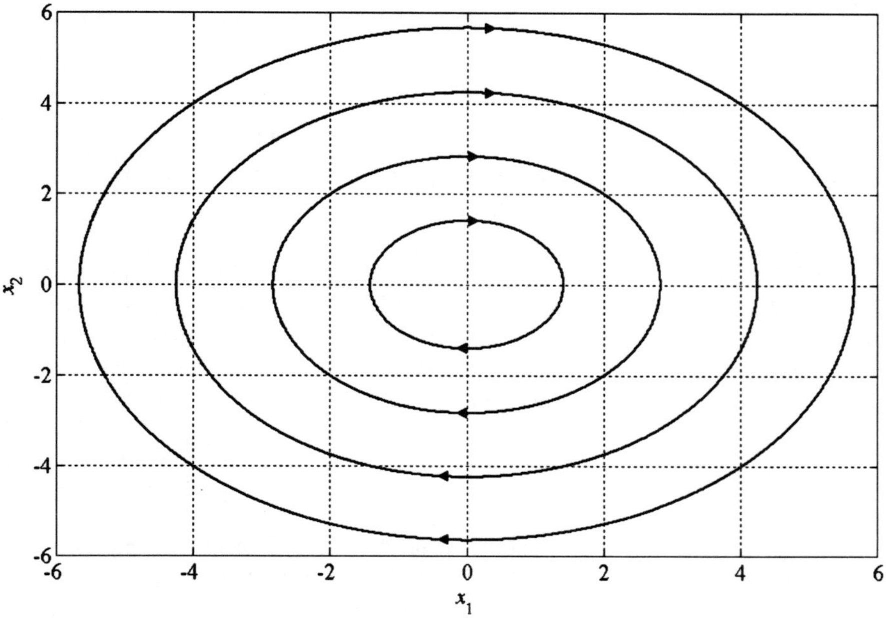

Fig. 1 Phase plane of the marginally stable system (1)

For given matrix A in (1), $\|e^{At}\|$ equals to 1 for all t and therefore one can write,

$$\|x(t)\| \leq \|e^{At}\|.\|x(0)\| = \|x(0)\|. \tag{3}$$

Equation (2) can be rewritten as follows as well:

$$x(0) = e^{-At}x(t). \tag{4}$$

Since $\|e^{-At}\| = 1$, the following relation is hold.

$$\|x(0)\| \leq \|e^{-At}\|.\|x(t)\| = \|x(t)\|. \tag{5}$$

From (3) and (5), it can be concluded that:

$$\|x(t)\| = \|x(0)\| \quad \forall t > 0. \tag{6}$$

This shows why the contours remain invariant in integer order system (1). For the given example, it also means that when the simulation starts from an initial condition, the 2-norm of the states remain constant for all times.

2.2 Analysis of Phase Plane in Marginally Stable LTI Fractional Order Systems

There are some different definitions for fractional order derivatives. In this paper, we use the Caputo definition because it has more physical meaning in the sense of initial conditions than the others. The Caputo definition of derivative with terminal value 0 is defined as follows:

$$D_t^\alpha f(t) = \frac{1}{\Gamma(n-\alpha)} \int_0^t \frac{f^{(n)}(\tau)}{(t-\tau)^{\alpha+1-n}} d\tau \quad n-1 < \alpha < n. \tag{7}$$

A fractional order LTI system is represented by the following relation:

$$D_t^\alpha x = Ax, \tag{8}$$

where D_t^α is the Caputo derivative of order $0 < \alpha < 1$ and $x \in R^n$. The instability region of fractional order system (8) is determined by sector $|\arg(s)| \leq 0.5\alpha\pi$ [6]. In other words, when all roots of characteristic equation of matrix A are placed anywhere outside this sector, (8) will be asymptotically stable. Now, we consider the following two-dimensional fractional order system:

$$D_t^\alpha x(t) = Ax(t), \quad A = \begin{bmatrix} 1 & \tan(0.5\alpha\pi) \\ -\tan(0.5\alpha\pi) & 1 \end{bmatrix}. \tag{9}$$

System (9) has poles on the stability margin for all values of α. Numerical simulations of the fractional order system in (9) show that in the phase plane representation of the system response there is a transient time before it reaches the semi-periodic response (we call it semi-periodic because there are still some small oscillations in amplitude) during which the amplitude of the response increases. If we start another simulation from a point on the resulted semi-contour, the next steady state semi-contour will have a distance from the first one. The same situation is experienced in all successive simulations (Fig. 2). The difference in phase plane trajectories between marginal stable integer and fractional order systems is, therefore, obvious. Our motivation in this paper is to analyze these phenomena.

To analyze the mentioned behavior one needs to have a closed form solution for system (9). This explicit solution may be determined by decomposing the system dynamics into two first order subsystems. To decompose a given system to independent first order subsystems, one may diagonalize its dynamic matrix A using its normalized eigenvector matrix V. For the given matrix A, the normalized eigenvector matrix, V, is of the following form:

$$V = \frac{1}{2}\begin{bmatrix} \sqrt{2} & \sqrt{2} \\ j\sqrt{2} & -j\sqrt{2} \end{bmatrix}.$$

Let $x^* = V^{-1}x$. Then,

$$D_t^\alpha x^*(t) = V^{-1}AVx^*(t) = \begin{bmatrix} 1 + j\tan(0.5\alpha\pi) & 0 \\ 0 & 1 - j\tan(0.5\alpha\pi) \end{bmatrix} x^*(t). \tag{10}$$

We define $\lambda_{1,2} = 1 \pm j\tan(0.5\alpha\pi)$. Therefore, the decomposed system will be as follows

$$\begin{cases} D_t^\alpha x_1^*(t) = \lambda_1 x_1^*(t), \\ D_t^\alpha x_2^*(t) = \lambda_2 x_2^*(t). \end{cases} \tag{11}$$

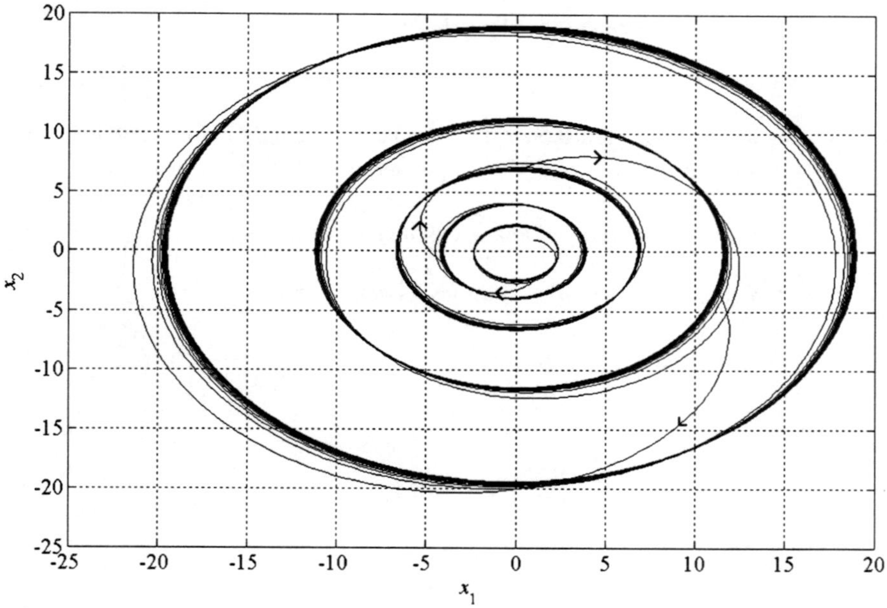

Fig. 2 Successive simulations of system (9) for $\alpha = 0.6$

It has been proved by Matignon [6] that the system of the form

$$D_t^{\alpha} x(t) = \lambda x(t), \tag{12}$$

has a fundamental solution of the following form

$$E_{\alpha}(\lambda, t) = \sum_{k=0}^{\infty} \frac{(\lambda t^{\alpha})^k}{\Gamma(\alpha k + 1)}. \tag{13}$$

This is the so called Mittag-Leffler function in one parameter. Use of the Mittag-Leffler function results in the following explicit solutions for systems in (11)

$$\begin{cases} x_1^*(t) = x_1^*(0) E_{\alpha}(\lambda_1, t), \\ x_2^*(t) = x_2^*(0) E_{\alpha}(\lambda_2, t). \end{cases} \tag{14}$$

The asymptotic behavior of the system in (9) is investigated through results of the following theorem.

Theorem 1 (See [6]) *When* $|\arg(\lambda)| \leq \alpha\pi/2$, *the asymptotic behavior of function is determined by the following relation*

$$\lim_{t \to \infty} E_{\alpha}(\lambda, t) \approx \frac{1}{\alpha} e^{\lambda^{1/\alpha} t}. \tag{15}$$

According to Theorem 1, we have

$$E_{\alpha}(\lambda_{1,2}, t) \approx \frac{1}{\alpha} e^{\lambda_{1,2}^{1/\alpha} t} = \frac{1}{\alpha} e^{(1 \pm j \tan(0.5\alpha\pi))^{1/\alpha} t}, \tag{16}$$

for large t. It is straightforward to show that:

$$\arg((1 \pm j\tan(0.5\alpha\pi))^{1/\alpha}) = \pm\pi/2. \tag{17}$$

In other words, $(1 \pm j\tan(0.5\alpha\pi))^{1/\alpha}$ is a pure imaginary number and therefore, the asymptotic expression of $E_\alpha(\lambda_{1,2}, t)$ possesses sinusoidal terms alone i.e.

$$E_\alpha(\lambda_{1,2}, t) \approx \frac{1}{\alpha}\{\cos(\omega t) \pm j\sin(\omega t)\} \tag{18}$$

for large t. ω, the frequency of oscillations, is determined by

$$\omega = |(1 \pm j\tan(\alpha\pi/2))^{1/\alpha}| = \frac{1}{(\cos(\alpha\pi/2))^{1/\alpha}}. \tag{19}$$

Also, we have

$$\lim_{t\to\infty} |E_\alpha(\lambda_{1,2}, t)| = 1/\alpha. \tag{20}$$

It can be easily shown that

$$\begin{cases} x_1(t) = c_1 E_\alpha(\lambda_1, t) + c_2 E_\alpha(\lambda_2, t), \\ x_2(t) = jc_1 E_\alpha(\lambda_1, t) - jc_2 E_\alpha(\lambda_2, t), \end{cases} \tag{21}$$

where $c_{1,2} = 0.5(x_1(0) \pm jx_2(0))$. Since $E_\alpha(\lambda_1, t)$ and $E_\alpha(\lambda_2, t)$ are complex conjugates, then:

$$\|x(t)\| = |E_\alpha(\lambda_1, t)|.\|x(0)\|. \tag{22}$$

According to (20) and (22),

$$\lim_{t\to\infty} \|x(t)\| = \frac{1}{\alpha}\|x(0)\|. \tag{23}$$

This shows us an interesting result which is confirmed through simulations as well.

- For $\alpha = 1$, the ultimate norm of the phase plane contours is equal to the initial norm and therefore, the contours in phase plane are invariant.
- For $0 < \alpha < 1$, the ultimate norm of the phase plane trajectories is larger than the initial norm and the ratio is given by $1/\alpha$. So as α tends to zero the ultimate value will be farther from the initial value and for $\alpha = 0$, this would go to infinity!

The given result is completely in agreement with our observation in the phase plane trajectories of the integer and fractional order systems (1) and (9). For the integer order system (1), each initial condition results in a trajectory that has its own invariant contour. Since there is no transient response it remains on its contour from the beginning. But for the fractional order system (9), the ultimate norm of a trajectory, while is constant, is larger than its initial condition norm that means there always be a transition from the initial condition that a trajectory starts from to its ultimate contour.

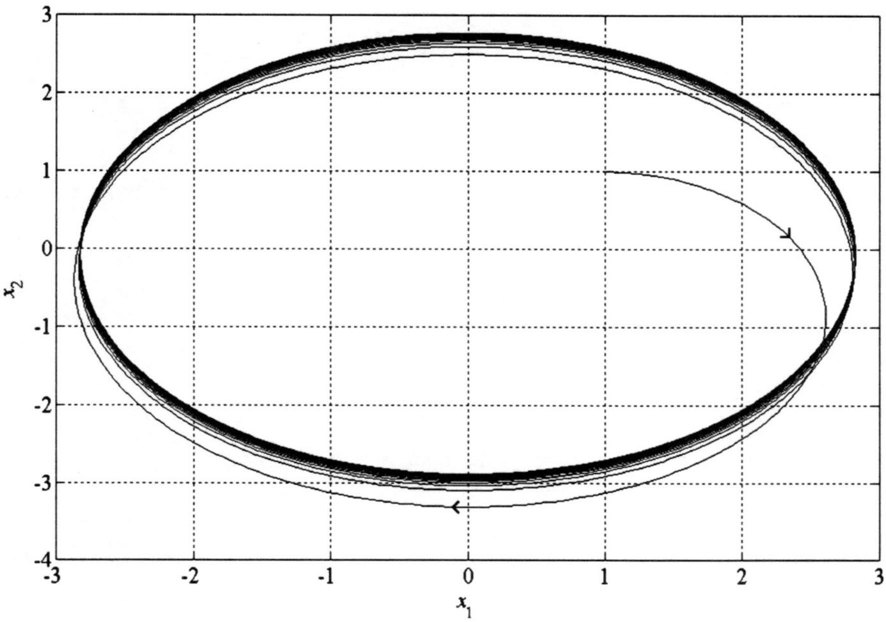

Fig. 3 Trajectory of system (9) for small values of t

In the next section some simulation results are presented to verify the above achievement.

3 Simulation Results

In this section, we provide some simulation results to confirm our analysis about the fractional order systems. We consider the system described by (9). When $\alpha = 0.5$, the phase plane for a trajectory that starts from initial condition $[1, 1]$ is shown in Fig. 3. It is expected that the drawn trajectory has final norm equal to $\sqrt{2}/\alpha$. Simulation result shows that the trajectory is asymmetrical for small t and it is inside the bound for some t and outside it for some others. However, for large simulation times the expected symmetry is appeared in the trajectory. As t grows and tends to infinity from theoretical point of view, the trajectory becomes more symmetric and it approaches to the limit we determined theoretically. Actually the trajectory becomes more like integer order case in steady state. According to (19), when α is 0.5, the frequency of steady state oscillations should be equal to 2 rad/s. Or the period of oscillations should be π s. Figure 4 confirms the equivalence between the calculated and simulation results.

In Fig. 2, we have plotted 5 successive trajectories such that the last point of each contour is the initial point for the next contour. If one does the same experiment in the integer order case one will see just one contour because all of them have the

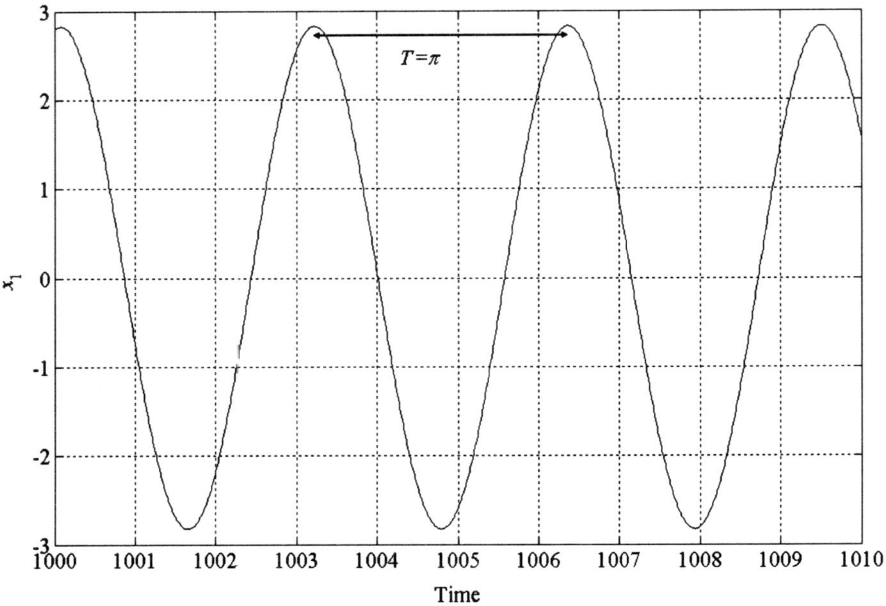

Fig. 4 Period of ultimate oscillations for $\alpha = 0.5$

same norm. The mentioned distance between the trajectories increases as α goes to 0, such that for orders near zero it tends to infinity.

4 Conclusion

Fractional order systems show more complex behavior in comparison to integer order systems. Undamped oscillations as one of linear fractional order system characteristics were investigated in this paper. An explicit solution for these systems was derived and its asymptotic behavior was analyzed. Also, we determined asymptotic amplitude and frequency of the oscillations. An interesting phase plane property of these systems was highlighted in our investigations. It has been shown that the trajectories of fractional order systems are not invariant in spite of integer order systems.

References

1. Y.A. Rossikhin, M.V. Shitikova, Acta Mech. **120**, 109 (1997)
2. N. Engheta, IEEE Trans. Antennas Propag. **44**, 554 (1996)
3. R.L. Bagley, R.A. Calico, J. Guid. Control Dyn. **14**, 304 (1991)
4. I. Podlubny, IEEE Trans. Autom. Control **44**, 208 (1999)
5. I. Podlubny, *Fractional Differential Equations* (Academic Press, San Diego, 1999)

6. D. Matignon, Stability results for fractional differential equations with applications to control processing, in *IMACS, IEEE-SMC*, Lille, France (1996)
7. D. Diethelm, N.J. Ford, J. Math. Anal. Appl. **265**, 229 (2002)
8. B. Bonilla, M. Rivero, J.J. Trujillo, Appl. Math. Comput. **187**, 68 (2007)
9. C. Bender, Syst. Control Lett. **54**, 671 (2005)
10. M.S. Tavazoei, M. Haeri, Physica D **237**, 2628 (2008)
11. M.S. Tavazoei, M. Haeri, Phys. Lett. A **367**, 102 (2007)
12. T.T. Hartley, C.F. Lorenzo, H.K. Qammer, IEEE Trans. Circuits Syst. I **42**, 485 (1995)

Application of Fractional Controllers for Quad Rotor

C. Lebres, V. Santos, N.M. Fonseca Ferreira, and J.A. Tenreiro Machado

Abstract This paper studies the application of fractional algorithms in the control of a quad-rotor rotorcraft. The development of a flight simulator provide the evaluation of the controller algorithm. Several basic maneuvers are investigated, namely the elevation and the position control.

Keywords Rotorcraft flight · Mathematical model · Position control · Fractional control · Nonlinear

1 Introduction

A full-scale four-rotor helicopter was built by *De Bothezat* in 1921. This idea of using four rotors is not new. Rotary wing aerial vehicles have distinct advantages over conventional fixed wing aircrafts on surveillance and inspection tasks, since they can take-off land in limited spaces and easily fly above the target [1, 2]. A quad-rotor is a four rotor helicopter, and are example is shown in Figs. 1 and 2. Helicopters are dynamically unstable and, therefore; suitable control methods is needed to make them stable [3]. Although an unstable dynamics is not desirable, it is good in the viewpoint of agility [4, 5]. The instability comes from the changing of the helicopter parameters and from the disturbances such as the wind [6]. A quadtrotor helicopter is controlled by varying the rotors speed, thereby changing the lift forces. It is an under-actuated dynamic vehicle with four input forces and six output coordinates [7]. One of the advantages of using a multi-rotor helicopter is the increased

C. Lebres · V. Santos · N.M.F. Ferreira (✉)
Institute of Engineering of Coimbra, Rua Pedro Nunes, 3031-601 Coimbra, Portugal
e-mail: nunomig@isec.pt

J.A.T. Machado
Institute of Engineering of Porto, Rua Dr. António Bernardino de Almeida, 4200-072 Porto, Portugal
e-mail: jtm@isep.ipp.pt

J.A.T. Machado et al. (eds.), *Nonlinear Science and Complexity*,
DOI 10.1007/978-90-481-9884-9_35, © Springer Science+Business Media B.V. 2011

payload capacity. It has more lift and before, heavier weights can be carried. The quadtrotors are highly maneuverable, which enables vertical take-off/landing, as well as flying into hard to reach areas, but the disadvantages are the increased helicopter weight and increased energy consumption due to the extra motors. Since the machine it is controlled with rotor-speed changes, it is more suitable to adapt electric motors. Large helicopter engines, that which a have slow response, may not be satisfactory without, incorporating a proper gear-box system [8, 9]. The main contribution of this study concerns the use of non-linear control techniques to stabilize and to perform output tracking control of the helicopter.

2 Helicopter Model

Unlike regular helicopters, that have variable pitch angles, a quad rotor has fixed pitch angle rotors and the rotor speeds are controlled in order to produce the desired lift forces. Basic motions of a quad rotor can be described using Fig. 1. In the first method, the vertical motion of the helicopter can be achieved by changing all of the rotor speeds at the same time. Motion along the x-axis is related to tilt around the y-axis.

This tilt can be obtained by decreasing the speeds of rotors 1 and 2 and by increasing speeds of rotors 3 and 4. This tilt also produces acceleration along the x-axis. Similarly y-motion is the result of the tilt around the x-axis.

It was tested another tilting method which consists in decreasing only the speed of rotor 1 and increasing his opposite rotor speed, the rotor 3 (for example) , however

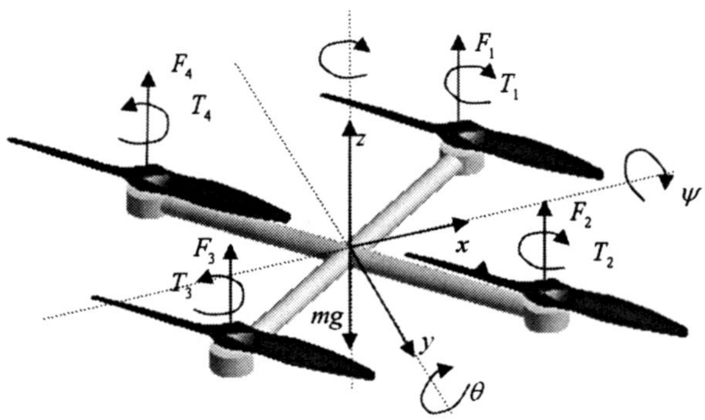

Fig. 1 The quad-rotor helicopter

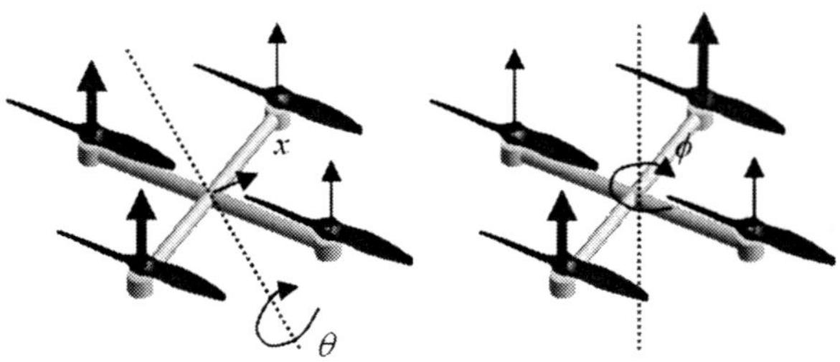

Fig. 2 The tilting method

this method led to less power to actuate the tilting motions, making the platform more difficult to control. The yaw motions are obtained using the moments that are created as the rotors spin. Conventional helicopters have the tail rotor in order to balance the moments created by the main rotor. With the four-rotor machine, spinning directions of the rotor are set to balance and to cancel these moments. This is also used to produce the desired yaw motions. To turn in a clock-wise direction, the speed of rotor's 2 and 4 should be increased to overcome the moments created by rotors 1 and 3. A good controller should be able to reach a desired yaw angle while keeping the tilt angles and the height constant. A body fixed frame is assumed to be at the center of gravity of the quad-rotor, where the z-axis is pointing upwards. This body axis is related to the inertial frame by a position vector (x, y, z) and three Euler angles, (θ, ψ, ϕ), representing pitch, roll and yaw, respectively.

A ZYX-Euler angle representation given in (1), has been chosen for the representation of the rotations.

$$R = \begin{bmatrix} C_\varphi C_\theta & C_\varphi S_\theta S_\psi - S_\varphi C_\psi & C_\varphi S_\theta S_\psi + S_\varphi C_\psi \\ S_\varphi C_\theta & S_\varphi S_\theta S_\psi - C_\varphi C_\psi & S_\varphi S_\theta C_\psi + S_\varphi S_\psi \\ S_\theta & C_\theta S_\psi & C_\theta C_\psi \end{bmatrix} \tag{1}$$

where C_θ and S_θ represent $\cos(\theta)$ and $\sin(\theta)$ respectively.

Each rotor produces moments as well as vertical forces. These moments were observed experimentally to be linearly dependent on the forces at low speeds. There are four input forces and six output states $(x, y, z, \theta, \psi, \phi)$ and, therefore the quad-rotor is an under-actuated system. The rotation direction of two of the rotors are clockwise while the other two are counter clockwise, in order to balance the moments and to produce yaw motions as needed.

The equations of motion can be written using the force and moment balance, yielding:

$$\ddot{x} = \frac{(\sum_{i=1}^{4} F_i)(C_\phi S_\theta C_\psi + S_\phi S_\psi) - K_1 \dot{x}}{m}, \tag{2}$$

$$\ddot{y} = \frac{(\sum_{i=1}^{4} F_i)(S_\phi S_\theta C_\psi + c_\phi S_\psi) - K_2 \dot{y}}{m}, \tag{3}$$

$$\ddot{z} = \frac{(\sum_{i=1}^{4} F_i)(C_\phi C_\psi) - mg - K_3 \dot{z}}{m}, \tag{4}$$

$$\ddot{\theta} = \frac{l(F_3 + F_4 - F_1 - F_2 - K_4 \dot{\theta})}{J_1}, \tag{5}$$

$$\ddot{\psi} = \frac{l(F_2 + F_3 - F_1 - F_4 - K_5 \dot{\psi})}{J_2}, \tag{6}$$

$$\ddot{\phi} = \frac{l(F_2 + F_3 - F_1 - F_4 - K_6 \dot{\phi})}{J_3}. \tag{7}$$

The factors K_i $(i = 1, 2, \ldots, 6)$ given above are the drag coefficients. In the following we assume the drag is zero, since drag is negligible at low speeds. By convenience, we will define the inputs to be:

$$u_1 = \frac{(F_1 + F_2 + F_3 + F_4)}{m}, \tag{8}$$

$$u_2 = \frac{(-F_1 - F_2 + F_3 + F_4)}{J_1}, \tag{9}$$

$$u_2 = \frac{(-F_1 + F_2 + F_3 - F_4)}{J_2}, \tag{10}$$

$$u_2 = C\frac{(F_1 - F_2 + F_3 - F_4)}{J_3}, \tag{11}$$

where J_1, J_2 and J_3 are the moment of inertia with respect to the axes and C is the force-to-moment scaling factor. The variables u_l represents a total thrust on the body in the z-axis, u_2 and u_3 are the pitch and roll inputs and u_4 is a yawing moment.

3 Fractional Control

In this section we present the Fractional Order algorithms inserted at the position loops (Fig. 3).

The mathematical definition of a derivative of fractional order α has been the subject of several different approaches [10, 11]. For example, we can mention the Laplace and the Grünwald-Letnikov definitions:

$$D^{\alpha}[x(t)] = L^{-1}\{s^{\alpha}X(s)\}, \tag{12}$$

$$D^{\alpha}[x(t)] = \lim_{h\to 0}\left[\frac{1}{h^{\alpha}}\sum_{k=1}^{\infty}\frac{(-1)^k\Gamma(\alpha+1)}{\Gamma(\alpha+1)\Gamma(\alpha-k+1)}x(t-kh)\right], \tag{13}$$

$$C(s) = K_p + \frac{K_i}{T_i s} + K_D s^{\alpha}, \quad -1 < \alpha < 1, \tag{14}$$

where Γ is the gamma function and h is the time increment. In our case, for implementing *FO* algorithms of the type:

$$C(z) \approx K\frac{a_0 z^k + a_1 z^{k-1} + \cdots + a_k}{b_0 z^k + b_1 z^{k-1} + \cdots + b_k} \tag{15}$$

we adopt a 4th-order discrete-time Pade approximation ($a_i, b_i, c_i, d_i \in \Re, k = 4$): where K_{Pi} are the position gains, respectively.

Table 1 Time response of the forces applied in the quad-rotor for the vertical motion in the z axis of the helicopter

i	K_p	K_d	α
1—Pitch control	15	25	0.95
2—Roll control	15	25	0.95
3—Yaw control	30	40	0.95
4—*X* control	10	57	0.95
5—*Y* control	10	57	0.95
6—*Z* control	25	102	0.95

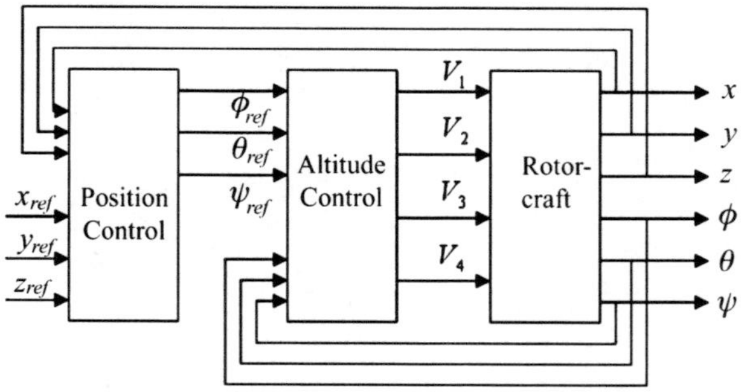

Fig. 3 The control diagram of the quad-rotor helicopter

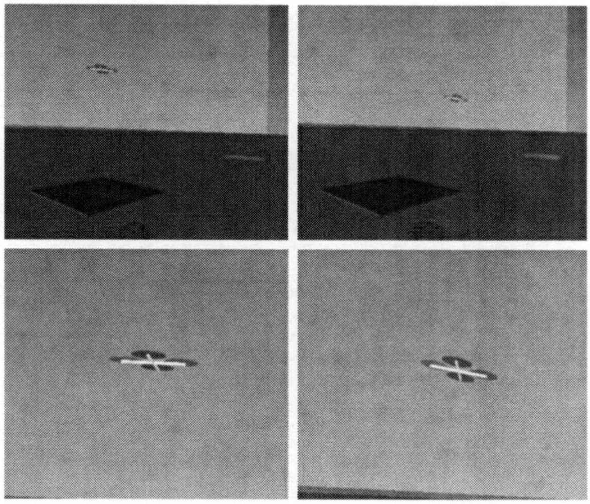

Fig. 4 The animation of the quad-rotor helicopter

Table 1 shows the PD^{α} tuning parameters implemented on the attitude and position controllers.

4 The Flight Simulator

We developed a flight simulator (Fig. 4) to provide a test bed for evaluating models. The simulator is written in *Matlab* and all the model parameters are stored in files. When using the simulator, the first thing that is obvious is how difficult it is to get the simulated quad-rotor helicopter to stop rising or falling. To get it to hover at one height you have to adjust the throttle until both velocity and acceleration in the z direction are zero. In a first phase we consider the vertical motion of the helicopter starting in $\{x, y, z\} \equiv \{0, 0, 10\}$ [m] to $\{x, y, z\} \equiv \{0, 0, 20\}$ [m]. In a second phase we consider the horizontal motion of the helicopter starting in $\{x, y, z\} \equiv \{0, 0, 10\}$ [m] up to $\{x, y, z\} \equiv \{10, 0, 10\}$ [m]. In a third phase we consider a circular trajectory centered at $\{x, y, z\} \equiv 0, 0, 10\}$ [m] with a 5 meter radius. The time responses show us that the quad-rotor it is very complex due to the several

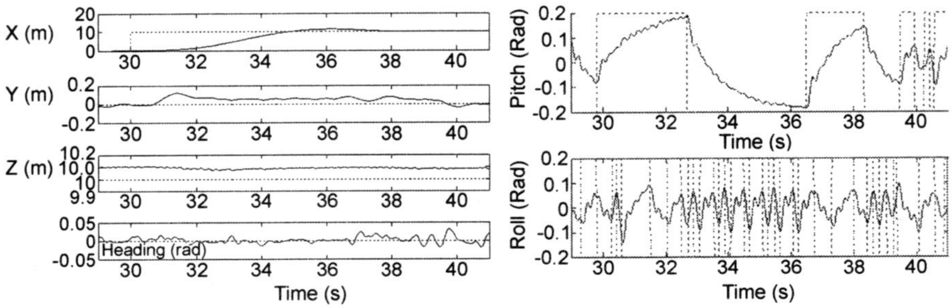

Fig. 5 Time response of the quad-rotor's position, considering horizontal motion in the x axis of the helicopter

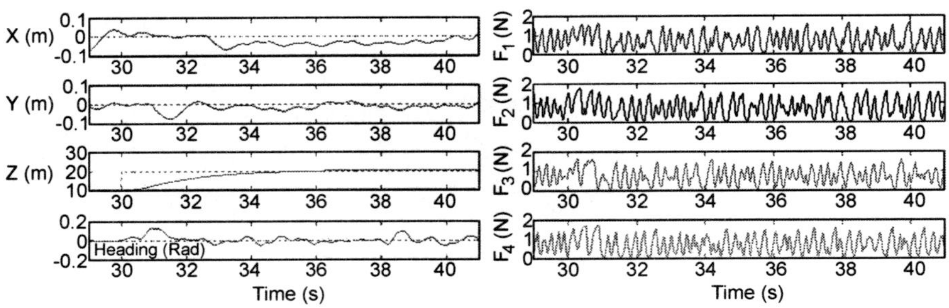

Fig. 6 Time response of the quad-rotor's position and the forces applied, considering vertical motion in the z axis of the helicopter

couplings effects caused by the several propellers drag moments. Nevertheless it reveals an high maneuverability, which enables a quick vertical take-off and landing.

Figures 5 and 6 show that the applied forces for the lifting are not the same for each motor due to the controller corrections required of the controllers to keep the quad-rotor near the desired references in order to compensate the coupling effects.

5 Conclusions

Our analysis has shown that the quad-rotor helicopter is a complex system. In this analysis we have developed a model of the quad-rotor and we tested some basic manoeuvres. We have explored the resulting forces and moments applied to the vehicle and through these, investigated their impact on elevation and position control.

References

1. B.W. McCormick, *Aerodynamics Aeronautics and Flight Mechanics* (Wiley, New York, 1995)
2. G. Leishman, *Principles of Helicopter Aerodynamics* (Cambridge University Press, Cambridge, 1995)
3. B. Etkin, L. Reid, *Dynamics of Flight-Stability and Control* (Wiley, New York, 1996)

4. S.N. Singh, A.A. Schy, Nonlinear decoupled control synthesis for maneuvering aircraft, in *Proceedings of the IEEE Conference on Decision and Control*, Piscataway (1978)
5. R.L.H. Romero, R. Benosman, Stabilization and location of a four rotors helicopter applying vision, in *American Control Conference (ACC'06)*, June 14–16, Minneapolis, MN, USA (2006)
6. R.L. Salazar, A. Palomino, *Trajectory Tracking for a Four Rotor Mini-Aircraft* (IEEE CDC/ECC, Sevilla, 2005)
7. S.T.A.K.R. Asep, F. Mora-Camino, An application of the nonlinear inverse technique to flight-path supervision and control, in *Proceedings of the 9th International Conference of Systems Engineering*, Las Vegas, NV (1993)
8. A.L.R. Castillo, P. Dzul, Real-time stabilization and tracking of a four rotor mini rotorcraft, in *European Control Conference ECC03*, Cambridge, UK, 1–4 September, 2003
9. L.R.G.P.A.P.P. Castillo, *Nonlinear Control of a Small Four-Rotor Rotorcraft: Theory and Real-Time Application. Nonlinear and Adaptive Control: Theory and Algorithms for the User* (Imperial College Press, London, 2005)
10. A. Oustaloup, *La Commande CRONE: Commande Robuste d'Ordre Non Entier* (Hermès, Paris, 1991)
11. J.A.T. Machado, FCAA J. Fract. Calc. Appl. Anal. **4**, 47 (2001)

Regularity of a Degenerated Convolution Semi-Group Without to Use the Poisson Process

Rémi Léandre

Abstract We translate in semi-group theory our regularity result for a degenerated convolution semi-group got by the Malliavin Calculus of Bismut type for Poisson processes.

Keywords Malliavin Calculus · Convolution semi-group

1 Introduction

Malliavin [10] proved again Hoermander's theorem by using the Brownian motion. Malliavin for that used an heavy apparatus of functional analysis: Bismut don't use this heavy apparatus in order to prove Hoermander's theorem by probabilistic methods [1].

Bismut's way of the Malliavin Calculus for diffusions was translated by Léandre in semi-group theory (See [6, 9] for reviews).

Bismut [2] considered Poisson processes and stochastic differential equations driven by them in order to state some regularity theorems for Markov semi-groups generated by integro-differential operators. This allowed Léandre to generalize for jump processes Hoermander's theorem [3–5]. Let us recall that jump processes are classically related to semi-groups associated to fractional powers of diffusion generators [11].

Léandre has translated Bismut's way of the Malliavin Calculus for jump processes [2] in semi-group theory [7, 8]. This allows [8] to prove again one of the results of Léandre [3–5] without to use the Poisson process.

R. Léandre (✉)
Institut de Mathématiques, Université de Bourgogne, 21000 Dijon, France
e-mail: Remi.leandre@u-bourgogne.fr

R. Léandre
Mittag Leffler Institute, Djursholm, Sweden

J.A.T. Machado et al. (eds.), *Nonlinear Science and Complexity*,
DOI 10.1007/978-90-481-9884-9_36,

Our goal is to recover another result of Léandre by using only the semi-group theory.

We will state our result, with simplified hypothesis in order to simplify the exposition.

Let $g_j(z)$ be m smooth functions from R^* into R^+ such that

$$\int_R (z^2 \wedge 1) g_j(z) dz < \infty. \tag{1}$$

Moreover we suppose that on a neighborhood of 0

$$g_j(z) = \frac{C}{|z|^{1+\alpha_j}} \tag{2}$$

for some $\alpha_j \in]0, 2[$.

We consider m smooth curves $\gamma_j(z)$ into R^d with bounded derivatives at each order such that

$$\gamma_j(0) = 0. \tag{3}$$

We do the following hypothesis:

Hypothesis H There exists a k such that $\bigcup_{j,l \leq k} \frac{d^l}{dz^l} \gamma_j(0)$ spans R^d.

We consider the Markov generator acting on $C_b^\infty(R^d)$

$$\begin{aligned} Lf(x) = \sum \int_R (f(x + \gamma_j(z)) - f(x) \\ - 1_{|\gamma_j(z)|<1} \langle \gamma_j(z), \mathrm{grad}_x f(x) \rangle) g_j(z) dz. \end{aligned} \tag{4}$$

L spans a convolution semi-group P_t. Our main result is the following

Theorem 1 (Léandre [3–5]) *Under hypothesis H, the semi group P_t has a smooth heat-kernel $y \to p_t(x, y)$ for $t > 0$*

$$P_t f(x) = \int_{R^m} p_t(x, y) f(y) dy. \tag{5}$$

The proof of this theorem is the translation in semi-group theory of the proof of the same theorem we got in [3] by using the Malliavin Calculus of Bismut type for Poisson processes.

We consider a smooth function $\nu(z)$ with compact support with values in R^* equal to z^4 in a neighborhood of 0. We consider the space $R^d \times M_d$ where M_d is the space of symmetric matrices on R^d. $(x, V) \in R^d \times M_d$. We consider the Malliavin generator acting on test functions $\hat{f}$ on this space

$$\begin{aligned} \hat{L}\hat{f}(x, V) = \sum \int_R (\hat{f}(x + \gamma_j(z), V + \nu(z)\langle \cdot, \gamma_j'(z) \rangle^2) - \hat{f}(x, V) \\ - 1_{|\gamma_j(z)|\langle 1} \langle \gamma_j(z), \mathrm{grad}_x \hat{f}(x, V) \rangle) g_j(z) dz. \end{aligned} \tag{6}$$

V is called the Malliavin matrix. The next theorem will allow us to prove the main theorem of this work.

Theorem 2 (Bismut [2]) *Let us suppose that the Malliavin condition is satisfied. This means that*

$$\hat{P}_t[V^{-p}](x,0) < \infty \tag{7}$$

for all even positive integer p and for all $t > 0$. Then the convolution semi-group P_t has a smooth heat-kernel.

For surveys on Malliavin Calculus interpreted in semi-group theory, we refer to [6, 9].

We thank the warm hospitality of the Mittag Leffler Institute where this work was done at the occasion of an activity about stochastic partial differential equations.

2 Proof of Bismut's Theorem Without to Use the Poisson Process

Since the proof is very similar to the proof of Theorem 1 of [8], we will give only the scheme. But since we consider a convolution semi-group, the algebra is much more simple. We will begin as in [8], part I, by elementary considerations.

Let $\hat{L}$ be the generator on $R^d \times R$ ($(x,u) \in R^d \times R$)

$$\hat{L}\hat{f}(x,u) = \sum_j \int_R (\hat{f}(x+\gamma_j(z), u+h(z)) - \hat{f}(x,u) \\ - 1_{|\gamma_j(z)|<1}\langle \gamma_j(z), \mathrm{grad}_x \hat{f}(x,u)\rangle) g_j(z)dz, \tag{8}$$

where $h(z) = -v'(z) - v(z)g_j'(z)/g_j(z)$ is supposed smooth.

We get as in [8] (13)

$$\hat{P}_t[fu](x,0) = \int_0^t P_{t-s}\left[\sum_j \int_R P_s[f](x+\gamma_j(z))h(z)g_j(z)dz\right]. \tag{9}$$

But the right side hand of (9) is equal to

$$\int_0^t P_{t-s}\left[\int_R \sum_j \langle \mathrm{grad}_x P_s[f](x+\gamma_j(z), \gamma_j'(z)\rangle v(z)g_j(z)dz\right]ds. \tag{10}$$

Therefore we get:

Proposition 1 (Bismut [3]) *We have the elementary integration by parts formula*

$$\hat{P}_t[fu](x,0) = \int_0^t P_{t-s}\left[\sum \int_R \langle \mathrm{grad}_x P_s[f](x+\gamma_j(z), \gamma_j'(z)\rangle v(z)g_j(z)dz\right]ds. \tag{11}$$

We put $E^l = R^d \times M_d \times R^{d_2} \times \cdots \times R^{d_l}$ $x^l = (x_1, v, x_2, \ldots, x_l)$ is the generic element of this big space. We consider a bounded map $g(z)$ from R^m into E^l. Its values in R^d are $\gamma(z) = \sum \gamma_j(z)$ and the others components have compact support and are bounded by $C|z|^2$. $d\mu(z) = \sum g_j(z_j)dz_j$.

We consider the generator

$$L^l f^l(x^l) = \int_{R^m} (f^l(x^l + g(z)) - f^l(x^l) - 1_{|\gamma(z)|<1} \langle \gamma(z), \mathrm{grad}_{x_1} f^l(x^l)\rangle) d\mu(z). \quad (12)$$

It generates a Markov semi-group P_t^l. The main remark is the following: if we consider a first order operator with constant coefficient D_{x^l}, D_{x^l} commutes with L^l. Therefore

$$D_{x^l} P_t^l[f^l](x^l) = P_t^l[D_{x^l} f^l](x^l) \quad (13)$$

which gives a simplified version of Proposition 7 of [8]. We suppose only in (13) that f^l has all derivatives bounded.

The second remark whose proof follows the same lines than the proof of Theorem 6 of [8] is the following theorem:

Theorem 3 *For a function f^l with polynomial growth*

$$P_t^l[|f^l|](x^l) < \infty. \quad (14)$$

The only change with respect of the proof of Theorem 6 of [8] is that $\gamma_j(z)$ are bounded.

In particular, if $f^l = h(x_1)h_1(x^l)$ where h_1 is a polynomial in the others component than x_1 and h is smooth with bounded derivatives in x_1, (13) is still true.

We put $g_1(z) = \gamma'(z)v(z)$ for the first component and for the others component $g_1(z) = g'(z)v(z)$. $g_1(z)$ satisfies still the hypothesis of Proposition 1: We consider the generator on $E^l \times E^l$

$$\tilde{L}^l f^l(x^l, V^l) = \int_{R^m} (f^l(x^l + g(z), V^l + g_1(z)) - f^l(x^l, V^l) - 1_{|\gamma(z)|<1} \langle \gamma(z), \mathrm{grad}_{x_1} f^l(x^l, V^l)\rangle) d\mu(z). \quad (15)$$

This generator is of type studied before.

We consider the generator on $E^l \times R$

$$\hat{L}^l \hat{f}^l(x^l, u) = \int_{R^m} (\hat{f}^l(x^l + g(z), u + h(z)) - \hat{f}^l(x^l, u) - 1_{|\gamma_j((z)|<1} \langle \gamma_j(z), \mathrm{grad}_{x^1} \hat{f}^l(x^l, u)\rangle) g_j(z) dz, \quad (16)$$

where $h(z) = -v'(z) - v(z)g_j'(z)/g_j(z)$. It generates a semi group $\hat{P}_t^l$.

We get the extension of Proposition 1.

Theorem 4 *We have the integration by parts formula*

$$\hat{P}_1^t[f^l u](x, 0) = \tilde{P}_t^l[\langle D_{x^l} f^l, V^l\rangle](x^l, 0^l) \quad (17)$$

if f^l has a polynomial growth in each variable.

Proof We remark first of all that $\tilde{L}^l$ commute with first order differential operators with constant coefficient. Therefore these differential operators commute with $\tilde{P}_t^l$: so $\tilde{P}_t^l$ defines a semi-group on affine functions of V^l. We consider the generator

$$\overline{L}f^l(x^l, V^l) = \int_{R^m} (f^l(x^l + g(z), V^l) \\ - f^l(x^l, V^l) - 1_{|\gamma(z)|<1}\langle\gamma(z), \text{grad}_{x_1} f^l(x^l, V^l)\rangle)d\mu(z) \quad (18)$$

and the generator acting on affine functions in V^l

$$\overline{L}^{l,1} f^l(x^l, V^l) = \int_{R^m} (f^l(x^l + g(z), g^1(z)) \\ - f^l(x^l + g(z), 0^l))d\mu(z). \quad (19)$$

We have clearly on affine functions in V^l:

$$\tilde{L}^l = \overline{L}^l + \overline{L}^{l,1}. \quad (20)$$

Therefore by Volterra expansion we get:

$$\tilde{P}_t^l[\langle D_{x^l} f^l, V^l\rangle](x^l, 0^l) \\ = \overline{P}_t^l[\langle D_{x^l} f^l, V^l\rangle](x^l, 0^l) \\ + \sum_n \int_{0<s_1<\cdots<s_n<t} \overline{P}_{s_1}^l \overline{L}^{l,1} \cdots \overline{L}^{l,1} \overline{P}_{t-s_n}^l [\langle D_{x^l} f^l, V^l\rangle](x^l, 0^l) ds_1 \cdots ds_n. \quad (21)$$

But $\overline{P}_t^l[\langle D_{x^l} f^l, V^l\rangle](x^l, V^l)$ is an affine function in V^l. In particular, if we apply $\overline{L}^{l,1}$ to this quantity, it does not depend on V^l. Therefore the Volterra expansion reads:

$$\tilde{P}_t^l[\langle D_{x^l} f^l, V^l\rangle](x^l, 0^l) = \overline{P}_t^l[\langle D_{x^l} f^l, V^l\rangle](x^l, 0^l) \\ + \int_0^t P_s^l \overline{L}^{l,1} P_{t-s}^l [\langle D_{x^l} f^l, V^l\rangle](x^l, 0^l) ds. \quad (22)$$

This last equality is nothing else but (11) in Proposition 1 but with this more complicated semi-group. □

Proof of Bismut's theorem We consider the Markov generator

$$L^1 f(x_1, V) = \int_{R^m} f(x_1 + \gamma(z), V + \sum \gamma'(z_j)v(z_j)\langle\gamma'(z_j), \cdot\rangle) - f(x_1, V) \\ - 1_{|\gamma(z)|<1}\langle\gamma(z), \text{grad}_{x_1} f(x, V)\rangle)d\mu(z). \quad (23)$$

We consider the function of Bismut type

$$(x_1, V) \to \langle D_{x_1} f(x_1), V^{-1} e_k\rangle \quad (24)$$

where e_k is the standard basis of R^d and we apply iteratively the previous integration by parts to this test function in order to get for all multi-index (α)

$$P_t[D^{(\alpha)} f](x) = P_t^l[f\theta](x,0) \tag{25}$$

where P_t^l is a semi-group of the type considered before and θ a polynomial in the extra-variables and in the inverse of the Malliavin matrix. It remains to apply (25) to Fourier exponentials. □

3 Proof of Léandre's Theorem Without to Use the Poisson Process

Let us recall the uniform Tauberian theorem of [3–5] which are an improvement of the Tauberian theorems of [2].

Let F be a set of smooth maps f from R into R equal to 0 in 0 such for all k

$$\sup_{f\in F, z\in R} |f^{(k)}(z)| < \infty \tag{26}$$

and such that there exists K such that

$$\inf_{f\in F}\left(\sup_{k\leq K} |f^{(k)}(0)|\right) > C > 0. \tag{27}$$

Let us define

$$\tau_f(\beta) = \int_R (\exp[-\beta|f(z)|] - 1)g(z)dz \tag{28}$$

where $\int_R (z^2 \wedge 1)g(z)dz < \infty$ and where $g(z) \geq 0$. Let us suppose that for some $\alpha \in]0,2[$

$$\underline{\lim}_{z\to 0}|z|^{1+\alpha} g(z) > 0. \tag{29}$$

Then

Theorem 5 *If* (26), (27), *and* (29) *are checked, there exists* $\alpha_1 > 0$ *such that*

$$\overline{\lim}_{\beta\to\infty}\left(\sup_{f\in F} \frac{\tau_f(\beta)}{\beta^{\alpha_1}}\right) < 0. \tag{30}$$

Let us recall that for all $p > 0$

$$\hat{P}_t[V^p](x,0) < \infty \tag{31}$$

such that is enough to show for all $p > 0$

$$\sup_{|\xi|=1} \hat{P}_t[\langle V,\xi\rangle^{-p}] < \infty. \tag{32}$$

We consider the semi-group P_t^ξ associated to the generator L^ξ on $R^d \times R$, $(x,u) \in R^d \times R$:

$$L^{\xi}\hat{f}(x,u)=\sum\int_R(\hat{f}(x+\gamma_j(z),u+\nu(z)\langle\xi,\gamma_j'(z)\rangle^2)-\hat{f}(x,u)$$
$$-1_{|\gamma_j(z)|<1}\langle\gamma_j(z),\mathrm{grad}_x\,\hat{f}(x,u)\rangle)g_j(z)dz. \quad (33)$$

We have as it was remarked by Bismut [2]

$$\hat{P}_t[\langle V,\xi\rangle^{-p}](x,0)=P_t^{\xi}[u^{-p}](x,0)=\Gamma(p)^{-1}\int_0^{\infty}\beta^{p-1}P_t^{\xi}[\exp[-\beta u]](x,0)d\beta. \quad (34)$$

We consider the generator L^{β} acting on functions on $R^d\times R\times R$ ($(x,u,v)\in R^d\times R\times R$)

$$L^{\beta}\hat{f}(x,u,v)=\sum\int_R(\hat{f}(x+\gamma_j(z),u+\langle\xi,\gamma_j'(z)\rangle^2\nu(z),v)-\hat{f}(x,u,v)$$
$$-1_{|\gamma_j(z)|<1}\langle\gamma_j(z),\mathrm{grad}_x\,\hat{f}(x,u,v)\rangle)g_j(z)dz$$
$$+\sum\int_R\langle\exp[-2\beta\nu(z)\langle\xi,\gamma_j'(z)\rangle^2]-1,\mathrm{grad}_v\,\hat{f}(x,u,v)\rangle g_j(z)dz. \quad (35)$$

Let us consider the function $\exp[-2\beta u-v]$. If we apply L^{β} to it, we find zero: Therefore

$$P_1^{\beta}[\exp[-2\beta u-v]](x,0,0)=1. \quad (36)$$

Since we have Markov semigroups, we get the following proposition proved in [2] by using martingales theory:

Proposition 2 (Bismut [2]) *For all positive* p

$$\hat{P}_t[\langle V,\xi\rangle^{-p}](x,0)\le\Gamma(p)^{-1}\int_0^{\infty}\beta^{p-1}(P_t^{\beta}[\exp[v]](x,0,0))^{1/2}d\beta. \quad (37)$$

But

$$P_t^{\beta}[\exp[v]](x,0,0)=\exp\left[\sum t\int_R(\exp[-2\beta\nu(z)\langle\gamma_j'(z),\xi\rangle^2]-1)g_j(z)dz\right] \quad (38)$$

and we deduce our results by the previous uniform Tauberian theorem.

References

1. J.M. Bismut, Z. Wahrscheinlichkeit **56**, 469–505 (1981)
2. J.M. Bismut, Z. Wahrscheinlichkeit **63**, 147–235 (1983)
3. R. Léandre, C. R. Acad. Sci., Sér. I **297**, 595–598 (1983)
4. R. Léandre, Une généralisation du théoreme de Hoermander a divers processus de sauts. These. Université de Franche-Comté, 1984
5. R. Léandre, Ann. Inst. H. Poincaré Probab. Stat. **21**, 125–146 (1985)

6. R. Léandre, Applications of the Malliavin calculus of Bismut type without probability, in *Simulation, Modelling and Optimization, C.D. (2006)*, ed. by A. Madureira. WSEAS transactions on Mathematics, vol. 5, pp. 1205–1210 (2006)
7. R. Léandre, Girsanov transformation for Poisson processes in semi-group theory, in *Int. Cong. Numer. An. Appl. Math 2007*, ed. by T.E. Simos. AIP Proceedings, vol. 936 (AIP, Melville, 2007)
8. R. Léandre, Malliavin calculus of Bismut type for Poisson processes without probability, in *Fractional Order Systems*. J.E.S.A, ed. by J. Sabatier et al., vol. 42, pp. 715–733 (2008)
9. R. Léandre, Far East J. Math. Sci. **30**, 1–26 (2008)
10. P. Malliavin, Stochastic calculus of variations and hypoelliptic operators, in *Stochastic Analysis*, Kinokuniya, Kyoto, ed. by K. Itô (1978)
11. K. Yosida, *Functional Analysis* (Springer, Heidelberg, 1977)

Part VII
Computational Techniques for Engineering Sciences

Image Processing for the Estimation of Drop Distribution in Agitated Liquid-Liquid Dispersion

L.M.R. Brás, E.F. Gomes, and M.M.M. Ribeiro

Abstract Image processing of particulate phases can provide an important assistance for the estimation of particle size and shape distributions in multiphase systems. The knowledge of these distributions is of major importance in the modeling of agitated liquid-liquid systems either for hydrodynamic and mass transfer (with or without chemical reaction) simulation. Often, obtaining these distributions implies visual/manual techniques for identifying, counting and measuring the particles. This implies high costs, intensive labor, weariness build-up and consequent high error rates. A largely automated computational approach presents a great potential for better performance. In this paper we describe our technique in image processing for shape discrimination and size classification for liquid drops in monochromatic digitized frames. We empirically evaluate our approach on examples of images.

Keywords Image processing · Liquid-liquid dispersions

1 Introduction

In this paper we describe our approach for the automatic estimation of drop size distribution by analyzing a set of images with photographs of an agitated dispersion.

L.M.R. Brás (✉)
Instituto Superior de Engenharia do Porto, Rua Dr. António Bernardino de Almeida, 431, Porto, Portugal
e-mail: luis-m-bras@telecom.pt

E.F. Gomes
GECAD, Instituto Superior de Engenharia do Porto,
Rua Dr. António Bernardino de Almeida, 431, Porto, Portugal
e-mail: efg@isep.ipp.pt

M.M.M. Ribeiro
CIGAR-SIPROM, Instituto Superior de Engenharia do Porto,
Rua Dr. António Bernardino de Almeida, 431, Porto, Portugal
e-mail: mgr@isep.ipp.pt

J.A.T. Machado et al. (eds.), *Nonlinear Science and Complexity*,
DOI 10.1007/978-90-481-9884-9_37, © Springer Science+Business Media B.V. 2011

These image frames were obtained at SIPROM (Modeling and Simulation of Multiphase Systems) Laboratory in ISEP (Instituto Superior de Engenharia do Porto) with a digital video camera + microscope + illumination set-up from a dispersion of toluene in water within a transparent mixing vessel [1].

Experimental conditions (mainly phase ratio and agitation speed) led to frames of deeply focused fields with partially overlapping drops and high background noise. Since non-invasive image acquisition and lighting were our uncompromising starting option, no significant improvements in the image quality were obtained although all available image acquisition and cleaning techniques were used within this fundamental constraint. Thus, partial images and ill-defined drop boundaries led to low-cost market-available image processing software packages requiring a high number of frames in order to secure statistically significant drop samples which, again, made for greatly time-consuming, unreliable procedures.

This motivated us to develop our own image processing software for which a preliminary test and calibration stage was performed on archive images and drop size distributions from a previous semi-automated procedure [2].

2 Description of the Method

We are working on an approach, implemented in Matlab, which, given one of the photographic images of a dispersion, automatically identifies the contour of existing drops and classifies them according to their diameter.

This work for the detection of the drops in an image was done in two distinct steps. In the first step, we detect the edges of the drops in the original image, manipulating the gradient and the descending thickness and creating an output image with those contours. In the second phase, we detect the drops in this contour image, using the Hough Transform.

2.1 Edges Detection

In Fig. 1 we represent the phases of the detection process of the drops. Thus, starting from the original image (leftmost), we detect the edges (centre) and, in the next step, we detect the contours of the drops using the Hough transform (rightmost). In our images, of weak quality, the drops have, in majority, dark edges.

In order to reduce the noise in the space frequency, the original image (Fig. 2) is smoothed using a Gaussian filter. By reducing noise, some false contours will be eliminated and the detection of false drops is reduced. As we can see in Fig. 3, the filtered image still has drops with dark edges, although smoother. The spatial frequency and the noise are lower.

Using the typical profile of the signal, the edge detection is made by evaluating the relation between the gradient descent (Gd) and its respective thickness (Ed). The filtered image, If, is derived, with kernels of convolution, indicated below,

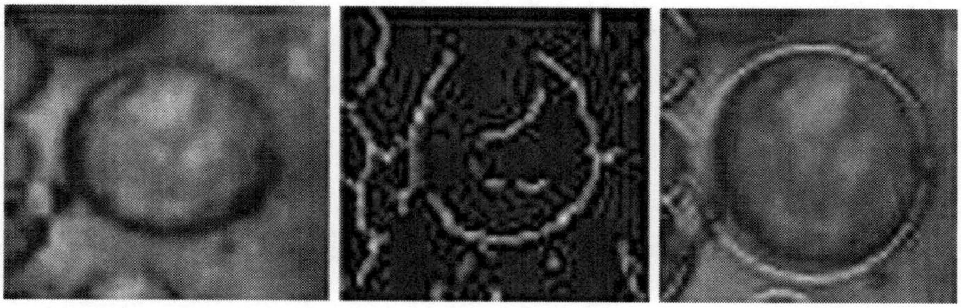

Fig. 1 Original image, the edge detection and the detection of the drop using the Hough transform

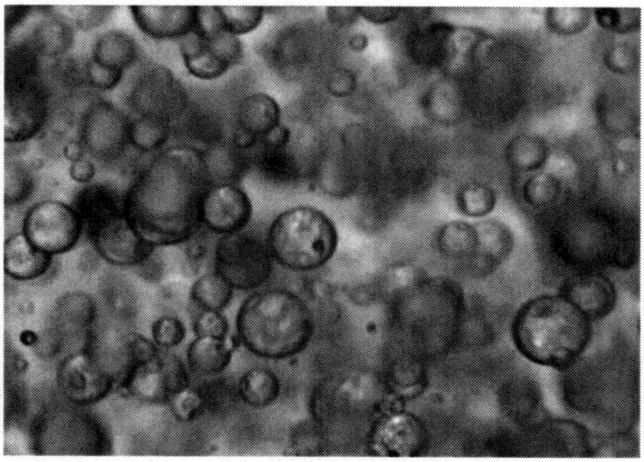

Fig. 2 Exemplary image, this one with very good quality. Although many drops are visually easy to identify, many others are *blurred* or *overlapped* (or *both*)

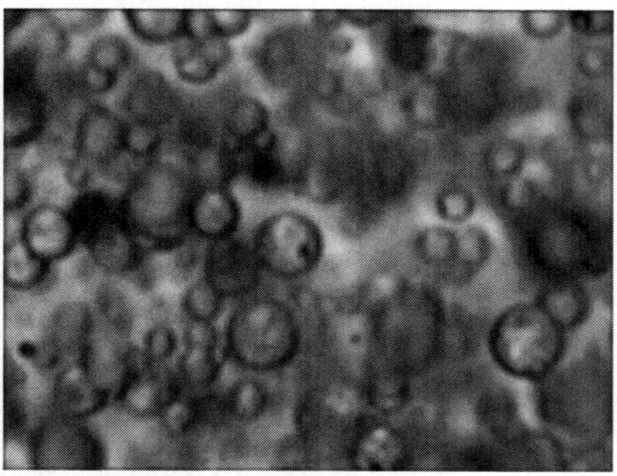

Fig. 3 The image after the Gaussian filter is applied. This filter reduces the noise

originating two images, Ix and Iy, corresponding to the partial derivatives dz/dx and dz/dy:

$$Ix = If * \begin{bmatrix} 1 & -1 \end{bmatrix}, \tag{1}$$

$$Iy = If * \begin{bmatrix} 1 \\ -1 \end{bmatrix}. \tag{2}$$

From the images Ix and Iy we obtain the average and the standard deviation of the descending gradient (μg, σg) and the average and the standard deviation of the descending thickness (μe, σe). The edge detection threshold is defined by the combination of *Gd* and *Ed*, where *Gd* takes negative values and *Ed* positive.

2.2 Detection of Drops in the Contour Image

In the second step of the work, we applied the Hough transform to the contour image (Fig. 4) to detect the contour of the drops for different radius. The Hough transform is widely used to detect lines and also to detect circles. A circle is specified with three numbers: the *X* and *Y* coordinate of its centre, and *R*, its radius [3, 4]. To find circles using a Hough transform, we need a three-dimensional accumulator array. Each edge element votes for all the points *X*, *Y* which are centres of the circles that it could lie on. The 3-D array is searched for peaks that give circle positions and radii. If the radius is known in advance, you only need a 2-D accumulator. In our case we have used a 2-D array accumulator, which, for each edge point in the image, given a radius r, votes for its possible centres. The detection process starts to treat the set of drops with radius between 17 and 48 pixels because these are the most frequent. This detection is accomplished with the erosion of the contour image in order to avoid the false drops with a higher radius. Then we treat the drops set with at least 48 pixels of radius. The third set to detect are the drops with radius between

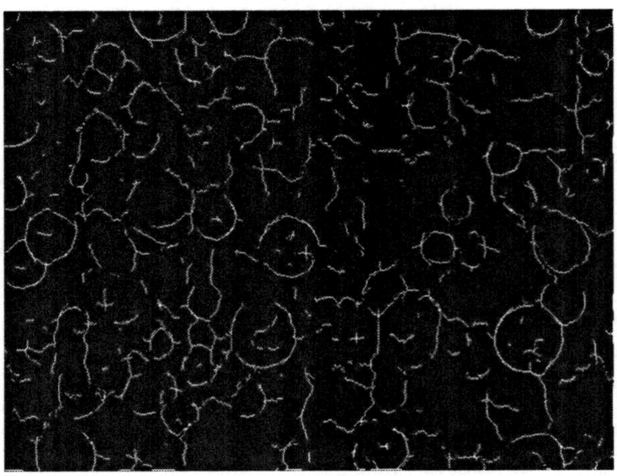

Fig. 4 The resulting contour image

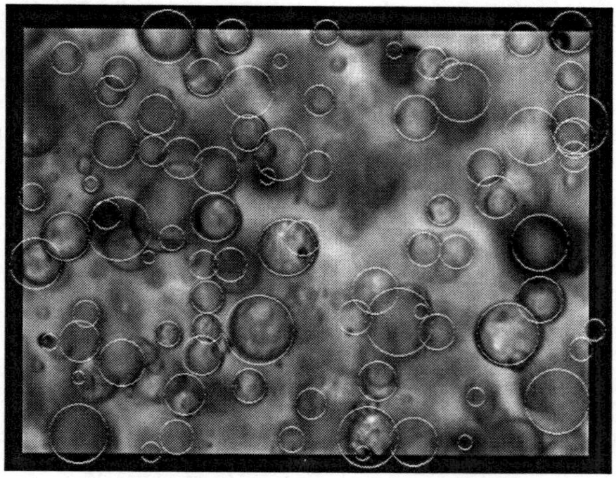

Fig. 5 The drops detected by our program. Each detected drop is marked with a *white circle on top* of the original image

8 and 16 pixels. The processing of the drop sets with higher radius previously to dealing with the drops of lower radius, allows reducing the probability of detecting false drops with a lower radius, motivates for agglomerates of pixels in the contour of the drops. However, the detection is not efficient for radius lower than 8 pixels. In Fig. 5 we can see the result obtained for a radius between 5 and 46 in pixels.

3 Results and Discussions

To evaluate more objectively our approach we have chosen two images with different levels of photographic quality, and compared the sets of drops obtained automatically with the drops manually identified by us. In Table 1 we show the results obtained with one of the images in terms of recall and precision values. We have calculated those values as defined in (3) and (4) where *TP* is the true positives (number of drops correctly identified by the program), *FN* the false negatives (actual drops not identified by the program) and *FP* the false positives (drops incorrectly identified by the program). In other words, recall measures the proportion of existing drops that the program was able to identify correctly, whereas precision is the proportion of drops identified by the program that are truly correct.

$$Recall = \frac{TP}{TP + FN} \tag{3}$$

$$Precision = \frac{TP}{TP + FP} \tag{4}$$

In the case of the first image, we have obtained the following results. For a radius less then 7 pixels the program cannot find any drop. For the radius 10, not in the table because recall $= 0$ and precision not defined, we have $TP = 0$, $FP = 0$ and $FN = 3$. For the radius 12, also not in the table, because recall $= 0$ and precision not defined, we have a $TP = 0$, $FP = 0$ and $FN = 1$. For the other values of the radius

Table 1 Precision and Recall for different radius on one image

Radius (pixels)	Recall	Precision	Radius (pixels)	Recall	Precision
7	0.25	1.00	22	0.60	1.00
8	0.50	1.00	23	0.67	1.00
9	0.57	1.00	24	0.33	1.00
11	1.00	0.67	25–26	1.00	0.75
13	0.67	1.00	28–30	1.00	1.00
14	0.80	1.00	32–33	1.00	1.00
15	0.67	1.00	34	1.00	1.00
16	0.75	0.75	36	1.00	1.00
17–19	1.00	0.90	39	1.00	1.00
20	0.75	1.00	44	1.00	1.00
21	0.60	0.60			

Table 2 Precision and Recall for different radius on one image

Radius (pixels)	Recall	Precision	Radius (pixels)	Recall	Precision
7	0.44	0.80	19	0.50	1.00
8	0.50	1.00	20	0.50	0.60
9	0.40	1.00	22	0.25	0.33
11	0.25	1.00	23	0.75	1.00
12	0.25	1.00	24–26	1.00	1.00
13	0.25	0.50	29	1.00	1.00
14	0.57	0.57	31	1.00	1.00
15	0.57	1.00	33	1.00	1.00
16	0.57	1.00	41	1.00	1.00
17	0.77	1.00	43	1.00	1.00
18	0.57	1.00			

not represented in the table, we have $TP = 0$, $FP = 0$ and $FN = 0$. As final results, taking into account all the values of radius, we have for this image a recall of 0.71 and a precision of 0.89.

For the second image, which has a lower photographic quality, we have worse results, having a total 0.55 for recall and 0.87 for precision, as we can see in Table 2. Nevertheless we have, for many values of radius in the image, maximum recall and precision. For several values of radius from 7 to 23 and for radius equal to 35, 40 and 45, recall and precision are not shown in the table, for the same reasons as in the previous image. We believe the cause for these worse results with respect to the previous image is the lower quality of this second image.

4 Conclusions and Future Work

In this paper we have presented a method for the automatic identification of drops in images taken from agitated liquid-liquid dispersion. The results obtained with two images lead to the conclusion that our program is able to detect a good percentage of the drops. In the case of a better quality image, the program recognized 71% of the drops. For the other image, with lower quality, only 55% were detected. We have also observed that the approach is less efficient for smaller values of the radius, since very small drops can be easily mistaken by noise. We think we can achieve yet better results if we can obtain a better contour image (with less noise) in the first step of the process, so we will do further work in this part of the process.

References

1. M.M.M. Ribeiro, Medição e Interpretação do Comportamento Hidrodinâmico de Dispersões Líquido-Líquido em Vaso Agitado em Regime Contínuo e Descontínuo, PhD Thesis, Universidade de Aveiro, 2002
2. M.M.M. Ribeiro, M.M.L. Guimarães, C.M.N. Madureira, J.J.C. Cruz-Pinto, Chem. Eng. J. **97**, 173–182 (2004)
3. C.R. Gonzalez, R.E. Woods, *Digital Image Processing Using MATLAB* (Prentice Hall, New York, 2006)
4. J.C. Russ, *Image Processing Handbook* (CRC Press, Boca Raton, 2002)
5. M. Shah, Fundamentals of computer vision, Computer Science Department, University of Central Florida, 1997
6. R.J. Schalkoff, *Digital Image Processing and Computer Vision* (Wiley, New York, 1989)
7. M.M.M. Ribeiro, C. Gonçalves, P.F. Regueiras, M.M.L. Guimarães, J.J.C. Cruz-Pinto, Chem. Eng. J. **118**, 47–54 (2006)
8. L.M.R. Brás, Image processing for drop detection in a dispersion, MS Thesis, Instituto Superior de Engenharia do Porto, Porto, Portugal, 2008

Music and Evolutionary Computation

Cecília Reis, Viriato M. Marques, and J.A. Tenreiro Machado

Abstract This paper presents a brief history of the western music: from its genesis to serialism and the Darmstadt school. Also some mathematical aspects of music are then presented and confronted with music as a form of art. The question is, are these two distinct aspects compatible? Can computers be of real help in automatic composition? The more appealing algorithmic approach is evolutionary computation as it offers creativity potential. Therefore, the Evolutionary Algorithms are then introduced and some results of GAs and GPs application to music generation are analysed.

Keywords Evolutionary computation · Evolutionary algorithms · Swarm intelligence · Computer music

1 A Brief History of the Western Music

What is music? Citing Edgar Varèse [1], music is *organized sound*. A well known definition from the late 19th century states that music is *the artful or pleasing organization of sound and silence*. However, the way sounds are organized has dramatically changed over the years and among different cultures. In fact, music is found in every culture and even in the most primitive and isolated societies. Perhaps invented in Africa at least 50,000 years ago, music spread all over the world giving rise to completely different forms of expression ranging from primitive tribal songs

C. Reis (✉) · J.A.T. Machado
Electrical Engineering Department, Institute of Engineering of Porto, Porto, Portugal
e-mail: cmr@isep.ipp.pt

J.A.T. Machado
e-mail: jtm@isep.ipp.pt

V.M. Marques
Computer Science Department, Institute of Engineering of Coimbra, Coimbra, Portugal
e-mail: viriato@isec.pt

J.A.T. Machado et al. (eds.), *Nonlinear Science and Complexity*,
DOI 10.1007/978-90-481-9884-9_38,

to sophisticated concerts, from Gregorian chant to serialism, also including Indian ragas, (Chinese) pentatonic scale based music, opera, jazz, The Beatles, etc.

The history of Western Music is typically divided into the following periods [2]: Medieval (500–1400); Renaissance (1400–1600); Baroque (1600–1750); Classical (1750–1800); Romantic (1800–1900); Modern and Contemporary (1900 up to now).

The Medieval period includes the Gregorian chant, the troubadours, the invention of polyphony, Ars Nova and styles like *ballade*, *virelai*, *rondo*, etc. The Renaissance period is characterized by forms like the *motet* (sacred) and the *madrigal* (secular); the *counterpoint*, also born in this period, essentially consists of sets of rules (organized in five *species*) aiming to conjugate different "independent" voices in the same musical piece. *Opera* is born in the Baroque period (Dafne by Jacopo Peri) where Bach, Handel and Vivaldi belong too. The Classical period is marked by the birth of the symphony, Haydn and Mozart. In the Romantic period music becomes more expressive and emotional, frequently showing some form of national appeal and/or a folkloric basis; some outstanding composers of this period are Beethoven, Schubert, Chopin, Liszt, Verdi and Puccini; the music chromatic complexity rises reaching a maximum (perhaps) with Wagner. After this period it seems that there's nothing more to explore in this direction. So, a search for completely different styles begins. The Modern period starts with Claude Débussy; in the XX century *atonalism* makes its appearance too; about 1920 Schoenberg creates *dodecaphony*, a composition method strongly based on rules that act as a means of ensuring that all 12 notes of the chromatic scale have equal importance. Extended to other musical aspects like rhythm, dodecaphony evolved to *serialism* divulgated by the Darmstadt School. Iannis Xenakis uses computers to write music based on statistical and probabilistic methods. John Cage (1912–1992), an experimentalist compositor, creates compositions such as 4′33″ where, during this lapse of time, the orchestra doesn't play a single note.

2 Music, Mathematics and Physics

There are very interesting relationships between music, mathematics and physics in what respects to the three fundamental characteristics of a sound: pitch, time duration and timbre.

The term *interval* designates the pitch difference between any two notes. A scale is an ordered sequence of intervals [3]. Western European music is primarily based on the *diatonic scale* composed by seven notes: C, D, E, F, G, A, B (C...) (Fig. 1a). An *octave interval* is defined between any note and the next one with the same name. In a diatonic scale the octave is divided into five intervals of one *tone* and two of *half-tone*. The half-tones occur between E-F and B-C. To position a note in an octave an index is used. The pitch of each note is defined by its (fundamental) frequency. For instance, A4 corresponds to 440 Hz. An ascending octave is defined by frequency duplication. So $f_2 = 2f_1$ where $f_2 =$ *frequency of the* X_n *note* and $f_1 =$ *frequency of* X_{n-1} *note*. Computing $3f_{C3}$ we get f_{G4}. So: $f_{G3} = \frac{3}{2} f_{C3}$.

Fig. 1 (**a**) A diatonic scale. (**b**) Some notes, rests, tuplets and their time duration

This $3/2$ ratio defines an interval of *fifth*. A *chromatic scale* consists of sequence of 12 notes all separated by a half-tone. Briefly, it contains all the ♯ (sharps) and ♭ (flats). As the first and the last notes are separated by an octave, the frequency relationship for any half-tone is defined by $f_2 = \sqrt[12]{2} f_1$.

The duration and absence of sounds is defined by *notes* and *rests* (Fig. 1b). In its basic form, the duration of each note is just $1/2$ of its precedent [3]. The duration of notes can also be altered by *dots*, *double dots*, *ties* and *tuplets* (Fig. 1b).

Timbre allows the distinction between different musical instruments and is the basis of the *instrumentation* phase in music composition [3]. It is determined by the harmonic contents of sound. Music is based on complex waveforms resulting from distinct harmonic. It is a matter of Fourier analysis, but a musical sound is also characterized by its *envelope* that comprises 4 phases named: Attack, Decay, Sustain and Release (ADSR).

3 Foundations *versus* Styles

While there are no sounds that can be described as inherently unmusical, musicians in each culture have tended to restrict the range of sounds they will admit [4]. Western music has been *tonal* from Baroque to the advent of dodecaphonism in the 20th century. The *tonal* system defines some hierarchical relationships between the notes of a scale. A chord is a combination of three or more notes played simultaneously. There are many chord types but the basic ones are obtained by building a *third interval* and a *fifth interval* on a fundamental pitch. When correctly combined with a melody, chords fulfil the musical ambience and give the listener the complete music sense and feeling. For centuries the Western European music was based on *major* and *minor* scales. The *C major* scale is shown in Fig. 1a. A piece based on a major scale tends to transmit happiness; a piece based on a minor scale evocates sadness or nostalgia. This is one of the things that allow us to recognize Western music. Dodecafonism states that there should not be any hierarchy between any of the 12 notes of an octave. A set of composition rules tries to ensure this principle. Emotions are, in great part, absent.

4 Music and Art

But music is, perhaps, above all, a form of art. In fact, [5] defines music as *an art concerned with combining vocal or instrumental sounds for beauty of form or emotional expression, usually according to cultural standards of rhythm, melody, and, in most Western music, harmony.* All of these aspects are evaluable by aesthetic considerations. *Beauty of form* refers to the beauty of melody and/or harmony as a whole. A melody is not a simple sequence of beautiful musical sounds. In fact, it can be divided into phrases that form a kind of dialogue, a succession of questions and answers.

And why do we like music based on diatonic scales? Why do they sound "well"? Briefly, traditional Western music deals with profound emotions, human consciousness and cognitive processes. Some *cultural standards of rhythm [and] melody* have already been presented. Finally, *harmony*, where *third* intervals have been considered dissonant until Renaissance are, nowadays, perhaps, the simplest form of two voices singing.

5 Music and Algorithms

As music is a product of the culture it belongs to, it is quite natural that the 20th century produces computer based music. Iannis Xenakis, Lejaren Hiller and John Cage are examples of composers that used computers to produce music, or as a tool for an *a posteriori* composition process. The work of Xenakis and Hiller was based on stochastic methods. Other approaches include neural networks, rule-based systems, knowledge-based systems, evolutionary computation and hybrid systems that combine two or more of the preceding ones. But, what tasks are these systems supposed to perform? We classify these tasks as follows: Melody generation, Harmony generation, Instrumentation and Variation/improvisation on a musical theme. The three former items implement the complete musical composition process. The fourth is of another kind. From the algorithmic point of view any of these tasks can be considered a search task: looking for solutions that are aesthetical appealing according to some restrictions that represent beauty and/or composition rules satisfaction. From the author's point of view, the more appealing subject is the *melody generation* process as it seems the more challenging and/or susceptible of innovation. The more appealing algorithmic approach is *evolutionary computation* as it offers creativity potential along with criticizing possibilities by means of the *fitness function*. It becomes evident the main role played by the fitness function that should, according to some patterns, correctly judge the quality of the solutions. The fitness function is the big challenge! Solutions based on heuristics, rules, neural networks and interactive have already been tried.

6 Evolutionary Computation

Evolutionary Computation (EC) is a subfield of artificial intelligence, namely computational intelligence. EC uses iterative progress, such as growth or development in a population. This population is then selected in a guided random search using parallel processing to achieve the desired end. Such processes are often inspired by biological mechanisms of evolution [6]. Therefore, EC is the general term for several computational techniques which are based, to some degree, on the evolution of biological life in the natural world. Examples of evolutionary techniques are the *Evolutionary Algorithms* (EAs): genetic algorithms, evolutionary programming, evolution strategy, genetic programming and learning classifier systems, and the *Swarm Intelligence* (SI): particle swarm optimization, ant colony optimization and the bees algorithm.

EAs differ from more traditional optimization techniques in that they involve a search from a "population" of solutions, not from a single point. The solutions with high "fitness" are "recombined" with other solutions by swapping parts of a solution with another. Solutions are also "mutated" by making a small change to a single element of the solution. Recombination and mutation are used to generate new solutions that are biased towards regions of the space for which good solutions have already been seen. SI is artificial intelligence based on the collective behavior of decentralized, self-organized systems. SI systems are typically made up of a population of simple agents interacting locally with one another and with their environment. The agents follow very simple rules and, although there is no centralized control structure dictating how individual agents should behave, local interactions between such agents lead to the emergence of complex global behavior. Natural examples of SI include ant colonies, bird flocking, animal herding, bacterial growth, and fish schooling.

6.1 Genetic Algorithms (GAs)

GAs are adaptive heuristic search algorithm based on the evolutionary ideas of natural selection and genetic. The basic concept of GAs is designed to simulate processes in natural system necessary for evolution, specifically those that follow the principles first laid down by Charles Darwin [7].

GAs were invented by John Holland in the sixties and were developed by Holland, his students and colleagues at the University of Michigan. Holland's GA is a method for moving from one population of *chromosomes* to a new population by using a kind of *natural selection* together with the genetics-inspired operators of crossover, mutation, and inversion. Processes loosely based on natural selection, crossover, and mutation are repeatedly applied to a population of binary strings which represent potential solutions. Over time, the number of above-average individuals increases, and highly-fit building blocks are combined from several fit individuals to find good solutions to the problem at hand. The individuals are represented by chromosomes and each chromosome consists of genes. The selection operator chooses those chromosomes in the population that will be allowed to reproduce. Crossover exchanges subparts of two chromosomes, roughly mimicking biological recombination between two organisms. Mutation randomly changes the values of some locations in the chromosome. Inversion reverses the order of a contiguous section of the chromosome.

6.2 Genetic Programming (GPs)

GP is an EA based methodology inspired by biological evolution to find computer programs that perform a user-defined task. It is a specialization of GAs where each individual is a computer program. It is a machine learning technique used to optimize a population of computer programs according to a fitness landscape determined by a program's ability to perform a computational task. John R. Koza is a main proponent of GP and has pioneered its application in various complex optimization and search problems. GP evolves computer programs, traditionally represented as tree structures. Trees can be easily evaluated in a recursive manner. Every tree node has an operator function and every terminal node has an operand, making mathematical expressions easy to evolve and evaluate. The main operators are crossover and mutation. Crossover is applied on an individual by simply switching one of its nodes with another node from another individual in the population. With a tree-based representation, replacing a node means replacing the whole branch. This adds greater effectiveness to the crossover operator. Mutation affects an individual in the population. It can replace a whole node in the selected individual, or it can replace just the node's information.

6.3 Particle Swarm Optimization (PSO)

Non-computer scientists did early research, which led into the theory of particle swarms. In these areas, the term *swarm intelligence* is well known and characterizes the case when a large number of individuals are able to accomplish complex tasks. Motivated by these facts, some basic simulations of swarms were abstracted into the mathematical field. The usage of swarms for solving simple tasks in nature became an intriguing idea in algorithmic and function optimization. Eberhart and Kennedy were the first to introduce the PSO algorithm [8], which is an optimization method inspired in the collective intelligence of swarms of biological populations. In the PSO, instead of using genetic operators, as in the case of GAs, each particle adjusts its flying according with its own and its companions experiences.

Each particle is a point in a D-dimensional space. The velocity and the position of the particles is dynamically updated and is divided in three parts, namely the momentum, the cognitive and the social parts. In the first part the velocity cannot be changed abruptly: it is adjusted based on the current velocity. The second part represents the learning from its own flying experience. The third part consists on the learning group flying experience. The algorithm starts with the random generation of the initial population. The initial velocity of each particle is initialized with zero. These particles have memory and each one keeps information of its previous best position and its corresponding fitness. The swarm has the of all the particles and the particle with the greatest fitness is the global best.

As particular cases of the PSO we can refer the Ant Colony Optimization [9], which is a class of optimization algorithm modeled on the actions of an ant colony and the Bees Algorithm [10], which is an algorithm that mimics the food foraging of swarms of honey bees.

7 Applications of EAs to Computer Music

Computational Intelligence techniques have been applied to musical problems across a wide range of areas, including algorithmic composition, artificial listening, musical cognition and sound synthesis [11]. In 1991 Gibson and Byrne employed a GA to produce and combine musical fragments and a neural network to evaluate their fitness. A GA is also a part of the improvisation and accompaniment system called GenJam which has been developed since 1993 by Al Biles. Since 1996 Rodney Waschka II has been using GAs for music composition. Thywissen (1996) developed a compositional system called GeNotator that applies GAs to several aspects of the composition. In 1997 Brad Johanson and Riccardo Poli developed the GP-Music System.

Burton presents a good survey on applications of genetic techniques to music up to 1999 [12]. Michael Towsey also presented a GA for melody generation (Fig. 2)

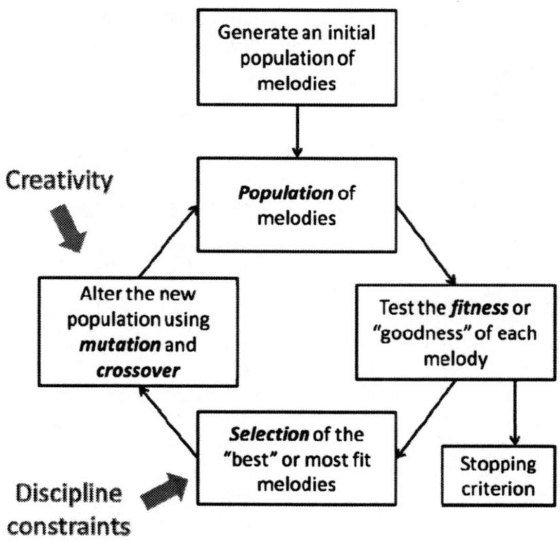

Fig. 2 A GA for melody generation [13]

[13]. CONGA is an interesting system by Tokui and Iba, which combines GAs with GP. Marques describes a GA for music composition [14]. Gartland-Jones and Copley focus on the fitness function problem—human critic and automatic fitness assessment [15]. New musical directions such as Blackwell's *Swarm Music* are also referred.

8 Conclusions and Future Work

EC seems the more promising tool when trying to algorithmically produce music. However, the term music designates an enormous and complex set of styles and tasks. Melody generation has called the author's interest as being one of the more challenging problems, particularly when considering Western tonal music that has the capability of evocating human emotions. From the EC point of view, the solution evaluation is made by the fitness function that must be able to evaluate beauty and the emotional strength of melodies. Therefore, a fitness function incorporating some kind of affective computing suggests further investigation. However, research may also progress in a completely different direction: as music is, as every art, a matter of aesthetical evaluation, computers may be called to produce completely new musical pieces based on unusual pitches, different harmonies and timbres.

Acknowledgement The authors would like to acknowledge the GECAD Unit.

References

1. E. Varèse, *The Liberation of Sound, in Contemporary Composers on Contemporary Music* (Da Capo Press, New York, 1998)
2. J. Stehman, *História da Música Europeia*, 2nd edn. (Bertrand, Lisboa, 1979)
3. J. Zamacois, *Teoria de la música, Liv. 1 e 2* (Ed. Labor, Barcelona, 1986)
4. *Encyclopædia Britannica*, 15th edn. (Encyclopædia Britannica, Inc, 2007)
5. Britannica online (2008), http://www.britannica.com/ebc/article-9372879
6. D. Goldberg, *Genetic Algorithms in Search, Optimization and Machine Learning* (Addison-Wesley, Reading, 1989)
7. M. Mitchell, *An Introduction to Genetic Algorithms* (MIT Press, Cambridge, 1997)
8. J. Kennedy, R.C. Eberhart, in *Particle Swarm Optimization, Proc. of the IEEE Int. Conf. Neural Networks* (1995), pp. 1942–1948
9. M. Dorigo, Optimization, learning and natural algorithms, PhD Thesis, Politecnico di Milano, Italy, 1992
10. D.T. Pham, et al., *The Bees Algorithm*. Technical Note, Manufacturing Engineering Centre, Cardiff University, UK (2005)
11. P.M. Todd, D.G. Loy (eds.), *Music and Connectionism* (MIT Press, Cambridge, 1991)
12. A.R. Burton, T. Vladimirova, Generation of musical sequences with genetic techniques. Comput. Music J. **23**(4), 59–73 (1999)
13. M. Towsey et al., in *Towards melodic extension using genetic algorithms*. Educ. Technol. Soc. **4**(2) (2001)
14. M. Marques et al., Music composition using genetic evolutionary algorithms. Proc. CEC-2000 **1**, 714–719 (2000)
15. A. Gartland-Jones, P. Copley, The suitability of genetic algorithms for musical composition. Contemp. Music Rev. **22**(3), 43–55 (2003)

Application of Computational Intelligence to Engineering

Viriato M. Marques, Luís Roseiro, Cecília Reis, and J.A. Tenreiro Machado

Abstract Computational Intelligence (CI) includes four main areas: Evolutionary Computation (genetic algorithms and genetic programming), Swarm Intelligence, Fuzzy Systems and Neural Networks. This article shows how CI techniques overpass the strict limits of Artificial Intelligence field and can help solving real problems from distinct engineering areas: Mechanical, Computer Science and Electrical Engineering.

Keywords Artificial intelligence · Computational intelligence · Evolutionary computation · Fuzzy systems · Neural networks

1 Introduction

Webster's New Collegiate Dictionary defines intelligence as "the ability to learn or understand or to deal with new or trying situations, the skilled use of reason and the ability to apply knowledge to manipulate one's environment or to think abstractly as measured by objective criteria (or tests)".

V.M. Marques (✉)
Computer Science Department, Institute of Engineering of Coimbra, Coimbra, Portugal
e-mail: viriato@isec.pt

L. Roseiro
Mechanical Engineering Department, Institute of Engineering of Coimbra, Coimbra, Portugal
e-mail: lroseiro@isec.pt

C. Reis · J.A.T. Machado
Electrical Engineering Department, Institute of Engineering of Porto, Porto, Portugal

C. Reis
e-mail: cmr@isep.ipp.pt

J.A.T. Machado
e-mail: jtm@isep.ipp.pt

J.A.T. Machado et al. (eds.), *Nonlinear Science and Complexity*,
DOI 10.1007/978-90-481-9884-9_39,

Artificial Intelligence (AI) is the area of computer science focusing on creating machines that can engage on behaviors that humans consider intelligent. The ability to create intelligent machines has intrigued humans since ancient times and today, with the advent of the computer and 50 years of research into AI programming techniques, the dream of smart machines is becoming a reality. Researchers are creating systems which can mimic human thought, understand speech, beat the best human chess player, and countless other feats never before possible. Artificial Intelligence has come a long way from its early roots, driven by dedicated researchers. The beginnings of AI reach back before electronics, to philosophers and mathematicians such as George Boole and other theorists on principles that were used as the foundation of AI logic. AI really began to intrigue researchers with the invention of the computer in 1943. The technology was finally available, or so it seemed, to simulate intelligent behavior. Alan Mathison Turing was one of the great pioneers of the computer field. He inspired the now common terms of "The Turing Machine" and "Turing's Test". As a mathematician he applied the concept of the algorithm to digital computers. His research into the relationships between machines and nature created the field of artificial intelligence.

Computational Intelligence is a successor of Artificial Intelligence. Eberhart defines computational intelligence as "a methodology involving computing that exhibits an ability to learn and/or to deal with new situations, such that the system is perceived to possess one or more attributes of reason such as generalization, discovery, association and abstraction". Computational Intelligence (CI) as defined by the IEEE Computational Intelligence Society includes four main areas: Evolutionary Computation (EC)(genetic algorithms and genetic programming), Swarm Intelligence, Fuzzy Systems and Neural Networks (NNs).

This article presents a brief introduction to CI main areas and describes their application to three engineering problems: (1) Neural Networks for identification of material constants (2) Fuzzy Logic for equipment fault diagnosis (3) Evolutionary Computation for digital circuit synthesis.

2 Identification of Material Constants with NNs

Among several NNs architectures, the feedforward type has been considered more suitable for the purpose of signature analysis. In this work the cross validation method is used: a group of data, chosen at random among the training values, is used in each training stage to evaluate an error function. When this function differs from the error obtained with the training samples, the process is early stopped and the node weighting and bias are frozen. During the testing stage, new data is provided as input and the output is calculated using the fixed node weights and bias. A feedforward type NN with four layers was considered and the training of the NNs was performed with a second order type algorithm, the Levenberg Marquardt [1].

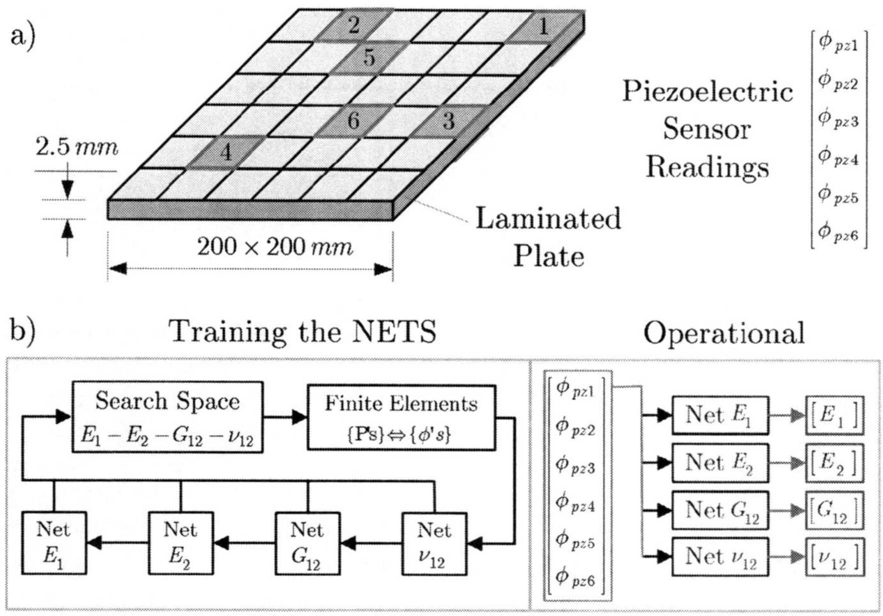

Fig. 1 (**a**) Laminated plate with piezoelectric sensors. (**b**) Neural networks model

2.1 Problem Definition and NN Model

Consider a laminated plate made of 12 glass/epoxy equal thickness layers having a stacking sequence $[0/+45/-45/90/-45/+45]_s$. Plate dimensions are $200 \times 200 \times 5$ mm. A finite element discretization with 36 equal elements is considered. Six piezoelectric sensors are placed on the plate surfaces (Fig. 1a). The plate is loaded with a 5N force in its center and the sensor readings are obtained with a finite element numerical model programmed in Matlab, based on a displacement field with nine independent variables given by:

$$\begin{cases} u_1(x,y,z,t) = u(x,y,t) + z\phi_1(x,y,t) + z^3\theta_1(x,y,t), \\ u_2(x,y,z,t) = v(x,y,t) + z\phi_2(x,y,t) + z^3\theta_2(x,y,t), \\ u_3(x,y,z,t) = w(x,y,t) + z\psi_3(x,y,t) + z^2\theta_3(x,y,t), \end{cases} \tag{1}$$

where u_1, u_2, u_3 are the displacement components at any point in the laminate space in the x, y, z directions, respectively, u, v, w are the displacements of a generic point on the reference surface, ϕ_1, ϕ_2 are the rotations of a normal to the reference surface about the y and the x axes, respectively, and $\theta_1, \theta_2, \theta_3, \psi_3$ are the higher order terms. The objective is to determine the mechanical properties E_1, E_2, G_{12} and ν_{12} of a target plate. The NNs are used to interpret the relationship between potential differences in sensor outputs and the mechanical properties of the plate. Thus, we define four NNs establishing a mapping between the potential differences and each

Table 1 Time and number of finite elements calculations

Model	Time [s]	Number of finite elements calculations
NNs Model	1458	110
Liu et al. (2002)	120040	3001

Table 2 Obtained results

Model		E_1 (GPa) (38.480)	E_2 (GPa) (9.380)	G_{12} (GPa) (3.410)	ν_{12} (0.292)
Neural networks model	Value	38.486	9.389	3.409	0.291
	Error	0.017	0.096	0.029	0.342
Liu et al.	Value	38.340	9.240	3.480	0.293
	Error	0.364	1.493	2.053	0.342

property. Figure 1b shows an outline of the model. In order to obtain a good compromise between the number of training patterns and the performance of the networks, an iterative methodology adopted.

2.2 Results

The NN model was built in Matlab. The obtained results were compared with those of Liu et al. [2]. The computation time and the number of finite elements calculations are shown in Table 1. The obtained values for the properties and the mean relative error for each one are shown in Table 2.

Identification of the material constants of a glass-epoxy plate has been carried out with satisfactory results. The NN iterative model revealed to be very accurate and fast. The numerical experiments showed that the iterative approach leads to small identification errors.

3 Equipment Fault Diagnosis with Fuzzy Systems

A Fuzzy Set [3] is a set to which each element belongs in a degree between 0 and 1, as given by the value of its *membership function*, $\mu(\mathrm{x})$. This concept allows the translation of linguistic terms. A possibility distribution function $\pi_x(u)$, that characterizes a possibility distribution π_X associated with a linguistic term, is numerically equal to the value of the membership function $\mu_{\tilde{F}}(u)$ that defines the linguistic term. A possibility distribution π_X is induced whenever the Relational Assignment Equation is used to associate a linguistic term $\tilde{F}$ with the value of the attribute A

of an object X. Given a fuzzy set $\tilde{A}$ and a possibility distribution function π_X, the *possibility measure* $\pi(\tilde{A})$ [3] is given by:

$$\text{poss}\{X is \tilde{A}\} \hat{=} \sup_{u \in U} \{\mu_{\tilde{A}}(u), \pi_X(u)\}. \tag{2}$$

3.1 Problem Definition and Fuzzy Model

The principles exposed were used in SADEX, a Case-Based-Reasoning (CBR) Fuzzy System for equipment fault diagnosis. SADEX makes use of a conceptual model of attribute values, relevance factors based on document retrieval techniques, combines information value and cost and can perform some case adaptation tasks based on taxonomic similarities. The operation cycle of a CBR system is described by Aamodt and Plaza [4]. However, CBR on its own is not enough, as technical staff know-how makes use of subjective experiences depending on visual inspection, noise, smell and approximate measurement of attribute values. The translation of this information is possible by means of fuzzy sets and fuzzy numbers or intervals.

Some examples of equipment fault descriptions follow: *It doesn't work; Temperature is* 10°C*; Temperature is low; Burned smell is evident.* These examples show that observations can be expressed in different ways. Nevertheless, in the CBR paradigm, and in order to allow cases to be compared, a normalized format is needed. Some definitions used in this process follow: (*i*) The *Absolute Domain* or *Possible Range* ($PR = [pr_l, pr_r]$) of an attribute is the set of values considered of interest for facts representation; (*ii*) The *Absolute Typical Range* ($A\tilde{T}R = (a, b, c, d)_{ATR}$) is a fuzzy set whose support is the set of values it can assume in *PR* in its "normal" state; (*iii*) The *Differential Range* ($DR = [dr_l dr_r]$) is a subinterval of $[-1, 1]$; (*iv*) The *Differential Typical Range* ($D\tilde{T}R = (a, b, c, d)_{DTR}$) is the fuzzy set that translates the linguistic term *Normal* in *DR*. The normalization process makes use of the important assumption that the fuzzy set $A\tilde{T}R$ corresponds to the linguistic term "normal" defined in $D\tilde{T}R$. The final results for the normalization process can be summarized as follows:

$$\begin{aligned} \alpha &= \min pr_l - A\tilde{T}R = pr_l - d_{ATR}, \\ \beta &= \max pr_r - A\tilde{T}R = pr_r - a_{ATR}, \end{aligned} \tag{3}$$

$$\theta = (pr_r - pr_l) + (d_{ATR} - a_{ATR}), \tag{4}$$

$$dr_l = \alpha/\theta, \qquad dr_r = \beta/\theta, \tag{5}$$

$$D\tilde{T}R = (a - d, b - c, c - b, d - a)_{ATR}/\theta. \tag{6}$$

Every time a system user makes an attribute definition by specifying some of the elements above, a weighted mean of its past definitions is calculated. The results are sets of linguistic terms that attempt to catch the concept *linguistic terms used to*

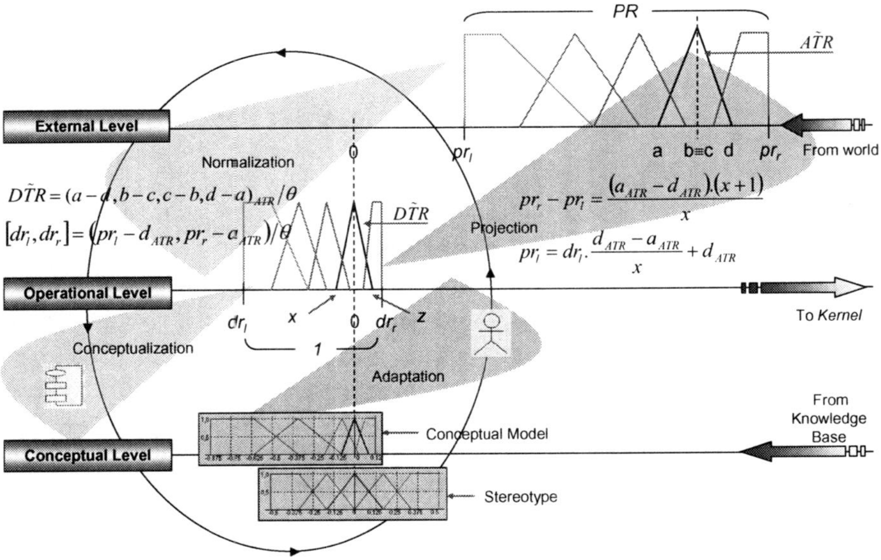

Fig. 2 A 3-Level model for attribute handling

describe attribute values, called *Conceptual Models*, that will act as templates for new definitions. The adaptation of templates to new definitions is called *Adaptation*. The *Projection* operation is symmetrical to *Normalization*. Therefore, attribute representation and handling can be conveniently supported by the 3-Level Model (3L Model) shown in Fig. 2. (*i*) External Level—models the real world; (*ii*) Operational Level—handles normalized attribute values; (*iii*) Conceptual Level—handles meta-models for future attribute definitions.

3.2 Results

A prototype was implemented according to the exposed principles and allowing Internet access. The system has been tested in the health-equipment maintenance field. The case library was initially loaded with work orders data from SUCH—Serviços de Utilização Comum dos Hospitais. The results were promising with a correct diagnosis near 100% in some cases.

4 Digital Circuit Synthesis with EC

Evolutionary Computation techniques consider the concept for automatic design of electronic systems[5]. Instead of using human conceived models, abstractions and techniques, EE employs search algorithms. GAs are adaptive heuristic search it algorithms based on the evolutionary ideas of natural selection and genetic [6]. GAs employ a population of individuals that undergo selection in the presence of operators such as mutation and crossover. A fitness function is used to evaluate individuals. In this case a GA strategy is adopted to design digital circuits, in particular combinational logic circuits.

4.1 Problem Definition and the Genetic Algorithm

The circuits are specified by a truth table and the goal is to implement a functional circuit with the least possible complexity. Four sets of logic gates were defined: Gset 6 = {AND, OR, XOR, NOT, WIRE}, Gset 4 = {AND, OR, XOR, NOT, NAND, NOR, WIRE}, Gset 3 = {AND, OR, XOR, WIRE} and Gset 2 = {AND, XOR, WIRE}, being Gset 2 the simplest one and Gset 6 the most complex gate set.

Logic gate named WIRE means a logical no-operation. The circuits are encoded as a rectangular matrix of logic cells. Three genes represent each cell: *<input1>* *<input2>* *<gate type>*, where *input1* and *input2* are the circuit inputs, if they are in the first column, or one of the previous outputs, if they are in other columns. The gate type is one of the elements adopted in the gate set. The chromosome is formed with as many triplets as the matrix size demands. The initial population of circuits as random generation. The search is then carried out among this population. The three different operators used are reproduction, crossover and mutation. Single point crossover is performed. The crossover point is only allowed between cells to maintain the chromosome integrity. The mutation operator changes the characteristics of a given cell in the matrix, meaning that a completely new cell can appear in the chromosome. Moreover, an elitist algorithm is applied and, consequently, the best solutions are always kept for the next generation.

To run the GA we have to define the number of individuals to create the initial population P. The crossover rate *CR* represents the percentage of the population P that reproduces in each generation. Likewise, the mutation rate *MR* is the percentage of the population P that can mutate in each generation.

The calculation of the fitness function F in (7a)–(7c) has two parts, f_1 and f_2, where f_1 measures the functionality and f_2 measures the simplicity. Firstly, we compare the output $\mathbf{Y}$ produced by the GA-generated circuit with the required values $\mathbf{Y}_R$, according with the truth table, on a bit-per-bit basis. Secondly, once the circuit is functional, the algorithm tries to generate circuits with the least number of gates. The index f_2, that measures the simplicity (the number of null operations), is increased by *one* (zero) for each wire (gate) of the generated circuit, yielding:

$$f_{10} = 2^{ni} \times no, \tag{7a}$$

$$f_1 = f_1 + 1, \quad \text{if } \{\text{bit } i \text{ of } \mathbf{Y}\} = \{\text{bit } i \text{ of } \mathbf{Y_R}\}, \quad i = 1, \ldots, f_{10}, \tag{7b}$$

$$f_2 = f_2 + 1, \quad \text{if gate type} = \text{wire}, \tag{7c}$$

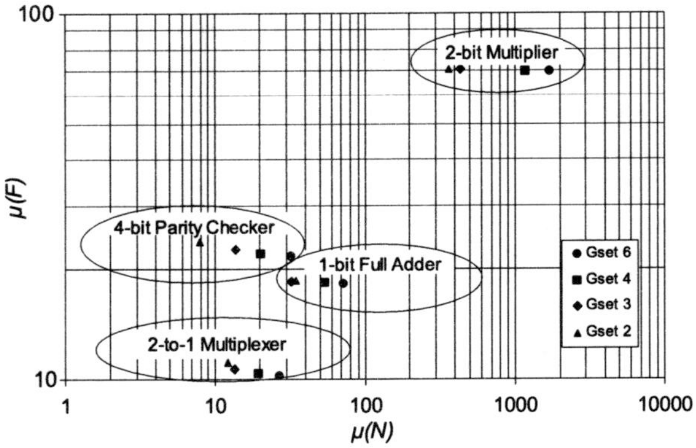

Fig. 3 Average of the fitness function $\mu(F)$ versus the average of the number of generations $\mu(N)$ to reach the solution, using the GA algorithm

$$F = \begin{cases} f_1, & F < f_{10}, \\ f_1 + f_2, & F \geq f_{10}, \end{cases} \tag{7d}$$

where *ni* and *no* represent the number of inputs and outputs of the circuit.

4.2 Results

Four different combinational logic circuits were studied, namely, a 2-to-1 multiplexer, a one-bit full adder, a four-bit parity checker and a two-bit multiplier, using the GA algorithm. Due to the stochastic nature of the GAs in order to evaluate its performance, for each gate set we perform 20 simulations. Figure 3 illustrates the average of the fitness function $\mu(F)$ *versus* the average of the number of generations μ(N) to reach the solution, for all the gate sets and for all the circuits.

It is possible to see the superiority of the gate sets 2 and 3 in terms of μ(N) and $\mu(F)$. Comparing the four studied cases, based on $\mu(N)$ to reach the solution and the $\mu(F)$ it is possible to conclude that, independent of the circuit complexity, the best results appeared with the reduced gate sets.

5 Conclusion

In this paper we show how CI techniques overpass the strict limits of AI field and help solving real problems from distinct engineering areas: Mechanical, Computer Science and Electrical Engineering. Still, as a relatively new disciplinary field, AI has a long way to go. "Take pride in how far you have come; have faith in how far you can go" (Anonymous).

Acknowledgements The authors would like to acknowledge the GECAD and the CEMUC Units.

References

1. T. Hagan et al., *Neural Network Design* (PWS, Boston, 1996)
2. Liu et al., An inverse procedure for determination of material constants of composite laminates using elastic waves. Comput. Methods Appl. Mech. Eng. **191**, 3543–3554 (2002)
3. R.E. Bellman, L.A. Zadeh, Local and fuzzy logics, in *Selected Papers by Lofti A. Zadeh, 1965–1996*. Advances in Fuzzy Systems—Applications and Theory, vol. 6 (World Scientific, Singapore, 1996)
4. A. Aamodt, E. Plaza, Case-based reasoning: foundational issues, methodological variations, and system approaches. Artif. Intell. Commun. **7**(1), 39–59 (1994)
5. R.S. Zebulum et al., *Evolutionary Electronics: Automatic Design of Electronic Circuits and Systems by Genetic Algorithms* (CRC Press, Boca Raton, 2001)
6. D. Goldberg, *Genetic Algorithms in Search, Optimization and Machine Learning* (Addison-Wesley, Reading, 1989)

Evolutionary Trajectory Optimization for Redundant Robots

Maria da Graça Marcos, J.A. Tenreiro Machado, and T.-P. Azevedo-Perdicoúlis

Abstract The trajectory planning of redundant robots is an important area of research and efficient optimization algorithms have been investigated in the last years. This paper presents a new technique that combines the closed-loop pseudoinverse method with genetic algorithms. In this case the trajectory planning is formulated as an optimization problem with constraints.

Keywords Redundant manipulators · Kinematics · Genetic algorithms · Trajectory planning

1 Introduction

Kinematic redundancy occurs when a manipulator possesses more degrees of freedom than the required to execute a given task. In this case the inverse kinematics admits an infinite number of solutions, and a criterion to select one of them is required. Most of the research on redundancy deals with the use of these extra degrees of freedom and is referred to in the literature as the resolution of redundancy [1].

Many techniques for solving the kinematics of redundant manipulators that have been suggested control the end-effector indirectly, through the rates at which the

M.G. Marcos (✉)
Dept. of Mathematics, Institute of Engineering, Polytechnic Institute of Porto, Porto, Portugal
e-mail: mgm@isep.ipp.pt

J.A.T. Machado
Dept. of Electrotechnical Eng., Institute of Engineering, Polytechnic Institute of Porto, Porto, Portugal
e-mail: jtm@isep.ipp.pt

T.-P. Azevedo-Perdicoúlis
Dept. of Mathematics, University of Trás-os-Montes and Alto Douro, Vila Real, Portugal
e-mail: tazevedo@utad.pt

J.A.T. Machado et al. (eds.), *Nonlinear Science and Complexity*,
DOI 10.1007/978-90-481-9884-9_40, © Springer Science+Business Media B.V. 2011

joints are driven, using the pseudoinverse of the Jacobian (see, for instance, [2]). The pseudoinverse of the Jacobian matrix guarantees an optimal reconstruction of the desired end-effector velocity—in the least-squares sense—with the minimum-norm joint velocity.

Baillieul [3] proposed a modified Jacobian matrix called the extended Jacobian matrix. The extended Jacobian is a square matrix that contains the additional information necessary to optimize a certain function. The inverse kinematic solutions are obtained through the inverse of the extended Jacobian.

One optimization method that is gaining popularity for solving complex problems in robotics is the Genetic Algorithm (GA). GAs are population-based stochastic and global search methods. Their performance is superior to that revealed by classical techniques [4] and has been used successfully in robot path planning.

Kubota et al. [5] studied a hierarchical trajectory planning method for a redundant manipulator using a virus-evolutionary GA. Nishimura et al. [6] proposed a motion planning method using an artificial potential field and a GA for a hyper-redundant manipulator whose workspace includes several obstacles. McAvoy and Sangolola [7] proposed an approach with GAs for optimal point-to-point motion planning of kinematically redundant manipulators. Their approach combines B-spline curves, for the generation of smooth trajectories, with GAs for obtaining the optimal solution. Peng and Wei [8] presented the ASAGA trajectory planning method of redundant manipulators by combining a stochastic search algorithm (simulated annealing algorithm) and a GA. Zhang et al. [9] proposed an algorithm to solve the inverse kinematics of a flexible macro-micro manipulator system which combines a GA and a neural network.

Having these ideas in mind, the paper is organized as follows. Section 2 introduces the fundamentals of the kinematics of redundant manipulators. Based on these concepts, Sect. 3 formulates the new closed-loop inverse kinematics method with genetic algorithms (CLGA). Section 4 presents the simulation results and, finally, Sect. 5 draws the main conclusions.

2 Kinematics of Redundant Manipulators

We consider a manipulator with n degrees of freedom whose joint variables are denoted by $\mathbf{q} = [q_1, q_2, \cdots, q_n]^T$. We assume that the class of tasks we are interested in can be described by m variables, $\mathbf{x} = [x_1, x_2, \ldots, x_m]^T$, $m < n$, and that the relation between $\mathbf{q}$ and $\mathbf{x}$ is given by the direct kinematics:

$$\mathbf{x} = f(\mathbf{q}). \tag{1}$$

Differentiating (1) with respect to time yields:

$$\dot{\mathbf{x}} = \mathbf{J}(\mathbf{q})\dot{\mathbf{q}} \tag{2}$$

where $\dot{\mathbf{x}} \in \mathbb{R}^m$, $\dot{\mathbf{q}} \in \mathbb{R}^n$ and $\mathbf{J}(\mathbf{q}) = \partial f\ (\mathbf{q})/\partial \mathbf{q} \in \mathbb{R}^{m \times n}$. Hence, it is possible to calculate a path $\mathbf{q}(t)$ in terms of a prescribed trajectory $\mathbf{x}(t)$ in the operational space.

Equation (2) can be inverted to provide a solution in terms of the joint velocities:

$$\dot{\mathbf{q}} = \mathbf{J}^{\#}(\mathbf{q})\dot{\mathbf{x}} \tag{3}$$

where $\mathbf{J}^{\#}$ is the Moore-Penrose generalized inverse of the Jacobian $\mathbf{J}$ [10].

3 Robot Trajectory Control

The Jacobian of a n-link planar manipulator (i.e., $m = 2$) has a simple recursive nature according with the expressions:

$$\mathbf{J} = \begin{bmatrix} -l_1 S_1 - \cdots - l_n S_{1\ldots n} & \ldots & -l_n S_{1\ldots n} \\ l_1 C_1 + \cdots + l_n C_{1\ldots n} & \ldots & l_n C_{1\ldots n} \end{bmatrix} \tag{4}$$

where l_i is the length of link i, $q_{i\ldots k} = q_i + \cdots + q_k$, $S_{i\ldots k} = \mathrm{Sin}(q_{i\ldots k})$ and $C_{i\ldots k} = \mathrm{Cos}(q_{i\ldots k})$, $i, k \in \mathbb{N}$.

In the closed-loop pseudoinverse (CLP) method the joint positions can be computed through the time integration of (5):

$$\Delta\mathbf{q} = \mathbf{J}^{\#}(\mathbf{q})(\mathbf{x}_{\mathrm{ref}} - \mathbf{x}) \tag{5}$$

where $\mathbf{x}_{\mathrm{ref}}$ is the vector of reference position in the operational space. Nevertheless, in a previous study, addressing the CLP method [11], we concluded that this method leads to unpredictable arm configurations and reveals properties resembling those that occur in chaotic systems.

Genetic algorithms (GAs) are a method for solving both constrained and unconstrained optimization problems, based on the mechanics of natural genetics and selection, that was first introduced by Holland [12]. A GA allows a population composed of many individuals to evolve under specified selection rules to a state that maximizes the fitness or the cost function. The GA modifies repeatedly the population of individuals (possible solutions). At each step, the GA selects individuals at random, from the current population, to be parents, and uses them to produce the offspring for the next generation. Over successive generations, the population evolves towards an optimal solution. The GAs can be applied to solve a variety of optimization problems that are not well suited for standard optimization algorithms, including problems in which the objective function is discontinuous, not differentiable, stochastic, or highly nonlinear.

Bearing these facts in mind, in this paper we propose a new method that combines the CLP with a GA, that namely the closed-loop inverse kinematics algorithm with genetic algorithms (CLGA). The optimal configuration is the one that minimizes the fitness function according to some specified criteria.

3.1 The CLGA Formulation

The CLGA adopts the closed-loop structure without requiring the calculation of the pseudoinverse. The CLGA uses an extended Jacobian matrix $\mathbf{J}^*$, $n \times n$, and an extended vector $\Delta\mathbf{x}^*$, as a way to limit the joint configurations for a given end-effector position.

The definition of $\mathbf{J}^*$ and $\Delta\mathbf{x}^*$ take the form:

$$\mathbf{J}^* = \begin{bmatrix} -l_1 S_1 - \cdots - l_n S_{1\ldots n} & \cdots & -l_n S_{1\ldots n} \\ l_1 C_1 + \cdots + l_n C_{1\ldots n} & \cdots & l_n C_{1\ldots n} \\ j_{31} & \cdots & j_{3n} \\ \cdots & \cdots & \cdots \\ j_{n1} & \cdots & j_{nn} \end{bmatrix}, \qquad \Delta\mathbf{x}^* = \begin{bmatrix} \Delta x_1 \\ \Delta x_2 \\ \Delta x_3 \\ \vdots \\ \Delta x_n \end{bmatrix} \tag{6}$$

where the matrix elements j_{ik} and Δx_i, $i = 3, \ldots, n$ and $k = 1, \ldots, n$, are values generated by the GA, satisfying the additional imposed constraints.

3.2 Representation and Operators in the CLGA

An initial population of strings, with dimension n_P, is constructed at random and the search is then carried out among this population. Each chromosome (string) is implemented by a matrix of $n_V = (n-2) \times n$ values (genes), represented as floating-point numbers, and initialized in the range $[\text{var}_{\text{min}}, \text{var}_{\text{max}}]$. The three different operators used in the GA are reproduction, crossover and mutation. In what respecting the reproduction operator, the successive generations of new strings are generated on the basis of their fitness function. In this case, it is used a rank weighting to select the strings from the old to the new population. For the crossover operator, the strings are randomly grouped together into pairs. Single crossover is then performed among pairs. Finally, for the mutation operator, one variable value is replaced with a new random one.

Begin
  $T = 0$
  $\Delta\mathbf{x} = \mathbf{x}_{\text{ref}} - \mathbf{x}_{\text{ini}}, \mathbf{J}$
  initialize random population P(T)
  get $\Delta\mathbf{q} = \mathbf{J}^{*-1}(\mathbf{q})\Delta\mathbf{x}^*$ *and* $\mathbf{q} = \int \Delta\mathbf{q}$
  evaluate P(T)
  repeat
    selection parents from P(T)
    crossover P(T)
    mutation P(T)
    form new generation P(T)
    get $\Delta\mathbf{q} = \mathbf{J}^{*-1}(\mathbf{q})\Delta\mathbf{x}^*$ *and* $\mathbf{q} = \int \Delta\mathbf{q}$
    evaluate P(T)
    $T = T + 1$
  until *termination condition is TRUE*
  get new $\mathbf{q}$
End

Fig. 1 Procedure for the CLGA

The CLGA procedure is shown in Fig. 1, where $\mathbf{x}_{ref}$ and $\mathbf{x}_{ini}$ are the vectors representing the reference and the initial position of the end-effecter in the operational space, respectively.

3.3 Optimization Criteria

The fitness function is designed according to the goal we want to achieve. Two criteria have been selected *a priori*. All constraints and criteria are translated into penalty functions to be minimized and that are defined in the sequel.

- The largest joint displacement between two adjacent robot configurations can be minimized through the fitness function:

$$f_1 = \max\{[q_j(k+1) - q_j(k)]^2\}, \quad j = 1, \dots, n \tag{7}$$

where k and $k+1$ are two consecutive sampling instants.

- The total level of joint velocities must be minimized at each configuration leading to the fitness function:

$$f_2 = \sum_{j=1}^{n} \dot{q}_j^2. \tag{8}$$

4 Simulation Results

This section presents the results of several simulations.

Without lacking generality, in the following experiments are adopted arms having identical link lengths, $l_1 = l_2 = \cdots = l_n$. The experiments consist in the analysis of the kinematic performance of a planar manipulator with $n = 3$ rotational joints ($3R$-robot) that is required to repeat a circular motion in the operational space with frequency $\omega_0 = 7.0$ rad sec^{-1}, center at $r = (x_1^2 + x_2^2)^{\frac{1}{2}}$, radius $\rho = 0.5$ and a step time increment of $\Delta t = 10^{-3}$ sec. The goal here is to position the end-effector of the $3R$-robot at a target location while satisfying a given optimization criterion.

If the robot's end-effector current position is $P_c = (x_c, y_c)$ and the desired final position is $P_f = (x_f, y_f)$, then the positional error, P_{error}, is defined as:

$$P_{\text{error}} = \sqrt{(x_c - x_f)^2 + (y_c - y_f)^2} \tag{9}$$

and the average of the positional error, $\overline{P}_{\text{error}}$, is defined as:

$$\overline{P}_{\text{error}} = \frac{\sum_{i=1}^{k} P_{\text{error}}}{k} \tag{10}$$

where k is the number of sampling points and is given by the expression:

$$k = \frac{2\pi}{\omega_0 \Delta t} n_C. \tag{11}$$

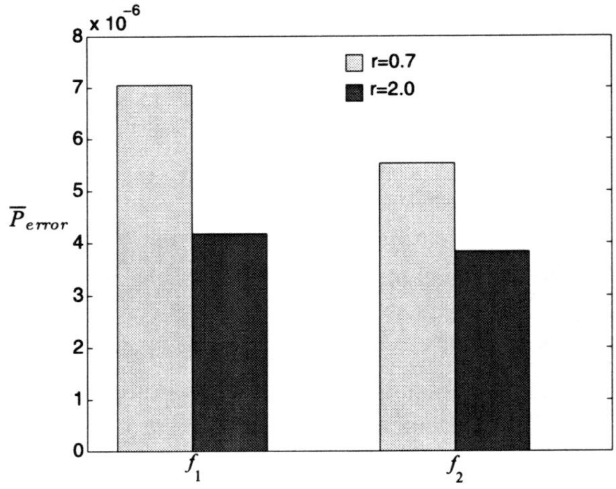

Fig. 2 $\overline{P}_{\text{error}}$ of the $3R$-robot, under the action of the CLGA for $n_P = 100$ string population, during $n_C = 100$ cycles for $r = \{0.7, 2.0\}$

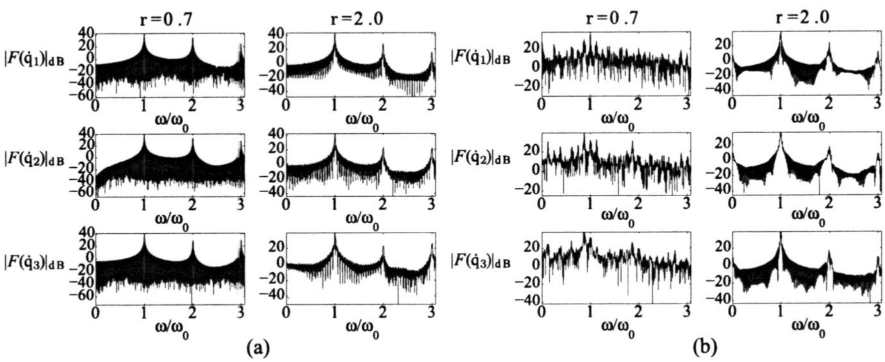

Fig. 3 $|F\{\dot{q}_i(t)\}|$ *vs* ω/ω_C of the $3R$-robot, under the action of the CLGA, during $n_C = 100$ cycles for $r = \{0.7, 2.0\}$ and the fitnesses (**a**) f_1 and (**b**) f_2

The CLGA algorithm adopts crossover and mutation probabilities of $p_c = 0.5$ and $p_m = 0.2$, respectively, a $n_P = 100$ string population, and the results are obtained for $n_G = 100$ consecutive generations.

The average of the positional error, $\overline{P}_{\text{error}}$, for $n_C = 100$ cycles and $r = \{0.7, 2.0\}$, is presented in Fig. 2, revealing that a high precision is achieved when using the CLGA with the fitness functions $\{f_1, f_2\}$.

The Fourier transform of the robot joint velocities are depicted in Fig. 3 revealing that:

(i) the signal energy is concentrated in the fundamental and multiple higher harmonics when we minimize the largest joint displacement through f_1;
(ii) the results depend on the radial distance when minimizing the total joint velocities through f_2.

We verify that the CLGA has a better performance than the CLP because, for example, when minimizing the largest joint displacement f_1, we get not only a good positioning but also a repetitive trajectory.

5 Conclusions

A CLGA scheme that combines the CLP with a GA scheme was presented. Several experiments were developed to study the performance of the CLGA when the manipulator is required to repeat a circular motion in the operational space while satisfying different optimization criteria.

The results show that the CLGA gives, in general, good results in what concerns the repeatability and positioning. The better result occurs when the CLGA minimizes the largest joint displacement between two adjacent configurations since not only we get a good positioning, but also the joint motion is repetitive.

References

1. D.N. Nenchev, Y. Tsumaki, Motion analysis of a kinematically redundant seven-DOF manipulator under the singularity-consistent method, in *IEEE Int. Conf. on Robotics and Automation* (2003), pp. 2760–2765
2. K.L. Doty, C. Melchiorri, C. Bonivento, A theory of generalized inverses applied to robotics. Int. J. Robot. Res. **12**, 1–19 (1993)
3. J. Baillieul, Kinematic programming alternatives for redundant manipulators, in *Proc. of the IEEE Int. Conf. on Robotics and Automation* (1985), pp. 722–728
4. D.E. Goldenberg, *Genetic Algorithms in Search Optimization, and Machine Learning* (Addison-Wesley, Reading, 1989)
5. N. Kubota, T. Arakawa, T. Fukuda, K. Shimojima, Trajectory generation for redundant manipulator using virus evolutionary genetic algorithm, in *IEEE Int. Conf. on Robotics and Automation* (1997), pp. 205–210
6. T. Nishimura, K. Sugawara, I. Yoshihara, K. Abe, A motion planning method for a hyper multi-joint manipulator using genetic algorithm, in *IEEE Int. Conf. on Systems, Man, and Cybernetics* (1999), pp. 645–650
7. B. McAvoy, B. Sangolola, Optimal trajectory generation for redundant planar manipulators, in *IEEE Int. Conf. on Systems, Man, and Cybernetics* (2000), pp. 3241–3246
8. Y. Peng, W. Wei, A new trajectory planning method of redundant manipulator based on adaptive simulated annealing genetic algorithm (ASAGA), in *IEEE Int. Conf. on Computational Intelligence and Security* (2006), pp. 262–265
9. Y. Zhang, Z. Sun, T. Yang, Optimal motion generation of a flexible macro-micro manipulator system using genetic algorithm and neural network, in *IEEE Conf. on Robotics, Automation and Mechatronics* (2006), pp. 1–6
10. B. Siciliano, Kinematic control of redundant robot manipulators: a tutorial. J. Intell. Robot. Syst. 201–212 (1990)
11. M.G. Marcos, F.B. Duarte, J.A.T. Machado, Complex dynamics in the trajectory control of redundant manipulators. Trans. Nonlinear Sci. Complex. 134–143 (2006)
12. J.H. Holland, *Adaptation in Natural and Artificial Systems* (University of Michigan Press, Ann Arbor, 1992), 2nd edn. MIT Press

Part VIII
Nonlinear Systems

Robust Communication-Masking via a Synchronized Chaotic Lorenz Transmission System

S. Poinsard and A. Loría

Abstract *Context*: a Lorenz transmitter produces a carrier signal to which a valuable signal of information is added; this may be recovered provided a receiver *synchronizes* with the transmitter. We present a comparative analysis of robustness for two chaotic Lorenz-based transmission systems. For the first we show that *ad-hoc* synchronization based on observer theory may be inefficient except for the specific case of binary information. Then, we show that by re-injecting the information to the dynamics of the transmitter the performance may be improved *at will*. We present an analytic study of robustness simulation results that corroborate the latter.

Keywords Chaotic systems · Synchronization · Secured transmission

1 Introduction

We say that the system

$$\dot{x}(t) = f_m(t, x(t)) \qquad x(t_\circ) \in \mathbf{R}^n t \geq t_\circ \geq 0 \tag{1}$$

is synchronized with

$$\dot{\hat{x}}(t) = f_s(t, \hat{x}(t)) \qquad \hat{x}(t_\circ) \in \mathbf{R}^n t \geq t'_\circ \geq 0 \tag{2}$$

A. Loría (✉) · S. Poinsard
LSS-Supelec, CNRS, 3 Rue Joliot-Curie, 91192 Gif sur Yvette, France
e-mail: loria@lss.supelec.fr

S. Poinsard
e-mail: s.poinsard@gmail.com

S. Poinsard
ISEP, 28 Rue Notre Dame des Champs, 75006 Paris, France

J.A.T. Machado et al. (eds.), *Nonlinear Science and Complexity*,
DOI 10.1007/978-90-481-9884-9_41,

if for any initial conditions $t_\circ$, $t'_\circ$, $x_\circ$, and $\hat{x}_\circ$ the respective motions of systems (1) and (2) satisfy:

$$\lim_{t\to\infty} \left|\hat{x}(t) - x(t)\right| = 0 \tag{3}$$

where $|\cdot|$ denotes the usual Euclidean norm. By a *motion* of a dynamical system, we consider the curve described by the points $(t, x) \in \mathbf{R} \times \mathbf{R}^n$ in the motion space.

In this article we concentrate on a specific form of synchronization: *master-slave* in which context, the master system, say (1), describes a motion unaffected by external forces while the slave system is controlled to describe, asymptotically, the *same* motion as the master system. Master-slave synchronization is of interest in different scientific communities, specifically, it is well-known from the works of Pecora and Carroll [1] and succeeding papers [2–4] that chaotic synchronization is useful for masking valuable information that needs to be transmitted from one end to another. The principle consists in building two Lorenz-type chaotic systems: one system (the transmitter) generates a chaotic *carrier* to which the information is added; another Lorenz oscillator receives the information masked in the chaotic signal. *Theoretically*, the information may be recovered by the receiver provided that the two circuits synchronize their motion perfectly [2, 4–6]. While some works rely on this *ad hoc* assumption, we shall see that this is not necessarily the case.

A particular way of synchronizing two chaotic systems is using *observers* theory [7]. An *observer* is an artificially constructed dynamical system with no particular physical meaning which is meant to recreate the *unmeasured* system's state based on a *measured* system's output. In this paper we address the problem of synchronization of chaotic systems using observer theory. Our study concentrates on studying the robustness of the approach that is, on establishing analytically how good the performance of master-slave synchronization for the purposes of information masking. This translates in the quality of information encoding and recovery. The main contribution of this article strives in the robustness analysis demonstrated and discussed on the basis of both a theoretical analysis and simulations.

2 The Lorenz System

We recall that the Lorenz system is defined by

$$\dot{x}_1 = \sigma(x_2 - x_1), \tag{4a}$$

$$\dot{x}_2 = rx_1 - x_2 - x_1x_3, \tag{4b}$$

$$\dot{x}_3 = x_1x_2 - bx_3, \tag{4c}$$

where r, σ and q are system's constant parameters. As is well-known the Lorenz system exhibits a chaotic behavior if $\sigma = 16$, $b = 4$ and $r = 45.6$ for different sets of initial states $x(t_\circ)$—cf. Fig. 1. An important characteristic to be noticed is that, since the system's trajectories converge to a strange attractor that may be strictly contained in a compact i.e. the phase trajectories are bounded for all $t \geq t_\circ \geq 0$.

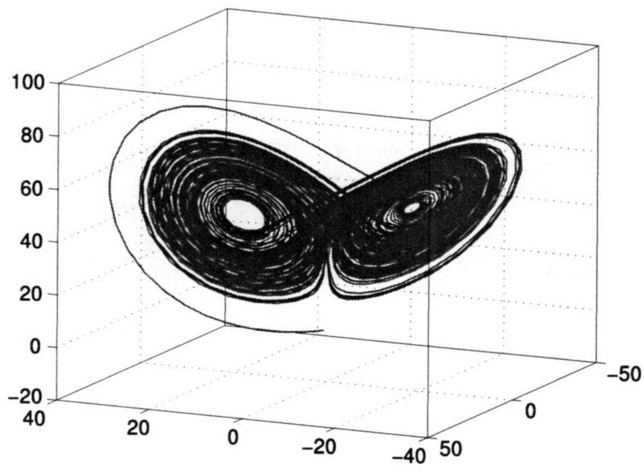

Fig. 1 Phase portrait of $x_1(t)$ vs. $x_2(t)$ for Lorenz system with $\sigma = 16$, $b = 4$, $r = 45.6$ and initial conditions: $x_1(0) = x_2(0) = x_3(0) = 10$

3 Master-Slave Synchronization Using a Luenberger-Type Observer

The first synchronization approach that we analyze is similar to that by [2] for a *particular* choice of the observer gain L. Our purpose is to study in fair generality the performance of a Luenberger-type observer for the Lorenz system, in the context of secure communication transmission.

Consider the Lorenz system (4a)–(4c); assume that only the variable x_1 is available for measurement; i.e., the matrix $C = [1\ 0\ 0]$. The Lorenz system may be rewritten as

$$\dot{x} = A(y^*)x, \qquad y^* := x_1 = Cx, \qquad A(y*) := \begin{bmatrix} -\sigma & \sigma & 0 \\ r & -1 & -y* \\ 0 & y^* & -b \end{bmatrix}. \tag{5}$$

Following basic observer theory [8] the receiver (slave) system is given by

$$\dot{\hat{x}} = A(y)\hat{x} - L(\hat{y} - y), \qquad y := x_1 + m. \tag{6}$$

It is important to stress that the measurable output is $y = Cx + m$ that is, one of the state variables of the transmitter plus the information. Typically, the information signal $m(t)$ has a much smaller amplitude than its carrier $x_1(t)$; based on which, it is claimed that the information may be properly retrieved [2]. As we shall see, this is not the case *in general*.

Also, note that the observer system (6) has the structure of the Luenberger observer for time-invariant systems however, the matrix A depends on the output which, in turn, includes the external time-varying (information) signal $m(t)$.

3.1 Robustness Study

Let us analyze the stability and robustness of the error dynamics which results from extracting (5) from (6). Using this and $\hat{x} = \bar{x} + x$ we have

$$\dot{\hat{x}} - \dot{x} = A(y)\hat{x} - L[\hat{y} - y] - A(y^*)x.$$

Recalling that $\hat{y} = C\hat{x}$, $y = Cx$ we see that

$$\dot{\hat{x}} - \dot{x} = A(y)\hat{x} - LC[\hat{x} - x] + Lm - A(y^*)x.$$

Define $A'(m)$ as

$$A(y) = A(y^*) + A'(m), \qquad A'(m) := \begin{bmatrix} 0 & 0 & 0 \\ 0 & 0 & -m \\ 0 & m & 0 \end{bmatrix}. \tag{7}$$

Since $CC^\top = 1$ we have $Lm = LCC^\top m$

$$\dot{\bar{x}} = A(y^*)\bar{x} + A'(m)\hat{x} - LC\big[\hat{x} - x - C^\top m\big]. \tag{8}$$

Using $\hat{x} = \bar{x} + x$, defining

$$\bar{A}(t, y^*) := A(y^*) - LC + A'(m(t)) \tag{9}$$

and regrouping some terms in (8) we obtain

$$\dot{\bar{x}} = \bar{A}(t, y^*(t))\bar{x} + LCC^\top m(t) + A'(m(t))x. \tag{10}$$

Strictly speaking, from an analytic view point, the state of the differential equation above is $\bar{x}$; the dynamics depends on time, through the measured output signal $y^*(t)$ and the information signal $m(t)$.

From a stability and robustness view point the information signal $m(t)$ is regarded as measurement noise; we wish to know whether the trivial solution $\bar{x} \equiv 0$ to the *nominal* system (i.e., without noise, $m \equiv 0$):

$$\dot{\bar{x}} = \bar{A}(t, y(t))\bar{x} \quad \text{where } \textit{here}, \ y = y^* \tag{11}$$

remains (robustly) stable in the presence of $m(t) \not\equiv 0$. To that end, we consider the Lyapunov function [8]

$$V(\bar{x}) = \frac{1}{2}|\bar{x}|^2. \tag{12}$$

Its total time derivative, with $\dot{\bar{x}}$ defined by (10), yields

$$\dot{V} = \frac{1}{2}\big(\bar{x}^\top[\bar{A} + \bar{A}^\top]\bar{x} + 2\bar{x}^\top A'(m)x + 2\bar{x}^\top Lm\big) \tag{13}$$

where, defining $L^\top := [l_1\ l_2\ l_3]$,

$$\begin{aligned}\bar{x}^\top[\bar{A} + \bar{A}^\top]\bar{x} &= -2(l_1 + \sigma)\bar{x}_1^2 + 2(r - l_2 + \sigma)\bar{x}_1\bar{x}_2 - 2l_3\bar{x}_3\bar{x}_1 - 2\bar{x}_2^2 - 2b\bar{x}_3^2 \\ &\le -[2(l_1 + \sigma) - r + l_2 - \sigma + l_3]\bar{x}_1^2 - [l_2 - r - \sigma + 2]\bar{x}_2^2 \\ &\quad - [l_3 + 2b]\bar{x}_3^2\end{aligned}$$

and the last two terms on the right-hand side of (13) are, in sum, equal to $[l_1\bar{x}_1 + (l_2\bar{x}_2 + x_3(t)\bar{x}_2) + (l_3\bar{x}_3 - x_2(t)\bar{x}_3)]m$. At this point, we observe that the master system's state trajectories $x_i(t)$ are bounded uniformly for all t—cf. Fig. 1. Collecting terms we find that

$$\dot{V}(\bar{x}) \le -2\alpha|\bar{x}|^2 + \beta \sup_{t\ge 0}|m(t)||\bar{x}| \tag{14}$$

where

$$\alpha := \frac{1}{4}\min\{2(l_1 + \sigma) - r + l_2 - \sigma + l_3, l_2 - r - \sigma + 2, l_3 + 2b\}, \tag{15}$$

$$\beta = |l_1| + \Big|l_2 + \sup_{t\ge 0}(|x_3(t)|)\Big| + \Big|l_3 + \sup_{t\ge 0}(|x_2(t)|)\Big|. \tag{16}$$

Let $\bar{m} := \sup_{t\ge 0}|m(t)|$ then, using the triangle inequality on $\beta\bar{m}|\bar{x}|$ we obtain

$$\dot{V}(\bar{x}) \le -\alpha|\bar{x}|^2 + \frac{\beta^2\bar{m}^2}{4\alpha}. \tag{17}$$

Defining $c_m := \frac{\beta^2\bar{m}^2}{4\alpha}$ and integrating on both sides of (17) we finally obtain

$$|x(t)|^2 \le |\bar{x}(t_\circ)|^2 e^{-\alpha(t-t_\circ)} + \frac{\beta^2\bar{m}^2}{2\alpha^2}. \tag{18}$$

Provided that one can find values of l_1, l_2 and l_3 such that $\alpha > 0$, we see from (18) that in the absence of information i.e., if $\bar{m} = 0$, the synchronization error tends to zero exponentially fast. Moreover, since α depends on the design parameters l_1, l_2, l_3 the convergence may be made arbitrarily fast.

When $\bar{m} \ne 0$ the master and slave systems synchronize up to an error of order $\frac{\beta^2\bar{m}^2}{2\alpha^2}$ where both α and β depend on the design gain L in the *same* order. As assumed (without analytical proof) in [2] one may conjecture that synchronization occurs if the information is relatively small with respect to the carrier (the x_1 variable) that is, if $\bar{m}$ is small (−20 dB). However, this is in general not sufficient since the ratio β/α is constant and may be large. The previous analysis includes but does not boil down to the design of [2] which may be recasted in the notations of this article by the choice $L = [0\ r\ 0]$.

The relatively poor performance of this approach is illustrated in Figs. 2 and 3. In the simulation, binary information of 3 kHz is transmitted and clearly, "recovered" if one is satisfied with simply detecting 'voltage changes' in the received signal;

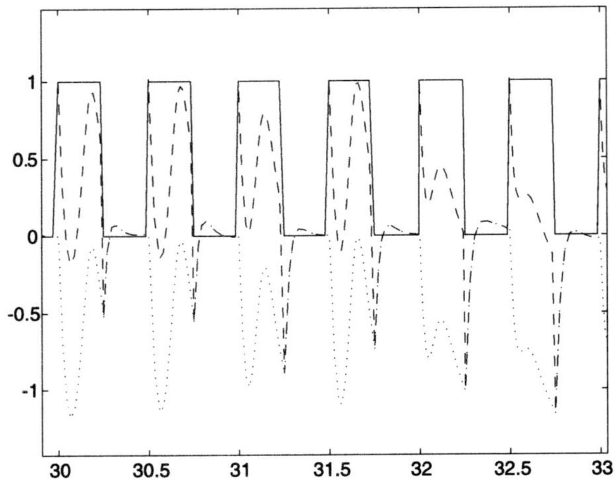

Fig. 2 Transmitted information $m(t)$—*solid line*; estimated extracted information $\hat{m}(t)$—*dashed line*; synchronization error $\hat{x}_1 - x_1$—*dotted line*

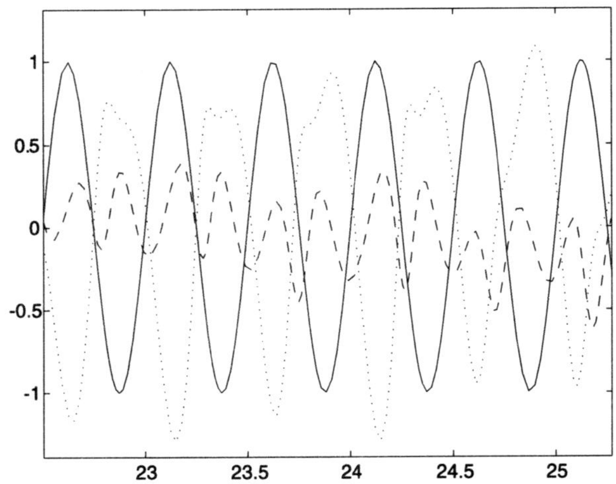

Fig. 3 Transmitted information $m(t)$—*solid line*; estimated extracted information $\hat{m}(t)$—*dashed line*; synchronization error $\hat{x}_1 - x_1$—*dotted line*

note that synchronization errors go to zero asymptotically only during the intervals where $m = 0$ (absence of information. For the transmission of sinusoids the results are clearly worse. In this simulation experiment we used $l_1 = 40$, $l_2 = 110$ and $l_3 = -5$ to comply with the stability condition that $\alpha > 0$.

4 Modified Lorenz-Based Observer

Next, we explore a slightly modified scheme which consists in injecting the information to the first dynamic equation of the Lorenz transmitter (4a)–(4c). As we shall see next, this slight modification leads to a considerable performance improvement. The transmission system is depicted in Fig. 4. The transmitter system is given by the equation

$$\dot{x} = A(y^*)x - L_e m \tag{19}$$

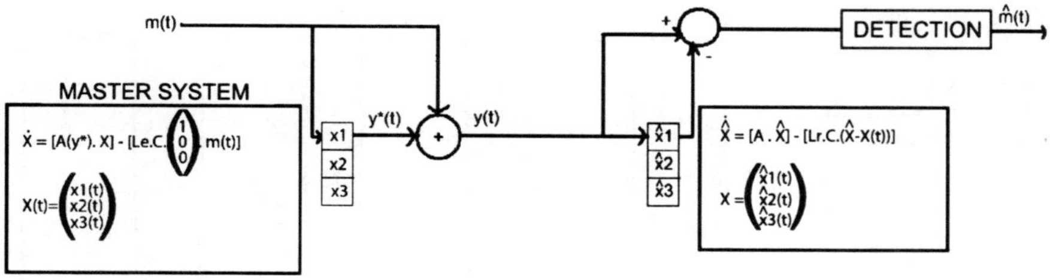

Fig. 4 Implementation scheme of modified Lorenz-based transmission system

where L_e is a design parameter to be defined. The receiver is given by (6) with $L = L_r$ to be defined as well. By a direct computation we obtain

$$\dot{\bar{x}} = A(y)\hat{x} - L_r C\big(\hat{x} - x - C^\top m\big) - A(y^*)x + L_e m$$

hence, by setting $L_r = -L_e$ for any choice of L_e we have

$$\dot{\bar{x}} = A(y)\hat{x} - L_r C\,\big(\hat{x} - x\big) - A(y^*)x. \tag{20}$$

Using $\hat{x} = \bar{x} + x$ and Identity (7) in the previous expression we obtain

$$\dot{\bar{x}} = [A(y) - L_r C]\bar{x} + [A(y) - A(y^*)]x$$

hence, defining $\bar{A}(t, y) := [A(y) - L_r C]$,

$$\dot{\bar{x}} = \bar{A}(t, y)\bar{x} + A'(m(t))x(t). \tag{21}$$

In the previous equation we see that the last term depends only on external *bounded* signals. Opposite to the previous approach the "perturbation" term does not depend on the design parameter L—see the second term in the right-hand side of (10). This allows to achieve arbitrarily perfect synchronization and therefore, arbitrarily perfect recovery of the masked information. To see this let us recall the Lyapunov function V defined in (12). Its time derivative along the trajectories of the error system (21) yields

$$\dot{V} = \frac{1}{2}\big(\bar{x}^\top[\bar{A} + \bar{A}^\top]\bar{x} + 2\bar{x}^\top A'(m)x\big)$$

which, following the computations from p. 360, satisfies (14) with α as in (15) and β redefined as

$$\beta := \sup_{t\geq 0}(|x_3(t)|) + \sup_{t\geq 0}(|x_2(t)|).$$

We conclude that the estimate (18) continues to hold which implies that

$$|x(t)| \leq |\bar{x}(t_\circ)|\mathrm{e}^{-0.5\alpha(t-t_\circ)} + \frac{\beta\bar{m}}{\sqrt{2}\alpha}.$$

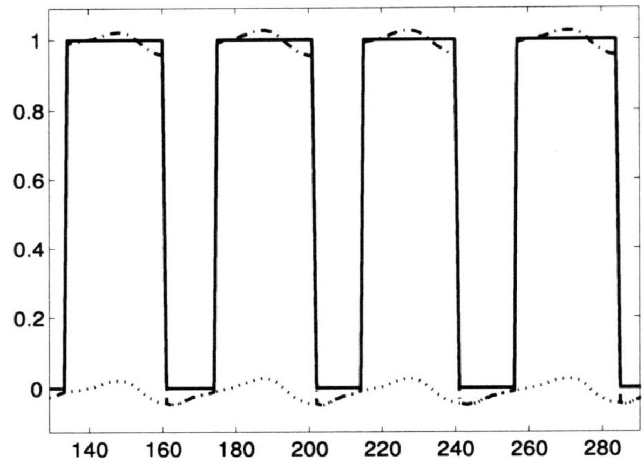

Fig. 5 Modified Lorenz-based observer: transmitted information—train pulse of frequency 3 kHz

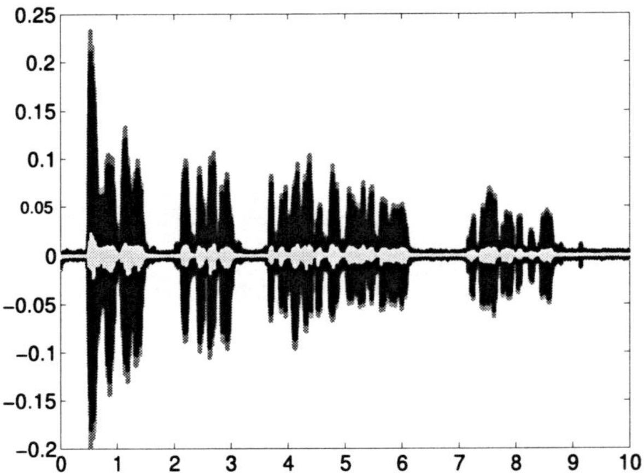

Fig. 6 Modified Lorenz-based observer with transmitted information—voice mail; synchronization error (*light grey*), estimated information (*grey*); transmitted information signal (*black*)

Since β does not depend on L_r by enlarging the latter we can enlarge α and therefore reduce the residual $\frac{\beta \bar{m}}{\sqrt{2}\alpha}$ i.e., synchronization can be arbitrarily exact and the information may be recovered with arbitrary accuracy.

For illustration we have performed a number of numeric simulations using SIMULINK™ of MATLAB™. For simplicity, we have considered $\sup_{t \geq 0}(|x_3(t)|) = 100$ hence $\beta = 200$ and we look for a value of α which makes the ratio (synchronization error)/(information amplitude) smaller than one, hence $\bar{x}/\bar{m} < 1$. To that end we need to pick $\alpha > 71$. We select the gains as $l_{r1} = l_{r2} = l_{r3} = 1000$ which lead to $\alpha = 235.6$. The performance of the transmission system is considerably improved as it may be appreciated from Figs. 5 and 6. As before, the binary signal is of frequency 3 kHz, the Lorenz spectrum ranges from 0 to 5 kHz which ensures an appropriate masking of the information (plots not shown here). Figure 5 presents simulation results for the transmission of voice; in this case, the Lorenz system is accelerated to cover higher natural voice frequencies (up to 20 kHz).

5 Conclusions

We presented two approaches for secured transmission of information using a chaotic Lorenz systems and classical observer theory for linear systems with focus on a robustness analysis vis a vis of the information amplitude. Future research is aimed at studying robustness with respect to transmission delays, parametric uncertainty, added measurement noise, etc.

References

1. L.M. Pecora, T.L. Carroll, Phys. Rev. A **64**, 821 (1990)
2. K.M. Cuomo, A.V. Oppenheim, S.H. Strogatz, IEEE Trans. Circ. Syst. I, Fundam. Theory Appl. **40**, 626 (1993)
3. Special issue on chaos synchronization, control, and applications. IEEE Trans. Circ. Syst. I, Fundam. Theory Appl. (1997)
4. G. Kolumban, M.P. Kennedy, L.O. Chua, IEEE Trans. Circ. Syst. I, Fundam. Theory Appl. **45**, 1129 (1998)
5. M. Boutayeb, M. Darouach, H. Rafaralahy, IEEE Trans. Circ. Syst. I, Fundam. Theory Appl. **49**, 345 (2002)
6. A.L. Fradkov, H. Nijmeijer, A. Markov, Int. J. Bifurc. Chaos **10**, 2807 (2000)
7. H. Nijmeijer, I. Mareels, IEEE Trans. Circ. Syst. I, Fundam. Theory Appl. **44**, 882 (1997)
8. H. Khalil, *Nonlinear Systems*, 2nd edn. (Macmillan, New York, 1996)

A Boundary Layer Problem in Power Law Fluids through a Moving Flat Plate

Chunqing Lu

Abstract The paper studies the bifurcation solutions for laminar flows governed by a third order nonlinear similarity equation $(|f''(\eta)|^{N-1}f''(\eta))' + f(\eta)f''(\eta) = 0$.

Keywords Bifurcation · Non-Newton flows · Shooting on an infinite interval

1 Introduction

The boundary layer on the moving porous flat plate with suction or injection of incompressible viscous fluid is governed by the nonlinear third order ordinary differential equation

$$(|f''(\eta)|^{N-1}f''(\eta))' + f(\eta)f''(\eta) = 0, \tag{1}$$

subject to boundary conditions

$$f(0) = -C, \qquad f'(0) = \xi, \qquad f'(+\infty) = 1. \tag{2}$$

Here η is the similarity variable, $f(\eta)$ denotes the dimensionless stream function, ξ is the velocity ratio between the plate and the fluid, and C is a constant related to suction if it is negative or injection if positive. If $\xi = 0$, then the flat plate does not move. The case $\xi > 0$ implies that the plate and fluid move in the same direction. When $\xi < 0$, they move in opposite directions. The number N comes from the assumption that the shear stress is a power of the velocity on the vertical direction.

The boundary layer flows have many applications in glacial advance, transport of coal slurries down conveyor belts, and several other geophysical, meteorology, oceanography, and industrial contexts [6, 9]. The case $N = 1$ and $\xi = C = 0$ is the Blasius problem. The first rigorous analysis of the Blasius problem was provided by Weyl [8] in 1942. Since then, many researches studied the model with different

C. Lu (✉)
Department of Mathematics and Statistics, Southern Illinois University, Edwardsville, USA
e-mail: clu@siue.edu

J.A.T. Machado et al. (eds.), *Nonlinear Science and Complexity*,
DOI 10.1007/978-90-481-9884-9_42,

values of ξ and C. Among them, Callegari and Friedman [1], Hussaini et al. [2, 3], and Soeweono et al. [7] considered the case for Newtonian fluids $N = 1$. For the non-Newtonian fluids $N < 1$, Nachman and Callegari [6] studied the case $\xi = 0$ and $C = 0$ and Zheng et al. [9] the case $\xi < 0$, with $C > 0$ for $0 < N < 1$ and $N > \frac{1}{2}$ if $C < 0$.

This paper continues the works of Lu [5] and Lu and Zheng [4]. It presents a proof of the bifurcation of solutions to the boundary value Problem (1)–(2) for $N \in (0, 1)$ and some $\xi < 0$ and $C > 0$. It seems that the power law Non-Newtonian and Newtonian flows have similar bifurcation behavior as the plate and the flow move in opposite directions, but their asymptotic behavior, as $\eta \to \infty$, are different. The rigorous proof of this result for the case $0 < N < 1$ by the shooting on the infinite interval and the explicit sufficient conditions for the bifurcation solutions have not been seen elsewhere. From the physics point of view, the two theorems in the paper show that for some C and ξ there may exist two different laminar flows depending on the $f''(0)$ while for some other values of C and ξ the laminar flow may not exist. This paper mainly studies the injection case $C > 0$. The suction case $C < 0$ for $N < 1$ will be reported in another paper.

The main results of the paper are the following theorems.

Theorem 1 *There exists at least two solutions of the boundary value problem* (1)–(2) *for some* $C \geq 0$ *and for some* $\xi < 0$ *given in Lemma* 12.

Theorem 2 *If* $\sqrt{C} > 2(1 + |\xi|)$ *and* $|C\xi| > 1$, *then the solution of the boundary value problem* (1)–(2) *does not exist.*

2 Proofs of Two Theorems

Consider the initial value problem

$$([f''(\eta)]^N)' + f(\eta)f''(\eta) = 0 \tag{3}$$

with

$$f(0) = -C, \qquad f'(0) = \xi, \qquad f''(0) = \alpha \tag{4}$$

where $\alpha > 0$ is a parameter. Our goal is to find an appropriate value of α depending on given values of C, N and ξ such that the corresponding solution $f(\eta)$ of (3)–(4) satisfies $f'(\infty) = \lim_{\eta\to\infty} f'(\eta) = 1$. In what follows, we always let C, N and ξ be fixed. It can be shown below that $f'' > 0$ and hence (1) is equivalent to (3).

Rewrite (3) as

$$f''' = -\frac{1}{N} f(f'')^{2-N}. \tag{5}$$

Multiply the integrating factor $e^{\int f(f'')^{1-N} ds}$ on (5), one gets

$$f''(\eta) = \alpha e^{-\frac{1}{N}\int_0^\eta f(f'')^{1-N} ds}. \tag{6}$$

This shows that $f'' > 0$ and the solution is well defined on the semi infinite interval $[0, \infty)$.

Lemma 1 *For given* $\xi \le 0$, $N \in (0, 1]$, $\alpha \ge 0$, *and* $C \in (-\infty, \infty)$, *if there exists a zero point of* $f'(\eta)$ *where* $f(\eta)$ *is the solution of the initial value problem* (3)–(4), *then* $\lim_{\eta\to\infty} f''(\eta; \alpha, \xi, C, N) = 0$.

Proof Assume that $\alpha > 0$ is given and there is a point $\eta = \eta_1 > 0$ with $f'(\eta_1) = 0$. It can be proved that there is a point $\eta_2 \ge \eta_1$ such that $f, f', f'' > 0$ for all $\eta > \eta_2$, the point at which $f \ge 0$ and $f' > 0$. From (5), $f''' \le 0$ for $\eta > \eta_2$ and hence f'' decreases as η increases for $\eta > \eta_2$. This shows that $\lim_{\eta\to\infty} f''(\eta; \alpha, \xi, C, N) = \beta$ exists and furthermore we can prove $\lim_{\eta\to\infty} f''(\eta; \alpha, \xi, C, N) = 0$.

Moreover, we can study the asymptotic behavior of f'' as $\eta \to \infty$. Without loss of generality, we can assume that $f'(\eta_2) = \xi_1 > 0$ because $f'' > 0$. Integrating $(f'')^{N-2} f''' = \frac{-1}{N} f$ with respect to the independent variable from η_2 to η, one obtains for $N < 1$

$$[f''(\eta)]^{N-1} = [f''(\eta_2)]^{N-1} + \frac{1-N}{N} \int_{\eta_2}^{\eta} f(s)ds. \tag{7}$$

Since $f'(\eta) \ge \xi_1$ and $f(\eta) \ge \xi_1(\eta - \eta_2)$ for $\eta \ge \eta_2$, we see that $\int_{\eta_2}^{\eta} f(s)ds \ge \frac{1}{2}\xi_1(\eta - \eta_2)^2$. It then follows from (7)

$$[f''(\eta)]^{1-N} \le \frac{2N}{\xi_1(1-N)}(\eta - \eta_2)^{-2} \le \frac{4N}{\xi_1(1-N)\eta^2} = \frac{A^{1-N}}{\eta^2}, \tag{8}$$

where $A = [\frac{4N}{\xi_1(1-N)}]^{\frac{1}{1-N}}$, provided $\eta > \eta_3 = 4\eta_2$. This shows that $f'' \le \frac{A}{\eta^{\frac{2}{1-N}}}$ for sufficiently large η. And $f'(\eta) \le \frac{1-N}{1+N} A(-\eta^{-\frac{1+N}{1-N}} + \eta_3^{-\frac{1+N}{1-N}}) + f'(\eta_3)$.

If $N = 1$, then $f, f', f'' > 0$ for all $\eta \ge \eta_2$ and $f'(\eta_2) = \xi_1 > 0$. Thus $f(\eta) \ge \xi_1(\eta - \eta_2)$, $\int_{\eta_2}^{\eta} f(s)ds \ge \frac{1}{2}\xi_1(\eta - \eta_2)^2$ for $\eta \ge \eta_2$,

$$f'' = f''(\eta_2)e^{-\int_{\eta_2}^{\eta} f ds} \le f''(\eta_2)e^{-\frac{1}{2}\xi_1(\eta-\eta_2)^2}, \tag{9}$$

which implies that $f''(\eta)$ approaches zero exponentially. □

From the asymptotic behavior of the solutions given by Lemma 1 one immediately gets the following lemma.

Lemma 2 *For given* $\xi \le 0$, $N \in (0, 1]$, $\alpha \ge 0$, *and* $C \in (-\infty, \infty)$, $\lim_{\eta\to\infty} f'(\eta; \alpha, \xi, C, N)$ *exists.*

By Lemma 1, we can define a function $F(\alpha) = F(C, N, \xi, \alpha)$ by $F(\alpha) = \lim_{\eta\to\infty} f'(\eta, C, \xi, \alpha)$. It is seen that $F(0) = \xi$ and that $F(C, N, \xi, \alpha)$ is well defined for all $\alpha \in [0, \infty)$ for any $C \in (-\infty, \infty)$ and $\xi \in (-\infty, 0]$.

Define a set $A = \{\alpha | f'(\eta, \alpha) = 0 \text{ for some } \eta \ge 0\}$. Then we see the set A is an open subset of $R^+ = (0, \infty)$. In the case $C \ge 0$, the set $A = (0, \infty)$. To see this, we begin with a solution f with $f(0) = -C \leq 0$, $f'(0) = \xi < 0$ and $f''(0) = \alpha > 0$. Then, $f < 0$ and $f' < 0$ for sufficiently small η, which implies $f''' > 0$. From the equation, we see that $f''' > 0$ as long $f' < 0$ in this case. Since $f'' > 0$, $f''' > 0$ if $f' < 0$, it follows that there must be a value of η at which $f' = 0$.

Lemma 3 *For given $\xi < 0$, $N \in (0, 1]$ and for any C, $F(\alpha)$ is a continuous function on the set A.*

Proof Suppose that for an $\alpha > 0$ the solution $f(\eta)$ of the initial value problem has a zero point. From the proof of Lemma 1, we use $f'(\eta) \leq \frac{1-N}{1+N} A(-\eta^{-\frac{1+N}{1-N}} + \eta_3^{-\frac{1+N}{1-N}}) + f'(\eta_3)$ for $\eta \geq \eta_3$ because the constant A continuously depends on α. On the interval $[0, \eta_3]$ we use the theorem that the solution continuously depends on the initial values. □

From this lemma, we see that in the case $C \geq 0$, $F(\alpha)$ is continuous for $\alpha \in (0, \infty)$.

Remark 1 The proof of continuity of $F(\alpha)$ for $N = 1$ in my previous work [5] should be given as above. If $C > 0$ and $\xi < 0$, the proof of continuity may not be as simple as in some previous works, [7, 9] since the continuity theorem of solutions depending initial values must be applied on a compact set. It should also be noticed that F at $\alpha = 0$ for $\xi < 0$ is discontinuous.

The next lemma shows that there exists at least a value α_0 of α such that $F(\alpha_0) > 1$.

Lemma 4 *For given $\xi < 0$, $N \in (0, 1]$, and $C \geq 0$, there exists at least a real number α_0 depending on ξ and C such that $F(\alpha) \geq 1$ for all $\alpha > \alpha_0$.*

Proof See Lu [5]. □

Lemma 5 *As $\alpha \to 0^+$, there exists at least one α such that $F(\alpha) > 1$ for any $C \geq 0$ and for any $\xi < 0$.*

Proof We can prove that as $\alpha \to 0$, $f''(\eta_1)$ becomes unbounded with the Crocco transformation, where η_1 is the zero point of f' similar to the proof in [7]. The rest proof is similar to the argument in my paper [5]. The case $C < 0$ can be proved similarly. □

Lemma 6 *Let $C > 0$, $\xi < 0$. There exists an α such that $F(\alpha) < 1$, if $N < 1$ and*

$$|\xi| \leq \frac{1}{2^{\frac{3-2N}{1-N}}(1 + \sqrt{1 + (k-1)^2})}, \tag{10}$$

$$|\xi|^{1-2N} < \frac{[k(k-2)]^N}{2^N} \frac{N}{k(1-N)} C^{-(1+N)}, \tag{11}$$

and

$$\frac{2^{\frac{1}{(1-N)^2}+\frac{1+N}{1-N}}}{(1+N)(1-N)^{\frac{N}{1-N}}}\frac{(1+\gamma)^{\frac{1+N}{1-N}}}{\gamma^{\frac{1}{1-N}}(k-1)^{\frac{2(1+N)}{1-N}}}C^{-\frac{1+N}{1-N}}|\xi|^{\frac{N}{1-N}}<\frac{1}{2}, \tag{12}$$

or $N=1$ *with* $(k+\frac{k\sqrt{k-2}}{2\sqrt{\gamma_0}}\sqrt{\pi})e|\xi|<1$ *and* $\frac{1}{2}\le\frac{2C^2}{(k-2)|\xi|}\le 1$, *where* $k\ge 3$.

Proof Let $C>0$. There must be a point η_1 such that $f'(\eta_1)=0$ and $f(\eta_1)<0$. Then, f becomes an increasing function with positive concavity as long as $f\le 0$. It turns out that there must be a point $\eta_0>\eta_1$ where f becomes zero. Noting that $f'''(\eta_0)=0$ and $f'''>0$ for $\eta<\eta_0$ and $f'''>0$ for $\eta>\eta_0$, we see that $f''(\eta_0)$ is the maximum of f'', i.e., $f''(\eta_0)>\alpha$.

Integrating the inequality $f''>\alpha$ once with respect to η, one gets $f'(\eta)>\alpha\eta-|\xi|$ as long as $f\le 0$, from which $\eta_1<\frac{|\xi|}{\alpha}$. Further integration of last inequality shows that

$$f(\eta)>\frac{\alpha}{2}\eta^2-|\xi|\eta-C. \tag{13}$$

Denote

$$g(\eta)=\frac{\alpha}{2}\eta^2-|\xi|\eta-C. \tag{14}$$

It is seen that f becomes zero before g does. The real zero point of g is $t_0=\frac{|\xi|+\sqrt{\xi^2+2\alpha C}}{\alpha}$. Thus, $\eta_0\le t_0$. Let $k=1+\sqrt{1+\frac{2\alpha C}{\xi^2}}$. Then, $\alpha=\frac{k(k-2)\xi^2}{2C}$, $t_0=\frac{k|\xi|}{\alpha}$ and $\eta_0<t_0$ where η_0 is the zero point of f. Integrating (13) with respect to η from 0 to η_0, we get

$$\int_0^{\eta_0}f(\eta)d\eta\ge\int_0^{t_0}g(\eta)d\eta=\frac{k^3|\xi|^3}{6\alpha^2}-\frac{k^2|\xi|^3}{2\alpha^2}-\frac{C|\xi|k}{\alpha}. \tag{15}$$

To make the estimate simpler, we choose $k\ge 3$ so that

$$\int_0^{\eta_0}fd\eta>-\frac{kC|\xi|}{\alpha}. \tag{16}$$

By (7), we find that if $N<1$,

$$f''(\eta_0)^{N-1}>\alpha^{N-1}-\frac{k(1-N)}{\alpha N}C|\xi|. \tag{17}$$

If we set a sufficient condition on C and ξ such that

$$\alpha^{N-1}-\frac{k(1-N)}{\alpha N}C|\xi|\ge\frac{1}{2\alpha^{1-N}} \tag{18}$$

which gives $f''(\eta_0)^{N-1}\ge\frac{1}{2\alpha^{1-N}}$. Note that $k\le\frac{\alpha^N}{2(1-N)C|\xi|}$. Then, the inequality (18) implies $k(1-N)C|\xi|<\frac{N\alpha^N}{2}\le N(\frac{k(k-2)\xi^2}{2C})^N$. This requires the first sufficient condition (11) on C and ξ, for $N<1$. We proceed the proof with (17) and (18). At this moment, we have $f''(\eta_0)<2^{\frac{1}{1-N}}\alpha$. Since $f''\le f''(\eta_0)$ for all $\eta>\eta_0$, it follows that $f''<2^{\frac{1}{1-N}}\alpha$ for all $\eta>\eta_0$. Thus, for all $\eta>0$,

$$f'<2^{\frac{1}{1-N}}\alpha\eta-|\xi|, \tag{19}$$

and $f(\eta) < 2^{\frac{N}{1-N}}\alpha\eta^2 - |\xi|\,\eta - C$.

Next, we find a lower bound for η_0, the zero point of f. Denote

$$h(\eta) = 2^{\frac{N}{1-N}}\alpha\eta^2 - |\xi|\,\eta - C.$$

Then we see that $g(\eta) \le f(\eta)$ and $g' \le f'$ hold only if $f \le 0$, but $f'(\eta) \le h'(\eta)$ and $f \le h$ for all $\eta \ge 0$. We observe that $f(\eta)$ becomes zero after $h(\eta)$ does. Let $t_1 = \frac{|\xi|+\sqrt{\xi^2+2^{\frac{1}{1-N}}\alpha C}}{2^{\frac{1}{1-N}}\alpha}$. Then, $h(t_1) = 0$, $\eta_0 > t_1$, $f > f'(\eta_0)(\eta - \eta_0) > f'(t_1)(\eta - \eta_0)$ because $f'' > 0$. Now, we have $t_1 \le \eta_0 \le t_0$, i.e., $\frac{|\xi|+\sqrt{\xi^2+2^{\frac{1}{1-N}}\alpha C}}{2^{\frac{1}{1-N}}\alpha} \le \eta_0 \le \frac{|\xi|+\sqrt{\xi^2+2\alpha C}}{\alpha}$. Noting that $g'(t_1) = |\xi|(\frac{1+\sqrt{1+2^{\frac{N}{1-N}}k(k-2)}}{2^{\frac{1}{1-N}}} - 1) = \gamma|\xi|$, where $\gamma = \frac{1+\sqrt{1+2^{\frac{N}{1-N}}k(k-2)}}{2^{\frac{1}{1-N}}} - 1 > 0$ for $k \ge 3$. Here, we set $t_1 = \frac{1+\gamma}{\alpha}|\xi|$. Since $f'(t_1) > g'(t_1)$ where the function g is defined by (14), we may choose α so that $f'(t_1) \ge \gamma|\xi|$ where $\gamma > 0$ is a constant. It follows that $f(\eta) > \gamma|\xi|(\eta - \eta_0)$ for all $\eta > \eta_0$. Again, by (7),

$$(f'')^{N-1} > f''(\eta_0)^{N-1} + \frac{1-N}{2N}\gamma\,|\xi|\,(\eta - \eta_0)^2$$

for all $\eta > \eta_0$. Hence,

$$f''(\eta) < \left[\frac{2N}{\gamma|\xi|(1-N)}\right]^{\frac{1}{1-N}} (\eta - \eta_0)^{-\frac{2}{1-N}} \tag{20}$$

for all $\eta > \eta_0$. From (19), $f'(2\eta_0) < 2^{\frac{2-N}{1-N}}\alpha\eta_0 - |\xi|$. Integrating (20) with respect to the independent variable from $2\eta_0$ to η and letting $\eta \to \infty$ yields

$$F(\alpha) < f'(2\eta_0) + \frac{1-N}{1+N}\left[\frac{2^{\frac{1}{1-N}}N}{\gamma|\xi|(1-N)}\right]^{\frac{1}{1-N}} \frac{1}{\eta_0^{\frac{1+N}{1-N}}}. \tag{21}$$

Noting $t_1 < \eta_0 < t_0 = \frac{|\xi|(1+\sqrt{1+(k-1)^2})}{\alpha}$, we get $f'(2\eta_0) < f'(2t_0) \le h'(2t_0) = 2^{\frac{2-N}{1-N}}\alpha\frac{|\xi|(1+\sqrt{1+(k-1)^2})}{\alpha} - |\xi|$,

$$F(\alpha) < 2^{\frac{2-N}{1-N}}(1 + \sqrt{1+(k-1)^2})|\xi| \\ + \frac{2^{\frac{1}{(1-N)^2}+\frac{1+N}{1-N}}}{(1+N)(1-N)^{\frac{N}{1-N}}} \frac{(1+\gamma)^{\frac{1+N}{1-N}}}{\gamma^{\frac{1}{1-N}}(k-1)^{\frac{2(1+N)}{1-N}}} C^{-\frac{1+N}{1-N}}|\xi|^{\frac{N}{1-N}}.$$

In order to have $F(\alpha) < 1$, we first choose a $k \ge 3$ large enough to make $\gamma > 0$. Then the conditions (10) and (12) imply $F(\alpha) < 1$. Those two inequalities together with (11) give a sufficient condition on C and ξ to guarantee $F(\alpha) < 1$. If $2N \le 1$, we see that for any given $C > 0$ there exists a $\xi_c < 0$, such that for $0 > \xi > \xi_c$, $F(\alpha, C, \xi) < 1$ for an α given above. Also, it can be observed that the larger C the smaller $|\xi_c|$. But, for $2N \geq 1$, the choice of C is not arbitrary any more because the inequality (11) may fail. But, there are still some C, ξ satisfy the inequalities, (10), (11) and (12). In other words, these three inequalities are feasible. The case $N = 1$ and $C > 0$ can be studied similarly. □

2.1 *Proof of Theorem 1*

From above lemmas, we see that for ξ, C, N given in Lemma 12 the function $F(\alpha)$ is continuous on $(0, \infty)$ and it maps the interval $(0, \infty)$ into an interval containing $[\delta, \gamma]$ where $\delta < 1$ and $\gamma > 1$. Since continuous functions preserve the connectedness, the rage of $F(\alpha) > 0$ must be an interval which has a positive minimum point. By Lemmas 4 and 5, we see that for those ξ, N, and C in Lemma 12, there exists at least two values $\alpha_1 < \alpha_2$ of α such that $F(\alpha_i) = 1$, for $i = 1, 2$. This proves the nonuniqueness of solutions to the boundary value problem.

2.2 *Proof of Theorem 2*

Denote the set $\{(C, \xi) | \sqrt{C} > 1 + |\xi| \text{ and } |C\xi| > 1\} = S_1$. It is sufficient to prove that $F(\alpha) > 1$ for all values of α and for $C > 0$ and $\xi > 0$ with (C, ξ) in S_1. If $\alpha \geq 1$, we consider a point $\eta_0 = \sqrt{\frac{C}{\alpha}}$. If $f(\eta_0) \geq 0$, then, by the mean value theorem, $f'(\eta_1) = \frac{f(\eta_0) - f(0)}{\eta_0} \geq \frac{0 - (-C)}{\eta_0} = \frac{C}{\sqrt{\frac{C}{\alpha}}} = \sqrt{\alpha C} > \sqrt{C}$. If $f(\eta_0) < 0$, then $f' > \alpha\eta - \xi$ for $f''' > 0$. Thus, $f'(\eta_0) > \alpha\sqrt{\frac{C}{\alpha}} + \xi = \sqrt{\alpha C} + \xi > \sqrt{C} + \xi$. In either case, $f'(\infty) > 1$ provided $\sqrt{C} > 1 + |\xi|$. The case $\alpha < 1$, the proof is similar to Lu [5]. The proof is complete.

References

1. A.J. Callegrari, M.B. Friedman, An analytical solution of a nonlinear, singular boundary value problem in the theory of viscous fluids. J. Math. Anal. Appl. **21**, 510–529 (1968)
2. M.Y. Hussaini, W.D. Lakin, Existence and Non-uniqueness of similarity solutions of a boundary-layer problem. Q. J. Mech. Appl. Math. **39**, 17–24 (1986)
3. M.Y. Hussaini, W.D. Lakin, A. Nachman, On similarity solutions of a boundary layer problem with an upstream moving wall. SIAM J. Appl. Math. **47**(4), 699–709 (1987)
4. C. Lu, L.C. Zheng, Similarity solutions of a boundary layer problem in power law fluids through a moving flat plate. Int. J. Pure Appl. Math. **13**(2), 143–166 (2004)
5. C. Lu, Multiple solutions for a boundary layer problem. Commun. Nonlinear Sci. Numer. Simul. **12**, 725–734 (2007)
6. A. Nachman, A. Callegari, A nonlinear singular boundary value problem in the theory of pseudoplastic fluids. SIAM J. Appl. Math. **38**(2), 275–281 (1980)

7. E. Soewono, K. Vajravelu, R.N. Mohopatra, Existence and non-uniqueness of solutions of a singular nonlinear boundary-layer problem. J. Math. Anal. Appl. **159**, 251–270 (1991)
8. H. Weyl, On the differential equations of the simplest boundary-layer problems. Ann. Math. **43**(2), 385–407 (1942)
9. L.C. Zheng, J.C. He, Existence and non-uniqueness of positive solutions to a non-linear boundary value problem in the theory of viscous fluids. Dyn. Syst. Appl. **8**, 133–145 (1999)

An Overview of the Behaviour of a Scattering Map for the Dynamics of Two Interacting Particles in a Uniform Magnetic Field

D. Pinheiro and R.S. MacKay

Abstract The interaction of two charges moving in $\mathbb{R}^3$ in a magnetic field $\boldsymbol{B}$ can be formulated as a Hamiltonian system with six degrees of freedom. A scattering map is defined for trajectories which come from and go to infinite separation along the field direction. It determines the asymptotic parallel velocities, guiding centre field lines, magnetic moments and gyrophases for large positive time from those for large negative time. In regimes where gyrophase averaging is appropriate, the scattering map has a simple form, conserving the magnetic moments and parallel kinetic energies (in a frame moving along the field with the centre of mass) and rotating or translating the guiding centre field lines. When the gyrofrequencies are in low order resonance, however, gyrophase averaging is not justified and transfer of perpendicular kinetic energy occurs. In the extreme case of equal gyrofrequencies there is typically also transfer between perpendicular and parallel kinetic energy.

Keywords Hamiltonian dynamics · Scattering map

1 Introduction

In [2] we analyse the interaction of two charged particles moving in three-dimensional space under the action of a uniform magnetic field and an interaction potential depending only on the distance between the particles. This problem is important for plasma physics and for atomic physics in magnetic fields. Apart from [3], where the separation of the centre of mass is treated in the quantum-mechanical setting, attention has tended to focus on some limiting regimes such as very strong

D. Pinheiro (✉)
CEMAPRE, ISEG, Universidade Técnica de Lisboa, Lisboa, Portugal
e-mail: dpinheiro@iseg.utl.pt

R.S. MacKay
Mathematics Institute, University of Warwick, Coventry CV4 7AL, UK
e-mail: R.S.MacKay@warwick.ac.uk

J.A.T. Machado et al. (eds.), *Nonlinear Science and Complexity*,
DOI 10.1007/978-90-481-9884-9_43,

magnetic field or plasmas with all the particles of the same kind (see [4–7]) or with one heavy particle idealized as fixed (the diamagnetic Kepler Problem, see [8–10]) or the case with charges summing to zero (see [11, 12]). The problem of Coulomb scattering of a charged particle by a fixed charge or two identical particles in a very strong magnetic field is treated in [13] in the quantum-mechanical setting. We study the dynamics of two charged particles in a uniform magnetic field without making restrictions on the sizes of the magnetic field, the charges or the masses, except that we assume that the particles behave classically and that their velocities and accelerations are small enough that we can neglect any relativistic and radiation effects. Although it is well known that non-uniformity of the magnetic field introduces further significant effects, we believe that there is value in establishing firm results for the uniform case first, which seems not yet to have been done in detail.

2 The Planar Problem

In [1] we made a detailed study of the problem of the interaction of two charged particles moving in a plane under the effect of a uniform magnetic field. We assumed that the interaction between the particles was given by a potential depending on the distance between the two particles and that the magnetic field was orthogonal to the plane of motion. That problem can be formulated as a Hamiltonian system with four degrees of freedom. We made extensive use of the symmetries in that Hamiltonian system to obtain a reduction in the dimension of the problem to two degrees of freedom. In the special case of same sign charges with equal gyrofrequencies (equal ratio of charge to mass) or on some special submanifolds we proved that this system is integrable.

We then specialized our analysis to the most physically interesting case of a Coulomb-like potential. Analysing the reduced systems and the associated reconstruction maps we provided a detailed description for the regimes of parameters and level sets of the conserved quantities where bounded and unbounded motion are possible and we identified the cases where close approaches between the two particles are possible. Furthermore, we identified regimes where the system is non-integrable and contains chaos by proving the existence of invariant subsets containing a suspension of a non-trivial subshift.

3 The Spatial Problem

In [2] we look at the same problem but with the particles now moving in $\mathbb{R}^3$. This system can be formulated as a Hamiltonian system with six degrees of freedom. Assuming that the magnetic field is uniform and the interaction potential has rotational symmetry we reduce this Hamiltonian system to one with three degrees of freedom; for certain values of the conserved quantities or choices of parameters, we obtain a system with two degrees of freedom. Furthermore, it contains the planar case as a subsystem. The reductions simplify the analysis of some properties of this system: we use the reconstruction map to obtain a classification for the dynamics in terms of boundedness of the motion and the existence of collisions.

We achieve these results by constructing a set of coordinates in which the system exhibits a reduction to three degrees of freedom, and two degrees of freedom when it applies. This reduction is an extension to three-dimensional space of similar reductions obtained for the planar case in [1] (and for a similar problem in [14]). In [2] the total change of coordinates that exhibits the reduction is computed. This change of coordinates is just the lift of a $SE(3)$ subgroup that, given the base dynamics of the reduced Hamiltonian systems, enables us to describe the full twelve-dimensional dynamics. The planar case is obtained as an invariant plane of the three-dimensional problem.

4 The Scattering Map for the Spatial Problem

The motion of one particle moving in three-dimensional space under the action of a uniform magnetic field is simple. It is the composition of two motions: a drift with constant velocity in the direction of the magnetic field and a uniform rotation in a plane orthogonal to the field about a fixed centre—the guiding centre, with constant radius—gyroradius, and angular velocity—gyrofrequency. Choosing the magnetic field to be vertical and oriented upwards, the motion in the circle is clockwise if the charge is positive and anticlockwise otherwise. We sign the gyrofrequency according to the direction of rotation. This problem can be formulated as a three degrees of freedom Hamiltonian system. It has symmetry under a four-dimensional subgroup of the Special Euclidean group of $\mathbb{R}^3$ (three-dimensional translations and a one-dimensional rotation). These symmetries induce conserved quantities for this system which is easily seen to be integrable.

One of our main goals in [2] is to study the scattering problem associated with the interaction of the two charges in the presence of a magnetic field and a Coulomb interaction potential:

$$V(R) = \frac{e_1 e_2}{4\pi \epsilon_0} \frac{1}{R},$$

where R denotes the distance between the two particles, e_1 and e_2 denote the values of the charges and ϵ_0 denotes the permittivity of the vacuum. If there is a large distance between the particles then the interaction is negligible and in this case the two particles move freely as described above. If the distance between the two particles is small then the strength of the interaction can not be neglected anymore and the particles interact. We look at the situation where the particles have initially a large vertical separation and both move freely towards each other so that the particles start interacting when they get closer and then start moving apart until both particles move again like free particles. The goal is to describe the changes in their trajectories due to this interaction. For background on scattering in classical mechanics see the review [15] and references therein.

In [2] we study the scattering map associated with this problem in the limit where the two particles' trajectories are widely separated. We separate our analysis into two cases: rationally independent gyrofrequencies and rationally dependent gyrofrequencies. If the two particles' gyrofrequencies are rationally independent we obtain that the magnetic moments of the particles are adiabatic invariants and under the adiabatic approximation the vertical kinetic energy in the centre of mass frame is unchanged and the guiding centres have the following dynamics:

(i) in the case of two charges whose sum is not zero, the guiding field lines rotate by some angle about a fixed field line during an interaction,
(ii) in the case of two charges which sum to zero, the guiding field lines translate by some amount in a direction determined by the conserved quantities.

If the two particles' gyrofrequencies are rationally dependent, however (or one goes beyond the above adiabatic approximation), some transfer can occur between the horizontal kinetic energies of the two particles; there can also be a weaker exchange between horizontal and vertical kinetic energy. Indeed we prove both such transfers are typically non-zero when the gyrofrequencies are equal, and the first occurs when they sum to zero. If the gyrofrequencies differ in absolute value, we bound any such transfer by the fourth or higher inverse power of the distance between the gyrohelices. Furthermore, we prove that in the case of "bouncing-back" behaviour (see [2] for more details), even if the vertical kinetic energy is conserved in the centre of mass frame, there is a transfer of vertical kinetic energy between the particles when the vertical centre of mass velocity is non-zero. Furthermore, we have made a numerical study of the scattering map without using the assumption that the two particles trajectories are widely separated. We observed regular behaviour for large energies and chaotic scattering for small positive energies.

5 Possible Applications

One application of our results might be to reassess the derivations of kinetic equations for the velocity distribution functions of charged particles in a magnetic field, describing the effects of two-body scattering. The standard derivations (e.g. [16, 17]) appear to ignore the possibility of resonant interaction such as we have found for interaction of particles with equal gyrofrequencies. Even if the resonance effects might be significant only for a small fraction of interactions (those with low order rational ratio of gyrofrequencies and small relative parallel velocity), their net effect might turn out to be larger than the standard answers. By changing the Coulomb interaction to a Debye-shielded version, the Balescu-Lenard version of plasma kinetic theory could also be addressed. Any significant resulting changes to standard plasma kinetic theory could be valuable to understand the scattering of particles into the loss cone in the magnetosphere or that of particles into and out of banana orbits in tokamak fields. In particular, the result could shed light on the generation of toroidal current in tokamaks by such transitions and might contribute to the understanding of anomalous perpendicular electron heat transport.

Acknowledgements D. Pinheiro's research was supported by FCT—Fundação para a Ciência e Tecnologia grant with reference SFRH/BPD/27151/2006 and CMUP—Centro de Matemática da Universidade do Porto.

References

1. D. Pinheiro, R.S. MacKay, Interaction of two charges in a uniform magnetic field: I. Planar problem. Nonlinearity **19**, 1713–1745 (2006)
2. D. Pinheiro, R.S. MacKay, Interaction of two charges in a uniform magnetic field: II. Spatial problem. J. Nonlinear Sci. **18**, 615–666 (2008)
3. J.E. Avron, I.W. Herbst, B. Simon, Separation of center of mass in homogeneous magnetic fields. Ann. Phys. **114**, 431–451 (1978)
4. D.H.E. Dubin, Collisional transport in non-neutral plasmas. Phys. Plasmas **5**, 1688–1694 (1998)
5. F. Anderegg, X.-P. Huang, C.F. Driscoll, E.M. Hollmann, T.M. O'Neil, D.H.E. Dubin, Test particle transport due to long range interactions. Phys. Rev. Lett. **78**, 2128–2131 (1997)
6. C.F. Driscoll, F. Anderegg, D.H.E. Dubin, D.-Z. Jin, J.M. Kriesel, E.M. Hollmann, T.M. O'Neil, Shear reduction of collisional transport: experiments and theory. Phys. Plasmas **9**, 1905–1914 (2002)
7. M. Psimopoulos, D. Li, Cross field thermal transport in highly magnetized plasmas. Proc. R. Soc. Lond. **437**, 55–65 (1992)
8. K. Efstathiou, R.H. Cushman, D.A. Sadovskií, Hamiltonian Hopf bifurcation of the hydrogen atom in crossed fields. Physica D **194**, 250–274 (2004)
9. M.C. Gutzwiller, *Chaos in Classical and Quantum Mechanics* (Springer, New York, 1990)
10. G. Tanner, K.T. Hansen, J. Main, The semiclassical resonance spectrum of hydrogen in a constant magnetic field. Nonlinearity **9**, 1641–1670 (1996)
11. P. Schmelcher, L.S. Cederbaum, Regularity and chaos in the center of mass motion of the hydrogen atom in a magnetic field. Z. Phys. D **24**, 311–323 (1992)
12. P. Schmelcher, L.S. Cederbaum, Intermittent chaos in Hamiltonian systems: the three-dimensional hydrogen atom in magnetic fields. Phys. Rev. A **47**, 2634–2639 (1993)
13. L.M. Tannenwald, Coulomb scattering in a very strong magnetic field. Phys. Rev. **113**, 1396–1405 (1959)
14. C. Grotta Raggazo, J. Koiller, W.M. Oliva, On the motion of two-dimensional vortices with mass. J. Nonlinear Sci. **4**, 375–418 (1994)
15. E. Ott, T. Tél, Chaotic scattering: an introduction. Chaos **3**, 417–426 (1993)
16. R. Balescu, *Transport Processes in Plasma, vol. I* (N. Holland, Amsterdam, 1988)
17. P. Helander, A.J. Sigmar, *Collisional Transport in Magnetized Plasmas* (Cambridge University Press, Cambridge, 2002)

A Generalised Entropy of Curves Approach for the Analysis of Dynamical Systems

Aldo Balestrino, Andrea Caiti, and Emanuele Crisostomi

Abstract This paper provides a new approach for the analysis and eventually the classification of dynamical systems. The objective is obtained by extending the theory of the entropy of plane curves to $\mathbb{R}^n$ space. Properties of a dynamical system are inferred by investigating how a curve connecting a set of initial conditions in the phase space evolves with time, according to its generalised entropy. In particular all linear dynamical systems are characterised by a constant zero entropy, while higher asymptotic values indicate nonlinear behaviours. An algorithmic procedure to evaluate the entropy at each time step is outlined and it proves to be very efficient to describe chaotic systems as well. In this case the generalised entropy is proved to be linked to other conventional indicators known from literature. The entropy based approach is extensively tested for the analysis of several benchmark dynamical systems.

Keywords Entropy of curves · Nonlinear systems · Chaotic systems

1 Introduction

This paper provides a new approach for the analysis and eventually the classification of dynamical systems. Some classification methods are already known from literature and are now a cornerstone of modern systems theory. For instance, it is possible to classify a linear system as stable, asymptotically stable or unstable, on

A. Balestrino (✉) · A. Caiti · E. Crisostomi
Department of Electrical Systems and Automation, University of Pisa, Via Diotisalvi 2, Pisa, Italy
e-mail: Aldo.Balestrino@dsea.unipi.it

A. Caiti
e-mail: Andrea.Caiti@dsea.unipi.it

E. Crisostomi
e-mail: Emanuele.Crisostomi@dsea.unipi.it

J.A.T. Machado et al. (eds.), *Nonlinear Science and Complexity*,
DOI 10.1007/978-90-481-9884-9_44,

the basis of the eigenvalues of its transition matrix. The same approach can be extended to nonlinear systems applying a linearisation procedure with respect to some nominal evolution, although now information can only be inferred about the particular evolution. A second approach based on Lyapunov Exponents is used to classify attractors into equilibria, cycles and chaotic sets [1, 2]. This second approach has the advantage of providing a general view on nonlinear systems, without restrictions to a particular equilibrium, but has the drawback that it can be applied only to dynamical systems that admit attractive sets. Other less conventional criteria to classify dynamical system have been proposed in [3, 4], although they are far from being general and systematic approaches.

This paper proposes a novel method that is based on a generalisation of the entropy of a plane curve [5]. Roughly speaking, the entropy of a curve is 0 when the curve is a straight line, and increases as the curve becomes more "wiggly". Starting from the seminal work of [5], a new theory called thermodynamics of curves was developed [6], with some analogies with thermodynamics. A drawback of such theory has been that defining the entropy of a curve only for plane curves has restricted its use to a few applications, as for instance [7]. This work extends the concept to higher dimensions, while preserving the main features. The generalised entropy is then used to evaluate the behaviour of dynamical systems and its main property is that all linear systems, include time-varying ones, are characterised by a constant zero entropy. On the other hand, an entropy larger than 0 is a clear symptom of a nonlinear behaviour. Generalisation of the entropy concept together with the algorithmic procedure to evaluate it for the analysis of dynamical systems is described in the next section. In the special case of chaotic systems, the entropy approach proves to share common properties with other well known chaotic indicators, and an interesting comparison is provided in the third section. In Sect. 4 several well-known dynamical systems are compared according to the asymptotic value of the generalised entropy. Finally in the last section we summarise our results and conclude the paper.

2 The Generalised Entropy of Curves and its Application

Full general theory of the thermodynamics of plane curves can be found in [5, 6], where starting from Steinhaus theorem [11] and Shannon's measure of entropy [12], the entropy of a plane curve Γ was defined as [7]

$$H(\Gamma) = \log\left(\frac{2L}{C}\right) \tag{1}$$

where L is the length of Γ and C is the length of the boundary of its convex hull. The temperature of a curve is generally defined as the inverse of $\beta(\Gamma)$ [6], where $\beta(\Gamma) = \log(\frac{2L}{2L-C})$. The main property of the previous entities is that only straight segments are represented by a temperature $T = \beta^{-1} = 0$, and then $H = 0$. This is in accordance with Nernst's thermodynamic assumption and provides the analogy with thermodynamics as in physics.

The main idea of this work is that of considering a set of N initial conditions (state points), chosen aligned along a straight line Γ in the phase space. The entropy of Γ is then 0 according to the previous definition of the entropy of a line. The equations of a dynamical system move the points in the phase space and the entropy of the line that connects sequentially the evolving points changes accordingly. However, we first introduce a generalised definition of the entropy of a curve (2), so that the approach reamins valid in the case of $\mathbb{R}^n$-dimensional spaces.

$$H = \frac{\log(\frac{L}{d})}{\log(N-1)}. \tag{2}$$

In (2) d is the diameter of the smallest hypersphere covering the curve connecting the N points. This definition circumvents the difficulty of defining the length of the convex hull perimeter C in higher dimensions, but preserves the property that the minimal entropy is associated only with straight segments. Indeed, if points are all aligned, then $L \equiv d$ and the entropy is 0. Moreover, since collinearity is preserved under affine transformations, a line evolving according to linear dynamics has constant zero entropy, while higher values of the entropy are a symptom of nonlinearity.

Assuming that a dynamical system evolves according to the discrete-time model $x(k+1) = f(x(k), k)$, where $x \in \mathbb{R}^n$, the previous procedure can be summarised in an algorithmic way.

Algorithm 1

1. **Initialisation:** $k = 0$
   a. *Choose* N *initial points* $x_1(0), \dots, x_N(0)$ *ordered sequentially along a straight line* $(H(0) = 0)$
2. **Evolution:** step k
   a. *Compute the next state* $x_1(k+1), \dots, x_N(k+1)$ *for each state point according to system dynamics*
   b. *Compute the length L of the line that connects sequentially all the points*
   c. *Compute the diameter d of the smallest hypersphere including all the points*
   d. *Compute* $H(k)$ *according to* (2)
   e. *Go to next step* $(k = k + 1)$.

Deterministic inputs can be included in the system dynamics without significant changes, and have not been considered here for sake of simplicity. The problem of computing the minimum covering sphere can be formulated as a convex quadratic programming problem [18], although here algorithm [19] was used, as it is expected to be faster on average, as more detailed in [10].

2.1 Properties of the Generalised Entropy

This section lists the main properties of the proposed generalised entropy indicator (2) when used for the analysis of dynamical systems. Due to space limitations, here all proves are omitted, but they can be found in [8–10].

Proposition 1 *If two curves are equally long, the more tortuous one has a higher entropy.*

Theorem 1 *The entropy of a curve is* 0 *if and only if the curve is straight and the points are ordered sequentially along it.*

Proposition 2 *The entropy range of a line is always between* 0 *and* 1.

Theorem 2 *The entropy is constantly* 0 *if the dynamical system is linear.*

Proposition 3 *The entropy of a line is insensitive to changes of scale, rotations and translations.*

Theorem 3 *The entropy is constantly* 0 *if the dynamical system is one-dimensional and the state function is monotonic.*

The case when the state space dimension is 1 is indeed a degenerate situation in the sense that all the points remain necessarily aligned along the only available dimension, so the only possibility for the entropy to be larger than zero is that points exchange their ordering.

The value of the proposed indicator can be computed either theoretically exploiting the previous properties and theorems, or by the algorithmic procedure introduced formerly. However, the algorithmic approach suffers from being numerically inefficient if all the points converge to an equilibrium, and all distances go to zero, or diverge, in which case distances go to infinity. The algorithmic procedure provides instead good results if applied to the analysis of chaotic systems, since the states evolve within attractor sets and their norms remain bounded.

3 A Comparison with Other Chaotic Indicators

Chaotic behaviours have been widely investigated in the recent years, and methods like Lyapunov Exponents (LE), the auto-correlation function and the power spectrum have become classic tools of chaos theory [1]. Other less conventional chaos indicators have been introduced in more recent literature, as for instance [13–15]. In particular, the approach proposed here is closely related to LEs and the d_∞ parameter of [15]. Similarly to the generalised entropy, d_∞ is computed following an algorithmic procedure. At the beginning, several pairs of initial conditions very close to each other are chosen randomly. Then the average distance between the pairs of trajectories is recorded at each time step and d_∞ represents the average asymptotic (for time that goes to ∞) distance. If the initial pairs of points are chosen along a line, then the asymptotic entropy, here called H_∞ for analogy, could be computed as a function of d_∞

$$H_\infty = \frac{\log(\frac{(N-1)d_\infty}{d})}{\log(N-1)} \tag{3}$$

where to avoid confusion we remind that d stands as usual for the diameter of the smallest hypersphere that encloses all the points. It should be noted that also d reaches an asymptotic value since the state points spread themselves so to cover the whole attractor. Therefore a proportional relationship between the logarithm of d_∞ and the indicator H_∞ gets established.

A difference is that LEs describe the stretching aspects of a dynamical system, while d_∞ and H_∞ take into account both the stretching and the folding effects. Since during the first steps of evolution the stretching effect is dominant over the folding one, an empirical way of estimating the maximum LE from the evolution of the asymptotic distance was suggested in [16]. A similar approach could be followed for the generalised entropy as well, however results are not as accurate as if other approaches were used (see for instance [17]).

As a further difference, the proposed generalised entropy indicator provides a nice extension for the investigation of unusual dynamical systems defined on a discrete state space, as for instance the Kaprekar routine [10] addressed in the next section, when it is not clear how to define Lyapunov exponents. Please note also that the previous relationships between the maximum LE, d_∞ and the generalised entropy indicator only hold when the dynamical system is chaotic. Otherwise, for instance, Lyapunov exponents might not distinguish chaotic systems from unstable linear systems [10].

4 Simulation Examples

Extensive simulations have been performed to study the behaviour of the proposed index in many benchmark problems. Figure 1 compares the entropy associated to well-known dynamical systems, such as Lorenz equations, Van Der Pol oscillator, Kaprekar routine and Arithmetic-Geometric Mean (AGM) function. Their dynamic equations can be found in [1], while the less conventional Kaprekar routine is described in [10].

In the next example, parameter a tunes the "amount of nonlinearity":

$$\begin{cases} x_1(k+1) = a \cdot x_1(k), \\ x_2(k+1) = 3.2 \cdot x_2(k) \cdot (1 - x_2(k)). \end{cases} \tag{4}$$

The first component of the system (4) evolves with linear dynamics, so its expected entropy is 0. The second component evolves according to a logistic equation with

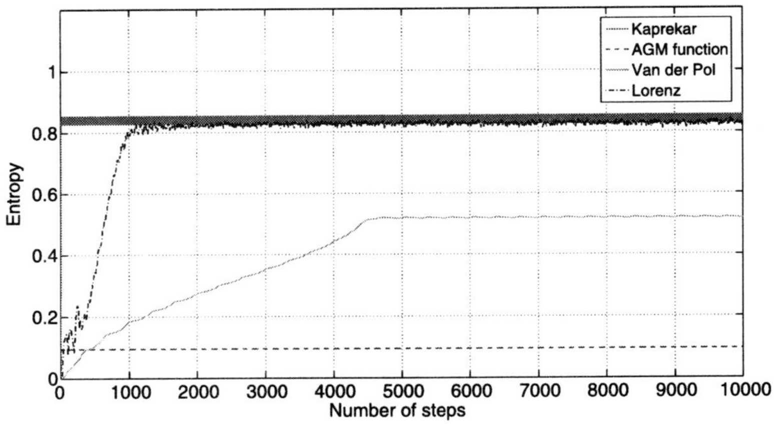

Fig. 1 Comparison of the entropy of some dynamical systems. Higher entropies are achieved respectively by the Kaprekar routine, Lorenz equations, Van Der Pol oscillator and AGM function

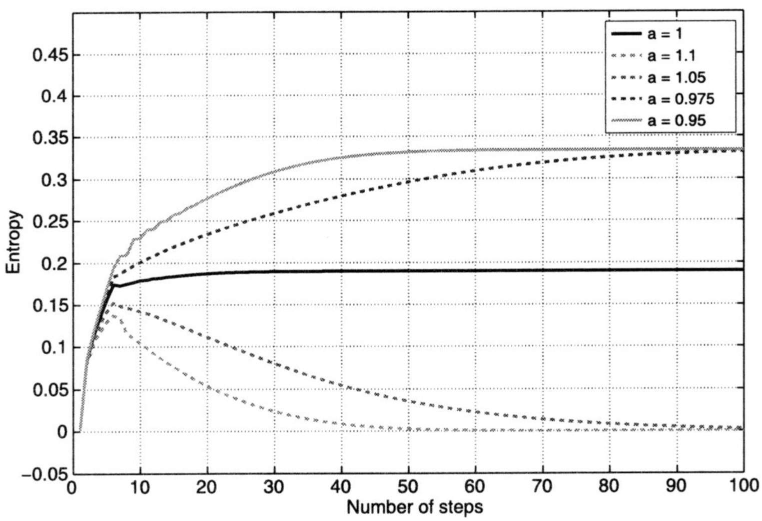

Fig. 2 Comparison of the entropy of a dynamical system as a function of a parameter that weighs the contribute of the linear and nonlinear component

parameter 3.2, in which case there are two equilibrium points and a higher entropy is expected due to the presence of nonlinear terms. In the example, when $|a| < 1$, the linear component goes to zero and the dynamical system reduces to the nonlinear part. On the contrary, when $|a| > 1$, the linear component overrides the nonlinear one. Only if $|a| = 1$ the two components have comparable values. The entropy of the system reflects this situation, and either assumes the value of the dominant component or an intermediate value in the last case, as can be shown in Fig. 2.

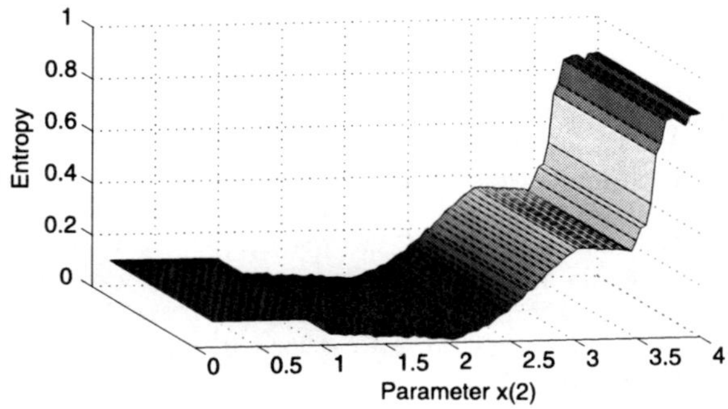

Fig. 3 Asymptotic value of the entropy in the case that the initial condition of the parameter x_2 varies from 0.1 to 4 with a 0.04 interval step

Not always dynamical systems have a unique behaviour in the whole state space. As an example, (5) generalises the logistic equation in the case that the state is extended to include the fixed rate parameter, so that the initial condition of the second state has a strong impact on the behaviour of the dynamical system.

$$\begin{cases} x_1(k+1) = x_2(k) \cdot x_1(k)(1 - x_1(k)), \\ x_2(k+1) = x_2(k). \end{cases} \tag{5}$$

Depending on the initial value of the parameter x_2, system (5) has one equilibrium when $x_2(0)$ is smaller than 3, it oscillates for parameter values between 3 and 3.57 (approximately), and shows a chaotic behaviour for values greater than 3.57 and smaller than 4. There is a so-called "island of stability" for values around 3.8 and finally it diverges for almost all initial conditions when the parameter is greater than 4. Thus, it can not be expected that the entropy of the dynamical system (5) can summarise all the possible behaviours with one only value, while it is sensible to compute the entropy as a function of the initial conditions, as is shown in Fig. 3 where the known behaviour of the logistic function is reproduced realistically. Further examples of the proposed entropy indicator can be found in [10].

5 Conclusions

This work provides a generalised definition of the entropy of a curve and applies it to the analysis and the classification of nonlinear systems. According to the proposed entropy based indicator all linear systems are characterised by zero entropy while higher values of the entropy always indicate nonlinear behaviours. The generalised entropy can be applied successfully to the analysis of chaotic systems as well, in which case there are common features with other known chaotic indicators. Moreover, it also extends the use of other chaotic indicators to the special case of not conventional dynamical systems defined over discrete state spaces. Finally, should the behaviour of the dynamical system under exam depend on the value of some parameters, a Monte Carlo based approach of the entropy indicator still recovers the general information, as described in the last example.

References

1. S.H. Strogatz, *Nonlinear Dynamics and Chaos* (Westview Press, Boulder, 2000)
2. E. Ott, *Chaos in Dynamical Systems* (Cambridge University Press, Cambridge, 2002)
3. L. Galleani, L. Lo Presti, A. De Stefano, Signals Process. **65**, 147–153 (1998)
4. M. Fliess, J. Levine, P. Martin, P. Rouchon, IEEE Trans. Autom. Control **44**, 922–937 (1999)
5. F.M. Mendès, Courrier Centre Nat. Rec. Sci. **51**, 5–9 (1983)
6. Y. Dupain, T. Kamae, F.M. Mendès, Arch. Ration. Mech. Anal. **94**, 155–163 (1986)
7. A. Denis, F. Crémoux, Math. Geol. **34**, 899–914 (2002)
8. A. Balestrino, A. Caiti, E. Crisostomi, in *IFAC Sympos. Nonlinear Control Systems*, Pretoria, South Africa (2007)
9. A. Balestrino, A. Caiti, E. Crisostomi, Chem. Eng. Trans. **11**, 119–124 (2007)
10. A. Balestrino, A. Caiti, E. Crisostomi, in *Conference on Nonlinear Science and Complexity*, Porto, Portugal (2008)
11. R. Moore, A. Van Der Potten, in *Conference on Geometry and Physics*, Canberra, Australia (1989)
12. C.E. Shannon, Bell Syst. Tech. J. **27**, 379–423, 623–656 (1948)
13. C. Skokos, J. Phys. A **34**, 10029–10043 (2001)
14. G. Lukes-Gerakopoulos, N. Voglis, C. Efthymiopoulos, Physica A **387**, 1907–1925 (2008)
15. A. Bonasera, M. Bucolo, L. Fortuna, M. Frasca, A. Rizzo, Nonlinear Phenom. Complex Syst. **6**, 779–786 (2003)
16. M. Bucolo, F. Di Grazia, F. Sapuppo, M.C. Virzí, in *Mediterranean Conference on Control and Automation*, Ajaccio, France (2008)
17. Y.B. Pesin, Russ. Math. Surv. **32**, 55–114 (1977)
18. D.J. Elzinga, D.W. Hearn, Manag. Sci. **19**, 96–104 (1972)
19. T.H. Hopp, C.P. Reeve, Natl. Inst. Stand. Technol. **5831** (1996)

Uncertainty on a Bertrand Duopoly with Product Differentiation

Fernanda A. Ferreira and Alberto A. Pinto

Abstract The conclusions of the Bertrand model of competition are substantially altered by the presence of either differentiated goods or asymmetric information about rival's production costs. In this paper, we consider a Bertrand competition, with differentiated goods. Furthermore, we suppose that each firm has two different technologies, and uses one of them according to a certain probability distribution. The use of either one or the other technology affects the unitary production cost. We show that this game has exactly one Bayesian Nash equilibrium. We do ex-ante and ex-post analyses of firms' profits and market prices. We prove that the expected profit of each firm increases with the variance of its production costs. We also show that the expected price of each good increases with both expected production costs, being the effect of the expected production costs of the rival dominated by the effect of the own expected production costs.

Keywords Game theory · Industrial organization · Optimization · Bertrand model · Uncertainty

1 Introduction

The Bertrand model is one of the cornerstones of the modern theory of oligopoly. In this model, firms' strategic variable is the price of the good that they produce. It is well-known that, if the firms produce a homogeneous product at a common constant marginal costs, the Bertrand competition leads to a price equal to the marginal cost (see [2]). The conclusions of the Bertrand model of competition are substantially

F.A. Ferreira (✉)
ESEIG, Instituto Politécnico do Porto, R.D. Sancho I, 981, 4480-876 Vila do Conde, Portugal
e-mail: fernandaamelia@eu.ipp.pt

A.A. Pinto
Departamento de Matemática, Universidade do Minho, 4710-057 Braga, Portugal
e-mail: aapinto1@gmail.com

J.A.T. Machado et al. (eds.), *Nonlinear Science and Complexity*,
DOI 10.1007/978-90-481-9884-9_45,

altered by the presence of either differentiated goods or asymmetric information about rival's production costs.

A number of well-know models have introduced firm market power by departing from the Bertrand model in various ways: product differentiation (see [10]), repeated games (see [6, 7]), capacity precommitment (see [12]). Also, Klemperer examines consumer switching costs (see [11]), and Allen and Thisse introduce consumer insensitivity to small price changes (see [1]). The incentives to share information in oligopoly is examined in a number of papers (see, for example, [8, 9, 13, 14, 17], and the references therein). Spulber analyzes the Bertrand competition in presence of asymmetric information about rivals' production costs (see [15]).

Let F_1 and F_2 be two firms, each producing a differentiated product. Both firms simultaneously choose the price for the corresponding good with the purpose to maximize their expected profit. We consider an economic model in which we suppose that each firm has two different technologies, and uses one of them according to a certain probability distribution. The use of either one or the other technology affects the unitary production cost. We suppose that firm F_1's unitary production cost is c_A with probability ϕ and c_B with probability $1-\phi$ (where $c_A > c_B$), and firm F_2's unitary production cost is c_H with probability θ and c_L with probability $1-\theta$ (where $c_H > c_L$). Both probability distributions of unitary production costs are common knowledge. In this work, we determine the prices in the Bayesian Nash equilibrium for the above model, and we analyze the advantages, for firms and for consumers, of using the technology with highest production cost *versus* the one with cheapest production cost. We prove that the expected profit of each firm increases with the variance of its production costs. We also show that the expected price of each good increases with both expected production costs, being the effect of the expected production costs of the rival dominated by the effect of the own expected production costs. Ferreira et al. studied different duopoly models under uncertainty production costs (see [3, 4]).

2 The Model and the Equilibrium

We consider an economy with a monopolistic sector with two firms, F_1 and F_2. Firm F_i produces a substitutable product i at a constant marginal cost, for $i \in \{1, 2\}$. The firms simultaneously choose prices, respectively, $p_1 \geq 0$ and $p_2 \geq 0$. The direct demands are given by

$$q_i = a - p_i + bp_j,$$

where q_i stands for quantity, $a > 0$ is the intercept demand parameter and $b \geq 0$ is a constant representing how much the product of one firm is a substitute for the product of the other (see, for example, [16]). For simplicity, we assume $b \leq 1$. These demand functions are unrealistic in that one firm could conceivably charge an arbitrary high price and still have a positive demand provided the other firm also charges a high enough price. However, this function is chosen to represent a linear approximation to the "true" demand function, appropriate near the usual price settings where the equilibrium is reached. Usually, in the case of complete information, the literature considers firm F_i's profit, π_i, given by

$$\begin{aligned}\pi_i(p_i, p_j) &= q_i(p_i - c)\\ &= (a - p_i + bp_j)(p_i - c),\end{aligned}$$

where $0 < c < a$ is the unitary production cost for both firms. Here, we suppose that each firm has two different technologies, and uses one of them following a certain probability distribution. The use of either one or the other technology affects the unitary production cost. The following probability distributions of the firms' production costs are common knowledge among both firms:

$$C_1 = \begin{cases} c_A & \text{with probability } \phi,\\ c_B & \text{with probability } 1-\phi,\end{cases}$$

$$C_2 = \begin{cases} c_H & \text{with probability } \theta,\\ c_L & \text{with probability } 1-\theta.\end{cases}$$

We suppose that $c_A > c_B$, $c_H > c_L$ and $c_A, c_B, c_H, c_L < a$. Moreover, we suppose that the highest unitary production cost of any firm is greater than the lowest unitary production cost of the other one, that is, $c_A > c_L$ and $c_H > c_B$.

Firms' profits, π_1 and π_2, are given by

$$\pi_1(p_1(c_1), p_2(c_2)) = (a - p_1(c_1) + bp_2(c_2))(p_1(c_1) - c_1),$$

$$\pi_2(p_1(c_1), p_2(c_2)) = (a - p_2(c_2) + bp_1(c_1))(p_2(c_2) - c_2),$$

where the price $p_i(c_i)$ depends on the unitary production cost c_i of firm F_i, for $i \in \{1, 2\}$.

Theorem 1 *Let $E(C_1) = \phi c_A + (1-\phi)c_B$ be the expected unitary production cost of firm F_1, and let $E(C_2) = \theta c_H + (1-\theta)c_L$ be the expected unitary production cost of firm F_2. For the Bertrand model with uncertainty costs considered, the Bayesian Nash equilibrium is*

$$\big((p_1^*(c_A), p_1^*(c_B)), (p_2^*(c_H), p_2^*(c_L))\big),$$

where

$$p_1^*(c_A) = \frac{2a(2+b) + (4-b^2)c_A + b^2E(C_1) + 2bE(C_2)}{2(4-b^2)}, \tag{1}$$

$$p_1^*(c_B) = \frac{2a(2+b) + (4-b^2)c_B + b^2E(C_1) + 2bE(C_2)}{2(4-b^2)}, \tag{2}$$

$$p_2^*(c_H) = \frac{2a(2+b) + (4-b^2)c_H + b^2E(C_2) + 2bE(C_1)}{2(4-b^2)}, \tag{3}$$

$$p_2^*(c_L) = \frac{2a(2+b) + (4-b^2)c_L + b^2E(C_2) + 2bE(C_1)}{2(4-b^2)}. \tag{4}$$

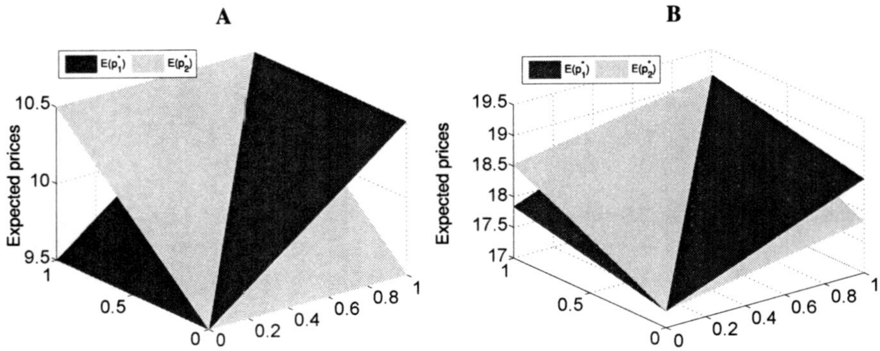

Fig. 1 Expected prices, $E(p_1^*)$ and $E(p_2^*)$, in the case of: (**A**) firms producing independent goods ($b=0$); and (**B**) firms producing differentiated goods ($b=0.9$). Other parameters values: $a=15$, $c_A=c_H=6$, $c_B=c_L=4$

This theorem is proved in [5]. In the following Corollary 1 and Theorem 2, we present an ex-ante analysis by giving the expected prices in the market and the profits that the firms can expect, before the knowledge of the production costs of both firms.

Corollary 1 *The expected price, $E(p_1^*)$, of the good produced by firm F_1 is given by*

$$E(p_1^*)=\frac{a(2+b)+2E(C_1)+bE(C_2)}{2(4-b^2)}$$

and the expected price, $E(p_1^*)$, of the good produced by firm F_2 is given by

$$E(p_2^*)=\frac{a(2+b)+bE(C_1)+2E(C_2)}{2(4-b^2)}.$$

The effect of the probabilities ϕ and θ over the expected prices is shown in Fig. 1, for some parameter region of the model. The expected prices are lower when both firms use their more expensive technologies with low probabilities, and it is higher when both firms use their more expensive technologies with high probabilities. Furthermore, the expected prices are lower in the case of independent goods than in the case of differentiated goods.

Theorem 2 *Let $V(C_i)$ be the variance of the firm F_i's unitary production cost, for $i\in\{1,2\}$. Firm F_1's expected profit $E(\pi_1^*)$ is given by*

$$E(\pi_1^*)=\frac{(a(2+b)-(2-b^2)E(C_1)+bE(C_2))^2}{(4-b^2)^2}+\frac{V(C_1)}{4}$$

and firm F_2's expected profit $E(\pi_2^*)$ is given by

$$E(\pi_2^*)=\frac{(a(2+b)+bE(C_1)-(2-b^2)E(C_2))^2}{(4-b^2)^2}+\frac{V(C_2)}{4}.$$

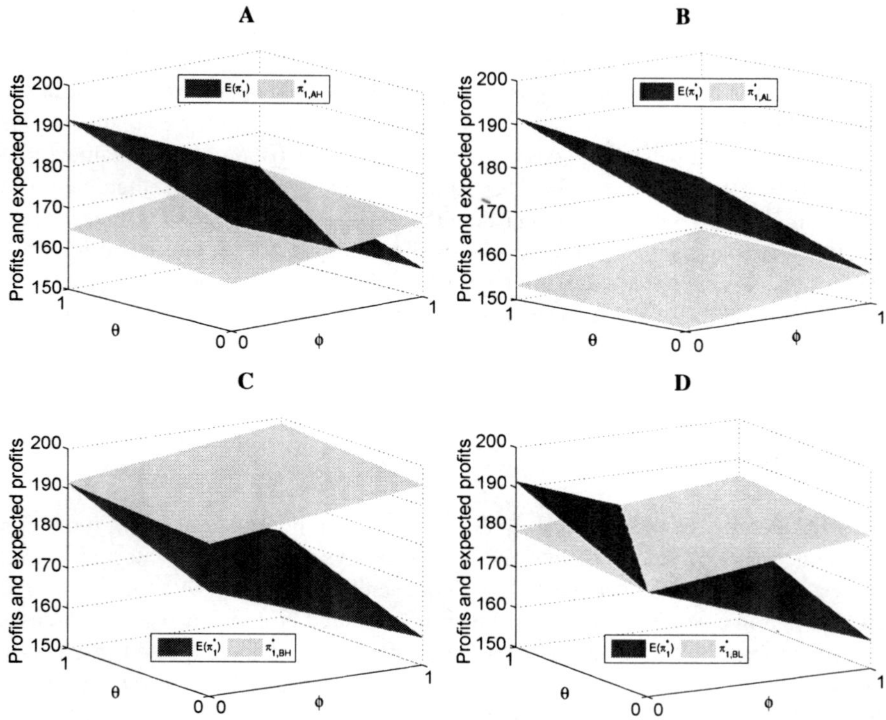

Fig. 2 Profits and expected profit of firm F_1, in the case of firms producing differentiated goods ($b = 0.9$). Other parameters values: $a = 15$, $c_A = c_H = 6$, $c_B = c_L = 4$. **(A)** Profit $\pi^*_{1,AH}$ in the case of $c_1 = c_A$ and $c_2 = c_H$; **(B)** Profit $\pi^*_{1,AL}$ in the case of $c_1 = c_A$ and $c_2 = c_L$; **(C)** Profit $\pi^*_{1,BH}$ in the case of $c_1 = c_B$ and $c_2 = c_H$; and **(D)** Profit $\pi^*_{1,BL}$ in the case of $c_1 = c_B$ and $c_2 = c_L$

This theorem is proved in [5].

Now, we are going to analyse the profits that the firms obtain after the observation of the production cots realization. The profits that the firms obtain, at equilibrium, are given by

$$\pi_1^* = \begin{cases} (a - p_1^*(c_A) + bp_2^*(c_H))(p_1^*(c_A) - c_A) & \text{if } c_1 = c_A \text{ and } c_2 = c_H, \\ (a - p_1^*(c_A) + bp_2^*(c_L))(p_1^*(c_A) - c_A) & \text{if } c_1 = c_A \text{ and } c_2 = c_L, \\ (a - p_1^*(c_B) + bp_2^*(c_H))(p_1^*(c_B) - c_B) & \text{if } c_1 = c_B \text{ and } c_2 = c_H, \\ (a - p_1^*(c_B) + bp_2^*(c_L))(p_1^*(c_B) - c_B) & \text{if } c_1 = c_B \text{ and } c_2 = c_L, \end{cases}$$

$$\pi_2^* = \begin{cases} (a - p_2^*(c_H) + bp_1^*(c_A))(p_2^*(c_H) - c_H) & \text{if } c_1 = c_A \text{ and } c_2 = c_H, \\ (a - p_2^*(c_L) + bp_1^*(c_A))(p_2^*(c_L) - c_L) & \text{if } c_1 = c_A \text{ and } c_2 = c_L, \\ (a - p_2^*(c_H) + bp_1^*(c_B))(p_2^*(c_H) - c_H) & \text{if } c_1 = c_B \text{ and } c_2 = c_H, \\ (a - p_2^*(c_L) + bp_1^*(c_B))(p_2^*(c_L) - c_L) & \text{if } c_1 = c_B \text{ and } c_2 = c_L, \end{cases}$$

where $p_1^*(c_A)$, $p_1^*(c_B)$, $p_2^*(c_H)$ and $p_2^*(c_L)$ are given by equalities (1)–(4). In Fig. 2, we show the plots of the expected profits and of the real profits, as functions of the probability distributions' parameters ϕ and θ, for some parameter region of the model. We see that if $c_1 = c_A$ and $c_2 = c_L$, then the profit that firm F_1 really obtains is lower than the ex-ante expected profit, for every pair (ϕ, θ) of the parameters of the probability distributions; for other hand, if $c_1 = c_B$ and $c_2 = c_H$, then the profit that firm F_1 really obtains is higher than the ex-ante expected profit, for every pair (ϕ, θ) of the parameters of the probability distributions; for the other two possibilities of production costs, there are pairs (ϕ, θ) for which the real profits of firm F_1 are higher than its ex-ante expected profits, and there are pairs (ϕ, θ) for which the real profits of firm F_1 are lower than its ex-ante expected profits.

3 Conclusions

We considered a Bertrand model with product differentiation and with production costs uncertainty. We did ex-ante and ex-post analyses of firms' profits and market prices. We proved that the expected profit of each firm increases with the variance of its production costs. We also showed that the expected price of each good increases with both expected production costs, being the effect of the expected production costs of the rival dominated by the effect of the own expected production costs. Furthermore, we saw that, for some production costs realizations, the firms' ex-post profits are either higher or lower than their ex-ante expected profits; and for others production costs realizations, to be higher or lower depend on the probability distributions' parameters.

Acknowledgements This research was partially supported by the Programs POCTI and POCI by FCT and Ministério da Ciência, Tecnologia e do Ensino Superior. F.A. Ferreira also thanks financial support from ESEIG/IPP and from Centro de Matemática da Universidade do Porto. A.A. Pinto acknowledges financial support from Centro de Matemática da Universidade do Minho.

References

1. B. Allen, J.-F. Thisse, Price equilibria in pure strategies for homogeneous oligopoly. J. Econ. Manag. Strategy **1**, 63–82 (1992)
2. J. Bertrand, Théorie mathématiques de la richesse sociale. J. Savants **68**, 303–317 (1883)
3. F.A. Ferreira, F. Ferreira, A.A. Pinto, Bayesian price leadership, in *Mathematical Methods in Engineering*, ed. by Kenan Tas et al. (Springer, Dordrecht, 2007), pp. 371–379
4. F.A. Ferreira, F. Ferreira, A.A. Pinto, Unknown costs in a duopoly with differentiated products, in *Mathematical Methods in Engineering*, ed. by Kenan Tas et al. (Springer, Dordrecht, 2007), pp. 359–369
5. F.A. Ferreira, A.A. Pinto, Bertrand model under incomplete information, in *Numerical Analysis and Applied Mathematics*, ed. by T.E. Simos et al. AIP Conference Proceedings, vol. 1048 (AIP, New York, 2008), pp. 209–212
6. J.W. Friedman, A non-cooperative equilibrium for supergames. Rev. Econ. Stud. **38**, 1–12 (1971)

7. J.W. Friedman, *Oligopoly and the Theory of Games* (North-Holland, Amsterdam, 1977)
8. E. Gal-Or, Information sharing in oligopoly. Econometrica **53**, 329–343 (1985)
9. E. Gal-Or, Information transmission: Cournot and Bertrand equilibria. Rev. Econ. Stud. **53**, 85–92 (1986)
10. H. Hotelling, Stability in competition. Econ. J. **39**, 41–57 (1929)
11. P. Klemperer, Markets with consumer switching costs. Q. J. Econ. **102**, 375–394 (1987)
12. D. Kreps, J. Scheinkman, Quantity precommitment and Bertrand competition yield Cournot outcomes. Bell J. Econ. **14**, 326–337 (1983)
13. W. Novshek, H. Sonnenschein, Fulfilled expectations and Cournot duopoly with information acquisition and release. Bell J. Econ. **13**, 214–218 (1982)
14. C. Shapiro, Exchange of cost information in oligopoly. Rev. Econ. Stud. **52**, 433–446 (1986)
15. D. Spulber, Bertrand competition when rivals' costs are unknown. J. Ind. Econ. **43**, 1–11 (1995)
16. J. Tirole, *The Theory of Industrial Organization* (MIT Press, Cambridge, 1994)
17. X. Vives, Duopoly information equilibrium: Cournot and Bertrand. J. Econ. Theory **34**, 71–94 (1984)

Price-Setting Dynamical Duopoly with Incomplete Information

Fernanda A. Ferreira, Flávio Ferreira, and Alberto A. Pinto

Abstract We consider a price competition in a duopoly with substitutable goods, linear and symmetric demand. There is a firm (F_1) that chooses first the price p_1 of its good; the other firm (F_2) observes p_1 and then chooses the price p_2 of its good. The conclusions of this price-setting dynamical duopoly are substantially altered by the presence of either differentiated goods or asymmetric information about rival's production costs.

In this paper, we consider asymmetric information about rival's production costs. We do ex-ante and ex-post analyses of firms' profits and market prices. We compare the ex-ante firms' expected profits with the ex-post firms' profits.

Keywords Game theory · Industrial organization · Optimization · Uncertainty

1 Introduction

Case studies find that in a wide variety of oligopolistic industries, such as the cigarette, steel, automobile, ready-to-eat-cereal and gasoline industries, new price announcements arrive in a sequential manner: price increases by one firm are followed immediately by its rivals.

In this paper, we consider the following price-setting dynamic model of duopoly (simultaneous decisions corresponds to Bertrand model, see [2]): There is a leading

F.A. Ferreira (✉) · F. Ferreira
ESEIG, Instituto Politécnico do Porto, R.D. Sancho I, 981, 4480-876 Vila do Conde, Portugal
e-mail: fernandaamelia@eu.ipp.pt

F. Ferreira
e-mail: flavioferreira@eu.ipp.pt

A.A. Pinto
Departamento de Matemática, Universidade do Minho, 4710-057 Braga, Portugal
e-mail: aapinto1@gmail.com

J.A.T. Machado et al. (eds.), *Nonlinear Science and Complexity*,
DOI 10.1007/978-90-481-9884-9_46,

firm that chooses the price for its good first, taking into account the follower's optimal response to its price choice. The follower then sets the price for its good based on the leader's choice. It is well-known that a second-mover advantage exists when firms are identical, under quite general conditions (see [7]). In the case of asymmetric information about rival's production costs, we will see that, the leading firm can have a higher expected profit than the one that follows. The uncertainty on the production costs is driven by considering that each firm has two different technologies, and uses one of them following a certain probability distribution. The use of either one or the other technology affects the unitary production cost. We suppose that firm F_1's unitary production cost is c_A with probability ϕ and c_B with probability $1-\phi$ (where $c_A > c_B$), and firm F_2's unitary production cost is c_H with probability θ and c_L with probability $1-\theta$ (where $c_H > c_L$). Both probability distributions of unitary production costs are common knowledge. We note that the leading firm does not know the unitary production costs of its rival, while the firm that plays second knows the unitary production costs of the leading firm by looking to the price set by this firm. In this work, we do ex-ante and ex-post analyses of firms' profits and market prices. We show that, in contrast to the case with complete information, in which case it is known that the firm that chooses its price in the second place is in advantage, the leading firm can profit more than the follower. We also show that the expected profit of the leading firm increases with the variance of its production costs, and the expected profit of the follower firm increases with both variances of production costs, being the effect of the variance of the rival's production costs dominated by the effect of the variance of the own production costs.

Van Damme and Hurkens studied a related question by considering that one firm has higher production cost, but in a game of complete information (see [3]). Amir and Stepanova also studied a second-mover advantage (see [1]). Ferreira et al. studied different duopoly models under uncertainty production costs (see [4, 5]).

2 The Model and the Equilibrium

We consider an economy with a monopolistic sector with two firms, F_1 and F_2. Firm F_i produces a substitutable product i at a constant marginal cost, for $i \in \{1,2\}$. We present a sequential-move model, with incomplete information, in which firms choose the prices for their goods. In a game of complete information, the players' payoff functions are common knowledge. In a game of incomplete information, in contrast, at least one player is uncertain about, at least, another player's payoff function. The timing of the game is as follows: (i) Firm F_1 (leader) chooses a price $p_1 \geq 0$ for its good; (ii) firm F_2 (follower) observes p_1 and then chooses a price $p_2 \geq 0$ for its good.

The direct demands are

$$q_i = a - p_i + bp_j,$$

provided that the quantities q_i are non-negative, with $i, j \in \{1,2\}$ and $i \neq j$, where $0 \leq b \leq 1$ reflects the extent to which firm F_i's product is a substitute for firm F_j's product, and $a > 0$ (see, for example, [8]). We suppose that each firm has two different technologies, and uses one of them following a certain probability distribution. The use of either one or the other technology affects the unitary production cost. The following probability distributions of the firms' production costs are common knowledge among both firms:

$$C_1(q_1) = \begin{cases} c_A\, q_1 & \text{with probability } \phi, \\ c_B\, q_1 & \text{with probability } 1-\phi, \end{cases}$$

$$C_2(q_2) = \begin{cases} c_H\, q_2 & \text{with probability } \theta, \\ c_L\, q_2 & \text{with probability } 1-\theta. \end{cases}$$

We suppose that $c_L < c_B < c_H < c_A < a$. Firm F_i's profits, π_i, are given by

$$\pi_i(p_i, p_j) = (a - p_i + bp_j)(p_i - c_i),$$

with $i, j \in \{1, 2\}$ and $i \neq j$. Firm F_1 should choose a price for its good, $p_1^*(c_A)$ or $p_1^*(c_B)$, depending on its unitary production cost, to maximize its expected profit; and firm F_2, knowing firm F_1's decision, should choose a price, $p_2^*(c_H|p_1^*(c_A))$, $p_2^*(c_L|p_1^*(c_A))$, $p_2^*(c_H|p_1^*(c_B))$ or $p_2^*(c_L|p_1^*(c_B))$, depending on its unitary production cost, to maximize its expected profit.

Theorem 1 *Let $E(C_2) = \theta c_H + (1-\theta)c_L$ be the firm F_2's expected unitary production cost. For the model presented above, the perfect Bayesian equilibrium is*

$$\big((p_1^*(c_A), p_1^*(c_B)), (p_2^*(c_H|p_1^*(c_A)), p_2^*(c_H|p_1^*(c_B)), p_2^*(c_L|p_1^*(c_A)), p_2^*(c_L|p_1^*(c_B)))\big),$$

where

$$p_1^*(c_A) = \frac{a(2+b) + (2-b^2)c_A + bE(C_2)}{2(2-b^2)}, \tag{1}$$

$$p_1^*(c_B) = \frac{a(2+b) + (2-b^2)c_B + bE(C_2)}{2(2-b^2)}, \tag{2}$$

$$p_2^*(c_H|p_1^*(c_A)) = \frac{a(4+2b-b^2) + (2b-b^3)c_A + (4-2b^2)c_H + b^2E(C_2)}{4(2-b^2)} \tag{3}$$

$$p_2^*(c_L|p_1^*(c_A)) = \frac{a(4+2b-b^2) + (2b-b^3)c_A + (4-2b^2)c_L + b^2E(C_2)}{4(2-b^2)}, \tag{4}$$

$$p_2^*(c_H|p_1^*(c_B)) = \frac{a(4+2b-b^2) + (2b-b^3)c_B + (4-2b^2)c_H + b^2E(C_2)}{4(2-b^2)}, \tag{5}$$

$$p_2^*(c_L|p_1^*(c_B)) = \frac{a(4+2b-b^2) + (2b-b^3)c_B + (4-2b^2)c_L + b^2E(C_2)}{4(2-b^2)}. \tag{6}$$

In the following Corollary 1 and Theorem 2, we present an ex-ante analysis by giving the expected prices in the market and firms' expected profits, before the knowledge of the production costs of both firms.

Corollary 1 *Let $E(C_1) = \phi c_A + (1-\phi)c_B$ be firm F_1's expected unitary production cost, and let $E(C_2) = \theta c_H + (1-\theta)c_L$ be firm F_2's expected unitary production cost. The expected market prices, $E(p_1^*)$ and $E(p_2^*)$, for the goods produced by firms F_1 and F_2 are, respectively,*

$$E(p_1^*) = \frac{a(2+b) + (2-b^2)E(C_1) + bE(C_2)}{2(2-b^2)}$$

and

$$E(p_2^*) = \frac{a(4+2b-b^2) + (2b-b^3)E(C_1) + (4-b^2)E(C_2)}{4(2-b^2)}.$$

Theorem 2 *Let $V(C_i)$ be the variance of firm F_i's unitary production cost, for $i \in \{1,2\}$. Firm F_1's expected profit $E(\pi_1^*)$ and Firm F_2's expected profit $E(\pi_2^*)$ are, respectively, given by*

$$E(\pi_1^*) = \frac{(a(2+b) - (2-b^2)E(C_1) + bE(C_2))^2}{8(2-b^2)} + \frac{(2-b^2)V(C_1)}{8}$$

and

$$E(\pi_2^*) = \frac{(a(4+2b-b^2) + b(2-b^2)E(C_1) - (4-3b^2)E(C_2))^2}{16(2-b^2)^2} + \frac{b^2 V(C_1)}{16} + \frac{V(C_2)}{4}.$$

We note that the expected profit of the firm F_1 increases with the variance of the correspondent production costs; and the expected profit of the firm F_2 increases with both variances of firm F_1's production costs and firm F_2's production costs.

All the above results are proved in [6]. When studying the expected profits for both firms as a result of the set of parameters chosen, we observe the existence of two possible outcomes when the firms are at equilibrium. So, we have two different regions of parameters, *region X* and *region Y*, determining different relative expected outcomes for firms F_1 and F_2. This is different from what is observed in the corresponding game with complete information where firm F_2 always has a higher profit than firm F_1. Let A be the region on the probability parameters space in which firm F_1 has a higher expected profit than firm F_2 (i.e., $E(\pi_1^*) > E(\pi_2^*)$). The region A decreases as the parameter b increases (see Fig. 1), and, depending on the other parameters, it can become empty (see Fig. 1A) or not (see Fig. 1B).

Now, we are going to analyse the profits that the firms obtain after the observation of the production costs realization. The profits that the firms obtain, at equilibrium, are given by

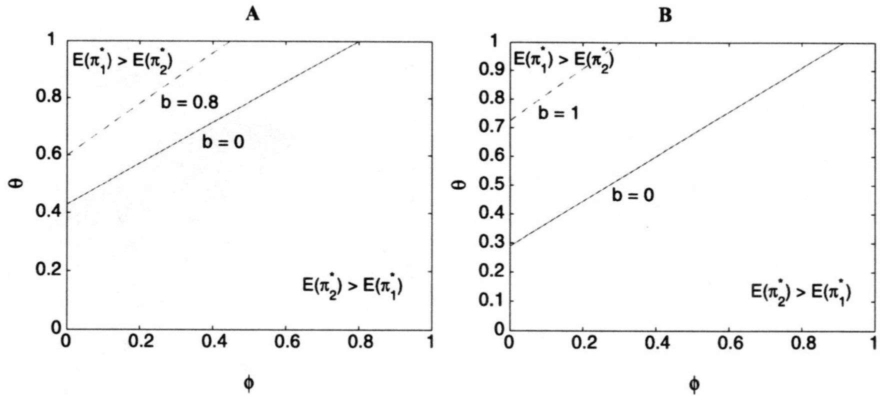

Fig. 1 Regions on the probability distributions' parameters in which the expected profit $E(\pi_i^*)$ of one firm is higher than the expected profit $E(\pi_j^*)$ of the other firm. The "*line*" and "*dashes*" correspond to the set of parameters, θ and ϕ, that give both firms the same expected profit for independent and substitutable goods, respectively. (**A**) Parameters values: $a = 8$, $c_A = 7$, $c_B = 4$, $c_H = 6$ and $c_L = 3$ (*region X*); (**B**) $a = 8$, $c_A = 7$, $c_B = 2$, $c_H = 6$ and $c_L = 1$ (*region Y*)

$$\pi_1^* = \begin{cases} (a - p_1^*(c_A) + bp_2^*(c_H|p_1^*(c_A)))(p_1^*(c_A) - c_A) & \text{if } c_1 = c_A \text{ and } c_2 = c_H, \\ (a - p_1^*(c_A) + bp_2^*(c_L|p_1^*(c_A)))(p_1^*(c_A) - c_A) & \text{if } c_1 = c_A \text{ and } c_2 = c_L, \\ (a - p_1^*(c_B) + bp_2^*(c_H|p_1^*(c_B)))(p_1^*(c_B) - c_B) & \text{if } c_1 = c_B \text{ and } c_2 = c_H, \\ (a - p_1^*(c_B) + bp_2^*(c_L|p_1^*(c_B)))(p_1^*(c_B) - c_B) & \text{if } c_1 = c_B \text{ and } c_2 = c_L, \end{cases}$$

$$\pi_2^* = \begin{cases} (a - p_2^*(c_H|p_1^*(c_A)) + bp_1^*(c_A))(p_2^*(c_H|p_1^*(c_A)) - c_H) \\ \quad \text{if } c_1 = c_A \text{ and } c_2 = c_H, \\ (a - p_2^*(c_L|p_1^*(c_A)) + bp_1^*(c_A))(p_2^*(c_L|p_1^*(c_A)) - c_L) \\ \quad \text{if } c_1 = c_A \text{ and } c_2 = c_L, \\ (a - p_2^*(c_H|p_1^*(c_B)) + bp_1^*(c_B))(p_2^*(c_H|p_1^*(c_B)) - c_H) \\ \quad \text{if } c_1 = c_B \text{ and } c_2 = c_H, \\ (a - p_2^*(c_L|p_1^*(c_B)) + bp_1^*(c_B))(p_2^*(c_L|p_1^*(c_B)) - c_L) \\ \quad \text{if } c_1 = c_B \text{ and } c_2 = c_L, \end{cases}$$

where $p_1^*(c_A)$, $p_1^*(c_B)$, $p_2^*(c_H|p_1^*(c_A))$, $p_2^*(c_L|p_1^*(c_A))$, $p_2^*(c_H|p_1^*(c_B))$ and $p_2^*(c_L|p_1^*(c_B))$ are given by equalities (1)–(6). In Fig. 2, we show the plots of the expected profits and of the real profits, as functions of the probability distributions' parameters ϕ and θ, for some parameter region of the model. We see that if $c_1 = c_A$ and $c_2 = c_L$, then the profit that firm F_1 really obtains is lower than the ex-ante expected profit, for every pair (ϕ, θ) of the parameters of the probability distributions; for other hand, if $c_1 = c_B$ and $c_2 = c_H$, then the profit that firm F_1 really obtains is higher than the ex-ante expected profit, for every pair (ϕ, θ) of the parameters of the probability distributions; for the other two possibilities of production costs, there are pairs (ϕ, θ) for which the real profits of firm F_1 are higher than its ex-ante

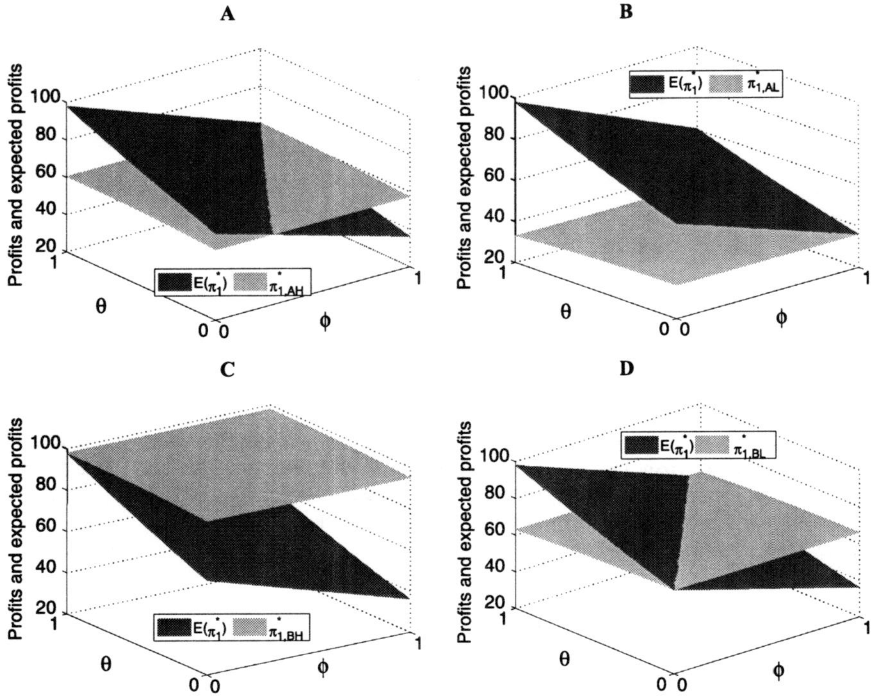

Fig. 2 Profits and expected profit of firm F_1, in the case of firms producing differentiated goods ($b = 1$). Other parameters values: $a = 8$, $c_A = 7$, $c_B = 2$, $c_H = 6$ and $c_L = 1$. (**A**) Profit $\pi^*_{1,AH}$ in the case of $c_1 = c_A$ and $c_2 = c_H$; (**B**) Profit $\pi^*_{1,AL}$ in the case of $c_1 = c_A$ and $c_2 = c_L$; (**C**) Profit $\pi^*_{1,BH}$ in the case of $c_1 = c_B$ and $c_2 = c_H$; and (**D**) Profit $\pi^*_{1,BL}$ in the case of $c_1 = c_B$ and $c_2 = c_L$

expected profits, and there are pairs (ϕ, θ) for which the real profits of firm F_1 are lower than its ex-ante expected profits.

3 Conclusions

We studied a price competition in a dynamic duopoly with substitutable goods, linear and symmetric demand and with unknown costs. We proved that, at equilibrium, the expected profit of the leading firm increases with the variance of its production costs, and the expected profit of the follower increases with the variance of both production costs. We saw that, depending on the relation between the demand parameters, the expected profit of one firm can be either higher or lower than the other one. Hence, to be the follower firm does not assure higher profits, and small changes in the parameters of the model can reverse the order of the firms' profits. Furthermore, we saw that, for some production costs realizations, the firms' ex-post profits are either higher or lower than their ex-ante expected profits; and for others production costs realizations, to be higher or lower depend on the probability distributions' parameters.

Acknowledgements We thank the Programs POCTI and POCI by FCT and Ministério da Ciência, Tecnologia e do Ensino Superior for their financial support. F.A. Ferreira and F. Ferreira gratefully acknowledge financial support from ESEIG/IPP and from Centro de Matemática da Universidade do Porto. A.A. Pinto also acknowledges financial support from Centro de Matemática da Universidade do Minho.

References

1. R. Amir, A. Stepanova, Second-mover advantage and price leadership in Bertrand duopoly. Games Econ. Behav. **55**, 1–20 (2006)
2. J. Bertrand, Théorie mathématiques de la richesse sociale. J. Savants **68**, 303–317 (1883)
3. E. van Damme, S. Hurkens, Endogenous price leadership. Games Econ. Behav. **47**, 404–420 (2004)
4. F.A Ferreira, F. Ferreira, A.A. Pinto, Bayesian price leadership, in *Mathematical Methods in Engineering*, ed. by Kenan Tas et al. (Springer, Dordrecht, 2007), pp. 371–379
5. F.A Ferreira, F. Ferreira, A.A. Pinto, Unknown costs in a duopoly with differentiated products, in *Mathematical Methods in Engineering.*, ed. by Kenan Tas et al. (Springer, Dordrecht, 2007), pp. 359–369
6. F.A Ferreira, F. Ferreira, A.A. Pinto, Price leadership competition under uncertainty (2009, in preparation)
7. E. Gal-Or, First mover and second mover advantages. Int. Econ. Rev. **26**, 649–653 (1985)
8. J. Tirole, *The Theory of Industrial Organization* (MIT Press, Cambridge, 1994)

Inductor-Free Version for Chua's Oscillator Based in Electronic Analogy

Guilherme Lúcio Damião Andrucioli and Ronilson Rocha

Abstract Although the literature presents several alternatives, an approach based in the electronic analogy was still not considered for the implementation of an inductor-free realization of the chaotic Chua's circuit. This work presents a new topology of Chua's circuit based on the electronic analogy, which has new and interesting features for many real applications. A simple, versatile and functional inductorless analogous circuit is designed and its implementation satisfactorily reproduces the chaotic behavior of Chua's oscillator.

Keywords Chaos · Chua's circuit · Electronic analogy

1 Introduction

The study of chaos in nonlinear electronic circuits has been a very active topic of research, principally in a circuit proposed by Leon O. Chua in the fall of 1983. This circuit was created in order to propose a system that can be realistically modeled, demonstrating the chaos as a robust physical phenomenon, and not merely an artifact of computer round-off errors. Since its initial proposal, this circuit has been intensely investigated and accepted as paradigm for study of important features of nonlinear systems. In spite of its simplicity and easy implementation, the Chua's circuit is robust and exhibits a very complex dynamical behavior, presenting a rich scenario formed by a large variety of bifurcations, homoclinic orbits, and distinct periodic and chaotic attractors.

G.L.D. Andrucioli · R. Rocha (✉)
EM/DECAT, Federal University of Ouro Preto, Campus Morro do Cruzeiro, 35400-000 Ouro Preto, MG, Brazil
e-mail: rocha@em.ufop.br

G.L.D. Andrucioli
e-mail: guilhermeandrucioli@yahoo.com.br

J.A.T. Machado et al. (eds.), *Nonlinear Science and Complexity*,
DOI 10.1007/978-90-481-9884-9_47,

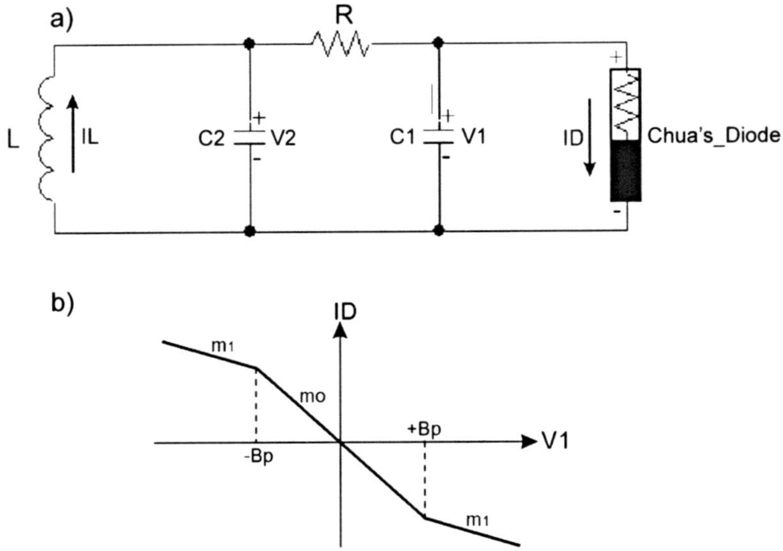

Fig. 1 Schematic Chua's circuit: (**a**) Chua's circuit (**b**) Characteristic of Chua's diode

The Chua's circuit is an autonomous system with chaotic behavior whose standard form is shown in Fig. 1(a). It is composed of a network of linear passive elements connected to a nonlinear active component known as Chua's diode. Its dynamics are described by three-coupled first-order nonlinear differential equations:

$$\begin{aligned} \dot{v}_1 &= -\frac{(v_1 - v_2)}{RC_1} - \frac{i_D(v_1)}{C_1}, \\ \dot{v}_2 &= \frac{(v_1 - v_2)}{RC_2} - \frac{i_L}{C_2}, \\ \dot{i}_L &= -\frac{v_2}{L} + \frac{ri_L}{L}, \end{aligned} \tag{1}$$

where r denotes the internal resistance of the inductor winding and $i_D(v_1)$ is the function of the Chua's diode, a nonlinear resistance with a negative current-voltage characteristic. Although the Chua's diode can be represented by any scalar function of one variable, generally a three-segment piecewise linear curve shown in Fig. 1(b) is chosen for convenience in synthesizing the physical circuit [1].

The success of an experimental implementation of the Chua's circuit basically depends on the realization of this nonlinear element. Several topologies to approach the Chua's diode are presented in the literature, considering its realization using circuit elements such as diodes [2], transistors [3], conventional voltage op amps (VOA) [4, 5], current feedback op amps (CFOA) [6, 7], and OTAs [8]. Another critical element in the implementation of the Chua's circuit is the inductor. Since commercial inductance values do not cover a very wide range, the inductor is assembled separately in most applications, resulting in a component with low accuracy and generally large dimensions if compared to other circuit elements. Furthermore, the internal resistance of the inductor must be relatively small to assure a correct operation of the Chua's circuit. Many alternatives to replace the physical inductor in the implementation of Chua's circuit have been presented in literature, such as the use of Wien-bridge [9] and inductance emulator circuits based in transistors [10], VOA's [11], CFOA's [6], OTAs [12, 13], and FTFN [14]. A comparative investigation about topologies of Chua's diode and inductance emulators, as well the combination of them to obtain inductorless Chua's circuits, is presented in [15].

Although the literature presents several alternatives for the implementation of an inductor-free realization of the Chua's circuit, an approach based on the electronic analogy was still not considered. Since the dynamics of Chua's circuit are governed by a set of differential equations, it can be electronically emulated using structures based on op-amp, which can realize several mathematical operations involving voltage signals: each first-order differential equation can be electronically implemented using a weighted analog integrator, and nonlinear functions can be approached splitting its curve into line segments generated by polarized diodes in an inverter amplifier [16]. However, a direct electronic implementation of an analogous Chua's oscillator is subjected to strict restrictions related to amplitude and frequency of the signals, which must be conditioned to stay within an operational range imposed by real implementation.

This work presents a new inductor-free topology of Chua's oscillator based in the electronic analogy, resulting in a simple, versatile and functional inductorless implementation of the Chua's circuit that offers new and interesting features for control and synchronization purposes. An analogous circuit is designed using this technique and its implementation reproduces satisfactorily the chaotic behavior of Chua's oscillator.

2 Electronic Analogy

Since the dynamics of a system are defined by a set of differential equations, it can be physically emulated using inexpensive and versatile analogous electronic circuits. These electronic circuits are based on op-amps, multistage amplifiers with differential inputs that can execute several mathematical operations involving voltage signals, such as multiplication by constant, subtraction, derivatives, weighted sum, weighted integration and nonlinear functions.

Since the original system variables and its derivatives are represented by voltage signals in the electronic analogy, they are subjected to strict limitations which can compromise the direct implementation of an analogous electronic circuit. The first limitation is the maximum voltage admissible by electronic devices, which is usually determined by the power supply. Another extreme limitation is the minimum voltage value, since noise or errors signals can corrupt low amplitude signals. The frequency of analog signals represents another limitation, since it cannot exceed the operational frequencies of electronic devices and/or measurement system. Thus, the scaling of the original model with appropriate factors can be necessary to restrict amplitudes and frequencies of the voltage signals, to assure the accurate reproduction of the dynamical system. The steps for the design of an analogous electronic circuit that reproduces the dynamical behavior of a system are [16]:

- To avail the dynamical model of the system. Since the weighted integrator output is inverted, it is advantageous to adopt the higher order term as negative for odd order equations aiming to simplify the electronic implementation.
- To estimate the expected range of value for each system variable, and, if necessary, to apply appropriate scale factors to restrict the signal amplitudes.
- To normalize the scaled system, dividing all set of equation by the greatest parameter value.
- To establish the dynamics for analogous implementation according to application.
- To assembler each first order equation of system using an analog weighted integrator and associated subcircuits related to other arithmetic operations. The inverse of normalized parameters corresponds to resistor values in p.u. (per unity) in the analogous implementation.

3 Design of an Analogous Chua's Oscillator

In this paper, the following parameters are considered for Chua's oscillator: $C_1 = 1$, $R = 1.47$, $C_2 = 1/9$, $L = 9/7$, $r = 0$, $m_0 = 1$, $m_1 = 0.5$, and $B_p = 1.4$. Figure 2 shows the simulated attractors of the original chaotic oscillator. The original equation system is scaled in amplitude to assure output voltage signals in the range of $9V$. Thus the original state variables are redefined as $V_1 = 0.76v_1$, $V_2 = 0.17v_2$ and $I_L = 0.82i_L$. The slopes m_0 and m_1 of the three-segment piecewise linear curve of the Chua's diode characteristic are not affected by amplitude scaling, but its break point B_p must be changed to 1.8.

Since the dynamics of the Chua's circuit is described by a set of three first order differential equations, the main cell of analogous electronic circuit is the analog inverting weighted integrator with op-amp. The transfer function of a weighting

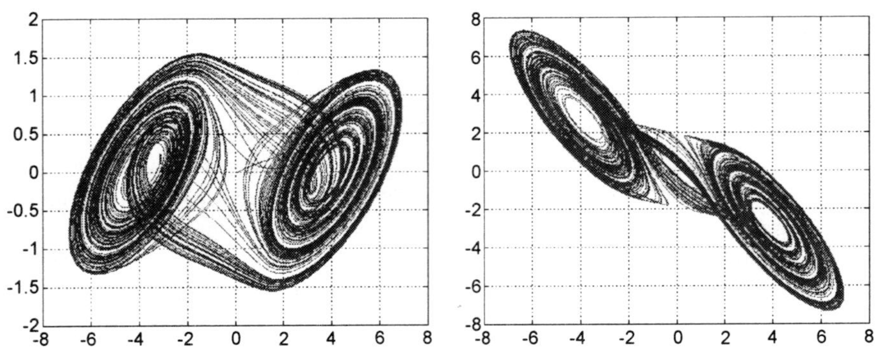

Fig. 2 Simulated Chua's attractors: projections V_1 vs V_2 and V_1 vs I_L

integrator is given by:

$$v_o = -\frac{1}{RC}\int\left(\frac{v_x}{R_{x\_}} + \frac{v_y}{R_{y\_}} + \frac{v_z}{R_{z\_}}\right)dt, \tag{2}$$

where v_o is the output voltage; v_x, v_y and v_z are the input voltages and $R_{x\_}$, $R_{y\_}$ and $R_{z\_}$ are the normalized values of the input resistances in p.u. (per unity). The dynamics of the analogous circuit are determined by the base resistance R and the integrator capacitance C. A comparison between the integrator transfer function and a normalized first order differential equation shows that the p.u. values of the input resistances correspond to the inverse of the respective differential equation coefficients.

The three-segment piecewise linear function of Chua's diode is created using an inverter amplifier based on an op-amp, whose gain is switched in the break point B_p by diodes connected in anti-parallel. In practice, these diodes can be the result of associations of rectifier, zener or LED diodes to obtain a voltage drop approximately equal to B_p. If the output voltage V_1 is within the range $\pm B_p$, both diodes are blocked and the gain of the inverter amplifier is given by $m_o = R_{m_o}/R$, else one

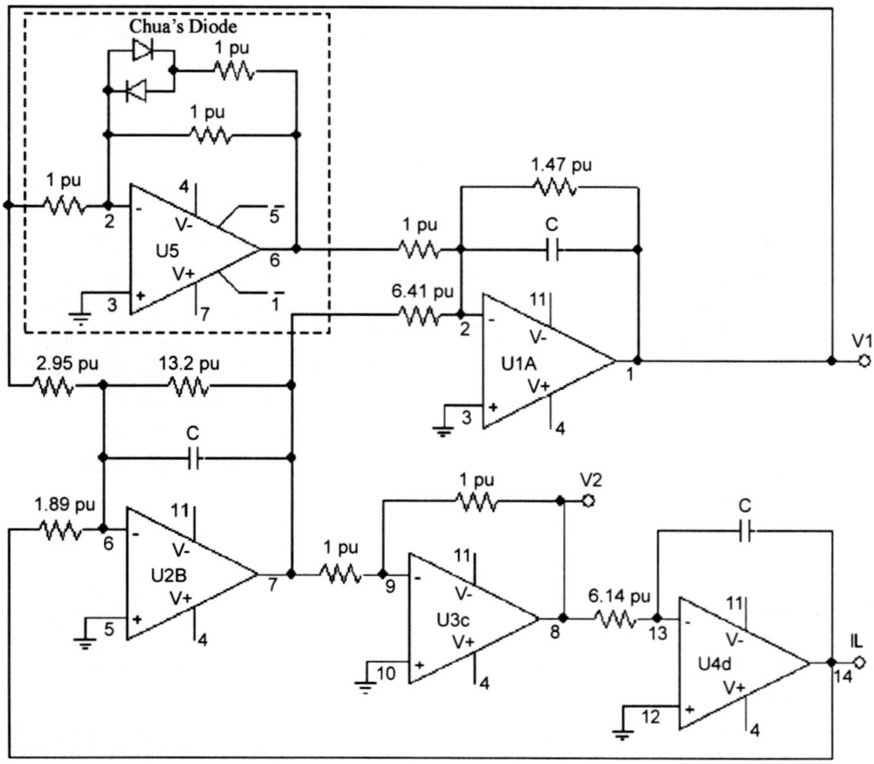

Fig. 3 Analogous Chua's oscillator

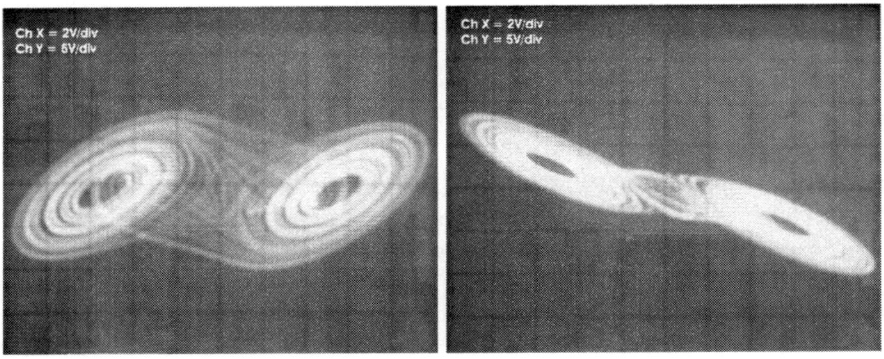

Fig. 4 Experimental attractors obtained from proposed analogous Chua's oscillator implementation: projections V_1 vs V_2 and V_1 vs I_L

of the diodes is switched to on, putting the resistor R_{m_1} in parallel with R_{m_o} and reducing the gain of the inverter amplifier to $m_1 = (R_{m_o} R_{m_1})/[(R_{m_o} + R_{m_1})R]$.

The analogous Chua's circuit that reproduces the normalized three-coupled first-order nonlinear differential equations of Chua's system is obtained through the adequate connection of the individual electronic weighting integrators and the analogous Chua's diode, as shown in Fig. 3. An extra inverting amplifier (U3c) is used to transmit a compatible polarity of the signal from V_2 integrator output (U2B) to I_L integrator input (U4d).

4 Experimental Implementation

This version of the Chua's circuit is experimentally implemented to verify its feasibility. The value of the base resistance is $10k\Omega$ and all capacitors are $1nF$. Texas Instruments IC's TL071 (single op-amp) and TL074 (quad op-amp) are utilized in the circuit implementation. Two red LEDs, whose individual voltage drop is approximately 1.8 V, are connected in anti-parallel to implement the Chua's diode. The Fig. 4 shows the projections of the attractor generated by experimental implementation observed in a 20 MHz analog oscilloscope on X–Y mode. The characteristic double-scroll strange attractor of the analogous circuit is restricted within ± 9 V range as established by amplitude scaling.

5 Conclusions

Herein, the design and implementation of a realization of the Chua's circuit based in the concept of electronic analogy is presented, an approach that can be always used to reproduce a certain set of differential equations. The proposed analogous circuit successfully reproduces the dynamical behavior of the original Chua's circuit, providing a versatile and functional inductor-free implementation of the chaotic Chua's oscillator. Although the number of components used to implement the analogous Chua's circuit is larger than other Chua's oscillator versions, it offers new and interesting features if compared with other alternatives proposed in literature. The three state variables are accessible and explicitly available in the analogous Chua's circuit as voltage signals (including the inductor current I_L), whose amplitude can be independently defined for each output in circuit design from scaling. Since the dynamics of analogous Chua's circuit can also be defined in the circuit design, very slow dynamics can be obtained, when necessary, establishing a large base resistance R and avoiding the use of electrolytic capacitors which reliability is low. In this context, both the amplitude and frequency of these signals can be matched according to application, allowing a design for operations with slow dynamics, as required for control purposes, or extremely fast oscillations, in order to create appropriate circuits for use in chaos-based communications. Since the input resistances of each analog integrator can be substituted by potentiometers, each coefficient of the coupled equation system of the Chua's circuit can be independently varied in the analogous Chua's circuit, allowing the experimental observation of other attractors. External control inputs can be easily included in the analogous Chua's circuit only introducing new input resistors in the integrators. Although an ideal inductance is considered in this analogous Chua's circuit, the effect of an inductor resistance can be easily incorporated introducing a feedback resistor R_{zz} in the I_L integrator. It is possible to adapt the analogous Chua's circuit considering negative values for inductance L and/or capacitances C_1 and C_2, which is impossible in the original Chua's circuit. Thus, the proposed analogous Chua's circuit is very flexible and allows the experimental observation of a surprisingly large number of topologically distinct chaotic attractors, presenting several interesting features for many real applications.

Acknowledgements The authors would like to thank Fundação de Amparo à Pesquisa do Estado de Minas Gerais (FAPEMIG), Coordenação de Aperfeiçoamento de Pessoal de Nível Superior (CAPES), Conselho Nacional de Desenvolvimento Científico e Tecnológico (CNPq) and Fundação Gorceix for their financial support in the undertaking of this project.

References

1. R. Brown, IEEE Trans. CAS **40**, 878–884 (1993)
2. T. Matsumoto, L.O. Chua, M. Komuro, IEEE Trans. CAS **32**, 797–818 (1985)
3. T. Matsumoto, L.O. Chua, M. Komuro, Int. J. Circuit Theory Appl. **14**, 117–146 (1986)
4. G.O. Zhong, F. Ayrom, Int. J. Circuit Theory Appl. **13**, 93–98 (1985)
5. M.P. Kennedy, Frequenz **46**, 66–80 (1992)
6. R. Senani, S.S. Gupta, IET Electr. Lett. **34**, 829–830 (1998)
7. A.S. Elwakil, M.P. Kennedy, IEEE Trans. CAS **47**, 289–306 (2000)
8. J.M. Cruz, L.O. Chua, IEEE Trans. CAS **39**, 985–995 (1992)
9. Ö. Morgül, IET Electr. Lett. **31**, 1424–1430 (1995)
10. T.P. Weldon, Am. J. Phys. **58**, 936–941 (1990)
11. L.A.B. Tôrres, L.A. Aguirre, IET Electr. Lett. **36**, 1915–1916 (2000)
12. J.M. Cruz, L.O. Chua, IEEE Trans. CAS **40**, 614–625 (1993)
13. A. Rodriguez-Vazquez, M. Delgado-Restituto, IEEE Trans. CAS **40**, 596–611 (1993)
14. R. Kiliç, U. Çam, M. Alçi, H. Kuntman, Int. J. Bifurc. Chaos **12**, 1429–1435 (2002)
15. R. Kiliç, Int. J. Bifurc. Chaos **13**, 1475–1493 (2003)
16. R. Rocha, L.S. Martins Filho, R.F. Machado, Int. J. Elect. Eng. Educ. **43**, 334–345 (2006)

Model Reduction of Nonlinear Continuous Dynamic Systems on Inertial Manifolds with Delay

Jia-Zhong Zhang, Li-Ying Chen, and Sheng Ren

Abstract In comparison with traditional Approximate Inertial Manifolds (AIMs), an Approximate Inertial Manifolds with Time Delay (AIMTDs) is constructed for the model reduction of nonlinear continuous dynamic system governed by partial differential equation with second order in time. By this method, the nonlinear continuous dynamic system is studied in the phase space, and the solutions of the governing equations are projected onto the complete space spanned by the eigenfunctions of the linear operator of the governing equations. Then, the nonlinear *Galerkin's* procedure combined with AIMTDs is used to approach the solutions. Finally, the method is applied to the dynamics buckling analysis of the shallow arch under impact, and the comparisons between traditional *Galerkin's* procedure, traditional AIMs, and AIMTDs are given. It can be concluded that the methods presented are effective for the model reduction of the nonlinear continuous dynamic systems with second order in time.

Keywords Model reduction · Inertial manifolds with delay · Dynamic buckling

1 Introduction

For an infinite dimensional dissipative dynamic system, it is well known that the asymptotic behavior will evolve to a compact set known as a global attractor, which is finite-dimensional, and the equilibrium position, periodic solution etc., are included in such global attractor [1, 2]. In other words, such kind of dynamic systems can be described by the deterministic flow on a lower dimensional attractor for the long-term behavior of the system. However, how to approach the global attractor becomes another object.

J.-Z. Zhang (✉) · L.-Y. Chen · S. Ren
School of Energy and Power Engineering, Xi'an Jiaotong University, Xi'an 710049, P.R. China
e-mail: jzzhang@mail.xjtu.edu.cn

J.A.T. Machado et al. (eds.), *Nonlinear Science and Complexity*,
DOI 10.1007/978-90-481-9884-9_48,

Fortunately, the theory of Inertial Manifolds (IMs) is a technique for approaching the global attractor, and it is mainly used and developed in fluid dynamics [3, 4]. It has been proved that the existence of IMs usually holds only under the very restrictive spectral gap condition. Consequently, in practical applications the concept of Approximate Inertial Manifolds (AIMs) has been introduced. An AIMs can be defined as a finite-dimensional Lipschitz manifold and a thin surrounding neighborhood into which any orbit enters in a finite time. In fact, global attractor must lie within a small neighborhood of AIMs. If the *Galerkin* procedure is used to approach such AIMs, the AIMs can be considered as the interaction between the high and low modes. By this method, it splits the infinite-dimensional phase space of the PDE into two complementary subspaces: a finite-dimensional one spanned by slowly contracting modes, and its complement spanned by the high modes. With the introduction of the Approximate Inertial Manifolds with Time Delay (AIMTDs) later, the relation between this two subspaces is the one with time delay, that is, the evolution of the high modes is not only relevant to the instantaneous low modes, but also to the past high modes.

Mathematically, the shallow arch under impact is an infinite-dimensional or continuous dynamic system in the point of view of dynamics. Normally, the traditional *Galerkin* method is used to approach the solutions of the governing equations, namely, the second order in time nonlinear partial differential equations. One question will arise: how many buckling modes should be included in the mode expansion. There are some studies on the dynamic buckling of shallow arch, but only few buckling modes are considered [5]. However, an important and well-known aspect of nonlinear dynamics is the sensitive dependence of the solution on the perturbations [6]. More precisely, the truncation of higher buckling modes will have a great influence on the dynamic buckling. On the other hand, if much more buckling modes are considered in the mode expansion, considerable computing time will be required due to the many degrees-of-freedom.

2 Inertial Manifolds with Time Delay

For the nonlinear dissipative autonomous and continuous dynamic system with second order in time, the governing equation can be described in the general form,

$$\ddot{w} + C\dot{w} + Kw = h(w), \tag{1}$$

where $h(w)$ is the nonlinear term, K is the linear operator. Denote by P_k the orthogonal projection in the Hilbert space H, which the solution lives in, onto the space spanned by the first k eigenfunctions of K, and then $Q_k = I - P_k$. For the sake of simplicity, hereafter P_kH and Q_kH are termed as low and high mode subspaces, respectively. Defining $p \equiv P_kw$ and $q \equiv Q_kw$, and applying P_k and Q_k to (1), a set of equations in the following form can be obtained,

$$\ddot{p} + C_1\dot{p} + K_1p = h(p+q), \tag{2}$$

$$\ddot{q} + C_2\dot{q} + K_2q = g(p+q). \tag{3}$$

Apparently, the traditional *Galerkin* method corresponds to setting $q = 0$ in (2). By this method, the interaction between low modes and high modes are neglected. Normally, the high modes will be decayed compared with low modes, due to the damping. Hence, one kind of the AIMs can then be constructed as the following,

$$K_2 q = g(p+q), \tag{4}$$

namely, $q = \Phi(p)$ which captures the behavior of high modes in terms of the low modes.

More precisely, one can obtain an AIMs under the assumptions that the high modes of the w can be negligible, as well $\dot{q}(t)$ and $\ddot{q}(t)$. In the situation that $|q(t)|/|p(t)|$, $|\dot{q}(t)|/|\dot{p}(t)|$ and $|\ddot{q}(t)|/|\ddot{p}(t)|$ are small in comparison with the other terms in (2) and (3), then (4) can be used to approach (3).

It is clear that AIMs treats the number of the modes involved strictly, and such restrictive condition depends on the implicit that the relation between high and low modes is instantaneous. Later, the Inertial Manifolds with Time Delay (IMTDs), which is a much more rational concept, is proposed, and it implies the relationship between the high and low modes is relevant to the history of the evolution of the modes,

$$q(t) = \Phi(p(t), q(t-T)) \quad (T \text{ is a proper time delay}). \tag{5}$$

As stated [2], IMTDs does not require a spectral gap condition and they can exist for general dissipative systems of infinite dimension, and it is suitable for the numerical computation.

3 Governing Equations of Shallow Arch under Impact

The method presented above will be applied to the dynamic buckling analysis, the typical nonlinear dissipative autonomous and continuous dynamic system with second order in time. Under some assumptions, the governing equation for shallow arch under impact and with simply boundaries can be derived as the following with one unknown,

$$\rho h \frac{\partial^2 w}{\partial t^2} + \frac{Eh^3}{12} w'''' + V(y_0'' - w'') + d\frac{\partial w}{\partial t} = 0, \tag{6}$$

where y_0 is the positions of the middle axis of shallow arch pre-impact, d the damping per area, ρ the density of the arch, and $V = \frac{1}{l}\int_0^l [\frac{Eh}{2}(y'^2 - y_0'^2)]dx$. The initial velocity could simulate the varying of the impact load, i.e. $t = 0$: $w = 0$, $\frac{\partial w}{\partial t} =$ cons. The initial shape or configuration of the shallow arch is assumed as $y_0(x) = -blx + bx^2$.

The linear operator of (6) together with the boundary conditions can be defined as $Lw = \frac{Eh^3}{12} w''''$. Then $\{\sin \frac{n\pi}{l} x, n = 1, \ldots, +\infty\}$ constitute the set of eigenfunctions of the operator, which span an orthogonal basis of the space which the solution of the governing equation will be projected onto. Following *Galerkin* procedure, yields,

$$\frac{Eh^3}{12}w_m\frac{(m\pi)^4}{2l^3}+\sum_{n=1}^{\infty}\left\{w_n^2\frac{(n\pi)^2}{2l}\frac{Eh}{2l}+\frac{2Ebh}{n\pi}w_n[(-1)^{n+1}+1]\right\}$$
$$\times\frac{2bl}{m\pi}[(-1)^{m+1}+1]$$
$$+\sum_{n=1}^{\infty}\left\{w_n^2\frac{(n\pi)^2}{2l}\frac{Eh}{2l}+\frac{2Ebh}{n\pi}w_n[(-1)^{n+1}+1]\right\}w_m\frac{(m\pi)^2}{2l}$$
$$+\frac{\rho hl}{2}\ddot{w}_m+\frac{ld}{2}\dot{w}_m=0,\quad m=1,\ldots,+\infty. \tag{7}$$

The Approximate Inertial Manifolds with Time Delay (AIMTDs) is applied to the high buckling modes, and the backward Euler and middle difference methods are used to approach $\dot{w}_m$ and $\ddot{w}_m$, respectively. Consequently, the higher buckling modes can be expressed in terms of lower buckling modes as the follows,

$$w^i{}_m=\Phi(w^i{}_n, w^{i-1}{}_m, w^{i-2}{}_m)\quad m=k+1,\ldots,2k; n=1,\ldots,k. \tag{8}$$

4 Numerical Examples

The dynamic buckling of a shallow arch with the following system parameters are studied with different buckling modes: $\rho = 7896$ kg/m^3, $E = 2.1 \times 10^{11}$ N/m^2, $d = 2500$ N·s/m^3, $l = 0.105$ m, $h = 0.00065$ m, $b = -2.72$.

The influence of high buckling modes on the long-term dynamic behaviors are presented, with some comparisons between the traditional *Galerkin* (TGM) and nonlinear *Galerkin* methods combined with AIMTDs. Hereafter, all of the phase portraits to be shown represent the response at the center of the arch.

For understanding the influence of modes truncation on the long-term behaviors of the system, more modes are considered in the mode expansion. Figure 1 are

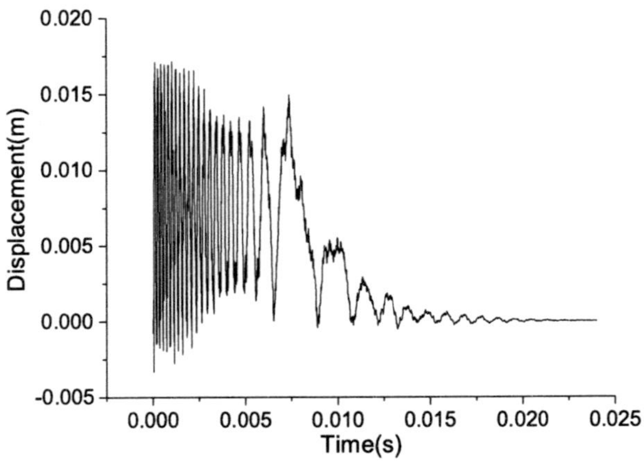

Fig. 1 Time history of system by TGM with initial velocity—180.0 m/s

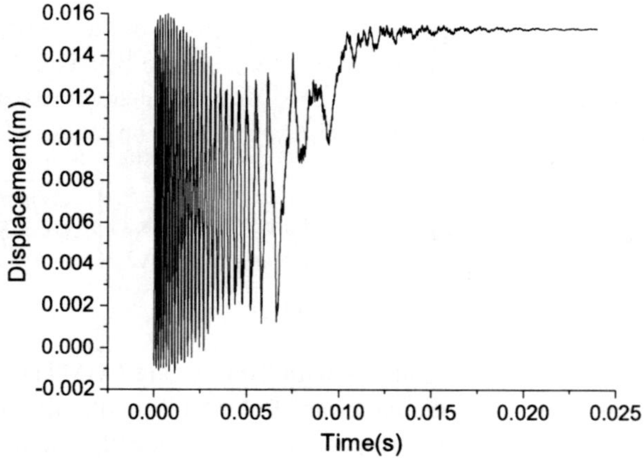

Fig. 2 Time history of velocity by TGM with initial velocity—180.0 m/s

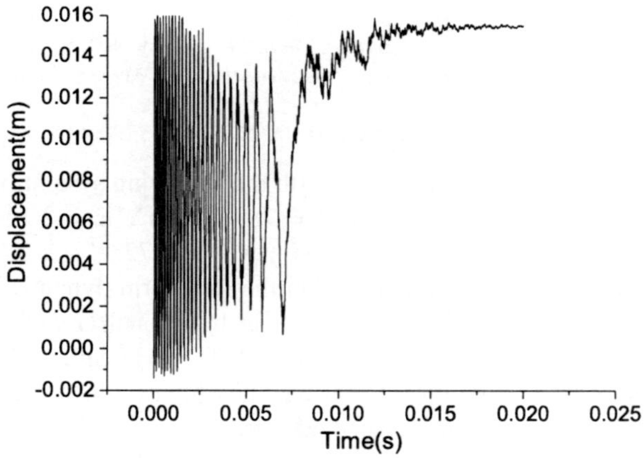

Fig. 3 Time history of velocity by AIMTDs with initial velocity—180.0 m/s

the results obtained from TGM with first 20 buckling modes considered and initial velocity—180 m/s. It is clear that it is finally reaching the initial situation after some periods of oscillation. Figures 2 and 3 are the results obtained from TGM and AIMTDs with first 100 buckling modes considered and initial velocity—180 m/s, respectively. It is clear that both of them are finally reaching a snap-through buckling, and the time histories are somewhat different each other. Obviously the time histories are distinctly different from Fig. 1.

5 Conclusions

Following Inertial Manifolds with Time Delay and Nonlinear *Galerkin* method, a numerical scheme is presented to reduce the governing equation of dynamic buckling of shallow arch under impact load. In comparison to the traditional *Galerkin* method, the presented method can improve the distance between the original and reduced systems on the long-term behaviors, since the interaction between the low and high buckling mode with time delay are considered, and requires less computing time.

Acknowledgements This work was supported by Program for New Century Excellent Talents in University in China, No. NCET-07-0685, and National Natural Science Foundation of China, No. 10772140.

References

1. R. Temam, *Infinite-Dimensional Dynamical System in Mechanics and Physics* (Springer, New York, 1997)
2. A. Debussche, R. Temam, Appl. Math. Lett. **8**, 21–24 (1995)
3. E.S. Titi, J. Math. Anal. Appl. **149**, 540–557 (1990)
4. F. Jauberteau, C. Rosier, R. Temam, Comput. Methods Appl. Mech. Eng. **80**, 245–260 (1990)
5. J.Z. Zhang, D.H. Campen, G.Q. Zhang, V. Bouwman, J.W. Weeme, AIAA J. **39**, 956–961 (2001)
6. J.Z. Zhang, Y. Liu, D.M. Chen, Appl. Math. Mech. **26**, 938–943 (2005)

A Fuzzy Crisis in a Duffing-Van der Pol System

Ling Hong and Jian-Qiao Sun

Abstract A crisis in a Duffing-Van del Pol with fuzzy uncertainties is studied by means of the fuzzy generalized cell mapping (FGCM) method. A crisis happens when two fuzzy attractors collide simultaneously with a fuzzy saddle on the basin boundary as the intensity of fuzzy noise reaches a critical point. The two fuzzy attractors merge discontinuously to form one large fuzzy attractor after a crisis. A fuzzy attractor is characterized by its global topology and membership function. A fuzzy saddle with a complicated pattern of several disjoint segments is observed in phase space. It leads to a discontinuous merging crisis of fuzzy attractors. We illustrate this crisis event by considering a fixed point under additive fuzzy noise. Such a crisis is fuzzy noise-induced effects which cannot be seen in deterministic systems.

Keywords Fuzzy dynamical systems · Fuzzy noise · Fuzzy bifurcation · Cell mapping methods

1 Introduction

Noise is ubiquitous in real-life physical systems and can be usually modeled as a random variable or a fuzzy set dependent on the available information about the noise [1–4]. Noise acting on nonlinear dynamical systems can be a source of new phenomena. It may qualitatively change the system behavior and induce bifurcations. This paper presents a method to analyze the response and bifurcation of nonlinear dynamical systems with fuzzy noise. We are interested in a nonlinear dynamical system whose response is a fuzzy process, and study how the fuzzy response

L. Hong (✉)
MOE Key Lab for Strength and Vibration, Xi'an Jiaotong University, Xi'an 710049, China
e-mail: hongling@mail.xjtu.edu.cn

J.-Q. Sun
School of Engineering, University of California, Merced, CA 95344, USA
e-mail: jqsun@ucmerced.edu

J.A.T. Machado et al. (eds.), *Nonlinear Science and Complexity*,
DOI 10.1007/978-90-481-9884-9_49,

changes as the fuzzy noise intensity varies. Specifically, our attention is focused on the analysis of crises in fuzzy attractors induced by fuzzy noise. It should be noted that few works dealing with this problem have been published to date.

In the theory of deterministic dissipative systems, sudden changes in chaotic attractors with parameter variation have been called crises [5, 6]. Three types of crisis can be distinguished according to the nature of the discontinuous change that the crisis induces in the chaotic attractor. In the first type, a chaotic attractor is suddenly destroyed as the parameter passes through a critical value. In the second type, the size of the attractor in phase space suddenly increases. In the third type, two or more chaotic attractors merge to form one chaotic attractor.

For fuzzy nonlinear dynamical systems, crisis analysis is difficult because the evolution of the membership function of the fuzzy response process is not readily obtained analytically. There is little study in the literature on the bifurcation of fuzzy nonlinear dynamical systems. There are studies of bifurcations of fuzzy control systems where the fuzzy control law leads to a nonlinear and deterministic dynamical system. The bifurcation studies are practically the same as that of deterministic systems [7, 8]. The work [9] deals with bifurcation of fuzzy dynamical systems having a fuzzy response. Numerical simulations are used to simulate the system response with a given parameter and fuzzy membership grade. The eigenvalues and the membership distribution are both used to describe the bifurcation. For a given membership grade, the bifurcation of the system is defined in the same manner as for the deterministic system. The authors have recently proposed a fuzzy generalized cell mapping (FGCM) method for the bifurcation analysis of fuzzy nonlinear dynamical systems and considered several very interesting scenarios of fuzzy bifurcations [10, 11]. The current paper studies a sudden change in a fuzzy attractor which is characterized by its global topology and membership function. Such a change is called a fuzzy crisis following Grebogi's definition of crisis in deterministic chaotic systems [5, 6]. We shall study a discontinuously merging crisis involving the collision of two fuzzy attractors with a fuzzy saddle on the basin boundary. The origin and evolution of a fuzzy attractor and saddle under fuzzy noise are also investigated. A fuzzy saddle develops from a saddle point into a complicated saddle pattern of several disjoint segments and plays an extremely important role in a fuzzy crisis. It leads to a discontinuously merging crisis of fuzzy attractors. We illustrate this crisis event by considering a fixed point under additive fuzzy noise.

The remainder of the paper is outlined as follows. In Sect. 2, we study a discontinuously merging crisis of two fuzzy attractors in the case of additive fuzzy noise. The paper concludes in Sect. 3.

2 A Fuzzy Crisis in a Duffing-Van der Pol System

We consider the Duffing-Van der Pol (DVP) equation driven by additive fuzzy noise

$$\begin{aligned} \dot{x}_1 &= x_2, \\ \dot{x}_2 &= 1.0x_1 + 0.2x_2 - x_1^3 - x_1^2 x_2 + S, \end{aligned} \tag{1}$$

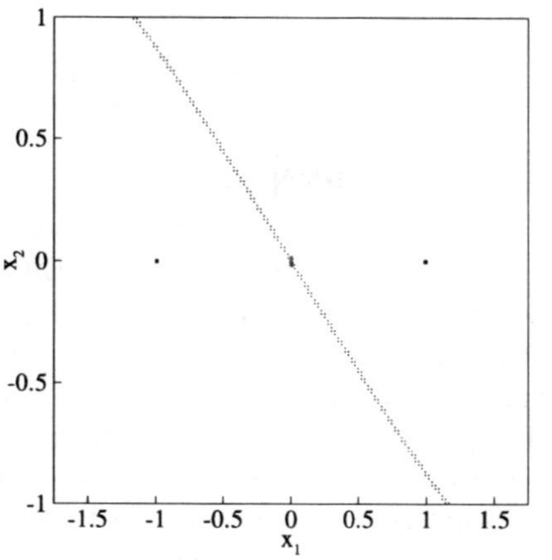

Fig. 1 Global phase portrait of the deterministic Duffing-Van der Pol equation. *Two dots* at $(-1, 0)$ and $(+1, 0)$ are stable fixed points and *the dot on the line* at $(0, 0)$ is a saddle. *The line* is the stable manifold of the saddle marking the boundary of the domains of attraction of the fixed points

where S is a fuzzy parameter with a triangular membership function,

$$\mu_S(s) = \begin{cases} [s - (s_0 - \varepsilon)]/\varepsilon, & s_0 - \varepsilon \leq s < s_0, \\ -[s - (s_0 + \varepsilon)]/\varepsilon, & s_0 \leq s < s_0 + \varepsilon, \\ 0, & \text{otherwise}, \end{cases} \tag{2}$$

$\varepsilon > 0$ is a parameter characterizing the intensity of fuzziness of S, and s_0 is the nominal value of S with membership grade $\mu_S(s_0) = 1$. The corresponding deterministic Duffing-Van der Pol oscillator with $S = 0$ is one of the most studied systems in nonlinear dynamics. Its local and global bifurcation behavior has been thoroughly investigated [12, 13]. The influence of stochastic noise on the Duffing-Van der Pol oscillator exhibiting codimension one and two bifurcations has also been studied [14–16]. For the case of fuzzy noise, to our knowledge, no attempt has been made regarding this problem. In the present work, we choose $\mu_1 = 1.0$, $\mu_2 = 0.2$ located in the upper right-hand quadrant of Fig. 7.3.7 in the literature [12]. When $\mu_1 = 1.0$, $\mu_2 = 0.2$, the deterministic DVP equation has two coexistent fixed point attractors and a saddle point on their basin boundary as shown in Fig. 1.

The domain $D = \{-1.75 \leq x_1 \leq 1.75, -1.0 \leq x_2 \leq 1.0\}$ is discretized into 141×141 cells when applying the FGCM method, 5×5 sampling points are used within each cell. The membership function is discretized into 201 segments ($M = 201$). Hence, out of each cell, there are 5025 trajectories with varying membership grades. These trajectories are then used to compute the transition membership matrix.

We fix $s_0 = 0$ and allow the fuzzy noise intensity ε to vary. As ε increases, two coexistent fuzzy fixed point attractors grow bigger simultaneously, and the saddle changes to a fuzzy saddle pattern with complicated structure. The global phase portrait of the deterministic system is shown in Figs. 2 and 3 when $\varepsilon = 0.2$ and $\varepsilon = 0.284$.

A merging crisis occurs when $\varepsilon = 0.284$ and $\varepsilon = 0.285$. In such a case, the fuzzy attractors collide simultaneously with a fuzzy saddle on the basin boundary, and

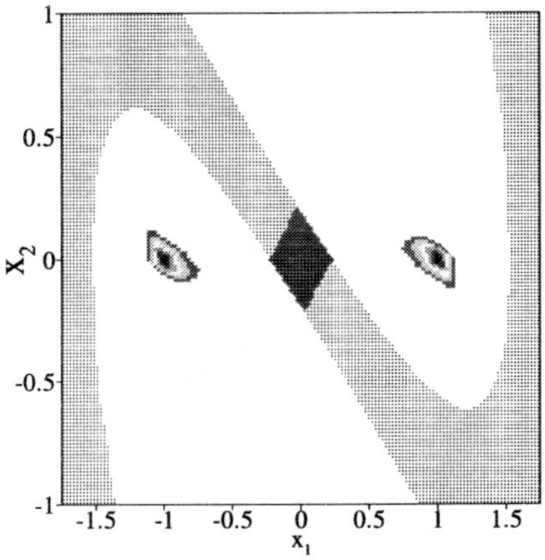

Fig. 2 (Color online) Global phase portrait of the noisy Duffing-Van der Pol equation with the fuzzy noise intensity 0.2. In the figure, the fuzzy attractors are marked by *the color symbol* ".". The membership distribution of fuzzy attractors is color-coded with *black* = 1.0, 0.8 < *red* < 1.0, 0.6 < *green* < 0.8, 0.4 < *yellow* < 0.6, 0.2 < *cyan* < 0.4, and 0.0 < *purple* < 0.2. The saddle is marked by *the blue symbol* "o". The basin boundary is marked by *the grey symbol* "."

Fig. 3 (Color online) A fuzzy crisis at a critical value of the fuzzy noise intensity 0.284 for the noisy Duffing-Van der Pol equation. Legends are the same as those in Fig. 2. Two fuzzy attractors are touching a fuzzy saddle with a complicated pattern of several disjoint segments

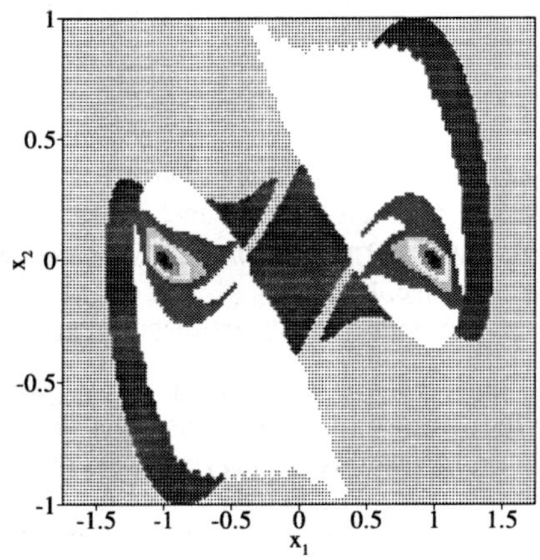

suddenly merge to form one large fuzzy attractor in the phase space after the crisis. The global phase portraits are shown in Figs. 3 and 4.

3 Concluding Remarks

In this paper, we have investigated fuzzy crises driven by fuzzy noise where a fuzzy saddle with a complicated structure leads two fuzzy attractors discontinuously to merge into a large one. A collision with a fuzzy saddle is the typical mechanism by which two fuzzy attractors can discontinuously merge. These fuzzy crises are

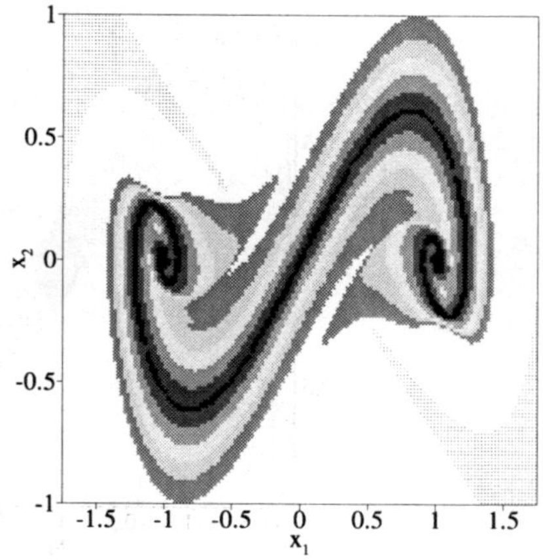

Fig. 4 (Color online) The expanded fuzzy attractor at post-crisis with the fuzzy noise intensity 0.285 for the noisy Duffing-Van der Pol equation. Legends are the same as those in Fig. 2

difficult to analyze with direct numerical simulations or analytical methods. The FGCM method is at present the only effective tool for bifurcation analysis of fuzzy nonlinear dynamical systems.

Acknowledgements This work is supported by the National Science Foundation of China under Grant No. 10772140 and the Scientific Research Foundation for the Returned Overseas Chinese Scholars, State Education Ministry.

References

1. F. Moss, P.V.E. McClintock, *Noise in Nonlinear Dynamical Systems* (Cambridge University Press, Cambridge, 1989)
2. G.J. Klir, T.A. Folger, *Fuzzy Sets, Uncertainty, and Information* (Prentice-Hall, Englewood Cliffs, 1988)
3. M. Bucolo, S. Fazzino, M.L. Rosa, L. Fortuna, Small-world networks of fuzzy chaotic oscillators. Chaos Solitons Fractals **17**, 557–565 (2003)
4. U. Sandler, L. Tsitolovsky, Fuzzy dynamics of brain activity. Fuzzy Sets Syst. **121**, 237–245 (2001)
5. C. Grebogi, E. Ott, Chaotic attractors in crisis. Phys. Rev. Lett. **48**, 1507–1510 (1982)
6. E. Ott, *Chaos in Dynamical Systems* (Cambridge University Press, Cambridge, 2002)
7. Y. Tomonaga, K. Takatsuka, Strange attractors of infinitesimal widths in the bifurcation diagram with an unusual mechanism of onset. Nonlinear dynamics in coupled fuzzy control systems. II. Physica D **111**, 51–80 (1998)
8. F. Cuesta, E. Ponce, J. Aracil, Local and global bifurcations in simple Takagi-Sugeno fuzzy systems. IEEE Trans. Fuzzy Syst. **9**, 355–368 (2001)
9. P.K. Satpathy, D. Das, P.B.D. Gupta, A fuzzy approach to handle parameter uncertainties in Hopf bifurcation analysis of electric power systems. Int. J. Electr. Power Energy Syst. **26**(7), 531–538 (2004)
10. L. Hong, J.Q. Sun, Bifurcations of fuzzy nonlinear dynamical systems. Commun. Nonlinear Sci. Numer. Simul. **11**(1), 1–12 (2006)
11. L. Hong, J.Q. Sun, Codimension two bifurcations of nonlinear systems driven by fuzzy noise. Physica D, Nonlinear Phenom. **213**(2), 181–189 (2006)
12. J. Guckenheimer, P.J. Holmes, *Nonlinear Oscillations, Dynamical Systems and Bifurcations of Vector Fields* (Springer, New York, 1983)

13. P. Holmes, D. Rand, Phase portraits and bifurcations of the non-linear oscillator. Int. J. Non-Linear Mech. **15**, 449–458 (1980)
14. N.S. Namachchivaya, Stochastic bifurcation. J. Appl. Math. Comput. **38**, 101–159 (1990)
15. K.R. Schenk-Hoppe, Bifurcation scenarios of the noisy Duffing-van der Pol oscillator. Nonlinear Dyn. **11**, 255–274 (1996)
16. N.S. Namachchivaya, Co-dimension two bifurcations in the presence of noise. J. Appl. Mech. **58**, 259–265 (1991)

Name Index

J.A.T. Machado et al. (eds.), *Nonlinear Science and Complexity*,
DOI 10.1007/978-90-481-9884-9,

ONE

Lightning Source UK Ltd.
Milton Keynes UK
18 November 2010

163088UK00006B/2/P

9 789048 198832